Health and Welfare of Brachycephalic (Flat-faced) Companion Animals

Health and Welfare of Brachycephalic (Flat-faced) Companion Animals

A Complete Guide for Veterinary and Animal Professionals

Edited by
Dr. Rowena M.A. Packer and Dr. Dan G. O'Neill

CRC Press
Taylor & Francis Group
Boca Raton London New York

CRC Press is an imprint of the
Taylor & Francis Group, an **Informa** business

First edition published 2022
by CRC Press
6000 Broken Sound Parkway NW, Suite 300, Boca Raton, FL 33487-2742

and by CRC Press
2 Park Square, Milton Park, Abingdon, Oxon, OX14 4RN

© 2022 Taylor & Francis Group, LLC

CRC Press is an imprint of Taylor & Francis Group, LLC

ISBN: 978-0-367-20741-0 (hbk)
ISBN: 978-0-367-20724-3 (pbk)
ISBN: 978-0-429-26323-1 (ebk)

Typeset in Times
by codeMantra

Access the Support Material: www.routledge.com/9780367207243
All Figures are available to download for free at https://www.routledge.com/9780367207243. These are for use in teaching and should not be used commercially without permission from the publisher.

"Man selects only for his own good: Nature only for that of the being which she tends."

– Charles Darwin, The Origin of Species

Contents

Foreword xi

Editors xiii

Contributors xv

Chapter 1   Introduction: How Can a Brachycephalic Boom Cause a Welfare Bust? 1

Rowena M. A. Packer and Dan G. O'Neill

PART I   Wider Viewpoints

Chapter 2   A Historical Perspective on Brachycephalic Breed Health and the Role of the
Veterinary Profession 7

Alison Skipper

Chapter 3   Flat-Faced Fandom: Why Do People Love Brachycephalic Dogs and Keep
Coming Back for More? 25

Rowena M. A. Packer

Chapter 4   Ethical Challenges of Treating Brachycephalic Dogs 41

Anne Quain, Paul McGreevy and Siobhan Mullan

Chapter 5   Discussing Brachycephalic Health with Current and Prospective Dog Owners 55

Zoe Belshaw and Sean Wensley

Chapter 6   Nurses and the Brachycephalic Patient – Practical Considerations and the Role
of Veterinary Nurses in Improving Brachycephalic Health 69

Kate Price

Chapter 7   The Epidemiology of Brachycephalic Health – Understanding the Science and
Exploring the Evidence on Demography, Disorder Frequency and Risk Factors 85

Dan G. O'Neill

Chapter 8   The Genetics of Brachycephaly, Population Genetics and Current Health
Testing for Brachycephalic Breeds 107

David Sargan

Chapter 9   International and National Approaches to Brachycephalic Breed Health
Reforms in Dogs 127

Brenda N. Bonnett, Monique Megens, Dan G. O'Neill, and Åke Hedhammar

PART 2   *Clinical Viewpoints*

Chapter 10 Brachycephalic Obstructive Airway Syndrome (BOAS) – Clinical Assessment
and Decision-Making 155

Jane Ladlow and Dr Nai-Chieh Liu

Chapter 11 Brachycephalic Obstructive Airway Syndrome – Surgical Management and
Postoperative Management 177

Mickey Tivers and Elizabeth Leece

Chapter 12 Ophthalmology in Practice for Brachycephalic Breeds 201

Màrian Matas Riera

Chapter 13 Dermatological Problems in the Brachycephalic Patient 219

Hilary Jackson and Deborah Gow

Chapter 14 Dental and Oral Health for the Brachycephalic Companion Animal 235

Fraser A. Hale

Chapter 15 Brain Disorders Associated with Brachycephaly 251

Clare Rusbridge and Susan Penelope Knowler

Chapter 16 Vertebral Malformations and Spinal Disease in Brachycephalic Breeds 291

Steven De Decker and Rodrigo Gutierrez-Quintana

Chapter 17 Obesity and Weight Management of Brachycephalic Animals 311

Eleanor Raffan

Chapter 18 Reproduction in Brachycephalic Companion Animal Species 329

Aoife Reid, Laura Cuddy, and Dan G. O'Neill

Chapter 19 Anaesthesia for the Brachycephalic Patient 353

Fran Downing and Rebecca Robinson

Chapter 20 Conclusions: Where Are We Now? And What Should or Could We Do Next? 377

Dan G. O'Neill and Rowena M. A. Packer

Index 387

Foreword

Health and Welfare of Brachycephalic Breeds: A Guide for Veterinary Professionals

They say love is blind, and the relentlessly increasing popularity of brachycephalic dogs is a perfect example of this. It is not difficult to understand why brachycephalic dogs such as the Pug and French Bulldog have climbed the popularity list of dog breeds at a phenomenal rate: they are generally friendly little dogs with 'cute faces', seen everywhere from on the arm of social media influencers to clothing, general advertising and greetings cards. But our love for the looks and behaviours of these dogs has blinded society to the suffering that many of these dogs endure.

It would be right to say that every dog breed is likely to have its own unique health problems: but for brachycephalic breeds, these problems are often more severe and impact directly on basic bodily functions, such as breathing. Their flattened faces can lead to a lifetime of struggling to breathe, while clinical problems such as snoring and snuffling become 'normalised' and accepted as 'part of the breed' even by veterinary professionals. In consequence, videos of brachycephalic dogs needing to fall asleep upright to be able to breathe go viral as being 'cute'. Even as the numbers of flat-faced dogs continue to rise, the list of other known significant health concerns in these breeds grows longer, including skin problems, spinal and neurological problems, and dental, birthing and eye problems. As demand increases, so does the monetary value of these breeds, leading to unscrupulous breeders seeing brachycephalic dogs as a way to make quick money. Humans blinded to the suffering of many brachycephalic breeds, but determined to own one of these cute, friendly dogs, just further perpetuate this cycle, and we find ourselves in a never-ending supply and demand loop, where the ultimate victims are the dogs.

But how do we have productive discussions about these popular breeds? Breeders who love their breed, and genuinely believe they are doing their best by them, feel like they are being attacked from all sides. Veterinary professionals and welfare professionals, who are often left to pick up the pieces of sick and suffering dogs, feel frustrated at fighting a losing battle against the tide of rising ownership. And owners of flat-faced dogs feel demonised simply for just loving their pets. Evidence, anecdote and personal belief often contradict each other, and conversations about brachycephaly rapidly turn difficult or stall: precisely because most people that interact with brachycephalic breeds want the 'best' for these animals but are looking at the issues from differing perspectives. Consequently, we cannot reach consensus on what is 'best'. Sometimes, evidence makes uncomfortable reading and challenges our perceptions and beliefs. Veterinary professionals may forget the human factor behind the brachycephalic boom, breeders may find it difficult to accept their breed is not as 'healthy' as they believe, and current or prospective owners may find it a challenge to accept that purchasing a brachycephalic dog is perhaps not (or was not) the right thing to do. We owe it to these dogs to ensure we work together to address the ethical, moral and health-related challenges associated with brachycephalic breeds.

Whoever you are, breeder, dog owner, veterinary professional or simply a caring human being, it is precisely because we all care about these dogs that you should read this book.

Daniella Dos Santos
British Veterinary Association President 2019–2020

Editors

Dr. Rowena M. A. Packer graduated from the University of Bristol, UK, with a first-class BSc in Animal Behaviour and Welfare (2009). She then went on to complete a PhD at the Royal Veterinary College (RVC), London, UK, (2013) exploring the impact of conformational extremes on canine health, focusing on brachycephalic breeds. This work culminated in a research impact event 'Building Better Brachycephalics' in 2013 and has since influenced international policy and legislation on the breeding of brachycephalic dogs. She has continued to work at RVC in the field of canine health and welfare research since finishing her PhD and was awarded a Biotechnology and Biological Sciences Research Council (BBSRC) Future Leader Fellowship in 2016. She is now a Lecturer in Companion Animal Behaviour and Welfare Science at RVC and leads a research team exploring diverse topics in this area. Rowena has authored over 60 papers on canine and feline health and welfare since 2012. In 2016 the Universities Federation for Animal Welfare (UFAW) recognised her achievements with its Young Animal Welfare Scientist of the Year Award. Rowena is co-leader of RVC's Brachycephalic Research Team and a founding member of the UK Brachycephalic Working Group.

Dr. Dan G. O'Neill graduated in veterinary medicine from Dublin in 1987. After 22 years in small and large animal general practice as well as in industry, he gained an MSc in epidemiology in 2009. He was awarded a PhD in 2014 at the RVC for developing the VetCompass™ Programme to evaluate breed effects on the health of dogs and cats. He is now a Senior Lecturer in Companion Animal Epidemiology at the RVC.

From 2012–2021 Dan has authored over 95 VetCompass™ papers that provide welfare evidence and benchmarks relevant to UK companion animals. He co-authored the third edition of the book *Breed Predispositions to Disease in Dogs and Cats* as well as a book on hamster health *Hamsters; in sickness and in health*. He was awarded a fellowship by RCVS in 2018 and the Blaine Award by BSAVA in 2019. His current research programme focusses on breed-related health in dogs and cats based on applying VetCompass clinical data shared from over 30% of UK vet practices.

Dan chairs the UK Brachycephalic Working Group and is co-leader of the RVC's Brachycephalic Research Team. Dan's ethos is that without good evidence, we are all just muddling around in the darkness.

Contributors

Dr. Zoe Belshaw, MA VetMB PhD CertSAM Dip ECVIM-CA AFHEA MRCVS: Zoe Belshaw is an EBVS and RCVS-recognised specialist in small animal internal medicine and was the 2020 recipient of the BSAVA PetSavers award for contributions to clinical aspects of small animal practice in its widest context. Her PhD from the Centre for Evidence-based Veterinary Medicine, University of Nottingham, UK, explored how veterinary surgeons and owners make welfare decisions about older dogs with osteoarthritis. Her ongoing research interests include quality of life assessment, owner experiences of looking after chronically ill pets, clinical research design and veterinary ethics. She combines clinical practice with a wide range of collaborative research and consultancy projects.

Dr. Brenda Bonnett, DVM, PhD: Brenda qualified as a veterinarian at the University of Guelph, in Canada. After many years as a tenured Associate Professor at the Department of Population Medicine, Ontario Veterinary College, University of Guelph in Ontario, Canada, she is now a Consulting Epidemiologist and CEO of the non-profit International Partnership for Dogs (IPFD).

Academic achievements include over 87 peer-reviewed publications and 4 book chapters in the fields of population-based and clinical epidemiology, theriogenology, human–animal interactions, veterinary education and communication, as well as population-based research using secondary data sources (most notably a large companion animal veterinary insurance database in Sweden) on dogs, cats and horses. She has supervised 12 doctoral and 10 Master's students and served on numerous other graduate student committees in several countries. Dr. Bonnett has been awarded an honorary doctorate by the Swedish University of Agricultural Sciences.

Dr Bonnett is a consultant on numerous welfare initiatives, including pet over-population in the USA, and works with cat and dog shelters, welfare and regulatory organisations. She facilitated a workshop on accessible veterinary care. As Lead Scientist at Morris Animal Foundation (2010), she assisted in development of the (now titled) Golden Retriever Lifetime Study. She is a frequent speaker at local, national and international venues to breeders, show judges, veterinarians, researchers and others.

She initiated the IPFD as a multi-stakeholder, collaborative organisation to enhance the health, well-being and welfare of dogs, and support human–dog interactions. Other developments include the International Dog Health Workshops which gather international decision leaders to address key issues in the dog world; and the IPFD Harmonization of Genetic Testing in Dogs on the IPFD platform DogWellNet.com. IPFD is facilitating international dialogue and actions to address ongoing and emerging issues relative to extreme conformation in dogs, among other challenges.

Dr. Laura Cuddy, MVB MS Diplomate ACVS-SA Diplomate ECVS Diplomate ACVSMR Diplomate ECVSMR MRCVS: Laura graduated from University College Dublin, Ireland, in 2008. She completed a rotating internship in small animal medicine and surgery, followed by an MSc and residency in small animal surgery at the University of Florida (UF), Gainesville, USA. In 2013, Laura joined the University College Dublin, Ireland, as an Assistant Professor in Small Animal Surgery. She was welcomed as a Diplomate of the American College of Veterinary Surgeons (ACVS) and European College of Veterinary Surgeons (ECVS) in 2014, and as a Diplomate of the American and European Colleges of Veterinary Sports Medicine and Rehabilitation (ACVSMR) in 2015. Laura is currently the director of Veterinary Specialists Ireland, a private specialty hospital located in County Meath, Ireland.

Dr. Steven De Decker, DVM, PhD, DipECVN, MvetMed, PGcert Vet Ed, FHEA, MRCVS: After graduating in 2005 from Ghent University in Belgium, Steven completed a small animal rotating internship and a 4-year PhD focussing canine cervical spondylomyelopathy ('wobbler syndrome') at the same institution. He completed his residency in Veterinary Neurology and Neurosurgery at the

Royal Veterinary College, University of London, UK. After becoming a board-certified specialist, he stayed at the RVC where he works currently as an Associate Professor and the Head of the Neurology & Neurosurgery service. His main research interests are chronic spinal conditions and neurosurgery.

Dr. Fran Downing, BVSc MSc DipECVAA MAcadMEd MRCVS: After graduating from the University of Liverpool, UK, in 2007, Fran spent 2 years working in a busy mixed practice in Lancashire. She then returned to the University of Liverpool to undertake an internship in Small Animal Anaesthesia. Since then, she has completed a residency in Veterinary Anaesthesia and spent time as a Lecturer at the Royal (Dick) School of Veterinary Studies in Edinburgh.

In 2013, she gained her European Diploma in Veterinary Anaesthesia and Analgesia, joining the Anaesthesia team at Davies Veterinary Specialists the same year; she became an RCVS-Recognised Specialist in Veterinary Anaesthesia in 2015.

Dr. Anne Quain, BA(Hons) BSc(Vet)(Hons) BVSc(Hons) MVS GradCertEdStud MANZCVS (Animal Welfare) DipECAWBM(AWSEL). Anne is a Lecturer at the Sydney School of Veterinary Science, Camperdown, Australia, and a companion animal veterinarian in private practise. She completed a Master's degree in small animal medicine and surgery through the Murdoch University, Perth, Australia, is a member by examination in the animal welfare chapter of the Australian and New Zealand College of Veterinary Scientists, and a Diplomate of the European College of Animal Welfare and Behaviour Medicine in Animal Welfare Science, Ethics and Law.

Anne co-authored the book *Veterinary Ethics: Navigating Tough Cases* with Dr Siobhan Mullan and is the author of numerous peer-reviewed journal articles and book chapters. She co-edited the sold-out *Vet Cookbook* (published by the Centre for Veterinary Education). She is a member of the Humane Society Veterinary Medical Association leadership council. Her research interests include professional ethics, ethical challenges faced by veterinary teams, the wellbeing of veterinary teams and veterinary primary care.

Dr. Deborah Gow, BVM&S PhD Dip. ECVD FHEA MRCVS: Following graduation from The Royal (Dick) School of Veterinary Studies (R(D)SVS) Edinburgh, Dr. Gow completed a small animal rotating internship at Glasgow Vet School, UK and then returned to Edinburgh to undertake a PhD in immunology at The Roslin Institute, UK, followed by a residency in veterinary dermatology. She obtained her Edinburgh Teaching Award in 2018, allowing her to become a Fellow of the Higher Education Association (FHEA) and gained her European Diploma in Veterinary Dermatology in 2020, becoming a European and RCVS Specialist in Veterinary Dermatology. She is currently Senior Lecturer in Dermatology at the R(D)SVS. She has published widely and has a particular interest in novel targets to control canine atopic dermatitis and genetic skin disease.

Dr. Rodrigo Gutierrez-Quintana, MVZ MVM, DipECVN, FHEA, MRCVS, European and RCVS recognised Specialist in Veterinary Neurology: After graduating from the National Autonomous University of Mexico in 2005, Rodrigo spent one year in small animal practice in Mexico before beginning an internship in 2006 at the National Veterinary School of Nantes, France. In January 2009, he started a residency in Veterinary Neurology at the University of Glasgow, and in 2012, he was awarded the European Diploma in Veterinary Neurology. He is currently a senior university clinician at the University of Glasgow.

Dr. Fraser Hale, DVM, FAVD, DiplAVDC: Fraser graduated from the Ontario Veterinary College, Canada, in 1984 and spent the next six and a half years in small animal general practice. In 1991, he started a mobile veterinary dental service, traveling from clinic to clinic. In 1994, he became a Fellow of the Academy of Veterinary Dentistry and established a stationary dental practice within a general practice in Guelph, Ontario. In 1997, Dr. Hale became a Diplomate of the American Veterinary Dental College, and in 2003, he opened Canada's first dental-only veterinary practice, also in Guelph.

Dr. Hale has been a member of the American Veterinary Dental Society (now The Foundation for Veterinary Dentistry) since 1987 and has served on several committees of the Academy of Veterinary Dentistry and the American Veterinary Dental College.

Dr. Hale was named Fellow of the Year by the Academy of Veterinary Dentistry in 2005 and received the American Veterinary Dental Society Research and Education Award in 2010. He has spoken at conferences large and small around the world, has published many articles, book chapters and an introductory textbook on veterinary dentistry, maintains an educational website and consults on the Veterinary Information Network.

Prof. Åke Hedhammar, DVM, PhD: Senior Professor in Internal Medicine – Small Animals and Diploma in Internal Medicine – CA: Åke is still involved in research mainly on the epidemiology and genomics of spontaneously occurring complex traits in dogs (diseases and behaviour) serving as models also for their human counterparts.

He has served as scientific advisor and veterinary consultant to the Swedish Kennel Club since 1978 and is the initiator of and project leader of the establishment of a Small Animal Epidemiology group at SLU 1994 and of the Swedish Dog Genetics group at Swedish University of Agricultural Sciences and Uppsala University, Sweden and The National Canine Biobank 2002.

Åke was the main applicant and coordinator of Strategic Funding SLU 4 MSEK/year on Canine Models of Complex Diseases and Behaviour Variation (2002–2006), and the node coordinator for SLU in the large-scale integrated project submitted for funding by EU seventh framework on Health 'Unravelling the molecular basis of complex human disorders using the dog as model system'.

He was the initiator of the first International Dog Health workshop held in Stockholm in 2012 and the founder of a national multi-stakeholder group on Canine Health and the International Working Group on Extreme Conformation.

Dr. Hilary Jackson, BVM&S ARPS DVD DipACVD DipECVD MRCVS: Hilary is a graduate of the Royal (Dick) School of Veterinary Studies in Edinburgh where she later returned to complete a residency in Veterinary Dermatology. Since then, she has worked in Academia (the University of Bristol, UK, and North Carolina State University, USA) and for the past 16 years has been in private practice at the Dermatology Referral Service in Glasgow. She has specific interests in canine atopic dermatitis and adverse food reaction and is a founder member of the International Committee for Atopic Diseases in Animals.

Dr. Susan Penelope Knowler, B.Sc. (Special Hons.) PG Teaching Diplomate, PhD: Susan (Penny) Knowler has researched canine Chiari-like malformation (CM) and secondary syringomyelia (SM) for over two decades and co-authored over 30 papers on the subject with Prof. Clare Rusbridge. Sourcing the genealogy of two toy dog breeds, she developed a computer database of worldwide pedigrees which played a pivotal role in establishing the heritability, prevalence and effectiveness of breeding guidelines to reduce the risk of SM. The database and accompanying collected DNA enabled the successful investigation of the genetic basis of CM/SM which suggested a complex multifactorial inherited condition. It also provided evidence of segregated traits and the possible involvement of recessive genes. Using sagittal magnetic resonance images, she developed a system of morphometric measurements of the skull and brain in small breed dogs which correlated with the risk of symptomatic CM and SM. In 2017, she was awarded a doctorate at Surrey University Veterinary Department for the characterisation of CM and secondary SM and its relationship with brachycephalic breeds.

Dr. Jane Ladlow, MA VetMB, Cert VR, Cert SAS, Dipl ECVS MRCVS: Jane has nearly 20 years' experience as a European Specialist in small animal surgery, and over the last 10 years, she has focused on upper airway disease in dogs. In 2005, she started researching brachycephalic obstructive airway syndrome (BOAS), and with Dr. Nai-Chieh Liu at the University of Cambridge, she devised a non-invasive method of assessing airway function in the Bulldog, French Bulldog and Pug and also introduced a clinical grading system for BOAS. During this time, she developed good relationships with breeders and worked with the breed clubs to introduce health schemes in these three breeds that included BOAS. The research has also led to the validation of some new surgical techniques (laser turbinectomy, laryngeal surgeries) that were shared by Prof Oechtering of Leipzig. The project led to the development of The Kennel Club/University of Cambridge Respiratory Function Grading (RFG) Scheme that was launched in 2019 to facilitate the breeding of healthy dogs in Bulldogs, French Bulldogs and Pugs. Jane is continuing this work by promoting the RFG scheme and now intends to start looking at some of the other brachycephalic breeds.

Dr. Elizabeth Leece, BVSc CertVA DipECVAA MRCVS: Liz is a European and RCVS Specialist in Veterinary Anaesthesia and Analgesia and also a past president of the Association of Veterinary Anaesthetists. After a residency in veterinary anaesthesia at the University of Liverpool, she moved to the Animal Heath Trust becoming head of anaesthesia in 2006. She has since worked in university and private referral hospitals while acting as a consultant in a number of clinics across Europe. She loves teaching both inside and outside the clinics, while her current clinical research interests are with the brachycephalics, neuroanaesthesia and chronic pain. She lectures to vets and nurses (and occasionally human anaesthetists) worldwide.

Dr. Nai-Chieh Liu, DVM MPhil PhD: Nai-Chieh Liu is a clinical research fellow at Cambridge BOAS Research Group. Nai-Chieh obtained her veterinary degree in Taiwan, followed by completing her Master's and PhD degrees from the University of Cambridge in 2016 under the supervision of Dr. Jane Ladlow where she focused on developing the non-invasive diagnostic test for BOAS using whole-body barometric plethysmography. Her research has led to a significant impact on BOAS research, including objective screening of BOAS, evaluation of surgical effectiveness and identification of the conformational risk factors. Her current research is focused on dynamic upper airway collapse and sleep apnoea in dogs.

Prof. Paul McGreevy, BVSc, PhD FRCVS: Paul is a veterinarian and behaviourist. He has trained dogs for agility trials, film and television. The author of 9 books and over 300 peer-reviewed articles, he has received numerous Australian and international awards for his research and teaching innovations. In 2017, he won the UK Kennel Club's International Lifetime Achievement Award for his contributions to dog behaviour and welfare.

Dr. Monique Megens, DVM: Monique has a passion for companion animal health and welfare. She dedicates much of her time to help raise awareness, to liaise between stakeholders and to help ensure companion animal health and welfare is high on the international agenda.

Graduated from Utrecht University, the Netherlands (1998), Monique started her career as a companion animal veterinary practitioner. In 2008, she sold her clinic to work as an independent project manager and consultant.

Monique has been involved in veterinary politics in Europe for many years, serving, among others, as the President of the Federation of European Companion Animal Veterinary Associations (FECAVA). On behalf of FECAVA, she was a member of the joint Union of European Veterinary Practitioners (UEVP) and Federation of Veterinarians of Europe (FVE) Animal Welfare Working Group. She was responsible for the European veterinary position paper on stray dogs, the position paper on (il)legal dog trade and the position paper on the responsible breeding of dogs. Monique has been a speaker on these topics at many international conferences.

Currently, she is a member of the World Small Animal Veterinary Association (WSAVA) Hereditary Disease Committee, a member of the Health Committee of the Dutch Kennel Club and in 2020, the COO of the International Partnership for Dogs (IPFD).

Prof. Siobhan Mullan, BVMS PhD DWEL DipECAWBM(AWSEL) MRCVS: Siobhan is Professor of Animal Welfare and Veterinary Ethics at UCD Dublin School of Veterinary Medicine. She has a longstanding interest in veterinary ethics which she has taught at both undergraduate and postgraduate levels. She instigated and ran the 'Everyday Ethics' column in the *In Practice* journal for 10 years. She is a European Veterinary Specialist in Animal Welfare Science, Ethics and Law and co-author of the book *Veterinary Ethics: Navigating Tough Cases.*

Miss Kate Price, BSc (Env. Science) RVN: Kate is a veterinary nurse with a range of experience in both small animal and wildlife practice. Having accomplished an honours degree in Environmental Science (Anglia Ruskin University, 1992), Kate secured her first position working at Shepreth Wildlife Park in Hertfordshire, followed by a position at Battersea Park's children's zoo in London. Eager to become a Qualified Veterinary Nurse, in 1995 she joined the RSPCA Animal Hospital in Putney where she went on to achieve her nursing qualification in 1998. In 2001, she moved to Norfolk to follow her passion for wildlife nursing where she spent 15 years nursing and rehabilitating seals and other wildlife at East Winch Wildlife centre.

In 2003, Kate bought her first pet, Olive, a black Pug from a UK show breeder. In 2009, after Olive experienced a plethora of conformation-related health issues, Kate started questioning and researching brachycephalic health and welfare, triggering over a decade of campaigning on this issue. To date, her campaigning has included online education, writing articles for magazines and TV appearances, including the programme 'Trust Me I'm a Vet', alongside her second Pug Ken, a rescue also affected by a range of health issues associated with his breeding. Kate's ardent determination to educate the public regarding brachycephalic health and welfare led her back to small animal practice in 2016 and the adoption of Ester the French Bulldog, an ex-breeding bitch that was signed over to the practice. In 2019, she began her dream role of Brachycephalic Clinic Nurse at the Animal Health Trust.

Dr. Eleanor Raffan, BVM&S PhD CertSAM DipECVIM-CA: Dr Raffan is a European and RCVS Specialist in small animal medicine. Having spent time in first opinion practice and undertaken specialist training at the vet schools in Liverpool and Cambridge, Eleanor now divides her time between clinical practice and research. She runs a clinical obesity referral service at the University of Cambridge, UK, where she uses her broad medical training to consider patients' problems in the round. Her research focusses on which genes are important in predisposing some individuals to obesity and how they exert their effect, focussing on dogs but with results also studied in humans.

Dr. Aoife Reid, MVB CertAVP(ECC) PGCertVetEd FHEA MRCVS: Aoife graduated from University College Dublin in 2001. She spent some time in equine, mixed practice and small animal practice. She entered small animal emergency practice in 2006 and has worked in this area ever since. She joined Vets Now Ltd. in 2009 and was appointed as the Head of Edge Programmes in January 2014. She has completed her CertAVP (ECC) and her postgraduation in Veterinary Education. She is an RCVS Advanced Practitioner in Emergency and Critical Care and a Fellow of the Higher Education Academy. Aoife enjoys teaching at graduate and post graduate level on various emergency and critical care topics. She has a special interest in canine dystocia and has co-authored several peer-reviewed and continuing education publications.

Dr. Màrian Matas Riera, DVM DipECVO PGCertVetEd FHEA MRCVS: Màrian is a recognised specialist in ophthalmology by the EBVS® after achieving the diploma of the European College of Veterinary Ophthalmology (DipECVO). She graduated from the Universitat Autònoma de Barcelona, Spain subsequently enrolling in a rotating internship in the same university and an ophthalmology internship at the Animal Health Trust (AHT). After finishing the ophthalmology internship, she undertook 3 years of training approved by the ECVO at the AHT and then moved to work at the Royal Veterinary College where she was a Lecturer in Ophthalmology and later appointed the Head of Service. After over 10 years living and working in highly recognised referral centres in England, she decided to move back home. She is now the Director of Memvet-Referral Centre in Palma, where she combines referral work with itineracy work of small animals and equine patients. She has multiple research articles published in peer-reviewed journals and lectures nationally and internationally aiming to share her ophthalmology passion.

Dr. Rebecca Robinson, BVSc MVetMed DipECVAA FHEA MRCVS, RCVS and EBVS® European Specialist in Veterinary Anaesthesia and Analgesia: Becky graduated from the University of Liverpool in 2008. After a short spell in first opinion small animal practice, she completed a 12-month rotating internship in the Small Animal Teaching Hospital at the University of Liverpool. Upon conclusion of this internship, she embarked upon a 3-year residency programme in veterinary anaesthesia and analgesia at the Royal Veterinary College, University of London. She passed her European Diploma in Veterinary Anaesthesia and Analgesia in 2013 and subsequently became an RCVS-recognised specialist in Veterinary Anaesthesia and Analgesia. Since then, she has spent time working as an anaesthesia clinician in a number of private, multi-disciplinary hospitals while also working as a Lecturer in Veterinary Anaesthesia at the Royal (Dick) School of Veterinary Studies, University of Edinburgh, UK. During the following years, she was awarded Fellowship of the Higher Education Academy (FHEA) in recognition of her teaching experience in higher education. She joined the anaesthesia team at Davies Veterinary Specialists in 2016. Over the years, she has authored a number of peer-reviewed research journal and CPD articles and book chapters while also contributing to both in-house staff training and externally organised CPD.

Prof. Clare Rusbridge, BVMS PhD DipECVN FRCVS: Prof Clare Rusbridge graduated from the University of Glasgow in 1991, and following an internship at the University of Pennsylvania, Philadelphia, Pennsylvania, USA, and general practice in Cambridgeshire, she completed a BSAVA/ PetSavers Residency and was Staff Clinician in Neurology at the Royal Veterinary College. She became a Diplomate of the European College of Veterinary Neurology in 1996, a RCVS Specialist in 1999 and Fellow of the Royal College of Veterinary Surgeons in 2016. In 2007, she was awarded a PhD from Utrecht University for her thesis on Chiari-like malformation & Syringomyelia and received the J. A. Wright Memorial Award from The Blue Cross Animal Welfare Charity in 2011. She joined Fitzpatrick Referrals and the University of Surrey, Guildford, UK, in 2013 – after operating a neurology and neurosurgery referral service in Wimbledon for 16 years. She has researched Chiari malformation and syringomyelia for 20 years. She has authored or co-authored over 140 scientific articles (over 50 on canine Chiari) in addition to several book chapters and has co-edited a human medical textbook on syringomyelia. Her other professional interests include neuropathic pain, neurological diseases related to conformation, epilepsy, polymyositis and rehabilitation following spinal injury.

Dr. David Sargan, MA PhD: David Sargan is a geneticist and molecular biologist working for much of his career in veterinary environments, with a particular interest in dogs both for their own sakes and as models for human disease. He obtained an MA in Genetics from the University of Cambridge, UK, went to University College London, UK, for his PhD and then held postdoctoral positions in Geneva and Houston before taking a position at the Royal (Dick) School of Veterinary Studies in Edinburgh initially to work on a sheep virus. Here, he became interested in the evolutionary plasticity of dogs, an interest he brought with him when he returned to Cambridge to head up a comparative genetics research group. He has aligned this interest in plasticity of form with a growing interest in inherited diseases of dog and their welfare implications. He has worked on BOAS as a disease for the last 7 or 8 years, as well as on several less complex defects.

Dr. Alison Skipper, MA Vet MB Cert VR MA MRCVS: Alison Skipper qualified from Cambridge Veterinary School and has worked in small animal general practice for many years. She has a long history of varied personal involvement with pedigree dog health initiatives, and has been one of the veterinary team at Crufts dog show since 2012. She is currently completing a PhD in History at King's College London, funded by the Wellcome Trust. This project explores changing attitudes to pedigree dog health and disease from 1890 to the present, with the aim of informing the current debate and improving effective communication between stakeholders. A co-founder of the Animal History Group, Alison is also increasingly interested in the potential role of veterinary humanities to inform and broaden clinical practice and sustain practitioner well-being.

Dr. Mickey Tivers, BVSc PhD CertSAS DipECVS MRCVS: Mickey is an RCVS and EBVS® European Specialist in Small Animal Surgery. He is Head of Surgery at Paragon Veterinary Referrals in Wakefield. Mickey has a particular interest in brachycephalic obstructive airway syndrome (BOAS), and he continually strives to improve outcomes for affected dogs and their owners.

Dr. Sean Wensley, BVSc MSc FRCVS: Dr Sean Wensley is Senior Veterinary Surgeon for Communication and Education at the UK veterinary charity, the People's Dispensary for Sick Animals (PDSA). Holding a Master's degree in Applied Animal Behaviour and Animal Welfare, he has contributed to animal welfare and conservation projects on five continents. Sean is an Honorary Lecturer in Animal Welfare at the University of Nottingham, UK, and a Guest Lecturer in Animal Welfare at Queen's University Belfast. He chairs the Federation of Veterinarians of Europe (FVE) Animal Welfare Working Group, was President of the British Veterinary Association (BVA) from 2015 to 2016 and is a Fellow of the Royal College of Veterinary Surgeons. In 2017, he received the inaugural World Veterinary Association (WVA) Global Animal Welfare Award for Europe.

1 Introduction

How Can a Brachycephalic Boom Cause a Welfare Bust?

Rowena M. A. Packer and Dan G. O'Neill
The Royal Veterinary College

We are in the midst of a brachycephalic dog welfare crisis. And this crisis is not just a UK phenomenon but is unfolding right across the world. But how can this be? Surely the public loves these adorable little flat faces, and surely this unbridled love can only be a good thing for these brachycephalic dogs? With this conundrum, we welcome you, our dear reader, to the complex world of brachycephalic animals. In this world, issues such as conformation, genetics, disease, human perceptions, veterinary care, breeding, social media along with a host of other factors all interact in a complex interplay to impact on the ultimate quality of the lives of these animals. This book will take you on a rollercoaster ride through the highs and lows of brachycephalism in companion animals and will help you to grasp the rich and fuller story of how our human drive to own and love flat-faced animals does not always equate with these animals living good lives, or even 'lives worth living'. Whether you have picked up this book as a veterinary professional, owner, student or breeder, we urge you to park your current views on brachycephalism at the front cover and to enter this book with an open mind. And we challenge you to have unchanged views by the time you have finished reading.

As the Internet age took hold some decades ago, some people were of the opinion that electronic media would herald the extinction of textbooks as we knew them then. Libraries would become museums. There was a view that the pace of knowledge accrual had become just too rapid and fluid for published books to keep up. But time has shown this view to be very wrong, and today, books are as useful and popular for summarising accumulated knowledge and for learning as ever. With this supportive background, we are pleased to present this book to you, and to share with you the accumulated knowledge and wisdom from 29 world-leading international experts on brachycephalic health and welfare. This book provides an unparalleled resource of factual information, scientific evidence and expert opinion across a range of topics relating to brachycephalism in companion animals, but it also aims to go deeper. In each chapter, the authors go beyond facts to offer their views and insights from their decades of experience working on their topic areas. We want to stimulate you to think deeply and openly about the issues that are raised in these chapters, and to challenge your current views on brachycephalism, with the aim that you move towards new and more considered beliefs.

But why is this book needed right now, given that many popular flat-faced breeds were invented in the 1900s? This book is needed because something odd is happening right now in the twenty-first century to put us in the midst of a brachycephalic crisis that is challenging the well-being of our dogs like never before. This brachycephalic crisis has pushed its way to the top of the global animal welfare agenda over the past decade and garnered the attention of the veterinary profession, welfare charities, government, kennel clubs, academia, media and wider public. Health issues of brachycephalic dog breeds have taken centre stage in such a dramatic fashion because this small group of dog breeds has become so phenomenally popular over a short span of time. Colloquially dubbed 'the brachy boom', this rapid popularity increase has magnified the welfare impacts from brachycephalism from affecting just a small proportion of dogs to now affecting huge swathes of the wider dog population. And the process of feeding an insatiable public demand to own some of these popular breeds has generated breeding and supply chains that often prioritise quick sales and large profits with little regard to the welfare of the animals or their future owners. This perfect storm has created the current paradox where a boom in popularity seems to have led to a bust in welfare.

The marked rise in numbers of brachycephalic dogs has been coupled with a rapid expansion in scientific evidence that is identifying more and more health challenges faced by brachycephalic breeds. In order to get a deeper understanding of these complex issues, this book will guide you through many facets of this 'brachycephalic issue'. Whether you read this book from cover to cover or choose to dip into specific sections that are pertinent to your specialism, profession or breed of interest, we hope this comprehensive guide to brachycephalic health and welfare will consolidate the veritable tsunami of information on brachycephalic breeds generated in recent years into an accessible, engaging and stimulating format. This is, however, no ordinary veterinary text book. The health problems of brachycephalic dogs do not only exist within the veterinary bubble, and acknowledging this, we have structured the book into two sections.

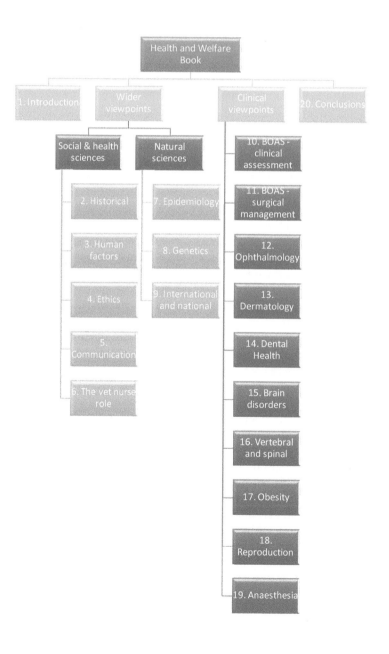

The first section covers important contextual topics that provide a solid grounding on the wider brachycephalic picture. The health and welfare of brachycephalic dogs is fundamentally a human issue; humans initially chose to harness those mutations that cause the flat-faced conformation, have perpetuated this phenotype over generations, and are currently proliferating brachycephalic dogs in numbers like never before. To achieve an understanding about options for the elusive 'way out' of the brachycephalic crisis requires a range of perspectives, including the human stories that have driven this trend. Chapter 2 explores over 100 years of historical perspectives on the breeding of brachycephalic dogs, including historical concerns for their health, stakeholder interactions and the roles of the veterinary profession. Such information is key to avoid repeating previously unsuccessful strategies aimed at improving the welfare of brachycephalic dogs that the current generation of veterinary professionals and animal welfare advocates may be unaware of. Despite being perceived as a 'dog problem', Chapter 3 explores human attraction and seeming 'addiction' to brachycephalic dogs that is fuelling the brachycephalic boom, questions whether humans are instinctually drawn to brachycephalic dogs and considers how owner (mis)perceptions of their dog's health may perpetuate the popularity of these breeds.

Veterinary professionals are a key stakeholder group in improving the health and welfare of individual brachycephalic dogs under their care and in driving reforms at a population level. Despite this seemingly positive role, the potential for ethical conflict when treating brachycephalic dogs, particularly in relation to veterinary involvement in reproductive procedures, and conformation-altering surgeries represent particular challenges for veterinary professionals. Some frameworks to navigate such issues are discussed in Chapter 4. Whether optimal welfare can be achieved for any individual brachycephalic dog will depend upon a variety of factors related to both their owner(s) (e.g. perception of disease severity and presence, financial constraints) and the veterinary team treating them (e.g. technical and communication skills). The potential for conflict may be high when veterinary surgeons manage brachycephalic cases but where owners do not perceive health problems in their much-loved pet, and therefore this chapter highlights the importance of good communication. Chapter 5 discusses opportunities and challenges for veterinary professionals associated with talking to current and prospective owners about common health issues encountered in brachycephalic dogs. In addition to veterinary surgeons, the role of the registered veterinary nurse (RVN) in welfare-focused patient care is increasingly recognised. Chapter 6 discusses opportunities for RVNs to help individual brachycephalic dogs by e.g. brachycephalic nurse clinics, as well as also exploring wider nursing educational and advocacy roles.

Understanding the extent of the brachycephalic crisis, and devising strategies to escape from it, requires evidence. The cold hard data of brachycephalic health is explored in Chapter 7, which offers cutting-edge epidemiological data on breed-specific and brachycephalic-wide disorder predispositions, demography and longevity. Chapter 8 explores the genetics of brachycephaly and common health disorders in brachycephalic breeds (termed 'the brachycephalic syndrome'), including the potential for genetic tests to help select towards healthier brachycephalic breeding stock in the future. Finally, public concern and even outrage concerning brachycephalic welfare have motivated regulators and government agencies to take diverse actions to protect brachycephalic welfare internationally. Chapter 9 describes issues and actions across regions and countries, especially Europe and North America, encompassing kennel and breed clubs, veterinary organisations, research institutions, regulators, humane groups and other non-profits.

The second section of this book moves to a more clinical setting. For the first time, this book brings together international veterinary experts to discuss the latest understanding on the health problems afflicting brachycephalic dogs. Taking a 'head-to-tail' approach, this section starts with the 'poster child' disorder of brachycephalic health, brachycephalic obstructive airway syndrome (BOAS; Chapters 10 and 11), and explores progressive methods of clinical assessment and surgical management of this complex disorder. Despite the commonly cited appeal of baby-like wide eyes in brachycephalic dogs, Chapter 12 discusses well-recognised common ophthalmic disorders in brachycephalic patients related to their conformation. Chapter 13 then reviews the main dermatological diseases related to brachycephaly in dogs, highlighting where these vary between common brachycephalic dog breeds and exploring how the extreme conformations typical of brachycephalic breeds are associated with complications of common dermatological disorders in dogs. Chapter 14 explores a neglected, but yet extremely important welfare topic, the dental and oral health of brachycephalic patients. Chapters 15 and 16 take us on a journey through the central nervous system of the brachycephalic dog. Chapter 15 explores the impact of brachycephaly on the nervous system, focusing on its disruption effects upon cerebrospinal fluid (CSF) movement and absorption, and the pathophysiology, clinical presentation, diagnosis and management of CSF disorders. Chapter 16 discusses common spinal problems in smaller brachycephalic breeds with an emphasis on vertebral malformations. Breed-specific differences are highlighted together with new insights into the prevalence, clinical relevance, diagnosis and treatment of these congenital malformations.

Obesity and overweight are growing welfare concerns for canine welfare, with even greater significance in brachycephalic animals due to the potentially exacerbating effects of obesity upon pre-existing conformation-related health problems such as respiratory disease and dystocia. Chapter 17 discusses obesity with particular reference to its impacts on brachycephalic animals, focussing on pathophysiology, prevention and treatment. Assisted breeding techniques, such as artificial insemination and caesarean section, are becoming more common in dogs. Increased popularity of some brachycephalic dog breeds has been associated with increased demand for these services. Chapter 18 discusses reproduction principles with special relevance for brachycephalic companion animals, including fertility, gestation, birth and postnatal care. The importance of tailored anaesthesia to facilitate surgical procedures to treat many of the disorders discussed in the second section of this book is increasing as brachycephalic breeds become more popular. Chapter 19 focuses on anaesthetic management during the entire perioperative period, including pre-anaesthetic assessment and stabilisation, induction and maintenance of general anaesthesia and recovery from anaesthesia and postoperative management.

We live at a pivotal moment in time for the future of our brachycephalic companion animals. We know more than ever about their health challenges, and more stakeholders than ever are invested in trying to protect their health and welfare. There is general consensus from many stakeholders in dog welfare that we all need to 'Stop and think before buying a flat-faced dog'. And yet, paradoxically, the numbers and popularity of brachycephalic animals continue to rise unabated. In a world where supply and demand shape the make-up of the companion animal populations that we own, exerting changes in human behaviour will be central to solving this brachycephalic crisis. In our new world of social media, how we can influence owners to actually stop and think before buying a flat-faced dog? Can hearts and minds be changed, or are brachycephalic dogs here to stay in ever-increasing numbers? If so, is a 'stick' rather than a 'carrot' approach needed to protect canine welfare by, for example, enacting legislation to enforce change? Can breeders be persuaded of the merits of breeding away from extremes of conformation such that a more moderate version of brachycephalism in these breeds can become compatible with acceptable health and welfare? Or can a dog with a brachycephalic head shape *ever* be considered healthy? These are just some of the many questions we hope you will be stimulated to reflect upon during your reading of this book. We hope that the combination of cutting-edge literature on key aspects of the brachycephalic crisis summarised by authoritative sources, alongside new and diverse perspectives on brachycephalic dog health and welfare, will encourage open-minded readers to reconsider their own preconceptions of this crisis and also perhaps inspire you to consider your own role in helping to shape a better future for companion animals.

Part I

Wider Viewpoints

2 A Historical Perspective on Brachycephalic Breed Health and the Role of the Veterinary Profession

Alison Skipper
King's College London

CONTENTS

Introduction..7
Note on Terminology and Scope..8
An Overview of Brachycephalic Health Activism, 1890 Onwards 8
    Early Brachycephalic Reshaping and Activism, c. 1900 8
    Mid-Century Veterinary Intervention, c. 1960 14
    Brachycephalic Activism in the Twenty-First Century 16
Discussion...17
Conclusions...19
References...20

INTRODUCTION

Since the turn of the twenty-first century, certain breeds of brachycephalic dogs, particularly French Bulldogs, Pugs and Bulldogs, have become increasingly fashionable (O'Neill et al., 2018, 2016, 2019). The recent explosion in the popularity of these breeds as pets and in contemporary visual culture has triggered a similarly dramatic increase in general concern for their welfare, in veterinary engagement with their diseases and in ethical debate over their breeding, as the need for this book itself demonstrates. Activists today often compare images of modern dogs with those of the past in an effort to demonstrate an indisputably progressive exaggeration in brachycephalic conformation over time, with an implicit assumption that poor health of brachycephalic breeds is also a new occurrence (Elegans, 2012). But brachycephalic diseases are not, as such comparisons might imply, a new development; the particular ailments of short-faced breeds have been repeatedly recognised and discussed for over a century (Skipper, 2020). The current controversy is actually the third time that these problems have attracted particular attention, with debate previously peaking in the years around 1900 and 1960 (Redwar, 1901b; Singleton, 1962b). Yet, despite this recurring concern, brachycephalic health issues remain a substantial challenge and a political minefield today (Atkinson, 2018).

    Informed by the aphorism that "[t]hose who cannot remember the past are condemned to repeat it" (Santayana, 1905), this chapter argues that the overlooked history of brachycephalic disease and welfare activism can both explain how the current situation arose and also inform more effective approaches to future engagement. Drawing on the observations and opinions of dog breeders and veterinarians recorded in British dog breeders' specialist newspapers and veterinary journals from the late nineteenth century to the present, it shows that those who favoured particularly exaggerated brachycephalic conformation have always been confronted by others who questioned their practices. Proponents of more extreme breeding have variously been motivated by commerce, fashion, aesthetics, 'love of the breed' and the demands of the show ring, while reformers – not always or only veterinarians – have challenged them with criticisms informed by their own understandings of science, welfare, aesthetics, breed heritage, and morality. Although the arguments used by each side have varied, the underlying constant has always been a conflict between two factions with fundamentally incompatible, yet deeply held, world views and agendas, whose interactions have therefore mostly been antagonistic, emotional and unproductive (Dangerfield, 1964).

Ample historical evidence demonstrates the inescapable truth that breeders – who ultimately decide which dogs to breed from, after all – are necessarily key to the success of any dog breeding health strategy (Skipper, 2016). Previous interventions in brachycephalic health have generally been unsuccessful because the relevant breeders were not persuaded by them, as this chapter will demonstrate. Despite some encouraging recent initiatives, communication between the various stakeholders in brachycephalic health remains hampered by mutual suspicion and hostility, often amplified and exacerbated by the bitter tribalism and soundbite rhetoric so characteristic of the wider zeitgeist in (inter)national politics and culture. In contrast, efforts to tackle genetic disease unrelated to conformation in other breed types have achieved far more successful cooperation, demonstrating that progress in canine health is possible with adequate buy-in from the important players (Lewis and Mellersh, 2019). However, both historical precedent and modern change theory show that the barriers to change in breeding practices are complex and hard to overcome; new 'human behaviour change' approaches are needed to defuse confrontation and facilitate productive debate (Bonnett and Sandøe, 2016). Offering history both as justification and a potential tool for meaningful engagement, this chapter draws key lessons from the past to inform more successful intervention in brachycephalic breeding in the future.

NOTE ON TERMINOLOGY AND SCOPE

The terms 'brachycephalic' and its opposite 'dolichocephalic' were not widely applied to canine skull morphology until the second half of the twentieth century, but other terms, such as 'short-faced' or 'noseless', held similar meanings beforehand.[1] The breeds included here – and the emphasis on Bulldogs – reflect those discussed in the historical sources, as does the terminology used, including the use of capitalised names for formalised breeds. 'Breeding' is here used as an inclusive term that covers both mate selection and reproductive husbandry. Pedigree dog breeding and health improvement is (and always has been) an international process, but the research which informs this chapter logistically focuses on Britain.

AN OVERVIEW OF BRACHYCEPHALIC HEALTH ACTIVISM, 1890 ONWARDS

Early Brachycephalic Reshaping and Activism, c. 1900

In Britain, the creation of the pedigree dog and the reshaping of canine bodies followed the rise of competitive dog shows under the overarching control of the Kennel Club (KC), which was founded in 1873 (Ritvo, 1987; Worboys, Strange, and Pemberton, 2018). KC regulation accelerated the separation of traditional types of dogs into distinct breeds, their desired appearances defined for the show ring by technical breed standards. By the end of the nineteenth century, the formalised 'dog fancy' had become a popular pastime, its emphasis on elite canine lineage mirroring the concurrent preoccupation with human class and heredity in wider society (Worboys, Strange, and Pemberton, 2018).

---

[1] The terms 'brachycephalic' and 'dolichocephalic' were coined by Anders Retzius, a nineteenth-century Swedish anatomist, anthropologist and veterinarian, to describe different human skull shapes, and were thereafter used in the doctrines of scientific racism (Gould, 1981; Larsell, 1924). They were later applied to dogs by the American medical eugenicist Charles Stockard (Stockard et al., 1941).

The most popular breeds tended to feature a distinctive appearance and an appealing breed mythology that resonated with the popular culture of the time. Brachycephalic breeds which fitted this pattern included (English) Bulldogs, Pekingese, Pugs and Toy Spaniels (now known as King Charles Spaniels). Bulldogs, long regarded as a particularly British breed, appealed to nationalistic Englishmen (Amato, 2015); 'Oriental' lapdogs, such as Pekingese and Japanese spaniels, were popular among wealthy women as part of a broader fashion for 'Oriental' arts (Cheang, 2006); Toy Spaniels had the elitist glamour of their supposedly royal antecedents (Ritvo, 1987). Their original breed standards emphasised their characteristic features, such as "a strikingly massive" head and "extremely short" face in the Bulldog, or a "short" muzzle, "large" head and "very large … prominent" eyes in the Pug, because these accentuated points were what distinguished them from other dogs (Dalziel, 1889). However, subsequent cumulative selection for these attributes progressively reshaped these breeds towards even more extreme conformation, in a process already well advanced before the First World War. The Bulldog of the 1870s resembled the modern Staffordshire Bull Terrier (Stonehenge, 1879), but by 1900, a new 'cloddy' type was common, which had short, twisted legs, a heavy build, a large head and a shorter face than before (Farman, 1899a). A few years later, Pekingese were similarly transformed. "Perfect flatness of face has become a fetish", one writer remarked, noting that rigorous selection had visibly altered the breed in just 6 years between 1904 and 1910 ('Lady Betty', 1911) (Figure 2.1).

These early changes were directly driven by show ring practices. Some judges prioritised extreme conformation, as when one described one Pug after another as "too long in face … not short enough in face … a nice short face" (Anon, 1896). Moreover, although breed standards had explicitly been developed to standardise judging decisions (Worboys, Strange, and Pemberton, 2018), some judges nevertheless still favoured more extreme morphology than the guidelines required. Objectors to the

Champion Palace Shi.
A study of shortness in face.

FIGURE 2.1 Pekingese showing "the ideal of 'shortness of face'" of 1904: by 1910, "noses [were] wiped out as superfluous". (Reproduced with permission of The Kennel Club Ltd from *The Kennel*, 1911, 438.)

1900 'cloddy' Bulldog complained that "the standard of points … in no way justifies this absurd exaggeration" (Hardcastle, 1902a). 'Screw' tails were then "expressly forbidden" by the Bulldog Club's breed standard (Farman, 1899b), yet the controversial 'cloddy' dogs, which often had screw tails, nonetheless achieved "any amount of winning", because fanciers and judges admired the type (Warland, 1898). By the 1920s, the explicit prohibition on screw tails had been removed from the breed standard (Leighton, 1924). This change responded to, rather than determining, show ring preferences, despite the original intention that breed standards should constitute blueprints of idealised conformation (Worboys, Strange and Pemberton, 2018). Eventually, of course, the 'screw' tail became ubiquitous in Bulldogs (which are now homozygous for the causal mutation (Mansour et al., 2018)). Yet at no time has the British breed standard explicitly demanded a screw tail, demonstrating that, throughout the history of the dog fancy, show ring fashions can override breed standard requirements and themselves drive conformational change (Vesey-Fitzgerald, 1948; Sutton, 1980; The Kennel Club, 1989). Moreover, in recent years some show Bulldog breeders have deliberately selected for less tightly screwed tails and claim to have "made massive strides" in "moving on" from producing dogs with the most extreme morphology (Collins-Natrass, 2020). Both show ring expectations and the canine bodies they shape thus remain plastic and subject to change today, despite reduced genetic variation and the weight of breed tradition.

The c. 1900 trend towards progressive exaggeration was accelerated by commercial demand. Dogs with more extreme conformation fetched higher prices, based on their show ring value, their breeding potential, or simply their desirability as possessions ('Old Dane', 1899). Advertisements often emphasised extreme features to lure buyers; for example, one Bulldog was offered for sale as the "largest-headed and heaviest-wrinkled dog living" (Lambert, 1898). One critic remarked that "certain so-called 'points' … increase the value of an animal until there seems to be no limit to the price which may be asked and paid for a dog" (Anon, 1898a). This was particularly true for the Bulldog which, many commentators agreed, "attained a greater popularity as a … creature of abnormal proportions than he had ever done", reflecting a wider fin-de-siècle cultural fascination with novelty (Anon, 1899; Denisoff, 2007). The fashionable aesthetic of the refashioned Bulldog combined synergistically with a lucrative American market for British pedigree dogs (Derry, 2003) and a rise in nationalism, linked to British military action in South Africa (Daly, 2015), to produce an explosion in the market for 'cloddy' Bulldogs. Top prices soared from £100 in 1898 to £1000 in 1900, just 2 years later – equivalent to roughly £120,000 in 2018 (Anon, 1898b, 1900)[2] (Figure 2.2). The trend towards flatter faces in lapdogs was similarly influenced both by 'Oriental' fashionable aesthetics (Cheang, 2006, 2007) and by the wider concurrent cultural innovation of 'cuteness' in toys and popular images, which were for the first time becoming visually distorted to emphasise juvenile features (Merish, 1996; Ngai, 2012). Pekingese, in particular, became "regarded as a sort of gold mine" ('Lady Betty', 1910a), with many "sold as good specimens … simply on the strength of a short face" (Tomkins, 1911).

However, more than a century ago, concern was already being expressed about health issues in these refashioned animals. In Bulldogs, some breeders and judges deplored the rise of physical features such as "teeth … mostly constructed to protrude or fall out" (Anon, 1898b), "nostrils … so small and pinched that it would be hard … to pass a tooth pick" (Crowther, 1893) and excessively undershot jaws (Beecher, 1902). They warned that 'cloddy' Bulldogs were sometimes "so exaggerated that a dog can only move with difficulty, and pants painfully with the least exertion" (Redwar, 1902a). They highlighted problems still recognised today, such as dystocia, which caused "terrible mortality amongst puppies and brood bitches" (Crowther, 1893), and short lifespan: 7 years was considered "quite an old age for show Bulldogs" (Anon, 1901). Bulldogs were also thought "peculiarly susceptible" to 'heat apoplexy', with famous prize winners regularly dying on the summer show circuit (Clarke, 1911; St John Cooper, 1900; W.G., 1900). In brachycephalic

---

[2] Calculation made using the Bank of England online inflation calculator (https://www.bankofengland.co.uk/monetary-policy/inflation/inflation-calculator), accessed 23 September 2019.

THE BULLDOG CHAMPION RODNEY STONE (K.C.S.B. 331c).
THE PROPERTY OF MR. W. F. JEFFERIES.

FIGURE 2.2   Rodney Stone, the first £1000 Bulldog. (Reproduced with permission of The Kennel Club Ltd from *The Kennel Gazette*, 1899, opposite 413.)

toy breeds, some commentators directly linked flat faces to health problems. One criticised the "snorty, watery troubles with which the short nosed, big eyed breeds are wont to plague themselves and their owners" (Dennis, 1911). Some Toy Spaniel breeders cautioned that extremely short faces were too often accompanied by cleft palates, screw tails and joined toes; one controversial leading breeder condemned such dogs as "noseless cripples" (Anon, 1903a, b; Lytton, 1911) (Figure 2.3). The few specialist canine veterinarians of the time also acknowledged that brachycephalic breeds were prone to dystocia, more risky to anaesthetise "due to the shape of the nose", and susceptible to eye conditions such as corneal ulceration, describing these problems in ways familiar to the modern clinician (Hobday, 1897; Skipper, 2020).

Nevertheless, there were few attempts to reform breeding practices before the mid-twentieth century. Not many veterinarians were even very interested in treating dogs, let alone in engaging with "the immoral ring of dog fanciers" ('Cave Canem', 1905; Skipper, 2020). The handful of specialists who served the dog fancy were preoccupied with combating distemper and rabies (Pemberton and Worboys, 2012), and probably also restrained by self-interest, particularly as professional ethics then deemed that a veterinarian's primary duty was to their clients, not their patients, except in cases of outright cruelty or neglect (Skipper, 2020; Woods, 2012b). Indeed, breeding was rarely conceptualised as a welfare issue at all. The few welfare campaigners (mostly upper-class women) who condemned dog shows focussed instead on "the infinite suffering to the animals exhibited" caused by show husbandry ('Ouida', 1897). Consequently, those who campaigned for breeding reform, such as deliberate outcrossing or a renewed focus on functionality, were mostly fanciers themselves, whose concern stemmed from their personal involvement with the breeds concerned (Redwar, 1901a).

Noseless Atrocity bred by the Author

FIGURE 2.3   Toy Spaniel puppy bred by Lady Wentworth, an outspoken opponent of 'noselessness'. (Reproduced with permission of The Earl of Lytton for the descendants of Judith Blunt-Lytton, Baroness Wentworth from *Toy Dogs and their Ancestors*, London: Duckworth and Co, 1911, 132.)

Critics of exaggerated conformation in 1900 understood inheritance and disease very differently from today. At that time, both biological scientists and wider Western society were obsessed with the idea that poor heredity and environment caused constitutional degeneration and racial decline, revealed by physical weakness or deformity (Denisoff, 2007; Pick, 1993). Informed by this general belief, reformers compared the "deplorable" physique of Toy Spaniels to people "in the idiot wards of any asylum or workhouse" (Anon, 1902) and attributed the exaggeration of the 'cloddy' Bulldog to racial degeneration from reckless inbreeding "concentrating … all the hereditary weaknesses … present in their closely-related ancestors" (Redwar, 1902a). This idea was so widespread that a cartoon showing a highly bred Bulldog, atavistically degenerated down the evolutionary ladder into a part-amphibian, even appeared in the popular magazine *Punch*[3] (Anon, 1906; Figure 2.4).

Although brachycephalic ailments were generally explained in terms of racial degeneration, they were perceived differently in different breeds. Lapdogs were regarded as "frail little beauties" ('Lady Betty', 1910b), "like a rare flower or a piece of china" (Lytton, 1911). Since they were expected to be delicate, breeders seldom argued over their health, demonstrating a 'culture of acceptance' of physical problems that normalised these issues among fanciers, paralleling the 'normal for the breed' mindset described within brachycephalic breed communities today (Packer, Hendricks and Burn, 2012). In contrast, the reshaping of the Bulldog triggered angry confrontation between different factions of enthusiasts. Admirers of the 'cloddy' Bulldog thought it "vastly improved" ('A Fancier', 1902). They considered stamina irrelevant – "he does not require to be so active, now that he is no longer used for bull-baiting" ('A Fancier', 1902), and praised the "healthy commercialism" which had "better[ed] the lot" of Bulldogs by increasing their "pecuniary worth" ('B', 1903).

---

[3] Cartoon originally found at https://terriermandotcom.blogspot.com/2011/03/development-of-bulldog-1906-punch.html, accessed 27 May 2019.

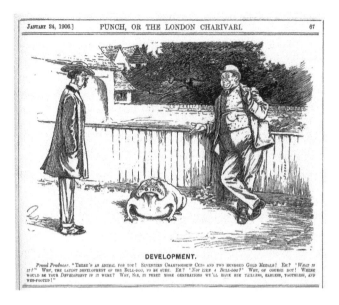

FIGURE 2.4   "The Latest Development of the Bull-dog" by Jaseff. (Reprinted with permission of Alison Skipper from *Punch* 130, 1906, 67.)

But reformers argued that such breeders betrayed the Bulldog's heritage (and, by implication, their country) by transforming "our poor old national breed" into "a 'freak'" (Hardcastle, 1902b), and basely seeking profit "in the same way as a purveyor of bacon or sausages" (Warland, 1902). One critic summed up their viewpoint:

> Our present show dog ... is ... a dog in which many of the typical points have been so exaggerated and travestied as to radically alter the character of the animal ... [J]udges ... have ... encouraged the breeding of a crippled dog, of enfeebled constitution, with little or no stamina, who has been produced by breeders ... merely to secure prizes and enhance the pecuniary value of their stock, without sufficient regard to the true interests and future well-being of the breed itself.
>
> *(Redwar, 1902b)*

Each side hurled impassioned abuse at the other, with 'cloddy' supporters calling their critics "canine faddists" ('A Fancier', 1902), and reformers claiming that 'cloddy' breeders held "effeminate" and "emasculated" views which had "prostituted" Bulldog breeding – vicious insults then (Hardcastle, 1902b; Warland, 1902). Some critics did try to temper their comments; "I do not desire to suggest that breeders are consciously inhumane. I do not think that we sufficiently realise the evils resulting from the course we are taking", one acknowledged (Redwar, 1901b). Reformers proposed a practical solution to brachycephalic problems – to choose less exaggerated and more distantly related dogs for breeding (Beecher, 1902; Redwar, 1901a, b) – which would have been effective and easily implemented, since outcrossing to unregistered dogs was then freely allowed (KC Registrations, 1910). But, unsurprisingly, the factions found no common ground. The reformers failed, and the popularity of the reshaped Bulldog continued to grow (Clarke, 1910). Flatter-faced conformation similarly continued to spread through other brachycephalic breeds, such as Toy Spaniels and Pekingese (Dean, 1913; Mallock, 1916). The trend was probably exacerbated by the First World War, which devastated the dog fancy through conscription of kennel staff, food shortages, public hostility to pet ownership, suspension of shows and, eventually, a KC ban on dog registrations, which together caused a catastrophic population bottleneck (Howell, 2013; Skipper, 2016).

Mid-Century Veterinary Intervention, c. 1960

There was minimal discussion of conformation-related disease during the interwar years, because the dog world was preoccupied with supporting research to develop a distemper vaccine (Bresalier and Worboys, 2014). The Second World War, like the First, then temporarily supplanted normal canine affairs (Kean, 2017). After the war, however, a surge in demand for pedigree dogs "dr[o] ve the mongrel off the market" (Hodgman, 1962) and increased KC registration numbers tenfold between 1941 and 1947 (The Kennel Club, 2009). This led to claims of increased indiscriminate breeding by opportunistic newcomers, with a subsequent rise in the incidence (and/or awareness) of breed-related disease (Anon, 1962a; Hodgman, 1962). Meanwhile, other fanciers were beginning to apply Mendelian genetics to improve pedigree dog health. Irish Setter breeders worked together to tackle progressive retinal atrophy (PRA) by implementing a highly successful breeder-led programme of pedigree analysis and test mating to identify and breed dogs free from the causal allele (Skipper, 2016). This project reflected the precepts of mid-century agricultural science, which advocated 'pure line' inbreeding to obtain a predictable strain of plants or animals of known (and, ideally, homozygous) genotype, intended to be free from hidden deleterious genes (Theunissen, 2014). The Setter PRA project became highly influential within the wider dog fancy. It catalysed a broader shift towards the 'pure line' approach by restricting the KC registration of Setters with unknown ancestry in order to keep the breed free from PRA, beginning a process which eventually culminated in the almost total closure of all KC breed registries (Skipper, 2016). Thus, the deliberately restricted gene pools that characterise modern pedigree dog breeding, which are now considered problematic barriers to genetic diversity and, consequently, breed health (Farrell et al., 2015), were originally introduced to improve and maintain genetic health – a deeply ironic reversal.

This new interest in breed-related disease also triggered breeders to begin debating conformation-related disease again, at first among themselves and later in response to an unprecedented wave of challenge from veterinarians. The default orthodoxy in the dog fancy was that selective breeding for the show ring led to the progressive improvement of pedigree dogs (Daly, 1954). By mid-century, however, some fanciers were questioning this assumption. They resurrected the 1900 concerns over extreme conformation in various brachycephalic (and other) breeds. Commentators were again particularly worried about the Bulldog, with the paradoxical difference that, since ongoing selection for the show ring had continued to alter the breed, the 'modern' winning Bulldogs of 1900 were, half a century later, nostalgically recalled as the 'old type' (Daly, 1955). One critic condemned the 1950s Bulldog as "[a] gross, misshapen, wheezing, snuffling, slobbering creature", blaming "generations of show breeders" for having "ruined" the "national breed" (Anon, 1953) and triggering vigorous opposition to this criticism from enraged Bulldog lovers (Daly, 1953).

In parallel, a largely separate debate was surfacing among veterinarians, where, by the 1950s, small animal work was becoming a mainstream professional interest for the first time, as evidenced by the launch of the British Small Animal Veterinary Association (BSAVA) in 1957 (Gardiner, 2014).[4] Clinicians at the veterinary schools and the Animal Health Trust (AHT), a pioneering centre for veterinary clinical research, were rapidly advancing knowledge in canine medicine (Joshua, 1958). Meanwhile, new discourses of farm and laboratory animal welfare were challenging the ethics of intensive husbandry and breeding practices (Kirk, Cantor, and Ramsden, 2014; Woods, 2012a). In this climate, some veterinarians began to question the ethics of the dog fancy (Knight, 1956).

Few veterinarians had previously engaged with the problems of dog breeding, apart from one or two who were also fanciers themselves. Early post-war veterinary efforts to improve pedigree dog health were largely brokered by John Hodgman, the founding director of the AHT's new canine section and a skilled political strategist, who served as President of both the British Veterinary Association (BVA) and the BSAVA (Ormond, 1973). Hodgman became a KC member in 1946

---

[4] This trend was driven by increased demand for small animal treatment, reduced competition from unqualified practitioners following the 1948 Veterinary Surgeons Act and the increasing number of female veterinarians, for whom companion animal practice was often a pragmatic career choice (Hipperson, 2018).

to forge links between the veterinary and breeding communities; his work facilitated a gradual strengthening of the relationship between the KC and the BVA during the 1950s (Anon, 1946; The Kennel Club, 1953). But some other veterinarians, increasingly concerned by pedigree dog diseases, found this progress too slow. Brian Singleton, a pioneer of brachycephalic soft palate surgery in his Knightsbridge practice, took a much more confrontational stance (Singleton, 1962a). Positioning himself as a moral critic of the show world rather than its technical support, he condemned "the folly of present day trends" and was "disgusted to see what certain breeders could do with certain breeds" (Singleton, 1962b).

Despite their different perspectives, Hodgman and Singleton together launched a powerful critique of pedigree dog breeding at the 1962 BVA Congress. Hodgman argued that "veterinary surgeons must take a firm stand" to discourage hereditary defects, but also emphasised the importance of constructive engagement with breeders, despite this being "a slow and often a thankless task" (Hodgman, 1962). Singleton, more outspoken, felt that "the present picture was quite ridiculous". He claimed that "during the past 3 decades anatomical monstrosities had been created in an ever-increasing number of breeds" and particularly highlighted brachycephalic problems;

> Why should Bulldogs be reared with concave thoracic cages, deformed respiratory passages, hip dysplasia, corkscrew tails? Why should Pekes have faces so flat that the hair on the facial folds caused constant irritation to the cornea, resulting in ulceration, pigmentation and blindness?
>
> *(Singleton, 1962b)*

Singleton strongly believed that veterinarians should challenge irresponsible breeding practices by reporting corrective surgery for conformational disease to the KC, explaining "to clients who bred abnormal stock the seriousness of what they were doing" and educating the puppy-buying public (Singleton, 1962b). He subsequently reinforced his campaign with vigorous public engagement on national television and radio, using similar strategies to his successors, the pedigree dog health activists of the twenty-first century (Anon, 1962b).

This revolutionary approach was incendiary to the dog fancy, because breeders seldom perceived veterinarians as superior authorities or moral arbiters, and so were enraged by what they saw as unjustified interference (Ellis-Hughes, 1962). Although willing to enlist veterinary support to identify inherited disease inapparent to the layman, such as PRA, they considered that dog breeding was otherwise essentially the preserve of dog breeders. In particular, many felt that extreme conformation was above criticism, because it "is what a fancy has for its purpose" (Anon, 1962a). Many fanciers dismissed Hodgman's and Singleton's comments as "sweeping generalisations" (Anon, 1962a). Some were personally offended, such as one elderly breeder deeply distressed by "the horrid things Mr Singleton has said about Pekingese on the TV" (Ashton Cross, 1962). Yet others – generally not those in the most controversial breeds – took a different view. "Why ... such a hostile attitude to something which we know to be true?" one asked (Stanier, 1962). These commentators increasingly began to question the ethics of extreme breeding practices, such as the "unkindness of breeding for flat 'noseless' faces" in Pekingese (Roche, 1963). "It seems to me both foolish and cruel to demand a standard which makes natural whelping impossible", one medically qualified pundit wrote of Bulldogs (Frankling, 1963).

Nevertheless, much of the show community remained hostile to this veterinary intervention. One leading fancier argued that Singleton's confrontational approach was "the least likely way to achieve results", urging "the two sides [to] stop thinking of themselves as contestants and start thinking of themselves as partners" (Dangerfield, 1964). Productive communication was hampered not only by mutual distrust but also by incompatible perspectives. In 1964, a joint KC/BVA committee (including both Singleton and Hodgman) produced a memorandum on inherited disease, intended to encourage breeders to cooperate with veterinarians (BVA, 1964b; The Kennel Club, 1963). This document prioritised five diseases considered "first-priority abnormalities" in pedigree dogs, chosen from among those reported by a BSAVA survey of veterinarians (Anon, 1963). It discussed hip dysplasia (HD), patella luxation, entropion, PRA and elongated soft palate. Veterinarians like Singleton made no distinction between the inadvertently inherited disorders on this list, such as PRA, and those directly linked to conformation, such as entropion. They simply urged breeders to select away from every type of physical defect. However, many breeders perceived "a clear line of demarcation between the two main categories of hereditary abnormalities" (Hinton, 1964). They were potentially willing to engage with efforts to tackle incidentally inherited diseases unconnected to show features, such as PRA, but problems linked to breed morphologies – such as brachycephalic soft palate troubles – were politically more sensitive, because they required breeders to fundamentally re-evaluate their breeding goals. Therefore, intervention pragmatically began where it seemed most likely to succeed, with formalised screening schemes for HD and PRA, rather than the trickier matter of conformation-related disease (BVA, 1964a, 1965).

Singleton's campaign did have some short-term impact on attitudes to brachycephalic disease. By the mid-1960s, some concerned Bulldog breeders had formed a Bulldog Reform Group, and even the mainstream Bulldog societies had acknowledged "a modicum of truth" in Singleton's criticism (Barnard and Lewis, 1965; Mitchell, 1965). Both fanciers and veterinary organisations discussed possible breed standard revisions to reduce extreme conformation (Anon, 1965; Ormond, 1967). However, these initiatives did not translate into effective implementation or lasting reform. The veterinary and breeding communities were both embittered by the legacy of their conflict. For many years, their subsequent engagement was largely confined to the formal channels of the KC/BVA health schemes and joint committee, where a few veterinarians with professional or personal links to the fancy continued to wrestle with breed standard wording and various other health initiatives (BVA, 1981; Stockman, 1983); meanwhile, polemicists like Singleton intermittently continued to criticise pedigree dog breeding (Anon, 1980).

BRACHYCEPHALIC ACTIVISM IN THE TWENTY-FIRST CENTURY

At the start of the twenty-first century, veterinary attention to pedigree dog health was largely directed to the relatively uncontroversial field of DNA gene testing. From the 1990s, new molecular technologies were used to develop breed-specific tests for undesirable alleles, enabling responsible breeders to avoid producing dogs affected by certain inherited disorders such as PRA or enzyme deficiencies (Petersen-Jones et al., 1995). Few practising vets were directly involved with these innovations, and veterinary engagement with pedigree dog disease was largely limited to clinical research and treatment, official health schemes and individual activism (Gough and Thomas, 2004; Milne, 2007).

This situation was abruptly disrupted by the dramatic catalyst, in 2008, of 'Pedigree Dogs Exposed' (PDE), a provocative documentary aired on primetime national television, directed by the journalist and activist Jemima Harrison (BBC, 2008). Harrison took a similar path to Singleton, 50 years before, in strongly criticising the dog fancy in general, and the KC in particular, for wilfully ignoring the adverse consequences of breeding for the show ring. Her documentary showcased a wide range of problems, ranging from incidental disease with variable inheritance, such as epilepsy, to issues exacerbated by progressive morphological exaggeration for the show ring, including brachycephalism. It also presented the opinions of many key contributors to the subsequent debate, including veterinarians, breeders, scientific experts, KC representatives and owner activists, thus opening up a wider discussion between a greater variety of stakeholders than before (subsequently facilitated and magnified by the growth in social media).

Despite some complaints of bias and sensationalism and a largely furious reaction from fanciers, the first weeks after the broadcast of PDE saw a groundswell of grassroots veterinary and public support for its claims (Anon, 2008; 'Chanticleer', 2008; Dudley, 2008a, b; Jeffrey, 2008; Milne, 2008; Watson, 2008). Space precludes a detailed discussion of the debate and initiatives which followed. However, a decade later, it is very apparent that PDE was a "landmark event … in animal welfare change" (O'Neill, 2014). Several independent organisations and charities subsequently produced high-profile reports on dog breeding (AGPAW, 2009; Bateson, 2010; Rooney and Sargan, 2009). PDE triggered new awareness of pedigree dog health among fanciers, veterinarians and (to some extent) the public. The KC reworded many breed standards, introduced new health strategies and initiated veterinary show checks, with these changes aimed to discourage excessively exaggerated conformation linked to disease (Anon, 2008, 2012; The Kennel Club, 2018). While many critics, ranging from animal rights activists to veterinary polemicists, remain deeply and vocally opposed to the dog fancy, some of its participants have made considerable efforts to improve pedigree dog health, despite inconsistent implementation of breed standard revisions by judges in the show ring and mixed cooperation from breeders (Hanson, 2016; Cage et al., 2017; Seath et al., 2012). Breed health coordinators and other breed health volunteers are contributing to their communities' efforts to develop breed-specific health plans and schemes (Collins-Natrass, 2020). Increased health awareness and changing culture among Bulldog fanciers means that exhibitors are now self-policing to discourage each other from presenting dogs with excessive respiratory noise in the show ring and are encouraging a less 'overdone' type than some years ago (Collins-Natrass, 2020). Even if such gains appear marginal, they can still incrementally result in real improvements in canine welfare.

But show ring trends are not the only cause of the current particular crisis with brachycephalic health. The last decade has seen an exponential population increase in some brachycephalic breeds, particularly the French Bulldog (O'Neill et al., 2013) and Pug (O'Neill et al., 2016), driven and exacerbated by their popularity with celebrity influencers, in social media and in advertising (Mills, 2018; Waters, 2017). Commentators argue that the 'cuteness' factor – including the human preference for neotenous flat faces famously described by Konrad Lorenz – both fuels this fashion and exacerbates its problems by encouraging the production of dogs with more extreme morphology (Hecht and Horowitz, 2015; Holland, 2019). The 'brachy boom' has caused a vast increase in ill-considered breeding by casual and commercial breeders and fuelled the dubious international puppy trade (Yorke, 2017). The show community seek to distance themselves from these opportunists, arguing, for example, that those who breed non-standard (but more expensive) colours, such as merle Bulldogs or French Bulldogs, are therefore less "responsible and ethical" than themselves (Bulldog Breed Council, 2010; French Bulldog Club of England, 2019).

This rise in brachycephalic numbers has, as during previous surges in this debate, driven increased concern over brachycephalic health, stimulating the formation of a new multi-stakeholder Brachycephalic Working Group (BWG) to address and respond to the issues of brachycephalic welfare (BWG, 2019). There has also been an explosion of research into Brachycephalic Obstructive Airway Syndrome (BOAS) and other brachycephalic disorders (Royal Veterinary College, 2019; University of Cambridge, 2020); we have a deeper technical understanding of their causes than ever before. However, unless this massed expertise leads to effective intervention which changes breeding or buying decisions in the brachycephalic marketplace, these problems will remain a largely unsolved challenge.

DISCUSSION

The history of engagement with brachycephalic disease offers some surprising insights for stakeholders today. As this chapter has shown, three bouts of intense controversy, each separated by roughly half a century, have all featured bitter antagonism between *proponents* of extreme brachycephalic conformation and *reformers* who have sought to reverse this trend. Each iteration of the debate has inevitably been shaped by its own current cultural and scientific context, and each has added further participants, with the 1900 debate confined to breeders, the 1960 version dominated by veterinarians, and the current incarnation also involving individual and organisational welfare activism. At the core of every dispute, however, has been a collision between two sides with incompatible yet deeply felt agendas and priorities, producing little effective engagement or deterrence to further brachycephalic exaggeration.

Consideration of this recurrent confrontation, although undoubtedly depressing, also illuminates the present controversy. From the outset, exaggerated brachycephalic conformation has been linked to disease, yet some breeders' choices, driven by breed standards, show ring practices, competition, economic gain and cultural influences, have nevertheless perpetuated and exacerbated the problem. While every stakeholder – breeders, vets, judges, the KC and the public – can contribute to brachycephalic health improvement, ultimately only breeders create the dogs of the future. Therefore, as Hodgman argued, "[t]he co-operation of breeders is an essential part of any attempt to control these defects … [O]ur efforts should be directed towards [t]his end" (Hodgman, 1962).

Breeders are not a homogenous group, however. The KC estimates that it only registers about 1/3 of UK-bred puppies; even among the subset of breeders who register their puppies, about 70% only ever register one litter (Bartrip, 2018). Many puppies are bred by opportunists whose transient involvement in a breed is driven by its popularity and who are unlikely to engage with any breed health initiatives, even if they can be reached at all. A large proportion of the UK puppy trade involves international puppy smugglers and 'grey market' imports, despite efforts at legislative reform (Dogs Trust, 2018; BBC, 2019). Some breeders, particularly those involved with certain breeds that are not currently recognised by the KC, such as 'American bullies', may deliberately choose ultra-extreme conformation and participate in illegal ear-cropping, presenting further major challenges to welfare and educational outreach that are beyond the scope of this chapter (Horton, 2021). Consequently, committed show and hobby breeders that register their puppies, although in the minority, are disproportionately important for productive engagement with breed health reforms. They are likely to be aware of health initiatives, through their involvement with breed clubs or on social media; because their dogs are almost always KC registered, there is the potential for accountability, however imperfect; their preferences influence wider expectations for a breed's appearance; and they tend to be the ultimate source of future breeding stock, so that their decisions largely shape a breed's biological legacy (Rooney and Sargan, 2009). Yet history clearly shows that simply telling those breeders who favour exaggerated conformation that they should breed less extreme brachycephalic dogs is an ineffective strategy, because proponents have fundamentally different perceptions and priorities from reformers. Veterinarians may dismiss breeders as – in the words of one 1970s vet – "scientifically illiterate … egotistical ogres" for ignoring what veterinarians consider obvious pathology (Archer, 1974). Breeders, however, may consider themselves as the specialist experts and "the veterinary surgeon as an interfering 'know it all'", as one dog-breeding vet wryly remarked many years ago (Joshua, 1947). The commonality is that neither group considers the other to hold the same authority with which they invest themselves. This is important, because a dictatorial educational approach does not incentivise or encourage those who see themselves as patronised to learn. It is especially critical to properly motivate (or at least, not to demotivate) the breeders who are already making efforts to improve brachycephalic health and who often have substantial power to influence others (Collins-Natrass, 2020).

Here, modern change theory provides useful insight; "people do not resist change, but they do resist being changed" (Senge et al., 1999). In human healthcare, establishing a framework which provides a supportive and no-blame environment, empowering people to make informed modifications to their own lifestyles, is more likely to lead to improved health outcomes than a top-down approach (Laverack, 2017). Similarly, BVA/KC screening schemes and gene tests for other conditions have empowered breeders by giving them the tools to choose healthier breeding stock themselves (Lewis and Mellersh, 2019). Consequently, most ethical breeders now view many of these incidentally inherited diseases as technical problems rather than shameful stigmata, translating into real improvements in canine health. Should current research on the genetics of BOAS decouple the pathology from the morphology by developing a targeted DNA test, then brachycephalic disease would similarly become a technical problem with an attractive solution for breeders (Ladlow et al., 2018). Otherwise, as suggested both by the historical evidence presented in this chapter and by change theory, progress will not be made unless the proponents of extreme breeding themselves decide to alter their morphological priorities, with appropriate education and technical support – or if they are compelled to do so by enforced legislation on conformational limitation, as recently imposed in the Netherlands (Limb, 2019). However, this is potentially problematic and, currently, seems unlikely in the UK, particularly as extant welfare legislation, which could already be used to curb extreme breeding, is seldom effectively enforced. Here, the new KC/Cambridge brachycephalic grading scheme potentially offers a useful route to achieve health improvement, because it enables breeders to select for better respiratory function alongside other physical attributes (The Kennel Club, 2019a). But, as with previous health initiatives in other breeds, widespread adoption of the scheme will take time. Grassroots veterinarians can help to expand its reach by promoting the scheme among breeders and owners, presenting respiratory dysfunction as a problem that can potentially be overcome through cooperative action, and by encouraging prospective owners determined to buy a brachycephalic puppy to at least choose one from tested parents. Breeders active in health work predict that breeder uptake of the scheme would increase if stimulated by greater public demand (Collins-Natrass, 2020). Consequently, pragmatic veterinary encouragement towards better brachycephalic breeding might drive improvement in canine welfare more effectively than condemnation of all things brachycephalic, which serves to demotivate breeders rather than raise their standards.

Another obstacle to brachycephalic health improvement is that even breeders willing to breed towards more moderate conformation currently have little scope to do so, because they are restricted by limited breed genetic diversity and the cultural fetishisation of the 'closed stud book', which prevents outcrossing (McGreevy and Nicholas, 1999; Pedersen et al., 2016). But, as discussed earlier, pedigree status only came to depend on a 'pure' gene pool in the middle of the last century, as part of the scientific orthodoxy of the time. Since then, understandings of heredity have shifted to consider genetic diversity as desirable, and the KC has even reopened its registries to allow controlled addition of unregistered dogs, although most breeders remain reluctant to pursue outcrossing (Lambert, 2019; The Kennel Club, 2019b). Here, history could be a tool for constructive engagement with breeders, opening a discussion based on a reappraisal of their own traditions rather than an externally imposed didacticism.

Although breeder engagement is essential to improving brachycephalic health, the process also entails educating a still remarkably ignorant puppy-buying public. Efforts to combat brachycephalic fashion on its home turf of social media, as with the recent BVA #breedtobreathe campaign, and to campaign against the use of brachycephalic breeds in advertising, may help to curb interest in these breeds (BVA, 2019, Brachycephalic Working Group, 2019). But such intervention wrestles with the 'cuteness' factor: research shows that purchasers find brachycephalic dogs appealing because of their appearance and behaviour and that owners can be surprisingly bad at recognising ill health in their pets (Packer et al., 2019). This is a complex problem, for 'cuteness' implies not only smallness and vulnerability, but physical distortion, clumsiness and even deformity. If "cuteness 'in distress' is the 'most affecting' cuteness" (Ngai, 2012), then meme-type images of brachycephalic dogs struggling with dyspnoea are popular not because their audience does not understand that the animals shown are distressed but actually because, unconsciously, they do. And if the appeal of cute dogs has such unsettling elements, then there is even more to overcome than the powerful effect of social media. The 'cuteness' aesthetic potentially elicits not just the obvious responses of affection and protectiveness, but also the darker ones of pity, control and power; perhaps, one commentator even suggests, cuteness "almost always involves an act of sadism" (Harris, 2000, Ngai, 2012). If so, no wonder that the demand for brachycephalic dogs is such a challenging and complicated dynamic to unravel and influence. Educational efforts such as breed-specific 'fun days' intended as outreach to link pet owners to breed communities and that acknowledge the aesthetic appeal of brachycephalic conformation while pointing out the associated problems may be less alienating than initiatives that are perceived as more judgemental by devotees of brachycephalism (Collins-Natrass, 2020).

CONCLUSIONS

It may be more than coincidence that half a century separates each iteration of the brachycephalic debate: two human generations is long enough for the previous cycle to be forgotten. Yet the past has much to teach us. The brachycephalic conundrum is a challenging and complex problem, but history clearly shows that antagonistic confrontation has previously tended to further entrench opposing viewpoints rather than lead to effective behaviour change. In contrast, successful veterinary communication with breeders can improve canine health and has done so before. The present situation will not last forever. Fashions will change; the brachycephalic 'boom' may already have peaked; smooth Dachshunds, with their own parallel problems, already show a worrying ascent (Clark, 2019). If, when the opportunists move on, new brachycephalic health initiatives can engage the core breeders productively, be it through gene tests, phenotype scores or even outcrossing, then perhaps a fourth cycle of brachycephalic outrage could be averted: but both proponents and critics of extreme brachycephalic conformation will need to accept the need for negotiation and compromise to make effective progress in reform.

REFERENCES

'A Fancier'. 1902. The Evolution of the Bulldog. *The Stock-Keeper and Fancier's Chronicle* 46:498.
AGPAW. 2009. *A healthier future for pedigree dogs: The report of the APGAW inquiry into the health and welfare issues surrounding the breeding of pedigree dogs.*
Amato, S. 2015. *Beastly Possessions: Animals in Victorian Consumer Culture.* Toronto: University of Toronto Press.
Anon. 1896. The pet dog show at the Aquarium. *Ladies' Kennel Journal* 3:332–336.
Anon. 1898a. Evolution of the modern dog. *The Field* 91:67.
Anon. 1898b. Dogs and their values. *The Field* 92:486.
Anon. 1899. Show of Bulldogs. *The Field* 93:691.
Anon. 1900. Whispers of Fancy. *The Stock-Keeper and Fancier's Chronicle* 42:359.
Anon. 1901. Whispers of Fancy. *The Stock-Keeper and Fancier's Chronicle* 44:93.
Anon. 1902. Toy dogs. *Illustrated Kennel News* 1:317.
Anon. 1903a. Toy dog notes. *The Illustrated Kennel News* 4:82.
Anon. 1903b. Toy dog notes. *The Illustrated Kennel News* 4:179.
Anon. 1946. Club and Kennel Notes: Mr S.F.J. Hodgman, M.R.C.V.S., Member of the Committee. *Kennel Gazette* 67:501.
Anon. 1953. Letter quoted anonymously in Macdonald Daly's weekly column. *Our Dogs* (November 13), 893.
Anon. 1962a. Save us from the candid friend. *Our Dogs* (September 28), 675.
Anon. 1962b. Vet attacks pedigree dog breeders. *Our Dogs* (November 16), 1039.
Anon. 1963. British Small Animals Veterinary Assoc. *Our Dogs* (April 26), 751.
Anon. 1965. Breed standards. *Our Dogs* (March 12), 471.
Anon. 1980. KC breed standards criticised. *Veterinary Record* 107:2–3.
Anon. 2008. Kennel Club launches review of UK pedigree dog breeds. *Veterinary Record* 163:464.
Anon. 2012. Veterinary checks take effect at Crufts. *Veterinary Record* 170:272–272.
Archer, W. 1974. A little learning is a dangerous thing. *Veterinary Practice* 6:5.
Ashton Cross, C. 1962. Hereditary abnormalities. *Our Dogs* (December 7), 1183.
Atkinson, M. 2018. How do we make breeding dogs better? *Veterinary Record* 183:698–699.
'B'. 1903. Mr Walter Jeffries and his Bulldogs. *The Illustrated Kennel News* 3:637.
Barnard, M., and Lewis, M. G. D. 1965. Bulldogs. *Our Dogs* (June 11), 1162.
Bartrip, S. 2018. KC registrations. Personal communication, May 1.
Bateson, P. 2010. *Independent Inquiry into Dog Breeding.* Suffolk: Micropress Ltd.
BBC. 2008. BBC - Press Office - Pedigree Dogs Exposed. http://www.bbc.co.uk/pressoffice/pressreleases/stories/2008/08_august/19/dogs.shtml (accessed September 30, 2019).
BBC. 2019. Lucy's Law: Puppy farm ban set to be confirmed. https://www.bbc.co.uk/news/uk-48249333 (accessed January 12, 2020).
Beecher, J. P. 1902. The Evolution of the Bulldog. *The Stock-Keeper and Fancier's Chronicle,* 8.
Bonnett, B., and Sandøe, P. 2016. Don't Know or Don't Care? How Beliefs and Attitudes about Dog Health and Welfare Limit Behaviour Change. https://dogwellnet.com/content/welfare-legislation/human-dog-interactions/how-beliefs-and-attitudes-about-dog-health-and-welfare-limit-behaviour-change-r406/ (accessed April 28, 2019).
Brachycephalic Working Group. 2019. Brachycephalic Working Group (BWG) Position Statement: The use of brachycephalic dogs in advertising and the media. http://www.ukbwg.org.uk/?page_id=453 (accessed January 3, 2020).

Bresalier, M., and Worboys, M. 2014. 'Saving the lives of our dogs': The development of canine distemper vaccine in interwar Britain. *The British Journal for the History of Science*, 47:305–334.

Bulldog Breed Council. 2010. Standard. http://www.bulldogbreedcouncil.co.uk/standard.html (accessed September 19, 2019).

BVA. 1964a. Scheme for the control of hip dysplasia. *Veterinary Record* 75:1474–1475.

BVA. 1964b. Joint memorandum from the British Veterinary Association and the Kennel Club on hereditary disease in dogs. *Kennel Gazette* 85:342–345.

BVA. 1965. Progressive retinal atrophy certification scheme. *Veterinary Record* 77:1592–1593.

BVA. 1981. Kennel Club/BVA standing committee. *Veterinary Record* 108:383.

BVA. 2019. BVA policy - brachycephalic dogs. https://www.bva.co.uk/news-campaigns-and-policy/policy/companion-animals/brachycephalic-dogs/ (accessed September 29, 2019).

BWG. 2019. Home – Brachycephalic Working Group. http://www.ukbwg.org.uk/?page_id=96 (accessed September 29, 2019).

Cage, A., et al. 2017. Improving pedigree dog health. *Veterinary Record* 180:309.

'Cave Canem'. 1905. Vaccination of dogs. *Veterinary Record* 17:714.

'Chanticleer'. 2008. Profession can no longer Know, Abhore and Ignore Pedigree Issue. *Veterinary Times* (October 20), 38.

Cheang, S. 2006. Women, Pets, and Imperialism: The British Pekingese Dog and Nostalgia for Old China. *The Journal of British Studies* 45:359–387.

Cheang, S. 2007. Selling China: Class, Gender and Orientalism at the Department Store. *Journal of Design History* 20:1–16.

Clark, K. 2019. Have brachy breeds fallen out of fashion? *Veterinary Record* 185:357–357.

Clarke, M. C. F. C. 1911. 'The Kennel' Bulldog section. *The Kennel* 2:262–264.

Clarke, S. 1910. Bulldogs in the South: Ten years' retrospect. *The Kennel* 1:63–66.

Collins-Natrass, V. 2020. Bulldogs and their health. *Interview*.

Crowther, F. W. 1893. Bulldogs. *The Kennel Gazette* 14:10–12.

Daly, M. 1953. Are we turning him into a jellyfish? The great bulldog controversy flares up. *Our Dogs* (November 27), 988.

Daly, M. 1954. Come along, you fireside critics, What's your opinion of this little lot? *Our Dogs* (December 17), 1157.

Daly, M. 1955. Bulldog men join critics of the modern trend. Is fat the big trouble? *Our Dogs* (January 21), 94.

Daly, N. 2015. Britain. In *The Fin-de-Siècle World*, ed. M. Saler, 117–130. London and New York: Routledge.

Dalziel, H. 1889. *British Dogs* (Vol. 2). London: L. Upcott Gill.

Dangerfield, S. 1964. A cool look at a burning topic. *Our Dogs* (July 17), 126.

Dean, R. L. 1913. Pekingese. *The Kennel Gazette* 34:15.

Denisoff, D. 2007. Decadence and Aestheticism. In *The Cambridge Companion to the Fin-de-Siècle*, ed. G. Marshall, 31–52. Cambridge: Cambridge University Press.

Dennis, M. L. 1911. Pomeranians. *The Kennel Gazette* 32:21.

Derry, M. E. 2003. *Bred for Perfection: Shorthorn Cattle, Collies and Arabian Horses since 1800*. Baltimore: Johns Hopkins University Press.

Dogs Trust. 2018. Puppy Smuggling: when will this cruel trade end? https://www.dogstrust.org.uk/puppy-smuggling/111018_puppy%20smuggling%202018_final.pdf (accessed January 12, 2020).

Dudley, J. 2008a. Pedigree documentary is 'missed opportunity'. *Veterinary Times* (September 1), 3.

Dudley. 2008b. Divisive dog documentary prompts double boycott. *Veterinary Times* (September 29), 4.

Elegans, C. 2012. 100 Years of Breed 'Improvement'. https://dogbehaviorscience.wordpress.com/2012/09/29/100-years-of-breed-improvement/ (accessed September 27, 2019).

Ellis-Hughes, T. 1962. Abnormalities in pedigree dogs. *Our Dogs* (September 28), 679.

Farman, E. 1899a. Bulldogs. *Kennel Gazette* 20:7–8.

Farman, E. 1899b. *The Bulldog: A Monograph*. Facsimile edition, Liss: Nimrod Book Services.

Farrell, L.L., Schoenebeck, J.J., Wiener, P., Clements, D.N. and Summers, K.M. 2015. The challenges of pedigree dog health: Approaches to combating inherited disease. *Canine Genetics and Epidemiology* 2:3 doi:10.1186/s40575-015-0014-9.

Frankling, E. 1963. Breeding and some of its problems. *Our Dogs* (December 27), 1331.

French Bulldog Club of England. 2019. Important Advice on Buying a Puppy. http://www.frenchbulldog-clubofengland.org.uk/important-advice-on-buying-a-puppy.html (accessed September 29, 2019).

Gardiner, A. 2014. The 'Dangerous' Women of Animal Welfare: How British Veterinary Medicine Went to the Dogs. *Social History of Medicine* 27:466–487.

Gough, A., and Thomas, A. 2004. *Breed Predispositions to Disease in Dogs and Cats*. Oxford: Blackwell Publishing Ltd.

Gould, S. J. 1981. *The Mismeasure of Man*. New York: WW Norton.

Jassef, 1906. The latest development of the bull-dog. Punch 130:67.

Hanson, M. 2016. Stop buying pedigree dogs. Stop breeding them. Stop these awful practices. https://www.theguardian.com/commentisfree/2016/mar/15/pedigree-dogs-breeding-crufts-german-shepherd-best-in-breed (accessed September 28, 2019).

Hardcastle, A. 1902a. The evolution of the bulldog. *The Stock-Keeper and Fancier's Chronicle*, 46:461.

Hardcastle, A. 1902b. The evolution of the bulldog. *The Stock-Keeper and Fancier's Chronicle*, 47:7–8.

Harris, D. 2000. *Cute, Quaint, Hungry and Romantic: The Aesthetics of Consumerism*. New York: Basic Books.

Hecht, J., and Horowitz, A. 2015. Seeing dogs: Human preferences for dog physical attributes. *Anthrozoös* 28:153–163.

Hinton, N. 1964. Public relations. *Our Dogs* (February 21), 344.

Hipperson, J. 2018. Professional entrepreneurs: Women veterinary surgeons as small business owners in inter-war Britain. *Social History of Medicine* 31:122–139

Hobday, F. T. G. 1897. Paper on canine and feline surgical operations. *Veterinary Record* 10:280–286.

Hodgman, S. F. J. 1962. Abnormalities of possible hereditary origin in dogs. *Veterinary Record* 71:1239–1247.

Holland, K. E. 2019. Acquiring a pet dog: A review of factors affecting the decision-making of prospective dog owners. *Animals* 9:124. doi:10.3390/ani9040124.

Horton, H. 2021. Designer 'ugly toad' bulldogs face life of pain, buyers warned. https://www.telegraph.co.uk/news/2021/01/23/social-media-trend-multicoloured-dogs-creating-extreme-breeds/ (accessed 20 March, 2021).

Howell, P. 2013. The dog fancy at war: Breeds, breeding, and Britishness, 1914–1918. *Society and Animals*, 21:546–567.

Jeffrey, N. 2008. We should campaign for end to current practices. *Veterinary Times* (October 6), 39.

Joshua, J. O. 1958. Current research in small animal diseases. *BSAVA Congress Proceedings* 1:13–18.

Joshua, J. O. 1947. Veterinary services at small animal shows. *Veterinary Record* 59:443–447.

KC Registrations. 1910. Bulldogs. *The Kennel Gazette* 31:127–128.

Kean, H. 2017. *The Great Cat and Dog Massacre: The Real Story of World War Two's Unknown Tragedy*. Chicago: University of Chicago Press.

Kirk, R. G. W., Cantor, D., and Ramsden, E. 2014. The invention of the 'stressed animal' and the development of a science of animal welfare, 1947–86. In *Stress, Shock, and Adaptation in the Twentieth Century*, ed. D. Cantor and E. Ramsden, 241–263. Rochester, N.Y.: University of Rochester Press.

Knight, G. C. 1956. Some effects on dogs of close breeding to type. *Veterinary Record* 68:1050–1051.

Ladlow, J., Liu, N.-C., Kalmar, L., and Sargan, D. 2018. Brachycephalic obstructive airway syndrome. *Veterinary Record* 182:375–378.

'Lady Betty'. 1910a. How to choose a puppy - toy dogs. *The Kennel* 1:32–34.

'Lady Betty'. 1910b. How to choose a puppy - toy dogs. *The Kennel* 1:60–62.

'Lady Betty'. 1911. The Pekingese: Notes and news. *The Kennel* 2:438–440.

Lambert, B. 2019. Unverified breeding registration. Personal communication, September 12.

Lambert, F. 1898. Dogs wanted and for sale: Bulldogs. *The Stock-Keeper and Fancier's Chronicle*, 38: (7 January), viii.

Larsell, O. 1924. Anders A. Retzius. *Annals of Medical History* 6:16–24.

Laverack, G. 2017. The challenge of behaviour change and health promotion. *Challenges* 8:25. doi:10.3390/challe8020025.

Leighton, R. 1924. *The Complete Book of the Dog*. London: Cassell and Company, Ltd.

Lewis, T. W., and Mellersh, C. S. 2019. Changes in mutation frequency of eight Mendelian inherited disorders in eight pedigree dog populations following introduction of a commercial DNA test. *PLOS ONE* 6:e0209864. doi:10.1371/journal.pone.0209864.

Limb, M. 2019. Dutch crackdown on brachycephalic breeds. *Veterinary Record* 184: 693.

Lytton, J. 1911. *Toy Dogs and Their Ancestors: Including the History and Management of Toy Spaniels, Pekingese, Japanese and Pomeranians*. London: Duckworth and Co.

Mallock, M. R. 1916. Toy Spaniels. *The Kennel Gazette* 37:42.

Mansour, T. A., Lucot, K., Konopelski, S. E., et al. 2018. Whole genome variant association across 100 dogs identifies a frame shift mutation in DISHEVELLED 2 which contributes to Robinow-like syndrome in Bulldogs and related screw tail dog breeds. *PLOS Genetics* 14:e1007850. doi:10.1371/journal.pgen.1007850.

McGreevy, P. D., and Nicholas, F. W. 1999. Some practical solutions to welfare problems in dog breeding. *Animal Welfare* 8:329–341.

Merish, L. 1996. Cuteness and commodity aesthetics: Tom Thumb and Shirley Temple. In *Freakery: Cultural Spectacles of the Extraordinary Human Body*, ed. R. Garland Thomson, 185–203. New York: New York University Press.

Mills, G. 2018. French bulldogs now the UK's top breed. *Veterinary Record* 182:705–705.

Milne, E. 2007. *The Truth about Cats and Dogs*. Brighton: Book Guild Ltd.

Milne, E. 2008. Profession must be vocal on pedigree dogs. *Veterinary Times* (October 6), 39.

Mitchell, J. 1965. Bulldog Reform. *Our Dogs* (February 5), 204.

Ngai, S. 2012. *Our Aesthetic Categories: Zany, Cute, Interesting*. Cambridge, MA: Harvard University Press.

'Old Dane'. 1899. Concerning Spaniels. *Field* (9 December), 903.

O'Neill, D. 2014. Progress in purebred dog health since the Bateson report of 2010. *Veterinary Record* 175:277–279.

O'Neill, D. G., Darwent, E. C., Church, D. B., and Brodbelt, D. C. 2016. Demography and health of Pugs under primary veterinary care in England. *Canine Genetics and Epidemiology*, 3:1–12.

O'Neill, D. G., Baral, L., Church, D. B., Brodbelt, D. C., and Packer, R. M. A. 2018. Demography and disorders of the French bulldog population under primary veterinary care in the UK in 2013. *Canine Genetics and Epidemiology*, 5:3.

O'Neill, D. G., Skipper, A. M., Kadhim, J., Church, D. B., Brodbelt, D. C., and Packer, R. M. A. 2019. Disorders of Bulldogs under primary veterinary care in the UK in 2013. *PLoS ONE*, 14:e0217928.

Ormond, A. 1967. Hereditary defects in dogs considered at the World Veterinary Congress. *Kennel Gazette* 88:397–398.

Ormond, H. 1973. Obituary: Mr.S.F.J. Hodgman, M.R.C.V.S. *Kennel Gazette* 94:7.

'Ouida'. 1897. *Dogs*. London: Simpkin, Marshall, Hamilton, Kent and Co., Ltd.

Packer, R.M.A., Hendricks, A., and Burn, C.C. 2012. Do dog owners perceive the clinical signs related to conformational inherited disorders as 'normal' for the breed? A potential constraint to improving canine welfare. *Animal Welfare* 21:81–93.

Packer, R.M.A., O'Neill, D.G., Fletcher, F., and Farnworth, M.J. 2019. Great expectations, inconvenient truths, and the paradoxes of the dog-owner relationship for owners of brachycephalic dogs. *PLoS ONE*. doi:10.1371/journal.pone.0219918.

Pedersen, N. C., Pooch, A. S., and Liu, H. 2016. A genetic assessment of the English bulldog. *Canine Genetics and Epidemiology* 3:6. doi:10.1186/s40575-016-0036-y.

Pemberton, N., and Worboys, M. 2012. *Rabies in Britain: Dogs, Disease and Culture, 1830–2000*. Basingstoke: Palgrave Macmillan.

Petersen-Jones, S. M., Clements, P. J. M., Barnett, K. C., and Sargan, D. R. 1995. Incidence of the gene mutation causal for rod-cone dysplasia type 1 in Irish Setters in the UK. *Journal of Small Animal Practice* 36:310–314.

Pick, D. 1993. *Faces of Degeneration: A European Disorder, c.1848-c.1918*. New York: Cambridge University Press.

Redwar, H. 1901a. The bulldog. *The Stock-Keeper and Fancier's Chronicle*, 44:178.

Redwar, H. 1901b. A plea for the bulldog. *The Stock-Keeper and Fancier's Chronicle*, 44:113.

Redwar, H. 1902a. The evolution of the bulldog. *The Stock-Keeper and Fancier's Chronicle* 46:443–444.

Redwar, H. 1902b. The evolution of the bulldog. *The Stock-Keeper and Fancier's Chronicle* 47:69–70.

Ritvo, H. 1987. *The Animal Estate: The English and Other Creatures in the Victorian Age*. Cambridge, MA: Harvard University Press.

Roche, G. 1963. Why these Handicaps? *Our Dogs* (December 27), 1334.

Rooney, N., and Sargan, D. 2009. *Pedigree dog breeding in the UK: a major welfare concern?* Independent report commissioned by the RSPCA. https://www.rspca.org.uk/adviceandwelfare/pets/dogs/puppy/pedigreedogs (accessed July 14, 2019).

Royal Veterinary College. Brachycephalic Research Team: Royal Veterinary College. 2019. Available from: https://www.rvc.ac.uk/research/focus/brachycephaly (accessed January 11, 2020).

Santayana, G. 1905. *The Life of Reason; or, The Phases of Human Progress*. New York: Charles Scribner's Sons.

Seath, I., et al. 2012. *The Kennel Club Breed Health Improvement strategy: A Step by Step Guide*. https://www.thekennelclub.org.uk/media/97934/bhcbreedhealthimpstrat.pdf (accessed September 28, 2019).

Senge, P. M., Roth, G., and Ross, R. B. 1999. *The Dance of Change: The Challenges to Sustaining Momentum in a Learning Organization*. New York: Currency.

Singleton, W. 1962a. Partial velum palatiectomy for the relief of dyspnoea in brachycephalic breeds. *Journal of Small Animal Practice* 3:215–216.

Singleton, W. B. 1962b. Discussion on abnormalities of possible hereditary origin in dogs at BVA congress. *Veterinary Record* 71:1244–1245.

Skipper, A. 2016. 'Night-Blindness' in the Irish Setter: Scientific eugenics and disease control in the pedigree dog, 1930–1950. MA diss., King's College London.

Skipper, A. 2020. The 'Dog Doctors' of Edwardian London: Elite canine veterinary care in the early twentieth century. *Social History of Medicine* 33, (2020), 1233–58. doi:10.1093/shm/hkz049.

St John Cooper, H. 1900. Dogs and railway journeys. *The Stock-Keeper and Fancier's Chronicle* 42:448.

Stanier, D. 1962. Abnormalities in dogs. *Our Dogs* (October 5), 734.

Stockard, C., Anderson, O.D., and James, W. T. (1941). *The Genetic and Endocrine Basis for Differences in Form and Behavior, as Elucidated by Studies of Contrasted Pure-Line Dog Breeds and their Hybrids.* Philadelphia, PA: Wistar Press.

Stockman, M. J. R. 1983. Kennel Club moves into a new era. *Veterinary Record* 113:286.

Stonehenge. 1879. *The Dog in Health and Disease.* London: Longmans, Green, and Co.

Sutton, C.G. 1980. Dog Shows and Show Dogs. Edlington: K & R Books Ltd.

The Kennel Club. 1953. Kennel Club committee meetings. *The Kennel Gazette* 74:455–457.

The Kennel Club. 1963. Club and kennel notes: Joint committee with B.V.A. *The Kennel Gazette* 84:367.

The Kennel Club. 1989. *The Kennel Club's Illustrated Breed Standards.* London: The Bodley Head.

The Kennel Club. 2009. Comparative table of registrations 1908–2009. Kennel Club library.

The Kennel Club. 2018. *Breed Watch: A guide for the health and welfare of show dogs.* https://www.thekennelclub.org.uk/media/341575/breed_watch_booklet.pdf (accessed September 28, 2019).

The Kennel Club. 2019a. The Kennel Club and University of Cambridge launch vital scheme to improve health of Pugs, French Bulldogs and Bulldogs. https://www.thekennelclub.org.uk/press-releases/2019/february/the-kennel-club-and-university-of-cambridge-launch-vital-scheme-to-improve-health-of-pugs-french-bulldogs-and-bulldogs/ (accessed September 29, 2019).

The Kennel Club. 2019b. Managing inbreeding and genetic diversity. https://www.thekennelclub.org.uk/health/for-breeders/inbreeding-and-genetic-diversity/managing-inbreeding-and-genetic-diversity/ (accessed September 29, 2019).

Theunissen, B. 2014. Practical animal breeding as the key to an integrated view of genetics, eugenics and evolutionary theory: Arend L. Hagedoorn. *Studies in History and Philosophy of Science Part C: Studies in History and Philosophy of Biological and Biomedical Sciences* 46:55–64.

Tomkins, A. C. 1911. Pekingese. *The Kennel Gazette* 32:21.

University of Cambridge. Cambridge BOAS research group, University of Cambridge. 2020. https://www.vet.cam.ac.uk/boas (accessed January 11, 2020).

Vesey-Fitzgerald, B. 1948. *The Book of the Dog.* London: Nicholas and Watson.

Warland, T. H. 1898. Bulldog points. *The Stock-Keeper and Fancier's Chronicle* 39:257–258.

Warland, T. H. 1902. The evolution of the bulldog. *The Stock-Keeper and Fancier's Chronicle* 46:520.

Waters, A. 2017. Brachycephalic tipping point: Time to push the button? *Veterinary Record*, 180:288.

Watson, S. 2008. We have 'done nothing' on pedigree dog issue. Veterinary Times (October 6), 39.

W.G. 1900. Dogs and Railway Journeys. *The Stock-Keeper and Fancier's Chronicle* 42:448.

Woods, A. 2012a. From cruelty to welfare: The emergence of farm animal welfare in Britain, 1964–71. *Endeavour* 36:14–22.

Woods, A. 2012b. The history of veterinary ethics in Britain, c. 1870–2000. In *Veterinary and Animal Ethics: Proceedings of the First International Conference on Veterinary Ethics*, 3–18. Chichester: Wiley Blackwell.

Worboys, M., Strange, J.-M., and Pemberton, N. 2018. *The Invention of the Modern Dog: Breed and Blood in Victorian Britain.* Baltimore: Johns Hopkins University Press.

Yorke, H. 2017. French bulldog craze is producing seriously ill puppies, leading vets warn. https://www.telegraph.co.uk/news/2017/02/14/dont-listen-celebrities-french-bulldog-craze-producing-seriously/ (accessed September 29, 2019).

3 Flat-Faced Fandom: Why Do People Love Brachycephalic Dogs and Keep Coming Back for More?

Rowena M. A. Packer
Royal Veterinary College

CONTENTS

Introduction 25
Why Are Humans Attracted to Brachycephalic Dogs? 26
The Baby Schema Effect 26
Brachycephalic Behaviour 28
Lifestyle Breeds 30
Social Influences on Breed Choice 30
What's Health Got to Do with It? 32
Normal for the Breed? 33
What is 'Good Health' for a Brachycephalic Dog? 33
Reacquisition Desire and the Development of 'Breed Loyalty' 35
Conclusions 35
References 36

INTRODUCTION

When reflecting on the paradoxical situation we face in 2020, where more than ever is known about the poor health of brachycephalic dogs, but yet their popularity is at an all-time high, the question of why owners are drawn to brachycephalic breeds, and even in the face of chronic or severe health problems, continue to show loyalty towards these breeds, is critically important to explore. Such understanding could inform our approaches to influencing decision-making in prospective owners considering buying brachycephalic breeds in the future, as well as discussions with the many thousands of current owners of brachycephalic breeds wishing to buy them again in the future. This chapter examines the growing literature on human attraction to brachycephalic animals, considering both biological explanations (e.g. *kindchenschema* and 'the cute phenomenon') and recent cultural explanations (e.g. fashion and lifestyle motivations). This chapter additionally explores owner experiences and perceptions of brachycephalic health that may perpetuate undertreatment of conformation-related disorders, while driving 'breed loyalty' (the desire to repeatedly acquire/own their breed) despite ill health in their own dog. In contrast to the vast literature describing the health challenges of brachycephalic breeds and studies to understand underlying causes and develop effective treatments for these disorders, literature understanding the 'human' component of the brachycephalic dog–owner dyad is in its relative infancy but may prove pivotal in turning the tide on the current 'brachycephalic crisis'.

WHY ARE HUMANS ATTRACTED TO BRACHYCEPHALIC DOGS?

What drives prospective dog owners, when faced with options from over 200 Kennel Club registered breeds along with a growing number of 'designer' crossbreeds, to choose a brachycephalic dog as their new companion is key in understanding their recent rise in popularity. Many factors have been identified that affect decision-making of prospective dog owners (Holland 2019), with physical features highlighted as an important consideration in several studies (Weiss et al. 2012, King, Marston, and Bennett 2009, Diverio et al. 2016, Hecht and Horowitz 2015, Packer, Murphy, and Farnworth 2017). Physical characteristics that owners consider important when choosing a dog breed include hair type and length (King, Marston, and Bennett 2009), body size (Diverio et al. 2016) and iris colour (Hecht and Horowitz 2015).

A recent study considered the importance of different canine characteristics to the owners of some popular brachycephalic (namely the Bulldog, French Bulldog and Bulldog) and non-brachycephalic breeds (including the Labrador Retriever and Cocker Spaniel) (Packer, Murphy, and Farnworth 2017). Owners were asked to rate the extent to which different factors influenced their decision to purchase their chosen breed. These factors included experiential factors (e.g. childhood experiences), breed health factors (e.g. generally healthy breed, long life expectancy) and lifestyle factors (e.g. breed size suited to lifestyle, breed is easy to take care of, cost). Of 15 potential influences, owners of brachycephalic dogs rated their chosen breed's appearance as the factor that most highly influenced their choice of breed. Appearance was significantly more influential for owners of brachycephalic breeds compared with owners of non-brachycephalic breeds, with traits such as breed health and longevity less influential upon owners of brachycephalic breeds than non-brachycephalic breeds (Packer, Murphy, and Farnworth 2017).

THE BABY SCHEMA EFFECT

So, what is it about the appearance of brachycephalic dogs that has their owners so captivated and that enable these owners to prioritise looks over their future dog's potential well-being? The defining feature of brachycephalic breeds, but also a key source of their poor health, is their facial appearance. In addition to a foreshortened muzzle giving them a distinctive 'flat-faced' appearance, their facial morphology commonly includes a large forehead, large and low-lying eyes, and bulging cheeks. These features potentially trigger instinctual human attraction to brachycephalic breeds as described in the so-called *Kindchenschema* (baby schema), hypothesised by Nobel prize winner and founding father of ethology, Konrad Lorenz (Lorenz 1971). *Kindchenschema* covers a set of infantile facial features that are described in classical ethology as 'social releasers', basic stimuli that evoke a stereotyped response from humans, in this case, nurturing. From an evolutionary perspective, this response would be essential for the survival of infants of species that rely on parental caregiving behaviour (e.g. protection, feeding) (Eibl-Eibesfeldt 2017). *Kindchenschema* is not limited to humans, or indeed even just mammals, with 'cute features' prevalent among mammals and birds, but are not shown by animals such as insect or reptile species that do not require parental care (Morreall 1993). Indeed, a recent study elegantly demonstrated that birds and reptile species that are dependent on parental care elicit greater caregiving intention from humans than other species within their class that do not rely on parental care (Kruger and Miller 2016). Human attraction to baby schema features has been found to extend beyond human infants, even to inanimate objects; cartoon characters (e.g. Mickey Mouse (Gould 1992)) and teddy bears (Hinde and Barden 1985) have become 'cuter' over time, with features that increasingly conform to the baby schema configuration.

Empirical studies using computer manipulation of real human infant photos (e.g. exaggerating width and length of face, size of forehead, eyes, nose and mouth to fit the baby schema more or less) have reported that 'high baby schema' infants (where facial features were manipulated towards the baby schema more closely) were rated as more cute, but also elicited a stronger caretaking motivation than 'low baby schema' infants (where facial features were manipulated away from the baby schema configuration) (Glocker et al. 2009). Using these manipulated images, the same team of researchers went on to test the effects of infant faces on the brain activity of 16 female participants who had not previously given birth to a child, measured with event-related blood oxygenation level-dependent (BOLD) fMRI (Glocker et al. 2009). The researchers found that the baby schema activated the nucleus accumbens, a key structure of the mesocorticolimbic system mediating reward processing and appetitive motivation, which the researchers suggested may be the neurophysiological mechanism by which baby schema promotes human caregiving, regardless of kinship (Glocker et al. 2009).

So how does the baby schema effect apply to dogs and specifically brachycephalic breeds? Lorenz noted that that breeds of dogs such as the brachycephalic Pekingese retain infant features into adulthood (Lorenz 1971) (Figure 3.1). Empirical studies have demonstrated that infantile features in cats, dogs and teddy bears increase their attractiveness and that women show higher ratings for pets with infant features than men (Archer and Monton 2011). Consequently, preference for dogs with infant features could form a major part of the initial attraction to brachycephalic breeds. The authors concluded that attraction to infant features, a general and phylogenetically ancient mechanism, is applied in an evolutionarily non-adaptive or maladaptive context in our human attraction to companion animals (i.e. it is not beneficial for our 'fitness' – reproductive success), facilitating the evolutionary paradox of pet keeping (Archer 1997, 2011). It has been argued that pets are essentially social parasites that manipulate the behaviour of their owners (hosts) to obtain one-sided fitness benefits (Archer 1997). So, are brachycephalic dogs the ultimate social parasites, given that the care they receive from humans has enabled them to become hugely successful in evolutionary terms? Some argue that who is being parasitised by whom within the pet–owner relationship is unclear, and the converse may be true. Given that humans potentially reap a range of physical and mental health benefits from dog ownership (Barker and Wolen 2008), while brachycephalic dogs are burdened with health problems related to their (human) desired body shape, it is possible that companion animals such as brachycephalic dog breeds are in fact the victims of human social parasitism, with 'cute' care-soliciting characteristics (e.g. neotenous morphological features, behavioural dependency, sociality, affection) being the product of selection for characteristics that enhance their ability to satisfy specific human needs (Archer 2011).

The impact of infant features on owners is not limited to the initial stages of our attraction to dogs. Attraction to infant features in dogs has the potential to have ongoing influences upon the dog–human relationship. Within Archer and Monton's (2011) study of attraction to infant features, the degree of attachment that owners reported to their pets (as measured by the 'pet attachment questionnaire', PAQ) was associated with their ratings of overall attractiveness of facial photographs of dogs and cats with

FIGURE 3.1   A human infant and a Pug exhibiting classic features of *Kindchenschema* including large eyes and a small nose. https://www.shutterstock.com/image-photo/funny-baby-boy-toddler-pug-dog-284016629.

infant features (Archer and Monton 2011). It should be noted that owners rated the attractiveness of unfamiliar animals rather than their own pets in this study. More recently, Thorn et al. (2015) extended this line of investigation, exploring whether owner perceptions of the cuteness of their own dog (rated on a 6-point Likert scale ranging from 1 (not cute at all) to 6 (really cute)) was associated with human–dog relationship quality (as measured by the Monash Dog Owner Relationship Scale (MDORS) (Dwyer, Bennett, and Coleman 2006) and the Dog Attachment Questionnaire (DAQ) (Archer and Ireland 2011)) (Thorn et al. 2015). This study found that owner perceived canine cuteness was associated with all measures of relationship quality that the authors termed the 'canine cuteness effect' (Thorn et al. 2015). Indeed, a recent investigation of the dog–owner relationship in brachycephalic breeds (specifically Bulldogs, French Bulldogs and Pugs) found that mean MDORS scores (Dwyer, Bennett, and Coleman 2006) for these breeds were higher across all three MDORS subscales than other recent studies utilising this tool in other breeds (Packer, Volk, and Fowkes 2017, Meyer and Forkman 2014, Rohlf et al. 2012). In the second part of Thorn et al.'s study of perceived cuteness, personality and relationship quality, participants rated the cuteness and perceived personality of photos of dogs' faces. Cute dogs were perceived to be significantly more amicable (e.g. friendly, safe and affectionate) than other dogs (Thorn et al. 2015), which has previously identified to be the most desirable trait in dogs (Ley, Bennett, and Coleman 2009). From this result, it is clear that owner interpretation of behaviour may be influenced by a dog's appearance. The influence of perceived brachycephalic 'personality' and behaviour will be considered next in this chapter.

The biological effects of the baby schema phenomenon go some way in explaining the international popularity of brachycephalic dogs. From the results explored here, it appears that the facial appearance of brachycephalic animals has the potential to innately predispose humans to be motivated to care for them, perceive their behaviour more positively and build stronger relationships with them. However, it is still unclear if, and how the baby schema effect applies for different brachycephalic breeds, given variation in key baby schema conformational traits such as eye shape and size, and face length. Further work needs to be done to better understand which human demographics are most sensitive to the 'cuteness' of brachycephalic dogs and to examine effects of viewing brachycephalic dogs (compared with breeds not conforming to the baby schema) upon brain activity. Whether less brachycephalic dogs are able to elicit similarly positive responses from humans may be key to encouraging the breeding and ownership of more moderate brachycephalic dogs.

BRACHYCEPHALIC BEHAVIOUR

As alluded to above, although the influence of brachycephalic behaviour (or indeed, perceived behaviour) may be an important element of human attraction to flat-faced breeds, this perspective has received little attention to date. Previous research has indicated that Pugs, Bulldogs and French Bulldogs are acquired partly based on perceived positive behavioural traits, namely making good companion dogs and being good breeds for households with children (Packer, Murphy, and Farnworth 2017). Breed standards of these three breeds describe dogs with an affectionate, happy nature (The Kennel Club 2018). Although breed standards only represent an idealised version of a breed, and there is likely to be great variation within breeds (and indeed overlap between breeds) in their behaviour, there is some empirical evidence supporting positive behavioural traits in these brachycephalic breeds.

Dogs with higher cephalic indices (shorter muzzles) are reported to be more affectionate, cooperative and interactive with unfamiliar humans than dogs with lower cephalic indices (longer muzzles) in both owner-reported and practical tests of dog behaviour (Stone et al. 2016, McGreevy et al. 2013). Dog–owner communication may be enhanced in brachycephalic dogs, with greater performance shown by brachycephalic dogs at utilising the human point gesture compared to dolichocephalic breeds (Gácsi et al. 2009). Furthermore, brachycephalic dogs display the longest looking times at human and dog portraits compared with mesocephalic breeds (with a skull of medium proportions e.g. Labrador Retrievers) and dolichocephalic breeds (with a relatively long

FIGURE 3.2   Heightened gazing behaviour may have enhanced dog–owner communication in brachycephalic breeds https://www.shutterstock.com/image-photo/closeup-studio-portrait-gazing-dog-black-795266767.

skull e.g. Greyhounds) (Bognár, Iotchev, and Kubinyi 2018). Rather than preference being purely based on their appearance, this supports the suggestion that the popularity of flat-faced breeds is also because humans might have preferred animals that looked at them for longer durations because this enhanced the effectiveness of communicative interaction (McGreevy, Grassi, and Harman 2004) (Figure 3.2).

Given experience plays an important role in the development of behaviour, including dog–owner interactions (e.g. Appleby, Bradshaw, and Casey 2002, Blackwell et al. 2008, Arhant et al. 2010), it is also possible that brachycephalic breeds are not intrinsically predisposed to behave differently to dogs with longer muzzles, but that owners consciously or unconsciously treat them differently, which results in differing reports in observed behaviours.

In a recent qualitative study of brachycephalic ownership that explored reasons why owners would recommend their breed, positive behavioural traits were the most dominant theme. Positive behavioural traits were often used to justify owners' ongoing allegiance to their breed despite other problems, for example chronic health issues. Valued elements of their dog's behaviour included loyalty, loving and affectionate natures, being comical or 'clown like', playful and easily trained (Packer et al. 2020). Owners often considered these traits to be unique to their breed, driving owner loyalty towards their breed. Perceived suitability for living with children has been identified as influential for acquisition of brachycephalic breeds (Packer, Murphy, and Farnworth 2017) and was recently identified as a core reason why Bulldog, French Bulldog and Pug owners recommend their breed for ownership to others (Packer et al. 2020). This may be a key factor perpetuating brachycephalic popularity in young families. In international studies of the 'ideal companion dog', safety with children was the most desirable behavioural characteristic, with over two-thirds of Australian (King, Marston, and Bennett 2009) and three-quarters of Italian (Diverio et al. 2016) owners considering it an 'extremely important' trait. With no conclusive evidence that brachycephalic breeds are safer breeds for households with children, and indeed, no robust evidence that breed is a risk or protective factor for dog bites (Newman et al. 2017), recommending brachycephalic dogs on this basis is unreliable and potentially irresponsible. Indeed, aggression was the thirteenth most common reason French Bulldogs were presented to UK veterinary surgeons in 2013 (O'Neill et al. 2018). As such, presenting a balanced view of brachycephalic behaviour is key to education initiatives around choosing suitable breeds, highlighting that within-breed differences in behaviour are common, influenced by genetics but also environment and experience (Mehrkam and Wynne 2014).

LIFESTYLE BREEDS

International data suggest that brachycephalic dogs represent 'lifestyle breeds', where their perceived characteristics fit with modern lifestyle needs of a companion dog. Trends in canine population morphology in Australia indicate that dogs with flatter faces, and breeds that are shorter and smaller have become relatively more popular between 1986 and 2013 (Teng et al. 2016). The authors postulated that this may be related to changes in human lifestyle in that period in Australia, with house purchases decreasing and apartment/flat purchases increasing, suggesting that living space available for potential dogs would have reduced. In a recent qualitative study exploring reasons why owners would recommend their brachycephalic breed, owner lifestyle and perception of their breed's ability to fit within it were influential in explaining why an owner would recommend their breed (Packer et al. 2020). Bulldogs, French Bulldogs and Pugs were promoted by owners as requiring little exercise or space, deeming them suitable for busy lifestyles, city-living and households considered unsuitable to other larger or more active breeds. These findings are in agreement with previous studies that identified brachycephalic breeds as being more likely than non-brachycephalic breeds to live in apartments and that breed size being suited to an owner's lifestyle was more influential in the decision-making on acquisition of brachycephalic than non-brachycephalic breeds (Packer, Murphy, and Farnworth 2017).

In Packer et al.'s (2020) qualitative study, brachycephalic breeds were also recommended as being 'lazy' low-energy dogs. Owners believed these perceived traits made them ideal dogs for owners unable to provide moderate to high levels of exercise for their dog for a variety of reasons, including disabled, 'low energy', working and elderly people. Others indicated that their brachycephalic breed would suit owners who wanted a dog but did not want to exercise them. Some owners tried to explain these perceived differences based on the 'purpose' these breeds were bred for, highlighting that they do not require 'dog-centric' tasks including exercise, as they were bred for companionship. Taking these results together, it appears that some brachycephalic breeds occupy an ownership niche in which traditional ownership activities such as dog walking are not highly valued, despite being one of the most common reasons (behind companionship) for acquiring dogs more generally (Herwijnen et al. 2018). Preconceptions of breed 'laziness' are inherently problematic. If individual dogs truly have a reduced motivation for exercise, this may be driven by pain or disease, for example osteoarthritis (Belshaw, Dean, and Asher 2020), or reduced exercise tolerance associated with airway obstruction common in brachycephalic breeds (Roedler, Pohl, and Oechtering 2013), which may go unrecognised and undertreated. If dogs are motivated to exercise but owners do not recognise this and consequently under-exercise them, dogs may be at risk of becoming overweight or obese (Robertson 2003), a commonly recognised problem in brachycephalic breeds such as the Pug (O'Neill et al. 2016).

SOCIAL INFLUENCES ON BREED CHOICE

As well as the biological effects described at the start of this chapter, owner perception of the attractiveness of the appearance of brachycephalic dogs is likely to be also socially reinforced. Previous studies of US dog registrations identified that changes in dog breed popularity follow boom-bust patterns of fashion, with social contagion highlighted as a major factor in determining dog breed preferences (Herzog, Bentley, and Hahn 2004). Media exposure can amplify breed popularity, for example, through their inclusion in films. This was prominently exemplified by the explosion in Dalmatian popularity in the US following the release of '101 Dalmatians' (Herzog 2006). Brachycephalic breeds have featured prominently in recent films, including the 2018 British film 'Patrick', where the title character is a Pug. Given that this film came after the UK popularity surge in Pugs, this film is more likely a consequence of Pug popularity rather than a driver of it. As highlighted by Herzog (2006), there is rarely a clear, causal relationship between a movie and a 'breed epidemic'; however, public exposure to brachycephalic breeds has been at an all-time high in the past decade, through a range of media including advertising and celebrity ownership.

The use of brachycephalic dogs in the advertising is pervasive, with flat-faced dogs used to promote products as diverse (and seemingly non-canine related) as soft drinks to mobile phones. This phenomenon has been campaigned against in the UK by both the veterinary profession (Waters 2017a, b) and animal welfare groups (e.g. Brachycephalic Working Group, CRUFFA: Campaign for the Responsible Use of Flat-Faced Dogs in Advertising). Celebrity endorsement has been considered an influence upon popularity for several breeds, for example the fad for miniature breeds such as Chihuahuas is thought to be driven by celebrity ownership by Paris Hilton in the early noughties (Redmalm 2014), and indeed, ownership of Bulldogs, Pugs and French Bulldogs appears extremely common in western celebrities. A recent study did not find celebrity ownership or endorsement a highly rated influence upon breed choice for owners of brachycephalic dogs (Packer, Murphy, and Farnworth 2017); however, it is possible these influences are not entirely conscious.

Although not yet studied in relation to dogs, the rise of 'influencer' culture via social media sites such as Instagram and YouTube has the potential to affect breed choice in young people. Pet Instagram accounts, where pet owners run Instagram accounts 'on behalf' of their pets are increasingly common (Maddox 2020). A prominent example of this is 'Doug the Pug', who at the time of writing has four million followers and over 3000 posted images. The US business magazine *Forbes* listed Doug #3 in their '10 Most Influential Pets on the Internet 2017', which also included another brachycephalic dog, "Manny the Frenchie" (#9) (Kocay 2017). In addition to pet Instagram accounts, a study commissioned by pet insurance company PetPlan suggested that over half of owners share photos of their pets on the internet (Schroeder 2011). Pets are commonly cited as part of the 'extended self'; online this can manifest as images of places, things or pets, with followers coming to know what matters to someone based on the images they share (Belk 2013). Influencers who post photos including their own brachycephalic dogs have the potential to increase desire for these breeds in young adults viewing these images and videos and wishing to emulate these individuals. Recent research found that 'celebrities' on Instagram have a strong influence on the purchasing behaviour of young female users, with non-traditional celebrities such as bloggers, YouTube personalities and 'Instafamous' profiles being more powerful, because young women finding them more credible and relatable than 'traditional' celebrities (Djafarova and Rushworth 2017). Studies on the demographics of brachycephalic ownership are limited by dog ownership not being a registered activity in many countries; however, international cross-sectional surveys have found that owners of brachycephalic breeds are predominantly young (mostly aged 25–34 years) and are younger than non-brachycephalic dog owners (Packer, Murphy, and Farnworth 2017, Packer et al. 2019). This was speculated to be due to this younger age group being more susceptible to social influence (fashion) (Figure 3.3).

The role of pets as part of one's identity, whether online or in the flesh, has been recently investigated. A qualitative study of dog owners used interview data to categorise owners into two distinct groups based on their motivation for pet ownership: 'intrinsic' or 'extrinsic' (Beverland, Farrelly, and Lim 2008). Intrinsically motivated owners were defined as those who preferred to achieve goals that are more innately satisfying and view their pets as unique beings possessing human characteristics that they value for the sake of the individual animal itself. In contrast, extrinsically motivated were defined as those owners who display behaviour that earns social acknowledgement and boosts 'status'; these owners view pets as possessions. The authors proposed the extrinsic owner group are more likely to acquire fashionable dog breeds as part of their 'personal identity project', commonly those breeds with distinctive physical appearance. A recent Danish study comparing the acquisition motivations of owners of French Bulldogs, Chihuahuas, Cavalier King Charles Spaniels and Cairn Terriers suggested that French Bulldog, Chihuahua and Cavalier King Charles Spaniel owners (all breeds previously considered brachycephalic (O'Neill et al. 2020)) might predominantly fall within the 'extrinsically motivated' group, compared with Cairn Terrier owners, on the basis that the Cairn

FIGURE 3.3   Brachycephalic breeds such as Pugs are popular across social media, in both 'Pet Instagram' accounts run 'on their behalf' and their owner's photographs https://www.shutterstock.com/image-vector/funny-pug-selfish-stick-on-blue-721140112.

Terrier group were more motivated by breed attributes such as health, and less motivated by appearance, in their breed choice (Sandøe et al. 2017).

Gleaning greater understandings into the digital world surrounding brachycephalic breeds, and its potential influence on ownership motivations in young people, is likely to be of importance if the next generation of dog owners are going to turn the tide on the current popularity crisis in brachycephalic breeds. Attempts to reduce the presence of brachycephalic dogs in traditional advertising are laudable and, indeed, have led to some companies changing their policies regarding animals in adverts (Brachycephalic Working Group 2019, British Veterinary Association 2018). However, whether this strategy is sufficient or targets the demographics most likely to desire a brachycephalic breed is questionable, and a greater understanding of social media influences on dog purchasing intent is needed.

WHAT'S HEALTH GOT TO DO WITH IT?

Despite being several pages deep into this chapter, there is an elephant in the room that has been little mentioned thus far – the health issues of brachycephalic dogs. Given this book was inspired by concerns over these health issues, and indeed, the second half explores several of them in great depth, the importance (or lack of importance) of health as an owner motivation towards or against owning a brachycephalic breed warrants discussion. Evidence suggests that health considerations are generally secondary to other physical and behavioural traits in many decisions to acquire dogs, particularly brachycephalic breeds. In a study of American Kennel Club registrations, health and longevity were not correlated with breed popularity, and on the contrary, the most popular breeds tended to have significant health problems (Ghirlanda et al. 2013). In a study of influences upon owner choice of breed, owners choosing brachycephalic dogs were reported to be less influenced by breed health and longevity than owners of non-brachycephalic, popular dog breeds (Packer, Murphy, and Farnworth 2017).

NORMAL FOR THE BREED?

Much of the initial attraction from prospective owners towards brachycephalic breeds appears to focus on the 'cute' external appearance of these dogs; however, continued appeal during owner-ship, particularly in the face of chronic health disorders, suggests strong influences and misalign-ment of other elements of owner psychology. Unconscious, or indeed conscious, fallacies in owner perceptions of their dog's true health status may perpetuate continuing satisfaction by owners with their brachycephalic choice. In a study of dogs presenting to a UK referral centre, over half (58%) of owners of brachycephalic dogs exhibiting clinical signs of BOAS did not recognise that their dog had a specific breathing 'problem', instead often justifying these signs as 'normal for the breed' (Packer, Hendricks, and Burn 2012). As such, owner expectations and beliefs of their dog's health may be influenced by their choice of breed, with greater levels of morbidity tolerated on average by brachycephalic breed owners. In a recent study of ownership experiences of Pug, French Bulldog and Bulldog owners, owners reported on the presence/severity of four common aspects of brachycephaly-associated airway impairments: breathing difficulty, heat intolerance, eating difficulties and sleep dysfunction (Packer et al. 2019). Owners then reported whether they perceived (yes/no) their dog to have a 'problem' with breathing, heat regulation, eating or sleeping. For all four areas of dysfunction, greater severity of signs was significantly associated with recog-nition of a problem in that domain, indicating that dogs may need to reach a critical level of clini-cal severity (a threshold) before owners consciously acknowledge their dog has a 'problem' that is somehow worse than just being 'normal for the breed'. Nearly 40% of dogs exhibited clinical signs indicative of airway obstruction, but when asked directly, only 17.9% of owners of all dogs were considered their dog to have a breathing problem. These contrasting and paradoxical results support the influence of 'normalisation' phenomenon whereby owners of brachycephalic dogs may be consciously aware that the dog is struggling to breathe but not consciously accept that this is a specific problem, instead considering it a 'normal' and therefore somehow acceptable feature of the breed (Packer et al. 2015). These normalisation and thresholding phenomena are of welfare concern because they may prolong suffering and worsen prognosis in affected dogs. For example, owners may delay seeking veterinary advice and/or intervention until secondary collapse of the airways occurs (Packer et al. 2019).

WHAT IS 'GOOD HEALTH' FOR A BRACHYCEPHALIC DOG?

In the same study described above, owners' perceptions of 'good health' were further revealed to be misaligned with what veterinary professionals may consider 'good health'. Owners reported which disorders their dog had previously been diagnosed with, the most common of which were allergies (27.0%), corneal ulcers (15.4%), skin fold infections (15.0%) and BOAS (11.8%). In addition, one fifth (19.9%) of owners reported that their dog had undergone one or more conformation-related surgeries. Despite relatively high levels of disease reported in this young population (mean age just 2.17 years), most owners in this population paradoxically perceived their dogs to be in the 'best health possible' (30.0%) or 'very good health' (40.9%). When further questioned on their own dog's health compared with the rest of their breed, owners most commonly reported that their dog was 'healthier than average for the breed', with just 6.8% of owners considering their dog 'less' or 'much less' healthy than average for their breed (Figure 3.4) (Packer et al. 2019). This deflection phenomenon may be driven by cognitive dissonance, where owners are aware of health problems in their dog's breed but find accepting these problems in *their own* dog psychologically uncomfortable, instead deflecting the issues to other individuals.

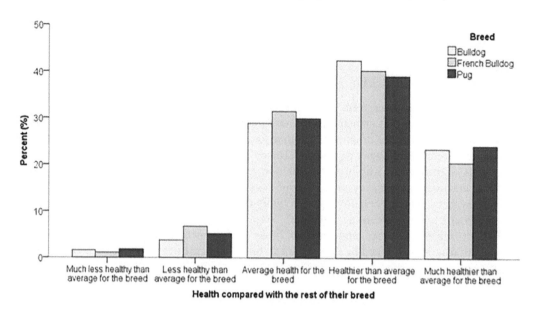

FIGURE 3.4   Owner ratings of their dog's current health from the worst to best health possible, in a population of Pugs (*n*=789), French Bulldogs (*n*=741) and Bulldogs (*n*=638), from Packer et al., 2019.

It is possible that poor health (and its associated caregiving requirements) is a desirable trait in their dogs for some owners. It has previously been demonstrated that brachycephalic dog owners, particularly females without children in the household, form strong emotional bonds with their dogs (Packer et al. 2019). This relationship may, to some extent, reflect a caregiver–infant relationship of owners seeking 'perpetual children' (Margolies 1999). Indeed, some owners in a recent qualitative study appeared to relish the dependency of their dog upon them and did not consider caring for their 'additional needs' as burdensome (Packer et al. 2020). High levels of caregiving are associated with high levels of attachment to dogs; caregiving activities may facilitate attachment development, which then cyclically motivates sustained caregiving (Kurdek 2008). This is likely not the case for all owners; indeed, in a Danish study, increasing number of owner-reported health and behaviour problems decreased the likelihood of owners being certain they would acquire a French Bulldog again, dropping from 31% for no health problems, to 20% for one problem reported and 12% for two problems reported (Sandøe et al. 2017). In an international study, owners who reported that their brachycephalic dog had a greater number of BOAS-related health problems or considered their dog to have health less than 'the best possible' had a reduced odds of wanting to own their brachycephalic breed again or recommending it to others, compared to those owning dogs with better health (Packer et al. 2020). Although these were significant effects, it should be noted that many dogs in that study population had significant health problems (Packer et al. 2019), yet the vast majority of owners (93.0%) would still elect to own their breed again, and thus, these effects are not as strong as might be expected. Understanding the presence and extent of 'caregiver burden' (Spitznagel, Solc, et al. 2019, Spitznagel et al. 2017, Spitznagel, Cox, et al. 2019) on owners of chronically ill brachycephalic dogs and potential influences of disease recognition could glean a better understanding of these results.

In summary, owners' perceptions of their dog's health may not align with assessments by veterinary professionals, with cognitive dissonance fuelling underrecognition or underestimation of the presence and/or severity of health issues. This could protect owners from psychological discomfort associated with buying an inherently unhealthy breed, along with potentially shielding owners from caregiver burden observed in the owners of dogs with other chronic illnesses.

REACQUISITION DESIRE AND THE DEVELOPMENT OF 'BREED LOYALTY'

As the popularity of brachycephalic breeds has already reached high levels in the UK (comprising 18.7% of all UK dogs (O'Neill et al. 2020)), in addition to understanding decision-making about initial acquisition of these breeds, there is an increasing need to understand aspects related to re-acquisition from existing owners. A recent study of Danish dog owners reported that nearly one-third of French Bulldog owners (29.2%) planned to acquire the same breed in the future, more than owners of any of the other small breeds studied (Chihuahua, 17.5%; Cairn Terrier, 20.1%; Cavalier King Charles Spaniel, 22.3%) (Sandøe et al. 2017). Dog breed-loyalty has received very little scientific attention, but a breed-loyal pet owner has been defined as "[An owner with] a positive attitude towards a specific breed of dog, [who will] buy that breed when compared with other existing breeds and have a continued allegiance to that breed over long periods of time" (Clark and Page 2009). If current brachycephalic dog owners become breed-loyal, then the current 'boom' phase of brachycephalic popularity (a phenomenon described in other breeds (Ghirlanda et al. 2013)) may plateau to remain at a high level, rather than dropping back to their pre-boom numbers. Given that 93.0% of brachycephalic dog owners indicated they would purchase their breed again in a recent study (Packer et al. 2020), unless their minds can be changed (or will 'naturally' change, if brachycephalic ownership is truly a fad and the boom will 'bust' (Herzog, Bentley, and Hahn 2004)), the flat-faced demographic may continue to dominate several countries' canine populations.

Several factors have been reported to increase the likelihood of owners wishing to purchase their chosen brachycephalic breed again, including the strength of the dog–owner relationship. In a study of Danish dog owners, owners reporting a stronger dog–owner (as quantified by the Lexington Attachment to Pets Scale (Johnson, Garrity, and Stallones 1992)) were more likely to be certain they would re-acquire a French Bulldog (Sandøe et al. 2017). In an international study of brachycephalic dog owners, the strength of the dog–owner relationship (as quantified by the MDORS (Dwyer, Bennett, and Coleman 2006)) increased the odds of reacquisition desire, specifically, an increased level of emotional closeness with their dog and a reduced perception of the costs of ownership (Packer et al. 2020). As discussed in the health section above, for some owners, poor health in their current dog reduces re-acquisition desire (Sandøe et al. 2017, Packer et al. 2020), but this is not a strong effect and poor health deters few owners. Owner expectations influence future intentions of breed ownership, with owners whose dogs behaved worse than they expected exhibiting reduced reacquisition desire than those whose behaviour met their expectations (Packer et al. 2020). Finally, prior ownership experience has an effect on reacquisition desire; first time owners had over twice the odds of wanting to own their breed again. This effect may be explained by differences in what first-time owners versus more experienced dog owners consider 'normal'. If a person's first ownership experience is with a brachycephalic breed, their baseline for 'normal' canine health and behaviour may become skewed. Signs of poor health may be considered 'normal' (Packer, Hendricks, and Burn 2012), and positive behavioural traits may be considered unique to their breed, even if available more widely. Combined, this may lead to more positive perceptions of their breed, fuelling the desire to continue owning the breed.

CONCLUSIONS

It is clear that relationships between humans and brachycephalic dogs are complex, which may explain why 'simple' educational messaging aimed at informing owners about their inherent health problems has apparently not lead to reductions in their popularity, with increases in brachycephalic ownership observed internationally. With millions of brachycephalic dogs currently being kept as companions, research aimed at improving their welfare (e.g. developing better diagnostic methods and treatments) remains extremely important; however, if we are to move beyond 'firefighting' the brachycephalic issue with veterinary interventions and move towards reducing the number of brachycephalic dogs bred and kept as companions internationally, understanding the human side of this issue is vitally important. Data such as those underlying the studies explored in this chapter and future social science research should be used to inform human behaviour change initiatives (Reed and Upjohn 2018), to influence owners and breeders to make better decisions in the acquisition and breeding of future dogs, to improve canine welfare in a lasting and meaningful way. If effective ways to reduce the acquisition of brachycephalic dogs cannot be found via behaviour change approaches, then legislative approaches may be require to limit their ownership and protect canine welfare.

REFERENCES

Appleby, D. L., J. W. S. Bradshaw, and R. A. Casey. 2002. "Relationship between aggressive and avoidance behaviour by dogs and their experience in the first six months of life." *Veterinary Record* 150 (14):434–438. doi:10.1136/vr.150.14.434.

Archer, J. 1997. "Why do people love their pets?" *Evolution and Human Behavior* 18 (4):237–259.

Archer, J., ed. 2011. Pet keeping: A case study in maladaptive behavior. Edited by C. Salmon and T. K. Shackelford, *The Oxford Handbook of Evolutionary Family Psychology*. Oxford, UK: Oxford University Press.

Archer, J., and J. L. Ireland. 2011. "The development and factor structure of a questionnaire measure of the strength of attachment to pet dogs." *Anthrozoös* 24 (3):249–261. doi:10.2752/1753037 11X13045914865060.

Archer, J., and S. Monton. 2011. "Preferences for infant facial features in pet dogs and cats." *Ethology* 117 (3):217–226. doi:10.1111/j.1439-0310.2010.01863.x.

Arhant, C., H. Bubna-Littitz, A. Bartels, A. Futschik, and J. Troxler. 2010. "Behaviour of smaller and larger dogs: Effects of training methods, inconsistency of owner behaviour and level of engagement in activities with the dog." *Applied Animal Behaviour Science* 123 (3–4):131–142.

Barker, S. B., and A. R. Wolen. 2008. "The benefits of human–companion animal interaction: A review." *Journal of Veterinary Medical Education* 35 (4):487–495. doi: 10.3138/jvme.35.4.487.

Belk, R. W. 2013. "Extended self in a digital world." *Journal of Consumer Research* 40 (3):477–500.

Belshaw, Z., R. Dean, and L. Asher. 2020. "Slower, shorter, sadder: A qualitative study exploring how dog walks change when the canine participant develops osteoarthritis." *BMC Veterinary Research* 16 (1):85. doi:10.1186/s12917-020-02293-8.

Beverland, M. B., F. Farrelly, and E. A. C. Lim. 2008. "Exploring the dark side of pet ownership: Status-and control-based pet consumption." *Journal of Business Research* 61 (5):490–496.

Blackwell, E. J., C. Twells, A. Seawright, and R. A. Casey. 2008. "The relationship between training methods and the occurrence of behavior problems, as reported by owners, in a population of domestic dogs." *Journal of Veterinary Behavior* 3 (5):207–217. doi:10.1016/j.jveb.2007.10.008.

Bognár, Z., I. B. Iotchev, and E. Kubinyi. 2018. "Sex, skull length, breed, and age predict how dogs look at faces of humans and conspecifics." *Animal Cognition* 21 (4):447–456. doi:10.1007/s10071-018-1180-4.

Brachycephalic Working Group. 2019. "Position Statement: The use of brachycephalic dogs in advertising and the media." accessed November. http://www.ukbwg.org.uk/wp-content/uploads/2020/03/BWG-Position-Statement-on-the-use-of-brachycephalic-dogs-in-advertising-and-the-media-190602-.pdf.

British Veterinary Association. 2018. "Policy Statement: Brachycephalic Dogs." accessed November. https://www.bva.co.uk/media/1183/bva-position-on-brachycephalic-dogs-full.pdf.

Clark, P. W., and J. B. Page. 2009. "Examining role model and information source influence on breed loyalty: Implications in four important product categories." *Journal of Management Marketing Research* 2:1.

Diverio, S., B. Boccini, L. Menchetti, and P. C. Bennett. 2016. "The Italian perception of the ideal companion dog." *Journal of Veterinary Behavior* 12:27–35.

Djafarova, E., and C. Rushworth. 2017. "Exploring the credibility of online celebrities' Instagram profiles in influencing the purchase decisions of young female users." *Computers in Human Behavior* 68:1–7. doi:10.1016/j.chb.2016.11.009.

Dwyer, F., P. C. Bennett, and G. Coleman. 2006. "Development of the Monash dog owner relationship scale (MDORS)." *Anthrozoös* 19 (3):243–256.

Eibl-Eibesfeldt, I. 2017. *Human Ethology*. Routledge, London.

Gácsi, M., P. McGreevy, E. Kara, and Á. Miklósi. 2009. "Effects of selection for cooperation and attention in dogs." *Behavioral and Brain Functions* 5 (1):31. doi:10.1186/1744-9081-5-31.

Ghirlanda, S., A. Acerbi, H. Herzog, and J. A. Serpell. 2013. "Fashion vs. function in cultural evolution: The case of dog breed popularity." *PLoS One* 8 (9):e74770.

Glocker, M. L., D. D. Langleben, K. Ruparel, J. W. Loughead, R. C. Gur, and N. Sachser. 2009. "Baby schema in infant faces induces cuteness perception and motivation for caretaking in adults." *Ethology* 115 (3):257–263. doi:10.1111/j.1439-0310.2008.01603.x.

Glocker, M. L., D. D. Langleben, K. Ruparel, J. W. Loughead, J. N. Valdez, M. D. Griffin, N. Sachser, and R. C. Gur. 2009. "Baby schema modulates the brain reward system in nulliparous women." *Proceedings of the National Academy of Sciences* 106 (22):9115–9119. doi:10.1073/pnas.0811620106.

Gould, S. J. 1992. *The Panda's Thumb: More Reflections in Natural History*. WW Norton & Company, New York.

Hecht, J., and A. Horowitz. 2015. "Seeing dogs: Human preferences for dog physical attributes." *Anthrozoös* 28 (1):153–163.

Herzog, H. 2006. "Forty-two thousand and one Dalmatians: Fads, social contagion, and dog breed popularity." *Society and Animals* 14 (4):383–397.

Herzog, H. A., R. A. Bentley, and M. W. Hahn. 2004. "Random drift and large shifts in popularity of dog breeds." *Proceedings of the Royal Society of London. Series B: Biological Sciences* 271 (suppl_5):S353–S356.

Hinde, R. A., and L. A. Barden. 1985. "The evolution of the teddy bear." *Animal Behaviour* 33 (4):1371–1373.

Holland, K. E. 2019. "Acquiring a pet dog: A review of factors affecting the decision-making of prospective dog owners." *Animals* 9 (4):124.

Johnson, T. P., T. F. Garrity, and L. Stallones. 1992. "Psychometric evaluation of the Lexington attachment to pets scale (LAPS)." *Anthrozoös* 5 (3):160–175.

King, T., L. C. Marston, and P. C. Bennett. 2009. "Describing the ideal Australian companion dog." *Applied Animal Behaviour Science* 120 (1–2):84–93.

Kocay, L. 2017. "Meet the Most Influential Pets: Grumpy Cat, Doug the Pug and More." accessed November. https://www.forbes.com/sites/lisakocay/2017/07/31/meet-the-most-influential-pets-grumpy-cat-doug-the-pug-and-more/?sh=bd0cd9b31a68.

Kruger, D. J. and S. A. Miller. 2016. "Non-mammalian infants dependent on parental care elicit greater kindchenschema-related perceptions and motivations in humans." *Human Ethology* 31:16–24.

Kurdek, L. A. 2008. "Pet dogs as attachment figures." *Journal of Social and Personal Relationships* 25 (2):247–266. doi:10.1177/0265407507087958.

Ley, J. M., P. C. Bennett, and G. Coleman. 2009. "A refinement and validation of the Monash Canine Personality Questionnaire (MCPQ)." *Applied Animal Behaviour Science* 116 (2–4):220–227.

Lorenz, K., ed. 1971. Part and parcel in animal and human societies. Edited by K. Lorenz. Vol. II, *Studies in Animal and Human Behaviour*. Cambridge, MA: Harvard University Press.

Maddox, J. 2020. "The secret life of pet Instagram accounts: Joy, resistance, and commodification in the Internet's cute economy." *New Media & Society*. doi:10.1177/1461444820956345.

Margolies, L. 1999. "The long goodbye: Women, companion animals, and maternal loss." *Clinical Social Work Journal* 27 (3):289–304. doi:10.1023/A:1022894320225.

McGreevy, P. D., D. Georgevsky, J. Carrasco, M. Valenzuela, D. L. Duffy, and J. A. Serpell. 2013. "Dog behavior co-varies with height, bodyweight and skull shape." *Plos One* 8 (12):e80529. doi:10.1371/journal.pone.0080529.

McGreevy, P., T. D. Grassi, and A. M. Harman. 2004. "A strong correlation exists between the distribution of retinal ganglion cells and nose length in the dog." *Brain, Behavior Evolution* 63 (1):13–22.

Mehrkam, L. R., and C. D. L. Wynne. 2014. "Behavioral differences among breeds of domestic dogs (Canis lupus familiaris): Current status of the science." *Applied Animal Behaviour Science* 155:12–27 doi:10.1016/j.applanim.2014.03.005.

Meyer, I., and B. Forkman. 2014. "Dog and owner characteristics affecting the dog–owner relationship." *Journal of Veterinary Behavior* 9 (4):143–150.

Morreall, J. 1993. "The contingency of cuteness: A reply to Sanders." *The British Journal of Aesthetics* 33 (3):283–286.

Newman, J., R. Christley, C. Westgarth, and K. Morgan, 2017. Risk factors for dog bites – An epidemiological perspective. Edited by D. Mills and C. Westgarth, *Dog Bites: A Multidisciplinary Perspective*. Sheffield: 5M Publishing.

O'Neill, D. G., L. Baral, D. B. Church, D. C. Brodbelt, and R. M. A. Packer. 2018. "Demography and disorders of the French Bulldog population under primary veterinary care in the UK in 2013." *Canine Genetics and Epidemiology* 5 (1):3. doi:10.1186/s40575-018-0057-9.

O'Neill, D. G., E. C. Darwent, D. B. Church, and D. C. Brodbelt. 2016. "Demography and health of Pugs under primary veterinary care in England." *Canine Genetics and Epidemiology* 3 (1):5. doi:10.1186/s40575-016-0035-z.

O'Neill, D. G., C. Pegram, P. Crocker, D. C. Brodbelt, D. B. Church, and R. M. A. Packer. 2020. "Unravelling the health status of brachycephalic dogs in the UK using multivariable analysis." *Scientific Reports* 10 (1):1–13.

Packer, R. M. A., A. Hendricks, and C. C. Burn. 2012. "Do dog owners perceive the clinical signs related to conformational inherited disorders as 'normal' for the breed? A potential constraint to improving canine welfare." *Anim Welfare* 21. doi:10.7120/096272812x13345905673809.

Packer, R. M. A., A. Hendricks, M. S. Tivers, and C. C. Burn. 2015. "Impact of facial conformation on canine health: Brachycephalic obstructive airway syndrome." *Plos One* 10 (10):e0137496. doi:10.1371/journal.pone.0137496.

Packer, R. M. A., D. Murphy, and M. J. Farnworth. 2017. "Purchasing popular purebreds: investigating the influence of breed-type on the pre-purchase motivations and behaviour of dog owners." *Animal Welfare* 26 (2):191–201. doi:10.7120/09627286.26.2.191.

Packer, R. M. A., D. G. O'Neill, F. Fletcher, and M. J. Farnworth. 2019. "Great expectations, inconvenient truths, and the paradoxes of the dog-owner relationship for owners of brachycephalic dogs." *Plos One* 14 (7):e0219918. doi:10.1371/journal.pone.0219918.

Packer, R. M. A., D. G. O'Neill, F. Fletcher, and M. J. Farnworth. 2020. "Come for the looks, stay for the personality? A mixed methods investigation of reacquisition and owner recommendation of Bulldogs, French Bulldogs and Pugs." *Plos one* 15 (8):e0237276.

Packer, R. M. A., H. A. Volk, and R. C. Fowkes. 2017. "Physiological reactivity to spontaneously occurring seizure activity in dogs with epilepsy and their carers." *Physiology & Behavior* 177:27–33 doi:10.1016/j.physbeh.2017.04.008.

Redmalm, D. 2014. "Holy bonsai wolves: Chihuahuas and the Paris Hilton syndrome." *International Journal of Cultural Studies* 17 (1):93–109.

Reed, K., and M. M. Upjohn. 2018. "Better Lives for Dogs: incorporating Human Behaviour change into a theory of change to improve canine Welfare Worldwide." *Frontiers in Veterinary Science* 5:93.

Robertson, I. D. 2003. "The association of exercise, diet and other factors with owner-perceived obesity in privately owned dogs from metropolitan Perth, WA." *Preventive Veterinary Medicine* 58 (1):75–83. doi:10.1016/S0167-5877(03)00009-6.

Roedler, F. S., S. Pohl, and G. U. Oechtering. 2013. "How does severe brachycephaly affect dog's lives? Results of a structured preoperative owner questionnaire." *The Veterinary Journal* 198 (3):606–610.

Rohlf, V. I., P. C. Bennett, S. Toukhsati, and G. Coleman. 2012. "Beliefs underlying dog owners' health care behaviors: Results from a large, self-selected, internet sample." *Anthrozoös* 25 (2):171–185.

Sandøe, P., S. V. Kondrup, P. C. Bennett, B. Forkman, I. Meyer, H. F. Proschowsky, J. A. Serpell, and T. B. Lund. 2017. "Why do people buy dogs with potential welfare problems related to extreme conformation and inherited disease? A representative study of Danish owners of four small dog breeds." *Plos One* 12 (2):e0172091. doi:10.1371/journal.pone.0172091.

Schroeder, S. 2011. "1 in 10 Pets Have a Social Networking Profile." accessed November. http://mashable.com/2011/07/13/pets-social-networking.

Spitznagel, M. B., M. D. Cox, D. M. Jacobson, A. L. Albers, and M. D. Carlson. 2019. "Assessment of caregiver burden and associations with psychosocial function, veterinary service use, and factors related to treatment plan adherence among owners of dogs and cats." *Journal of the American Veterinary Medical Association* 254 (1):124–132.

Spitznagel, M. B., D. M. Jacobson, M. D. Cox, and M. D. Carlson. 2017. "Caregiver burden in owners of a sick companion animal: A cross-sectional observational study." *Veterinary Record* 181 (12):321–321. doi:10.1136/vr.104295.

Spitznagel, M. B., M. Solc, K. R. Chapman, J. Updegraff, A. L. Albers, and M. D. Carlson. 2019. "Caregiver burden in the veterinary dermatology client: Comparison to healthy controls and relationship to quality of life." *Veterinary Dermatology* 30 (1):3-e2.

Stone, H. R., P. D. McGreevy, M. J. Starling, and B. Forkman. 2016. "Associations between domestic-dog morphology and behaviour scores in the dog mentality assessment." *PloS One* 11 (2):e0149403. doi:10.1371/journal.pone.0149403.

Teng, K. T., P. D. McGreevy, J.-A. L. M. L. Toribio, and N. K. Dhand. 2016. "Trends in popularity of some mor-
    phological traits of purebred dogs in Australia." *Canine Genetics and Epidemiology* 3 (1):2. doi:10.1186/
    s40575-016-0032-2.

The Kennel Club. 2018. "Breed Standards Information: Dog Breeds & Groups: The Kennel Club."

Thorn, P., T. J. Howell, C. Brown, and P. C. Bennett. 2015. "The canine cuteness effect: Owner-perceived
    cuteness as a predictor of human–dog relationship quality." *Anthrozoös* 28 (4):569–585. doi:10.1080/08
    927936.2015.1069992.

van Herwijnen, I. R., J. A. M. van der Borg, M. Naguib, and B. Beerda. 2018. "The existence of parenting
    styles in the owner-dog relationship." *Plos One* 13 (2):e0193471. doi:10.1371/journal.pone.0193471.

Waters, A. 2017a. "At last! Action to stop brachy dogs ads." *The Veterinary Record* 181 (24):635.

Waters, A. 2017b. "Brachycephalic tipping point: Time to push the button?" *The Veterinary Record*
    180 (12):288.

Weiss, E., K. Miller, H. Mohan-Gibbons, and C. Vela. 2012. "Why did you choose this pet? Adopters and pet
    selection preferences in five animal shelters in the United States." *Animals* 2 (2):144–159.

4 Ethical Challenges of Treating Brachycephalic Dogs

Anne Quain and Paul McGreevy
University of Sydney

Siobhan Mullan
University College Dublin

CONTENTS

Ethical Challenges in Veterinary Practice 41
Why Do Brachycephalic Breeds Present an Ethical Challenge for Veterinarians? 42
The Ethical Challenges of Treating Brachycephalic dogs 43
Contractarianism 46
Virtue Ethics 46
Consequentialism and Utilitarianism 46
Deontology 47
Recommendations and Conclusion 51
Acknowledgement 51
References 51

ETHICAL CHALLENGES IN VETERINARY PRACTICE

Ethically challenging situations, often described as ethical or moral dilemmas, are common in veterinary practice and are often stressful for veterinarians and their co-workers (Batchelor and McKeegan, 2012, Crane et al., 2015, Kipperman et al., 2018, Moses, 2018, Lehnus et al., 2019). They may give rise to moral distress, defined as

> the experience of psychological distress that results from engaging in, or failing to prevent, decisions or behaviours that transgress, or come to transgress, personally held moral or ethical beliefs

(Crane et al., p. 6, 2013)

or

> the psychological disequilibrium experienced from being constrained from following the perceived correct moral path

(Arbe Montoya et al., p. 1, 2019)

Moral distress in veterinary team members is topical because it is an occupational stressor that may contribute to psychological morbidity and mortality (Crane et al., 2015, Kipperman et al., 2018, Moses et al., 2018, Rollin, 2011, Foster and Maples, 2014, Arbe Montoya et al., 2019). There is merit in discussing ethically challenging situations that lead to moral distress because such discussions may disclose opportunities to improve aspects of veterinary care, including animal welfare, or to identify and address misplaced concerns (Camp and Sadler, 2019).

WHY DO BRACHYCEPHALIC BREEDS PRESENT AN ETHICAL CHALLENGE FOR VETERINARIANS?

The World Small Animal Veterinary Association Animal Welfare Guidelines Group identified the involvement of veterinarians when treating and facilitating reproduction of animals with extreme phenotypic traits as a 'common moral problem' in companion animal practice because such traits can contribute to 'significant welfare problems' (Figure 4.1) (Ryan et al., 2019). The strong correlation between the popularity of dog breeds and those breeds' predisposition to disorders suggests that prospective dog breeders and their clients prioritise appearance over health, longevity and behavioural characteristics (Ghirlanda et al., 2013).

Farrow and colleagues found that the majority of veterinarians (89.8%) believed that inherited disorders in dogs were a significant issue (Farrow et al., 2014). Concerns about the welfare of brachycephalic dogs have mobilised veterinary individuals and organisations around the globe in ways that other ethical issues in companion animal practice have not. Indeed, this issue has prompted veterinary associations to unite with animal protection organisations in public education campaigns increasing awareness of the consequences of brachycephaly, calling for an end to the use of brachycephalic dogs in advertising (British Veterinary Association, 2017, Latter, 2017) and for urgent change to breed standards (for example, see *Love is Blind*, https://www.loveisblind.org.au/) (Anon., Undated).

As discussed elsewhere in this volume, brachycephalic breeds suffer from a range of conformation-related conditions that compromise welfare, primarily through respiratory distress and air hunger, thermal and physical discomfort (Figure 4.1) and behavioural restriction (Fawcett et al., 2018). These impacts are concerning not only because of their severity but also their chronicity, and the prospect that they may be lifelong for affected dogs. In the most severe cases, affected dogs have 'a life worth avoiding', "because they experience suffering that cannot be meaningfully outweighed during those animal's relatively short (and generally unpleasant) lives" (Yeates, 2018). The growing popularity of brachycephalic breeds, particularly those with extreme brachycephalic phenotypes such as Pugs, French Bulldogs and English Bulldogs, means that these problems affect increasing numbers of dogs, owners and veterinary team members.

In this chapter, we outline the ethical challenges associated with treating brachycephalic dogs and argue the need for veterinarians to actively discourage the breeding of animals for conformation that compromises welfare.

FIGURE 4.1 A French Bulldog exhibiting signs of heat stress (the dog's rectal temperature was 40.1°C). (Image: Anne Quain.)

THE ETHICAL CHALLENGES OF TREATING BRACHYCEPHALIC DOGS

The key ethical challenge in relation to brachycephalic dogs can be summarised thus: are veterinary professionals complicit in perpetuating welfare problems associated with extreme brachycephalic phenotypes? If so, what practical steps can members of the profession take to address welfare problems associated with extreme brachycephalic phenotypes so that they behave in alignment with their values on animal welfare?

To answer these questions, we need to clarify the role of veterinarians in relation to companion animals. A moral philosophical analysis undertaken by Grimm and colleagues found that the moral foundation of companion animal practice is the principle of working "in the animal's best interest" (Grimm et al., 2018). It concluded that companion animal practice operates according to two key norms: the first aims to restore patient health; the second is to respect patient quality of life (Grimm et al., 2018). Members of the veterinary profession violate these norms if they are complicit in perpetuating welfare problems including those associated with extreme brachycephalic phenotypes.

But how are vets complicit in perpetuating welfare problems associated with extreme brachycephalic phenotypes?

Veterinarians may directly assist in breeding dogs with extreme brachycephalic conformation (for example by performing procedures such as artificial insemination or caesarean sections) or indeed breeding affected dogs themselves. Additionally, veterinarians may choose not to address brachycephaly as a veterinary public health issue. It is speculated that this may occur because individual veterinarians fear alienating their clients and breeders by criticising the very features that some clients find most endearing about their companion animals (Fawcett et al., 2018). The British Veterinary Association addresses this prospect as follows "…highlighting animal welfare problems may sometimes be viewed as a threat by veterinary surgeons, especially if done in a way that alienates animal-owning clients" (British Veterinary Association, 2016).

Veterinary professionals who draw an income from involvement in breeding or treating dogs with extreme brachycephalic conformation have a potential conflict of interest between the interests of the animal and their own interests in drawing an income. It is this conflict of interest which one anonymous veterinarian in the UK press blamed for the failure of most veterinarians to publicly acknowledge the negative health and welfare impacts of brachycephaly:

> …the vast majority of [veterinarians] work in general practice and our income is based on mending people's animals and getting paid for it, and, like it or not, a large number of those clients have brachycephalic dogs. In my practice alone we have a number of pug, shih-tzu and bulldog breeders and dozens of owners with squashed-nose pets…If I stood up and told the truth about these breeds, I would immediately alienate them and they would up sticks and move to the neighbouring practice where the vet was not as outspoken. Vets in general practice simply cannot afford to be honest and to speak out. You would be hard-pushed to find a general practitioner who likes the concept of a brachycephalic dog but you would be equally hard-pushed to find one being openly critical of them because this would put their livelihood on the line.
>
> *(Anonymous, 2016)*

There are two key implications of this passage. The first, let's call it A, is that when treating brachycephalic breeds without being honest with clients about the animal welfare implications of their dog's phenotype, the veterinarian prioritises financial interest over their primary professional concerns, that is, the interests of the animal and the client.

The second implication, B, is that in telling 'the truth' about these breeds, the veterinarian could alienate owners to the extent that they go to a different clinic and lose future opportunities to influence the current dog's health and welfare or the acquisition of dogs with a similar extreme phenotype by that client in the future. It presupposes that truth-telling is optional for veterinarians.

We will address each of these in turn.

With regard to A, the implications of this argument are concerning. Consider parvovirus, a disease associated with high morbidity and mortality in dogs. Veterinary costs associated with treatment of affected puppies vastly outweigh veterinary costs associated with vaccination. Veterinarians stand to earn more income from treating a puppy with parvovirus than they might from a series of vaccinations preventing parvovirus (Kelman et al., 2019). Yet a veterinarian who chooses not to speak up about the risks of parvovirus because they stand to earn more income from treating affected animals than providing from vaccinations would be considered unethical. Similarly, it would be unethical for a veterinarian who provided treatment to clinically affected animals not to proactively promote an effective vaccine that would prevent spread of the disease.

With regard to B, is telling 'the truth' about the impact of an animal's phenotype on current and future health and welfare optional for veterinarians? If a veterinarian believes that his or her role is to address a client's concerns about the health and welfare of their animal, he or she may prioritise managing the animal's current clinical condition, alleviating its suffering and dealing with the client's feelings as sensitively as possible. Of course, revealing the degree of current and future health and welfare impacts of the brachycephalic phenotype may cause clients who have deliberately selected or bred these dogs to feel responsible for causing suffering, to an animal they have bonded with. This may be further exacerbated because the emotional bond between that owners have with brachycephalic dogs may be stronger when compared to other dog–owner dyads (Packer et al., 2019). Indeed, James Serpell argues that owners, breeders, veterinarians and anyone involved in maintaining or perpetuating brachycephaly are harming others, albeit indirectly, for their own interests (Serpell, 2019). He concludes that the way many of us cope with this moral dissonance is to deny or minimise welfare problems or categorise inherited conformational disorders as normal or acceptable for the breed (Serpell, 2019).

People, in this case veterinarians, are more likely to speak up if they believe that doing so will have an impact (Smith and Kouchaki, 2018). So, the veterinarian presented with a brachycephalic patient may feel that there is little to be gained and everything to be lost by telling 'the truth'. The dog, after all, has already been acquired, and the owner(s) have already bonded. Thus, the veterinarian may feel helpless.

Hernandez and colleagues (Hernandez et al., 2018) argue that veterinarians have a professional duty to speak up about animal welfare issues, despite the risk of having uncomfortable conversations. In a similar vein, Coghlan (2018) maintains that companion animal veterinarians are ethically obliged to act as strong patient advocates, prioritising the interests of the patient they are treating. To optimise patient welfare, veterinarians must ensure appropriate and timely treatment or advocate appropriate management, if treatment is not an option. So, they must ensure that the clients appreciate the degree of suffering associated with extreme brachycephalic phenotypes, the relevant pathophysiology, and the steps needed to minimise suffering and to improve health and welfare. Ideally, those clients would be able to provide the best care for their dogs and may educate others about the welfare costs of brachycephalic breeds. They may be less likely to choose an extreme brachycephalic breed in the future. Indeed, one study found that clients who recognised more BOAS health issues in their dog were less likely to want to own a dog of the same breed in the future (Packer et al., 2020). On the other hand, clients who fail to understand this information may be less likely to seek appropriate treatment for their dog (Packer et al., 2012).

Veterinary oaths and codes of conduct give veterinarians obligations beyond the consultation room. For example, veterinarians in the UK currently swear to uphold their responsibilities to "the public, my clients, the profession, and the Royal College of Veterinary Surgeons" (One Welfare Portal, 2017). Such oaths suggest that simply serving the interests of the client, the individual patient and themselves and/or employer does not abrogate veterinarians' professional responsibilities to the wider animal population and community. They imply that professional organisations and the wider community expect veterinarians to not only optimise the welfare of the individual animal presented to their care, but also evaluate and act on systemic animal welfare issues at a broader level.

The British Veterinary Association's 2016 Animal Welfare Strategy states that veterinary professionals have opportunities to act on three levels:

1. Through individual vets direct to animal keepers and owners
2. Through veterinary practices to the surrounding community
3. Through professional associations to achieve political impact and change societal norms (British Veterinary Association, 2016).

According to the British Veterinary Association, failure of veterinarians to "speak out about systemic welfare problems", or to do so only in the face of a groundswell of public concern, can leave the profession vulnerable to "accusations of weak morality and, worse, complicity in animal welfare problems" (British Veterinary Association, 2016). Addressing breed-related health problems as they arise is an important role of veterinarians, but to do so without addressing the systemic issue of brachycephalic conformation is to do only part of the job (Fawcett, 2017).

Similarly, veterinarians who facilitate breeding of brachycephalic dogs by performing elective or emergency caesarean sections (Figure 4.2), and accept fees without taking steps to address the underlying problem, for example by reporting caesarean sections in pedigree dogs to a national canine register (where one exists) or by recommending spaying of these animals, are perpetuating poor welfare, and, in the words of the British Veterinary Association, "simply facilitating the status quo" (British Veterinary Association, 2016). But this is not enough. Owners rely on veterinarians to provide them with accurate, expert information about the health and welfare of their animals. The possibility of alienating a client by discussing their pet's health should not discourage veterinarians from doing so. Anticipation that telling clients 'the truth' will alienate them is simply an assumption. Such concerns highlight the need for veterinarians to develop their communication skills, rather than abrogate duties to avoid potentially uncomfortable situations.

Veterinarians who advocate for the best interests of their patients are likely to experience more moral stress/distress, according to Springer and colleagues:

> We found that veterinarians are particularly likely to face moral challenges when they are not able to comply with patient-centered demands …. We see a danger that a moralizing approach to small animal practitioners claiming that they *must* give priority to the best interest principle during animal patient care will simply increase levels of stress within the profession without really helping any patients.
>
> (Springer et al., p.13 2019)

This concern underscores the need for veterinary professional associations to proactively educate members of the public about the health and welfare costs of extreme phenotypes, rather than leaving it to the individual veterinarian to act as the 'moral hero'. Veterinary associations can advance animal welfare by influencing animal welfare legislation, breed standards and societal norms through advertising policies which specifically prohibit the use of breeds with phenotypes that compromise animal welfare. Veterinary associations can support individual veterinarians by promoting the role of the veterinary professional as someone who tells clients what they *need* to hear, rather than what they *want* to hear.

That alone may not be enough. According to Bradshaw and Hiby, "Despite the mounting evidence that inbreeding is progressively worsening the welfare of dogs, neither the breeders nor the dog-buying public seems to be sufficiently motivated to reverse this trend, which could be readily resolved by carefully planned outbreeding (merging of unrelated breeds)" (Bradshaw and Hiby, 2018).

To improve the welfare of dogs, research into the welfare impacts of extreme phenotypes must be combined with research on human behaviour change and evidence-based strategies to achieve it. Where possible, clinical veterinarians should be actively involved in this research, as many behaviour change strategies are likely to involve them.

How might different ethical frameworks assist in navigating ethical challenges associated with extreme brachycephalic phenotypes?

Ethical frameworks can help veterinarians make decisions when faced with ethical challenges and may assist in justifying their position to other parties including companion animal owners. In the following section, we provide a broad outline of four established frameworks that can be applied to ethical challenges associated with extreme brachycephalic phenotypes. Specifically, these are contractarianism, virtue ethics, consequentialism and utilitarianism, and deontology.

CONTRACTARIANISM

Contractarianism is an ethical framework based on the assumption that people are fundamentally self-interested and that acting ethically, fairly and justly is in one's self-interest. It is therefore rational for us to cooperate with others according to a social contract to ensure that we benefit through cooperation with others, without being harmed by them imposing their interests on us. In general, this theory applies to humans who are thought to be exclusively capable of rational agreement with others. Contractarianism has been criticised for excluding animals and some groups of humans (historically, women, some racial groups and minority groups) (Mullan and Fawcett, 2017).

According to this framework, the autonomy of persons must be respected where possible. One could argue that breeders have a right to breed animals as they see fit, and prospective owners have a right to choose an animal that appeals to them. But it seems harder to argue that private contracts between veterinarians and breeder or veterinarians and pet-owning client reduce the obligation to act on ethical challenges more broadly (Green, 2007). After all, veterinary professionals make a contract with professional organisations and society to uphold relevant oaths and adhere to codes of Professional Conduct (Susskind and Susskind, 2015).

If we consider domestication as a process of animals entering such a social contract, forfeiting their autonomy in return for the protection and care from people, then breeding that promotes suffering is surely exploitation and a breach of that contract.

VIRTUE ETHICS

Professionals are expected to be of 'good character' (Susskind and Susskind, 2015), a concept supported by oaths, codes of conduct and registration bodies. This concept aligns with virtue ethics. Virtues are morally relevant, reliable traits expected in persons of good character. Beauchamp and Childress (Beauchamp and Childress, 2013) identified six focal virtues as important for medical professionals, which can be applied to the treatment of brachycephalic patients, as outlined in Table 4.1 (adapted from Beauchamp and Childress, 2013, pp. 37–44). An earlier version of this table was published in Fawcett et al. (2018).

Importantly, a virtue ethics approach requires veterinarians to treat patients and clients (including breeders) compassionately and communicate with them honestly and openly, including declaring and managing their own conflicts of interest. It also requires them to improve animal health and welfare at the level of the individual patient and the broader population. Veterinarians may be regarded as role models of excellent husbandry and animal welfare, and should avoid portraying dogs with extreme brachycephalic phenotypes as normal (for example in practice marketing materials). One could also argue from a virtue ethics perspective that this applies to veterinarians in their personal lives, 'outside' of being a veterinarian – for example in the selection and care of their own companion animals, or on social media.

CONSEQUENTIALISM AND UTILITARIANISM

Generally speaking, consequentialism holds that the rightness or wrongness of an action must be judged by its consequences. Utilitarianism, of which there are a number of subtypes, generally holds

TABLE 4.1

Beauchamp and Childress' Focal Virtues for Medical Professionals, and How These May Manifest in the Veterinarian Presented with the Brachycephalic Patient

Virtue	Manifestation
Care	The veterinarian has an emotional commitment to, and willingness to act on behalf of, patients (brachycephalic dogs in this case) and clients (dog owners, breeders and clients requesting pre-purchase consultation).
Compassion	The veterinarian has an active regard for both the current and future dog's and owner's welfare, with imaginative awareness and sympathy, tenderness and discomfort at another's suffering. This implies the ability to identify and motivation to address suffering, including that caused by conformation-related conditions, in the short and long terms.
Discernment	The veterinarian can identify their potential conflicts of interest and make appropriate judgements and decisions without undue influence of fears, personal attachments or inducements.
Trustworthiness	The veterinarian always gives an honest, informed opinion about the dog's condition, potential causes and contributing factors, and prognosis, and declares any conflicts of interest.
Integrity	The veterinarian is faithful to their moral values and will communicate and defend these when necessary.
Conscientiousness	The veterinarian works conscientiously to do what is right: to provide the best possible care to the dog and to future patients by remaining up-to-date with scientific evidence. The conscientious veterinarian strives to prevent disease at the level of the individual as well as the population.

that the rightness or wrongness of an action is judged by its ability to maximise well-being and/or to minimise suffering. The best course of action may be judged on a cost-benefit analysis.

A utilitarian framework that would be familiar to most veterinary professionals is Russell and Burch's '3Rs' of replacement, reduction and refinement (Russell and Burch, 1959), used to assess proposed use of animals in research and teaching (National Health and Medical Research Council, 2013).

According to the 3Rs, which have been widely adopted in legislation globally, animals may be used in research and teaching only if welfare costs to animals are minimised while scientific benefits are maximised, and such proposed benefits must not be trivial. If we apply the 3Rs to companion animals and posit that animals are to be 'used' as companions with various benefits to humans, welfare costs or harms should be minimised to the greatest degree possible. Potential costs and benefits can be drawn up, as, for example, in Table 4.2.

The costs of veterinarians speaking up are highest at the first tier, but reduce at the second and third tiers, illustrating the need for the support of organisations and professional associations for veterinarians and practices.

Extreme brachycephalic dogs bear extensive welfare costs over a prolonged period, while the benefits to humans of owning an aesthetically appealing animal are relatively trivial and may be transient. While there are potential monetary benefits in perpetuating extreme brachycephalic breeds to breeders (from selling and showing dogs) and veterinarians (from treating affected dogs), there are also potential benefits from exploring alternative sources of income or maximising income by marketing improved animal welfare standards. In addition, the benefits to veterinarians may be short term. If extreme brachycephalic dogs erode public interest in dog ownership *per se*, veterinarians will have fewer dogs to treat in the future.

DEONTOLOGY

Rollin argued that our predominant social ethic combines utilitarianism with deontology, an ethical approach that judges rightness and wrongness against moral rules or norms. It has been argued that

TABLE 4.2

Costs and Benefits of Acting at Different Levels to Speak Up and Minimise Welfare Costs Associated with Extreme Brachycephalic Phenotypes

Level/Tier of Action	Potential Costs	Potential Benefits
Through individual veterinarians direct to animal keepers and owners	Loss of income from treatment of illnesses associated with brachycephalic conformation. Loss of opportunity to influence clients or breeders who feel alienated. Reduced income for breeders if demand for these breeds drop. Owners may feel guilty that they have chosen an animal which has more risk of suffering and risk of denial (Serpell, 2019). Vets speaking out may have no impact as some owners were aware of welfare problems but had other motivations for obtaining dogs with extreme phenotypes which outweigh these (Steinert et al., 2019). Owners may feel judged, alienated by veterinarian. Breeders may experience reduced income and reduced adoptions if demand is compromised. Credibility may be jeopardised (in one survey, many respondents believed that animal suffering is commonly accepted by breeders as necessary to fulfil breed standards (Steinert et al., 2019)). May harm reputation of breeders (in one survey many respondents believed that for pedigree breeders, the appearance of the dog was more important than its personality or health (Steinert et al., 2019)).	Reduced lifelong welfare improvements in dogs. Reduced veterinary fees for owners associated with treating brachycephalic-related health conditions. Reduced emotional costs for owners of managing an animal with chronic health and welfare issues or reduced longevity. Owners who appreciate the nature and potential consequences of conditions such as BOAS may be more likely to seek treatment (Packer et al., 2012). Owners may feel motivated and empowered to maximise their dog's welfare. Owners may feel motivated to ensure that they do not contribute to the problem further by breeding from the animal (Ladlow et al., 2018).
Through veterinary practices to the surrounding community	Veterinarians and community groups may clash.	Interaction with local kennel clubs and breed association provides veterinarians and organisations with opportunity to promote good welfare and improve overall credibility of members.
Through professional associations to achieve political impact and challenge societal norms	Veterinary professional associations may clash with breeders and breed associations. Within membership of professional associations, there may be divisions about appropriate stance.	Improved credibility in alignment with public expectations regarding animal welfare advocacy (BVA, 2015). In line with the expressed desire of members of associations, for example the British Veterinary Association, for animal welfare to be a top lobbying priority (BVA, 2015).

the "five freedoms of animal welfare" (freedom from hunger and thirst; discomfort; pain, injury and disease; fear and distress; and freedom to express normal behaviour) are, for all intents and purposes, a form of rights for animals (McCausland, 2014). In other words, it is generally held that, at a minimum, animals kept and used by humans should have these basic rights recognised.

Because extreme brachycephalic dogs do not enjoy freedom from discomfort, freedom from pain, injury and disease, nor freedom to express normal behaviour, it could be argued that their rights are violated. Therefore, according to a deontological approach, the breeding of dogs with extreme phenotypes, including brachycephaly, cannot be justified, and further, it must be discouraged because of our duties to companion animals. This is the basis of legislation in countries such as Austria, Switzerland and Germany; anti-qualzucht ('torture breeding') clauses in animal welfare legislation are designed to prevent people from knowingly breeding animals likely to experience pain, distress or harm (RIS, 2017, Bundesamt für Justiz, 2018, Federal Assembly of the Swiss Confederation, 2017).

According to a deontological framework, it can be argued that veterinarians have a duty to promote the interests of patients and must be honest with clients about the health and welfare impacts of brachycephaly despite the risk of offending or alienating them.

It is useful to see examples of these frameworks applied to situations that veterinarians are likely to encounter in practice settings.

Case Example 1: Caesarean Section

A veterinarian is presented with a brachycephalic bitch with dystocia due to fetopelvic disproportion (see Figure 4.2). Should the veterinarian insist on spaying the animal after performing the C-section if it is safe to do so?

Response: The question of whether desexing dogs with congenital defects should be mandated is not new (Roger, 2009, Green, 2007). The utilitarian justification (based on a harm-minimisation, well-being maximisation strategy) is that it will prevent future suffering of the dam if dystocia due to conformational factors recurs, and reduce potential suffering of future generations of brachycephalic dogs that are likely to be affected by disorders associated with their conformation. The risks to the dam or sire of desexing are arguably outweighed by the benefit of eliminating the suffering of potential puppies. Indeed, this is the basis of the UK legislation which states that "no dog may be kept for breeding if it can reasonably be expected, on the basis of its genotype, phenotype or state of health that breeding from it could have a detrimental effect on its health or welfare or the health or welfare of its offspring" (Government of the United Kingdom, 2018).

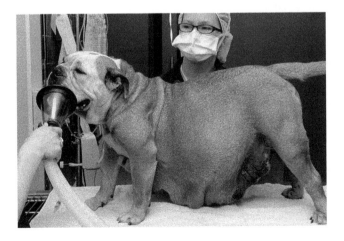

FIGURE 4.2   Bulldog presented for an elective caesarean section. (Image: Anne Quain.)

Mandatory desexing of dogs with congenital defects may also have wider benefits, by reducing the number of breeds with conformational defects and promoting good conformation. At the level of the individual vet (Tier 1), the veterinarian should recommend spaying the animal and explain the reasons for doing this, including benefits to the dam and future puppies. The veterinarian may even write out a list of costs and benefits of spaying the bitch versus not doing so. The owner has a right to decline but, in doing so, understands the welfare costs their decision exposes the animal and its potential descendants to. In countries where reporting of surgeries is allowed, caesareans should be notified to kennel clubs or breed societies if the dog is a pedigree. At the practice level (Tier 2), the practice can develop their own informed policy on whether or not they perform elective caesareans. Additionally, the practice can educate community members by discussing the risks and expenses associated with caesarean sections and explain their policy regarding caesareans. At the level of the profession (Tier 3), the veterinarian can notify their professional association about their concerns, and encourage their organisation – if it isn't already doing so – to engage with breeding societies to modify breed standards and encourage judicious outcrossing (Bradshaw and Hiby, 2018).

Case Example 2: Veterinarians Breeding, Selling and Performing Surgery on Dogs with Extreme Brachycephalic Phenotypes

You work in a busy practice where you perform, among other procedures, surgery in dogs with BOAS to reduce clinical signs associated with an extreme brachycephalic phenotype (Figure 4.3a and b). A colleague breeds French Bulldogs and justifies this on the grounds that she is improving the breed by selecting for a less extreme brachycephalic conformation. However, most of the animals she has bred have required surgery to widen stenotic nares (Figure 4.3a and b) and resection of the soft palate.

Response: Your colleague has a conflict of interest in drawing an income from selling the dogs she has bred, as well as drawing an income from treating those animals. This is particularly concerning because these animals are born with a phenotype that your colleague should expect. Additionally, she is role-modelling and normalising ownership of dogs with an extreme brachycephalic phenotype, potentially contributing to demand. There is clearly a gap between her stated intention and the intended outcome, which needs to be pointed out.

A virtue ethics approach may be helpful here. Your colleague may feel that she is being a good veterinarian by contributing to incremental improvement of an in-demand breed. Breeding these animals while registered as a veterinarian may give her credibility with other breeders who may be more likely to listen to her advice.

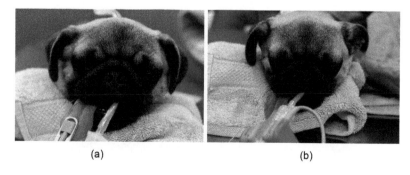

(a)   (b)

FIGURE 4.3   (a) A brachycephalic dog with stenotic nares prior to undergoing wedge resection to open the nares. (b) A brachycephalic dog following wedge resection of stenotic nares.

A discerning, honest, conscientious colleague would nonetheless raise the conflicts of interest with their colleague (Tier 1). You discuss the possibility of your colleague performing the surgeries on dogs she has bred at cost price (removing the financial conflict of interest in performance of the surgery) if their new owners agree to have their animals neutered at the same time, ensuring that those animals with extreme phenotypes do not breed. As per case example 1, all owners should be counselled about the welfare costs of brachycephaly. You should encourage your colleague to consider judicious outcrossing (Bradshaw and Hiby, 2018). At the practice level (Tier 2), the practice should avoid promoting dogs bred by your colleague, and be up-front (perhaps letting clients know in newsletters or via social media) about the costs and risks associated with surgical and medical management of BOAS. At the level of the profession (Tier 3), veterinarians can ask their professional associations to recognise ethical and responsible breeders. As attempts to ban breeds may lead to unintended consequences, e.g. sourcing of dogs from countries where breeding of companion animals is less regulated (Limb, 2019), quality assurance schemes can be introduced so that prospective owners can identify breeders who undertake informed, appropriate outcrossing to address conformation-related health and welfare issues, and improve quality of life (McGreevy, 2007).

RECOMMENDATIONS AND CONCLUSION

An ethical analysis of challenges associated with brachycephalic breeds and the role of the veterinarian suggests that veterinarians have professional and moral obligations to work to prevent or, at very least, minimise the health and welfare impacts of extreme morphology and inherited disorders at the level of the individual patient and more broadly. Specific strategies that describe how veterinarians can work at the level of the patient, their communities, and with breed and professional associations have been detailed by Belshaw and Wensley in this book, and elsewhere (Fawcett et al., 2018, Atkin, 2020).

Because of the well-documented health and welfare costs of extreme brachycephalic phenotypes, including but not limited to BOAS, coupled with the growing popularity of brachycephalic dogs, individual veterinarians, veterinary practices and veterinary professional organisations must take a strong position against perpetuating these phenotypes.

ACKNOWLEDGEMENT

The authors would like to thank Dr Magdoline Awad, Chief Veterinary Officer, Petsure, for providing access to deidentified insurance data.

REFERENCES

Anon. Undated. *Love is Blind* [Online]. A Joint Initiative of the RSPCA and the AVA. Available: https://www.loveisblind.org.au/ [Accessed 9 September 2020].

Anonymous. 2016. Pugs are anatomical disasters. Vets must speak out - even if its bad for business. *The Guardian*.

Arbe Montoya, A. I., Hazel, S., Matthew, S. M. & MCARTHUR, M. L. 2019. Moral distress in veterinarians. *Veterinary Record*, 185, 631.

Atkin, H. 2020. *#BreedtoBreathe: 10-point plan for veterinary practices* [Online]. British Veterinary Association. Available: https://www.bva.co.uk/news-and-blog/blog-article/breedtobreathe-10-point-plan-for-veterinary-practices/ [Accessed September 10 2020].

Batchelor, C. E. M. & Mckeegan, D. E. F. 2012. Survey of the frequency and perceived stressfulness of ethical dilemmas encountered in UK veterinary practice. *Veterinary Record*, 170, 19.

Beauchamp, T. L. & Childress, J. F. 2013. *Principles of Biomedical Ethics*, New York, Oxford: Oxford University Press.

Bradshaw, J. & Hiby, E. 2018. The paradoxical world of the dog. In: Butterworth, A. (ed.) *Animal Welfare in a Changing World*. Wallingford: CAB International.

British Veterinary Association 2016. Vets speaking up for animal welfare: BVA animal welfare strategy. BVA.

British Veterinary Association. 2017. *Information on advertising policy re bulldogs, French bulldogs and pugs* [Online]. BVA. Available: http://veterinaryrecord.bmj.com/pages/wp-content/uploads/sites/50/2017/04/Vet-record-Letter-for-advertisers-amended-for-website.pdf [Accessed 19 January 2018].

Bundesamt für Justiz 2018. Animal Protection Act § 11b.

BVA 2015. VetFutures: Voices from the future of the profession: BVA, RCVS.

Camp, M. & Sadler, J. 2019. Moral distress in medical student reflective writing. *AJOB Empir Bioeth*, 10, 70–78.

Coghlan, S. 2018. Strong patient advocacy and the fundamental ethical role of veterinarians. *Journal of Agricultural and Environmental Ethics*, 31, 349–367.

Crane, M., Bayl-Smith, P. & Cartmill, J. 2013. A recommendation for expanding the definition of moral distress experienced in the workplace. *Australasian Journal of Organisational Psychology*, 6. doi:10.1017/orp.2013.1.

Crane, M. F., Phillips, J. K. & Karin, E. 2015. Trait perfectionism strengthens the negative effects of moral stressors occurring in veterinary practice. *Australian Veterinary Journal*, 93, 354–360.

Farrow, T., Keown, A. J. & Farnworth, M. J. 2014. An exploration of attitudes towards pedigree dogs and their disorders as expressed by a sample of companion animal veterinarians in New Zealand. *N Z Vet J*, 62, 267–73.

Fawcett, A. 2017. Everyday ethics: Brachycephalic dogs and honesty with clients. *In Practice*, 39, 45.

Fawcett, A., Barrs, V., Awad, M., Child, G., Brunel, L., Mooney, E., Martinez-Taboada, F., Mcdonald, B. & Mcgreevy, P. 2018. Consequences and management of canine brachycephaly in veterinary practice: Perspectives from Australian veterinarians and veterinary specialists. *Animals*, 9, 3.

Federal Assembly of the Swiss Confederation 2017. Animal Protection Act. In: Confederation, F. A. O. T. S. (ed.) 455. The Federal Council: The Portal of the Swiss Government.

Foster, S. M. & Maples, E. H. 2014. Occupational stress in veterinary support staff. *Journal of Veterinary Medical Education*, 41, 102–110.

Ghirlanda, S., Acerbi, A., Herzog, H. & Serpell, J. A. 2013. Fashion vs. function in cultural evolution: The case of dog breed popularity. *PLoS One*, 8, e74770.

Government of the United Kingdom 2018. The Animal Welfare (Licensing of Activities Involving Animals) (England) Regulations 2018. *UK Statutory Instruments2018 No. 486 SCHEDULE 6*. legislation.gov.uk.

Green, R. 2007. Everyday ethics: Charity's stance on hereditary defects. *In Practice*, 29, 624.

Grimm, H., Bergadano, A., Musk, G. C., Otto, K., Taylor, P. M. & Duncan, J. C. 2018. Drawing the line in clinical treatment of companion animals: recommendations from an ethics working party. *Veterinary Record*, 182, 664.

Hernandez, E., Fawcett, A., Brouwer, E., Rau, J. & Turner, V. P. 2018. Speaking up: Veterinary ethical responsibilities and animal welfare issues in everyday practice. *Animals*, 8, 15.

Kelman, M., Ward, M. P., BARRS, V. R. & Norris, J. M. 2019. The geographic distribution and financial impact of canine parvovirus in Australia. *Transbound Emerg Dis*, 66, 299–311.

Kipperman, B., Morris, P. & Rollin, B. 2018. Ethical dilemmas encountered by small animal veterinarians: Characterisation, responses, consequences and beliefs regarding euthanasia. *Veterinary Record*, 182, 548.

Ladlow, J., Liu, N.-C., Kalmar, L. & Sargan, D. 2018. Brachycephalic obstructive airway syndrome. *Veterinary Record*, 182, 375.

Latter, M. 2017. AVA moves away from brachycephalic breeds in advertising. *Australian Veterinary Journal*, 95, N4.

Lehnus, K. S., Fordyce, P. S. & Mcmillan, M. W. 2019. Ethical dilemmas in clinical practice: A perspective on the results of an electronic survey of veterinary anaesthetists. *Veterinary Anaesthesia and Analgesia*, 46, 260–275.

Limb, M. 2019. Dutch crackdown on brachycephalic breeds. *Veterinary Record*, 184, 693.

Mccausland, C. 2014. The five freedoms of animal welfare are rights. *Journal of Agricultural and Environmental Ethics*, 27, 649–662.

Mcgreevy, P. D. 2007. Breeding for quality of life. *Animal Welfare*, 16, 125–128.

Moses, L. 2018. Another experience in resolving veterinary ethical dilemmas: Observations from a veterinarian performing ethics consultation. *American Journal of Bioethics*, 18, 67–69.

Moses, L., Malowney, M. J. & Wesley Boyd, J. 2018. Ethical conflict and moral distress in veterinary practice: A survey of North American veterinarians. *Journal of Veterinary Internal Medicine*, 32, 2115–2122.

Mullan, S. & Fawcett, A. 2017. Making ethical decisions. In: Mullan, S. & Fawcett, A. (eds.) *Veterinary Ethics: Navigating Tough Cases*. Sheffield: 5M.

National Health and Medical Research Council 2013. Australian Code for the Care and Use of Animals for Scientific Purposes 8ed.

One Welfare Portal 2017. *Veterinary Oaths* [Online]. Centre for Veterinary Education. Available: http://onewelfare.cve.edu.au/veterinary-oaths [Accessed June 10 2017].

Packer, R. M. A., Hendricks, A. & Burn, C. C. 2012. Do dog owners perceive the clinical signs related to conformational inherited disorders as 'normal' for the breed? A potential constraint to improving canine welfare. *Animal Welfare*, 21, 81–93.

Packer, R. M. A., O'Neill, D. G., Fletcher, F. & Farnworth, M. J. 2019. Great expectations, inconvenient truths, and the paradoxes of the dog-owner relationship for owners of brachycephalic dogs. *PLoS One*, 14, e0219918.

Packer, R. M. A., O'Neill, D. G., Fletcher, F. & Farnworth, M. J. 2020. Come for the looks, stay for the personality? A mixed methods investigation of reacquisition and owner recommendation of Bulldogs, French Bulldogs and Pugs. *PLoS One*, 15, e0237276.

Ris, R. D. B. 2017. *Animal Protection Act TSchG § 5* [Online]. www.ris.bka.gv.at. Available: https://www.ris.bka.gv.at/Dokument.wxe?Abfrage=Bundesnormen&Dokumentnummer=NOR40192428 [Accessed December 3 2018].

Roger, P. 2009. Everyday ethics: Bulldog caesarian. *In Practice*, 31, 95–95.

Rollin, B. E. 2011. Euthanasia, moral stress, and chronic illness in veterinary medicine. *Veterinary Clinics of North America-Small Animal Practice*, 41, 651–659.

Russell, W. M. S. & Burch, R. L. 1959. *The Principles of Humane Experimental Technique*. London: Methuen.

Ryan, S., Bacon, H., Endenburg, N., Hazel, S., Jouppi, R., Lee, N., Seksel, K. & Takashima, G. 2019. WSAVA animal welfare guidelines. *Journal of Small Animal Practice*, 60, E1–E46.

Serpell, J. A. 2019. How happy is your pet? The Problem of subjectivity in the assessment of companion animal welfare. *Animal Welfare*, 28, 57–66.

Smith, I. H. & Kouchaki, M. 2018. Moral humility: In life and at work. *Research in Organizational Behavior*, 38, 77–94.

Springer, S., Sandøe, P., Bøker Lund, T. & Grimm, H. 2019. "Patients' interests first, but … "–Austrian veterinarians' attitudes to moral challenges in modern small animal practice. *Animals*, 9, 241.

Steinert, K., Kuhne, F., Kramer, M. & Hackbarth, H. 2019. People's perception of brachycephalic breeds and breed-related welfare problems in Germany. *Journal of Veterinary Behavior*, 33, 96–102.

Susskind, R. & Susskind, D. 2015. *The Future of the Professions*. Oxford: Oxford University Press.

Yeates, J. 2018. Better to have lived and lost - the concept of a life worth living. In: Butterworth, A. (ed.) *Animal Welfare in a Changing World*. Proquest Ebook Central. Wallingford: CAB International.

5 Discussing Brachycephalic Health with Current and Prospective Dog Owners

Zoe Belshaw
EviVet Research Consultancy

Sean Wensley
People's Dispensary for Sick Animals (PDSA)

CONTENTS

Why Do People Buy Brachycephalic Dogs? 56
Veterinary Responsibility to Discuss Brachycephalic Health..56
Improving the Health and Well-being of Future Generations of Dogs 57
Pre-purchase Consultations...59
Improving the Health and Well-being of the Current Generation of Brachycephalic Dogs 60
    General Principles of Communicating with Existing Owners about Brachycephaly 60
Talking about Brachycephaly during the Puppy Vaccine Consult 62
    Respiratory 62
    Weight 62
    Exercise and Thermoregulation...63
    Eyes 63
    Skin 63
    Veterinary Checks 63
    Insurance 63
    Neutering...63
Talking to Owners of Brachycephalic Dogs with Existing Problems.............................64
Conclusions...65
References...66

In 2018, the French Bulldog topped the list as The UK Kennel Club's most registered breed for the first time (The Kennel Club 2018). Bulldogs, Pugs, Cavalier King Charles Spaniels, Chihuahuas, Boxers, Shih Tzus and Boston Terriers also featured in the top 20 most registered breeds, suggesting a shift in popularity towards dogs with brachycephalism. This change towards shorter and smaller breeds has also been recognised in other regions including Australia (Teng et al. 2016), Scandinavia (Nordic Kennel Union 2017) and America (American Kennel Club 2018). Brachycephaly is championed and selected for by breed interest groups, and brachycephalic dogs have featured in many advertising campaigns, television programmes and celebrity social media accounts internationally, with little recognition of their inherent health and welfare problems. This can make for challenging conversations in the consulting room between owners and veterinarians, with owners who may be very positive about brachycephalic breeds and unaware of the health and welfare challenges these dogs can face.

The health of brachycephalic dogs can be improved by reducing the negative impacts of brachycephaly on the current generation and by breeding healthier dogs for future generations. This chapter will cover veterinary communications relating to both, as well as the responsibility of veterinary professionals (veterinary surgeons and veterinary nurses) to have these discussions. We also look at why owners may be drawn to brachycephalic dogs in the first place. Where specific relevant research is available, we have made reference to it. Where none exists, we have drawn from other fields or provided opinions based on our experience from veterinary practice, policy development and research.

WHY DO PEOPLE BUY BRACHYCEPHALIC DOGS?

Recent research has started to explore owners' reasons for buying brachycephalic breeds. This information is useful in understanding the motivation of people who consider purchasing a brachycephalic dog and may subsequently present them for veterinary treatment.

In 2015, Packer and colleagues (Packer et al. 2017) surveyed owners of breeds that were then in The Kennel Club's top ten most registered breeds. The study identified that owners of French Bulldogs, Pugs and Bulldogs were younger in age and more likely to be owning a dog of that breed for the first time, compared to owners of non-brachycephalic breeds. Owners of all breeds (both brachycephalic and non-brachycephalic) ranked the importance of 'being a good companion' as the top reason for choosing a specific breed, but owners of brachycephalic dogs were more likely to rate the breeds' appearance, their suitability to spend time with children and their size as important additional influences. In contrast, they were less likely to rate good breed health or longevity as important influences, compared with the owners of non-brachycephalic breeds. Perhaps surprisingly, given speculation that celebrity ownership may be driving the rise in brachycephalic popularity, celebrity endorsement and recommendations from friends were not given as important factors by owners in their choice of breed.

A 2017 survey of French Bulldog, Pug and Bulldog owners conducted via social media (Packer et al. 2019) identified that almost 20% of the 2168 dogs involved had already undergone conformation-altering surgery, and many had suffered other illnesses common in brachycephalic breeds. While many of the owners recognised that their dog had one or more ongoing problems associated with brachycephaly such as heat intolerance, breathing, sleeping and eating difficulties, the majority felt their dog was either in the best health possible or in very good health. Similarly, most owners agreed that their dog was healthier than the average dog of their breed. The authors concluded that owners involved in this survey had perceptual errors in their beliefs about the health of their dogs compared to reality. In general, owners reported being very emotionally close to their dogs, with owners of Pugs rating this most highly. They may have deliberately chosen their breed based on their size, appearance and their personality, but may not have had a dog of that breed before. They may recognise that their dog has health problems and that they may face challenges associated with heat, sleeping and eating, but may not perceive them to be as bad as those of other dogs of the same breed. Further analysis of the same dataset (Packer et al. 2020) revealed that despite acknowledging their inherent health problems, high cost of ownership and negative behavioural attributes, over 90% of current owners stated they would acquire another dog of the same breed again. Reasons included their positive behavioural attributes for a companion dog, their sedentary nature and their compatibility with children. Understanding these perspectives that lead to breed loyalty is key to working alongside owners of brachycephalic dogs, to maintain or improve the welfare of their pets and to spread the message about the problems inherent in these breeds.

VETERINARY RESPONSIBILITY TO DISCUSS BRACHYCEPHALIC HEALTH

The veterinary and veterinary nursing professions are *animal welfare-focused*, providing leadership on animal welfare in society (BVA 2016). Being animal welfare-focused (prioritising the best interests of animals in our decision-making and advocacy) is distinct from being *client-focused* (prioritising our clients' interests, even when they do not align with improving animal welfare; for example, withholding advice on health problems linked to brachycephaly because it could cause client offence) or *vet-focused* (prioritising our own interests or those of our employer; for example, recommending surgery for a brachycephalic animal because it would be professionally interesting to perform or would generate more revenue than a conservative approach) (Yeates 2013). The interests of each group are important, but working with our clients and being economically viable are enablers for veterinary professionals to improve our primary goal of improving animal welfare. Overall, the veterinary professions have a dual animal welfare responsibility – to treat animals presented to us, while concurrently advocating for solutions to known root causes of common problems (BVA and RCVS 2015). Our responses to brachycephaly-related problems give a good example. Performing indicated airway surgery on a brachycephalic dog with Brachycephalic Obstructive Airway Syndrome (BOAS) is in that animal's best interests, but we have an additional responsibility to speak up about the selective breeding that causes a high incidence of BOAS within brachycephalic breeds, to stimulate and facilitate societal change.

Our opportunities to advocate for the best interests of animals exist at the levels of individuals (e.g. through our consultations with animal owners), communities (e.g. veterinary practices undertaking educational and outreach activities), nationally (e.g. veterinary associations and charities stimulating and contributing to public and political debate) and internationally (e.g. partnerships between national and international veterinary associations, and other international bodies and institutions) (BVA 2016, AVMA/FVE/CVMA 2020).

The UK Brachycephalic Working Group has described the veterinary professions' roles at these levels, with examples of actions that can be taken at each. At the national level in the UK, the British Veterinary Association (BVA) has run a high profile awareness-raising campaign, under the hashtag #BreedToBreathe (BVA 2018). This has included lobbying large companies to discontinue use of brachycephalic breeds in their marketing, resulting in positive commitments from brands including Costa Coffee, Heinz and HSBC Bank. Similarly, internationally, the Australian Veterinary Association and RSPCA Australia have run a 'Love is Blind' campaign since 2016 (AVA 2016), the New Zealand Veterinary Association made brachycephaly a priority issue in 2017 (NZVA 2017) and two new policy positions on breeding for brachycephaly were published in 2018, by the Humane Society Veterinary Medical Association in the US (HSVMA 2018) and in Europe jointly by the Federation of European Companion Animal Veterinary Associations and Federation of Veterinarians of Europe (FECAVA 2018). These veterinary campaigns have all served to raise the profile of brachycephalic breed health and provided supporting resources for veterinary practice teams.

IMPROVING THE HEALTH AND WELL-BEING OF FUTURE GENERATIONS OF DOGS

Position statements and campaigning by veterinary associations can reassure individual veterinary surgeons and veterinary nurses that it is appropriate and necessary to initiate what can be challenging conversations with prospective or current owners of brachycephalic dogs. This can drive action at the individual and community levels.

Community-level advocacy can be driven by the outreach activities of veterinary practices, engaging, for example, with local media, politicians, schools and through their in-practice displays and materials. In general, the practice management and culture should support the improvement of animal welfare in the local area, addressing the root causes of common welfare problems such as those linked to brachycephaly, beyond the bounds of animals under their direct care. Employees should be supported to have honest animal welfare-focused conversations with clients. Approached sensitively, such interactions should not normally lead to clients complaining, but support for clinicians by management staff is important if they do. Support for new and recent graduates in this respect may be particularly important, to help ensure they are not dissuaded from future advocacy by a negative reaction.

A specific ethical and practical challenge for effective community-level advocacy may be the existence of veterinary practices marketing themselves as 'brachy-friendly'. While we are unaware of a definition of such a practice, anecdotally they have a high caseload of brachycephalic dogs and are favoured by breeders for their welcoming, unquestioning approach to brachycephalic patients. While such practices may acquire a high level of expertise in treating conditions linked to brachycephaly, they may hamper efforts to tackle root causes of these conditions. Employees of such practices who are motivated to advocate brachycephalic health may feel disempowered, particularly if they weren't aware of the 'brachy-friendly' ethos when they joined, and the advocacy efforts of other local practices may be undermined if a single practice is taking a neutral or welcoming stance.

A list of advocacy activities that practices can undertake is given in the BVA #BreedToBreathe ten-point plan for veterinary practices (Table 5.1). It can be particularly effective when practices allocate a staff member as a dedicated 'champion' to lead on practical animal welfare activities like these, to take ownership and to motivate and assist their colleagues. Depending on the size of the practice or practice group, this can be coupled with an overarching 'task force', made up of staff members from various job roles, with a collective remit to regularly consider and audit how the practice can further advance animal welfare and ethics (Wensley et al. 2020).

TABLE 5.1

British Veterinary Association (BVA) #BreedToBreathe Ten-Point Plan for Veterinary Practices (Reproduced with Permission)

1. Offer pre-purchase consultations	You can use the PDSA 'Which pet?' consultation framework, with prospective dog owners. The potential health problems of brachycephalic conformation can be clearly outlined in these consultations
2. Strongly advise against breeding	If a dog is suffering from BOAS or requires conformation altering surgery – consider neutering (where best practice allows) to prevent further litters with extremes of conformation that negatively impact on their health and welfare
3. Promote the Puppy Contract	The Puppy Contract (comprising the Puppy Information Pack and contract for sale) can be promoted through the practice communication channels, e.g. website, social media, waiting room displays, newsletters, and in local print and broadcast media
4. Promote and actively participate in available health schemes	These can include the BVA/KC Health Schemes but also those for brachycephalic breeds that currently exist among Bulldog, French bulldog and Pug breed clubs
5. Carry out exercise tolerance test (ETT) and functional grading	ETT and functional grading can be conducted for brachycephalic breeds as part of their annual health assessment
6. Enrol the practice in clinical surveillance programmes	So as to contribute to data gathering and evidence generation. Examples include VetCompass and SAVSNET
7. Develop a practice communication strategy	Make sure to repeatedly, clearly and consistently communicate the health problems experienced by dogs with brachycephalic conformation through the practice communication channels
8. Maintain discipline in your practice communication strategy	Develop practice policy to ensure that practice communication channels (particularly social media and advertising materials) do not portray dogs with brachycephalic conformation as cute, humorous or appealing
9. Ensure practice policy supports staff	Everyone needs to appropriately convey evidence-based information and advice to owners of dogs with brachycephalic conformation
10. Support local breed clubs and representatives	Primarily in the development and implementation of plans to improve the health of dogs with brachycephalic conformation

PRE-PURCHASE CONSULTATIONS

One approach to improving the welfare of future generations of dogs, highlighted in the BVA #BreedToBreathe ten-point plan for veterinary practices, is to offer pre-purchase consultations, such as the People's Dispensary for Sick Animals (PDSA)'s 'Which Pet?' consultation framework (PDSA 2020a). These consultations with prospective pet owners can be an opportunity to outline the health problems linked to brachycephaly and to discuss ownership of healthier breeds.

Pre-purchase veterinary consultations are in their relative infancy. In 2019, 7% of current pet owners took advice from a veterinary professional before getting their pet, up from 5% in 2018 (PDSA 2018a, PDSA 2019). In 2018, 13% of veterinary practices offered free, dedicated pre-purchase clinics and 2% offered paid-for dedicated pre-purchase clinics (PDSA 2018a).

Effective marketing will be required to change cultural perceptions of what a veterinary practice can offer. Anecdotally, some prospective pet owners had not thought about visiting their local veterinary practice for pre-purchase advice, some thought there may have been a charge that they didn't wish to pay and some thought veterinary practices were too busy caring for sick animals. Practice marketing should aim to counter perceptions such as these, making it clear that prospective owners would be very welcome. Pre-purchase consultations may offer benefits to practices - of those currently offering free pre-purchase clinics, veterinary surgeons and veterinary nurses estimated that 71% of potential pet owners who were not existing clients of the practice go on to join the practice after receiving the consultation (BVA and BVNA 2018). At least one large veterinary practice group in the UK has introduced free pre-purchase consultations across its practices.

Frameworks help structure the available time in a consultation. PDSA's 'Which Pet?' consultation framework structures pre-purchase conversations around three areas. First, relevant aspects of the prospective owner's lifestyle and circumstances are discussed using the 'PETS' acronym:

Place – how suitable for a pet is the place where you live?
Exercise – how much exercise are you willing and able to do with a pet?
Time – how much time do you have for care activities like walking, training and cleaning?
Spend – are you aware of, and able to afford, the realistic costs of pet ownership?

This narrows the list of potentially suitable species and breeds. Then, the Five Welfare Needs (as set out in the UK Animal Welfare Acts; Table 5.2) are discussed for species and breeds of interest – what they are and how they would be provided for. The Health need offers a natural prompt to discuss the problems linked to brachycephaly.

Finally, responsible sourcing is discussed; for example, by promoting the Puppy Contract and Puppy Information Pack – a freely downloadable toolkit that guides puppy buyers through the questions they should ask a breeder before purchasing a puppy (Puppy Contract 2020). A practice can

---

TABLE 5.2

The Five Welfare Needs

The passing of the Animal Welfare Act (2006), the Animal Health and Welfare (Scotland) Act (2006) and the Welfare of Animals Act (Northern Ireland) (2011) updated animal protection legislation in the UK. These Acts retain an offence of causing unnecessary suffering but also introduced a legal duty of care based around an animal's welfare needs. A person responsible for an animal's welfare must ensure that the following needs are met:

1. **Environment** – a suitable environment
2. **Diet** – a suitable diet
3. **Behaviour** – to be able to exhibit normal behaviour patterns
4. **Companionship** – to be housed with, or apart from, other animals
5. **Health** – to be protected from pain, suffering, injury and disease

recommend local, reputable rescue and rehoming organisations, as well as reputable local breeders who are members of The Kennel Club Assured Breeders Scheme. A puppy should always be seen interacting with their mother, to reduce the risk of inadvertently purchasing a puppy from a puppy farm.

If during a pre-purchase consultation, an owner is insistent on acquiring a brachycephalic dog breed, providing specific information on how to source the healthiest possible individual from these high-risk breeds is important in helping to guide their purchasing process. You can help them to identify appropriate breeders, and the healthiest puppies available. Recent efforts have been made to provide advice to the public on this topic. In 2018, the UK Brachycephalic Working Group worked with the producers of a Disney movie, *Patrick the Pug*, to help mitigate risks of the film further increasing the popularity of brachycephalic dogs. A cinema flyer was produced to accompany the film, and included the following additional sourcing advice (agreed by all members of the Group), for those committed to buying a brachycephalic breed:

- Look for a puppy with a relatively long muzzle and nice wide open nostrils and parents with similar attributes.
- Check that the puppy's nose and eyes are not obscured by a large over-the-nose skin wrinkle.
- Ask if the parents have been health-tested[1] and ask to see the Health Certificate. Being health tested isn't a guarantee that a dog won't develop health problems but shows that breeders are trying to address some breed-related problems.

When first introducing pre-purchase consultations at a practice, it is recommended they should be piloted, followed by reflection and any improvements made before a full launch. It is important that all members of the practice team that are involved with marketing and delivering them should be fully briefed and trained on their purpose and delivery. Online marketing may be particularly effective in driving uptake, as over a third of dog owners looked online before they chose their pet (PDSA 2018a). A free online PDSA 'Which Pet?' toolkit includes an e-learning module, practice posters, a guide for the veterinary team and a client booklet, endorsed by BVA (PDSA 2020a). Each has space for a practice's logo and details to be added.

IMPROVING THE HEALTH AND WELL-BEING OF THE CURRENT GENERATION OF BRACHYCEPHALIC DOGS

GENERAL PRINCIPLES OF COMMUNICATING WITH EXISTING OWNERS ABOUT BRACHYCEPHALY

Until pre-purchase consultations have been further mainstreamed, the first interaction many of us will have with owners of brachycephalic dogs is after their purchase.

It is imperative that you form a good relationship with owners, whatever you may think personally about their choice of breed. While it can be tempting to blame them for making a bad choice, it is too late and will be counterproductive. At best, you are likely to make them feel guilty; at worst, they will feel patronised and judged, and may never set foot in your practice again. Some owners will have rescued their pet, and the presenting problems may pre-date their ownership, so it pays to be sensitive.

You could reel off a list of depressing health statistics to owners, but we suggest you don't. The evidence from fields such as climate change and vaccine denialism suggests reciting facts is likely to be ineffective, if not counterproductive (Trevors et al. 2016). It can harden existing attitudes

---

[1] Health testing could include Respiratory Function Grading, run by the Kennel Club and University of Cambridge Veterinary School. This scheme assesses Bulldogs, French Bulldogs and Pugs for BOAS, advises owners if their dog is affected by BOAS and gives guidance to breeders on how to lower the risk of producing affected puppies.

and make owners think you don't understand their perspectives. In addition, owners' beliefs about their freedom to choose a brachycephalic breed, or a distrust in you as an expert, could be aligned to those of their friends, their social group and their political or ideological beliefs. This phenomenon of 'motivated cognition' has been identified in studies exploring a wide range of other beliefs (Kahan 2016), but has yet to be tested among dog owners.

Instead, the evidence suggests that we need to frame messages positively, and we should strive to educate owners about practical steps they can take to keep their pet as well as they can. We could try to start the conversation by asking owners whether they are aware of the health problems their breed might be prone to. This open question allows them to tell you what they already know. Some may be unexpectedly expert, some may have no idea and others may be confident, but wrong. Hopefully, this will allow you to introduce the topic sensitively and without creating hostility.

If problems related to brachycephaly are not the reason they booked to see you, check that it is a good time to talk to them about these issues – they may be distracted by needing to get to work or make the school run, or they may not be the key decision-maker for that dog (Belshaw et al. 2018, Brunet 2020). If now doesn't work for them, tell them what you'd like to talk about and why it matters and book them yourself for another consultation at a more convenient time. Include in clinical records what you planned to say in case you don't see them next time, although continuity of care is likely to be very beneficial here.

Terms such as 'brachycephaly', 'stenotic nares', 'soft palate' and 'BOAS' will be incomprehensible to many owners. Describing the dog as 'flat faced' and explaining what you mean is likely to be more user-friendly. Use pictures, and point to relevant features on the dog in front of you whenever possible. Ensure that owners can see what you mean and understand what you're talking about by regularly stopping to ask if they have questions. Pausing for a few seconds after they have finished speaking to ensure they have nothing further to add and summarising back to them what they have said are good habits (Fawcett 2013). Always discuss the potential impact on the quality of life of the dog, and the prognosis – owners may be terrified that a minor ailment will be life-threatening or may be inappropriately unconcerned.

Empathy, compassion and time are keys to successful communication (Fawcett 2013). It's likely that your client, their wider family and friendship group, are very attached to this dog, and they may be worried or feel guilty about what you are telling them. This may come across as them being defensive. Try to find common ground – for example, you both want the best thing for the dog, and it would be awful if they started to suffer. Tell them that you want to work with them to prevent that. Research in other fields suggests that it is useful to remove any sense of guilt. Phrasing such as the following may help: *It's ok – you didn't know any of this before you came in, and it can be hard to find this information when you're doing your research, but now you do know. I can help, and we can work together to do all we can to prevent future problems.....* While this may feel like you are letting owners off the hook, it can be a more successful way to get them on board than blaming them. It may be worth asking your reception team to book longer first consultation appointments with brachycephalic breeds. Dedicated 'Brachycephalic Health Assessments' have also been suggested, where discussing the dogs' recognised health problems and quality of life provides the context for the visit, helping to make these topics less thorny and unexpected. Such assessments could be particularly useful for practices with high numbers of brachycephalic dogs registered.

Printed or online resources can be gathered together to help you and other practice staff explain your key messages to clients. One of the UK's largest veterinary practice groups has produced its own 'Guide to Brachycephalic Breeds' for educating owners about health issues and supplied these to its 800-plus practices (IVC 2019).

Websites from veterinary charities such as PDSA and Blue Cross, providing information for owners of brachycephalic dogs, (Blue Cross 2018, PDSA 2018b) can be useful in backing up what you have said and providing a reliable resource that owners can access later, in their own time. There's evidence that owners will look at a specific website if you write the link down for them – this is termed giving an 'information prescription' (Kogan et al. 2014). There may be a whole range of friends, relatives and dog walkers who are involved in looking after the dog and may be contributing to any problems – takeaway literature and web links will be helpful for them too.

It's important not to expect every owner to be on board with your advice. Anything you can do to educate the owner, keep the dog slim, get them insured or treat their problems will represent a positive benefit to that dog. Don't be too hard on yourself when dealing with these cases, and don't take this as a personal failure. Praise any small improvement, even if the dog has kept their weight stable or lost 0.5 kg when they need to lose 8 kg. Ask owners what, if anything, they are finding hard with any instructions you've given them if you don't see any changes, and work with them to find practical, bespoke solutions that work for them. In order to make any change, owners need to have the capability, opportunity and motivation to be able to do it (Michie et al. 2011).

TALKING ABOUT BRACHYCEPHALY DURING THE PUPPY VACCINE CONSULT

Many of us will first meet new brachycephalic dogs when presented for their initial course of vaccinations or for a new dog check. Our experience is that many owners of new brachycephalic breed puppies have little or no idea that their breed is associated with health problems. They may be horrified to learn that the cute puppy that may have cost them thousands of pounds, and who a breeder may have sold as being in excellent health, could be less than perfect. Using the general communication principles described above, this is a good time to start teaching owners about problems to look for and things they can prevent.

Some owners may think the vaccination consultation is just for giving a vaccine and nothing else (Belshaw et al. 2018), so introduce the idea at the outset that you're going to check the puppy over and chat with them about how to keep him or her healthy. Advise owners that you want to tell them about a few common things that can happen to this breed because catching problems early can make sure they don't get too serious, and some of them are preventable. Tell them you don't want to scare them, but in your experience, these are things that you see quite regularly. By doing this, you are positioning yourself as working alongside them as part of their team.

After your general clinical examination and after you've answered any questions the owner might have, go through some of the common health problems, demonstrating on the dog where the problem might arise, what signs they would spot if they occurred, and giving them reassuring information about how those problems might be fixable. Suggested common problems to cover include the following.

RESPIRATORY

Advise that noisy breathing sounds such as snorting, snuffling and snoring can be signs of obstructed breathing and shouldn't be considered 'normal for the breed', particularly if exhibited during the day (Packer et al. 2012). Show them their puppy's nostrils, and if they look stenotic, talk about why that can be a problem and what options might be available to help. Teach them what it means if their dog starts to snore, the risks of progressive problems if this is left and what could be done surgically if that happens.

WEIGHT

Emphasise the benefits and importance of keeping their dog a healthy weight; there is good evidence that obesity contributes significantly to BOAS (Liu et al. 2017) and was the most prevalent disorder in Pugs recorded by UK veterinary surgeons working in primary care during 2013 (O'Neill et al. 2016). Use Puppy Growth Curves to monitor healthy weight gain and demonstrate Body Condition Scoring; a body condition score chart specifically designed for Pugs is free to download from the Cambridge Veterinary School website (Cambridge BOAS Research Group 2017).

Exercise and Thermoregulation

Advise on healthy levels of exercise, including the need to avoid over-exercise, especially in hot weather. Teach the importance of keeping these breeds cool in warmer weather and being especially vigilant when they are in cars. Recent research suggests flat-faced dogs are at increased risk of heat stroke even from apparently innocuous activities such as sitting outside in hot weather (Hall et al. 2020).

Eyes

Show them where cherry eyes can pop up and what they can do if they see that happening. Teach them how to check the eyes for ulcers and the importance of taking their dog to see a veterinary surgeon promptly if they show clinical signs of an ulcer, as well as emphasising the potential risks of using eye drops purchased from non-veterinary websites.

Skin

Teach owners how to check their dog's ears for signs of infection or inflammation, and to encourage owners to positively train puppies to accept examination of their ears from a young age (see, e.g., Blue Cross 2020 and PDSA 2020b as useful resources for owners on reward-based training). If the dog has multiple skin folds, show them how they can inspect those and what products are available for reducing the risk of infections. Emphasise the importance of this regular home checking and, where necessary, cleaning.

Veterinary Checks

Advise regular checking by a veterinary surgeon. Explain that a veterinary surgeon may recommend treatment such as weight loss or specialist surgery to help them if breathing is obstructed.

Insurance

Discuss the benefits of insuring these breeds, given the unfortunate chance that one or more of these problems may develop. Remember that not every brachycephalic dog will suffer from any or all of these problems and emphasise the message of 'better safe than sorry'.

Neutering

Some owners may ask at this point whether or when they should neuter their dog. This is a useful time to discuss why you would advocate not breeding from brachycephalic dogs, particularly those with conformation-related disease. If they don't introduce the topic, you could signpost that you'll discuss it when they return for their second or third vaccination. This may prompt the owner to think about it in the interim, and the conversation should not be a surprise when it occurs.

Advise owners to read the handouts and websites you have given them, after their first vaccination consultation, and that you can answer any questions or queries in their second vaccination consultation.

Hopefully, these owners will now be on board with what they need to look for and do. Ensure they know that they can always come in with any questions or worries, even if they think it's something insignificant. Some owners may be reassured by having 6 monthly checks to help ensure they aren't missing anything; others will be happy to take matters into their own hands. As with any of these conversations, you won't convert everyone and some people won't be interested, but if even a few owners pick up some useful tips, then you will have helped improve the welfare of their dogs.

TALKING TO OWNERS OF BRACHYCEPHALIC DOGS WITH EXISTING PROBLEMS

Existing health problems can broadly be grouped into two types:

1. 'Presenting problems' are problems that the owner has noticed and that led them to book the appointment. Typically, these presenting problems create clinical signs that are easy for the owner to detect and have been recognised by the owner as something abnormal or bothersome to them and/or the dog.
2. 'Non-presenting problems' are ones that you pick up during the course of your clinical examination or that the owner mentions in the consulting room when the dog has been booked in for a different reason. Vaccination consultations can be a classic time for this to happen.

Relatively, these latter consultations can be a lot more difficult; you may need to raise an issue about which the owner is unaware or may consider to be normal. In our experience, facial and soft palate conformation problems, and obesity, can often fall into this category, with owners thinking that snoring or excessive panting is normal for their breed. Similar unexpected surprises for owners may arise when discussing the anaesthetic risks associated with brachycephaly as an adjunct to routine surgery, or when explaining reasons that their dog has recurrent corneal ulceration.

Where minor issues related to brachycephaly are not the reason for presentation, it can be tempting to ignore them, or leave them for someone else to deal with during a subsequent consultation. Concerns about how the conversation will go, and the very real-time pressures faced during veterinary consultations make this an understandable standpoint. In some instances, booking a future consultation to discuss the issue may be the best thing to do. However, our responsibility is to the welfare of the animal in front of us and should not be limited to telling owners what they wish to hear (Fawcett 2017, Fawcett et al. 2018). Fawcett (2017) proposes that a 'virtue ethics' approach can help with honest conversations with clients, structured around five focal virtues important for medical professionals.

There is a risk of seeming to be complicit in poor welfare if we do not speak up, and owners may ask why we said nothing, at a later date. As a minimum, we think emphasising the importance of keeping the dog within a normal body condition score, avoiding excessive heat and advising owners of common problems to watch for should be mentioned whenever possible.

Conversations about more severe problems related to brachycephalic conformations, particularly those about which the owner appears unaware, can be very challenging. Hughes et al. (2018) identified that owners of a range of species valued veterinary surgeons who "cared for their animals, knew their stuff, and took them seriously". Thinking about these key attributes of a good veterinary surgeon may form a useful way to have a conversation about an issue such as BOAS, in combination with our behavioural change attributes of capability, opportunity and motivation. Start by framing the conversation in terms of the animal's welfare – there is something that you've seen/heard/felt that you are worried about. Then, involve the owner by soliciting their viewpoint – is this something they were aware of, and were they concerned by it? This allows you to see where they stand on the issue, and it may help you to identify a reason that they could be motivated to do something about it. For example, owners of a dog with a loud snore may find it annoying or may even be having disturbed sleep as a result of it (Packer et al. 2020). You can then go on to describe what concerns you have and why, what the implications might be for the welfare of their dog and what might be done about it. Again, frame this as being a discussion you want to have because it's in the best interests of their pet. Introduce the fact that it is related to the dog's conformation, sensitively – they may have had no idea when they acquired the dog that they were prone to these problems, and blaming them for their bad decision will not help. As we described earlier, reassure owners that it was ok that they didn't know before, but now that they do, you can both do something about it – provide them with that clear opportunity. Not every owner will be capable of affording major surgery for their dog's airway, but they may well be capable of helping them to lose weight or be able to pay for their nares to be widened if these are relevant contributory factors. Discuss the options and find one that is practical for them. The key to any of these discussions is to tread carefully and sensitively. Ensure that you read the owner's level of engagement and their emotions on the day, and if you don't think it's the right time, then leave it as long as the dog's welfare permits, but write a note to discuss it next time.

It is possible that even with your best attempts, the owner may become upset or angry. Fawcett (2013) describes three broad reasons for this: it is not the outcome owners expected; expectations are misaligned; or one of the parties is behaving inflexibly. These may be underpinned by a wide range of reasons from feeling that their concern has not been addressed (they were worried about their dog's umbilical hernia, but you are talking about the dog's weight, for example), that they don't feel their perspectives have been taken into account and that they are worried about the costs of what you are discussing, or things outside your control such as they are having a bad day for other reasons. If you can sense rising emotions, stop and ask them what they think about what you are saying, allowing them the space to fully explain.

Again, not every owner will be on board, and this may leave you with a dilemma as to when to respect client autonomy versus intervene. The UK Animal Welfare Acts enshrine in law owners' legal duty to meet their pet's Five Welfare Needs (see the 'Pre-purchase Consultations' section), including the animal's need to be protected from pain, suffering, injury and disease. It is likely that many owners will be unaware of these legal duties (a quarter of UK dog, cat and rabbit owners have still never heard of the Animal Welfare Acts (Wensley et al. (2021)), so even advising them there is legislation to protect the welfare of their pet may be useful to all parties, but should not be used as a veiled threat. This legislation is enforced by the Royal Society for the Prevention of Cruelty to Animals (RSPCA), Scottish Society for the Prevention of Cruelty to Animals (SSPCA) and Animal Welfare Officers in Northern Ireland. Veterinary professionals have a responsibility to identify suspected instances of abuse, including neglect, and report this to the appropriate bodies. This will sometimes require us to break client confidentiality; in the UK, guidance on doing so, to act on animal welfare concerns, is laid out in the Royal College of Veterinary Surgeons (RCVS) Code of Professional Conduct and Supporting Guidance.

One specific situation where client confidentiality can be breached is in the reporting by veterinary surgeons to The Kennel Club of a caesarean section performed on a Kennel Club-registered dog. The Kennel Club registration specifically permits this, and veterinary surgeons should undertake such reporting to provide population data to The Kennel Club, to inform future health strategies. While a client's permission is not required to submit this information, it is nevertheless usually recommended to inform the client as part of good professional relations.

Breeders have additional legal responsibilities to protect animal welfare. Under the Animal Welfare (Licensing of Activities Involving Animals) (England), Regulations 2018, Schedule six states that "No dog may be kept for breeding if it can reasonably be expected, on the basis of its genotype, phenotype, or state of health that breeding from it could have detrimental effect on its health or welfare or the health and welfare of its offspring". This legislation could be a valuable additional tool for tackling the welfare problems associated with breeding for brachycephaly, but there has not been a test case at the time of writing.

CONCLUSIONS

Some owners of brachycephalic dogs will be naïve to their wide range of health problems, while others may be aware but still retain strong breed loyalty. Understanding owners' motivations to acquire these dogs is important in working alongside them. There are multiple time-points where you can sensitively discuss brachycephalic health and welfare, from the pre-purchase consultation to the routine consultation for another reason. Rather than making assumptions, ascertain what owners already know and how they feel about any current or potential issues and aim to educate and motivate, rather than blame them. Trying to improve brachycephalic health and encourage owners to consider alternative breeds can feel like an uphill challenge at times. Remember, you have multiple opportunities to make a difference – from your direct interactions with owners, to influencing your practice's marketing policies, to joining and supporting your professional bodies and their public awareness campaigns. Don't be disheartened – even if it takes a long time, you are likely to have a positive impact.

REFERENCES

American Kennel Club. 2018. Most Popular Dog Breeds–Full Ranking List 2018. https://www.akc.org/expert-advice/news/most-popular-dog-breeds-full-ranking-list/. (accessed September 2, 2020)

Australian Veterinary Association (AVA). 2016. Love is Blind resources https://www.ava.com.au/love-is-blind/. (accessed September 21, 2020)

American Veterinary Medical Association (AVMA), Federation of Veterinarians of Europe (FVE), Canadian Veterinary Medical Association (CVMA). 2020. Joint AVMA-FVE-CVMA statement on the roles of veterinarians in promoting animal welfare. https://www.fve.org/cms/wp-content/uploads/FVE-AVMA-CVMA-position-statement-on-animal-welfare-Clean-Version.docx.pdf. (accessed September 23, 2020).

Belshaw Z., Robinson N.J., Dean R.S., and Brennan M.L. 2018. Owners and veterinary surgeons in the United Kingdom disagree about what should happen during a small animal vaccination consultation. *Veterinary Sciences* 5, 7. doi:10.3390/vetsci5010007.

Blue Cross. 2018. Things to think about before buying a flat-faced (brachycephalic) dog https://www.bluecross.org.uk/pet-advice/things-think-about-buying-flat-faced-dog. (accessed September 2, 2020)

Blue Cross. 2020. Your puppy's first vet visit: introducing your puppy to vet handling training. https://www.bluecross.org.uk/pet-advice/your-puppys-first-vet-visit. (accessed September 2, 2020).

British Veterinary Association (BVA). 2018. All animals should be bred for health over looks. www.bva.co.uk/take-action/breed-to-breathe-campaign/. (accessed September 21, 2020).

British Veterinary Association (BVA). 2016. Vets Speaking Up for Animal Welfare – BVA Animal Welfare Strategy

British Veterinary Association (BVA) and Royal College of Veterinary Surgeons (RCVS). 2015. Vet Futures: A Vision for the Veterinary Profession for 2030

Brunet M. 2020. Chapter 8: Tending the garden. In: *The GP Consultation Reimagined: A Tale of Two Houses*, 65–77. Banbury: Scion Publishing Ltd.

BVA and BVNA. 2018. Voice of the Profession survey, reported in PDSA Animal Wellbeing (PAW) Report 2018. www.pdsa.org.uk/pawreport. (Accessed March 16, 2020)

Cambridge BOAS Research Group. 2017. Body condition score (BCS) in pugs. https://www.vet.cam.ac.uk/boas/about-boas/Pug_health_scheme_BCS_v2.jpg. (accessed September 2, 2020).

Fawcett A. 2013. Dealing with difficult clinical encounters. *In Practice*. doi:10.1136/inp.f5574.

Fawcett A. 2017. Brachycephalic dogs and honesty with clients. *In Practice*. doi:10.1136/inp.i6329.

Fawcett A., Barrs V., Awad M., Child G., Brunel L., Mooney E., Martinez Taboada F., McDonald B., and McGreevy P. 2018. Consequences and management of canine brachycephaly in veterinary practice: Perspectives from Australian veterinarians and veterinary specialists. *Animals*, 9(3), 1–25. doi:10.3390/ani9010003.

Federation of European Companion Animal Veterinary Associations (FECAVA). 2018. Brachycephalic issues: shared resources. https://www.fecava.org/policies-actions/healthy-breeding-3/. (accessed September 21, 2020)

Hall E.J., Carter A.J., and O'Neill D.G. 2020. Dogs don't die just in hot cars—exertional heat-related illness (heatstroke) is a greater threat to UK dogs. *Animals*, 10(8), 1324. doi:10.3390/ani10081324.

Humane Society Veterinary Medical Association (HSVMA). 2018. Policy statement on brachycephalic dogs. https://www.hsvma.org/policy_statements#brachycephalicdogs. (accessed September 21, 2020)

Hughes K., et al. 2018. 'Care about my animal, know your stuff and take me seriously': United Kingdom and Australian clients' views on the capabilities most important in their veterinarians. *Vet Record*. doi:10.1136/vr.10498.

Independent Vetcare (IVC) My Family Pet. 2019. A guide to brachycephalic breeds. https://issuu.com/ivcevidensiadigital/docs/brachy_breed_guide. (accessed September 21, 2020).

Kahan D. 2016. The expressive rationality of inaccurate perceptions. *Behavioral and Brain Sciences*, 40. doi:10.1017/S0140525x15002332.

Kogan L. et al. 2014. Information prescriptions: A tool for veterinary practices. *Open Veterinary Journal*, 4(2), 90–95.

Liu N.-C., Troconis E.L., Kalmar L., Price D.J., Wright H.E., Adams V.J., Sargan D.R., and Ladlow J.F. 2017. Conformational risk factors of brachycephalic obstructive airway syndrome (BOAS) in pugs, French bulldogs, and bulldogs. *PLoS One*. doi:10.1371/journal.pone.0181928.

Michie S., van Stralen M.M., and West R. 2011 The behaviour change wheel: A new method for characterising and designing behaviour change interventions. *Implementation Science* 6, 42.

Nordic Kennel Union. 2017. Statements and proposals regarding respiratory health in brachycephalic dogs. https://www.skk.se/globalassets/nku-en/documents/brachyreport.pdf. (accessed September 2, 2020).

New Zealand Veterinary Association (NZVA). 2017. When Beauty is Pain. https://www.nzva.org.nz/news/347325/When-Beauty-is-Pain.htm. (accessed September 21, 2020).

O'Neill D.G., Darwent E.C., Church D.B., and Brodbelt D.C. 2016. Demography and health of Pugs under primary veterinary care in England. *Canine Genetics and Epidemiology* 3, 5.

Packer R.M.A., Hendricks A., and Burn C.C. 2012. Do dog owners perceive the clinical signs related to conformational inherited disorders as 'normal' for the breed? A potential constraint to improving canine welfare. *Animal Welfare*, 21(S1), 81–93.

Packer R.M.A., Murphy D., and Farnworth M.J. 2017. Purchasing popular purebreds: Investigating the influence of breed-type on the pre-purchase attitudes and behaviour of dog owners. *Animal Welfare*, 26, 191–201

Packer R.M.A., O'Neill D.G., Fletcher F., and Farnworth M.J. 2019. Great expectations, inconvenient truths, and the paradoxes of the dog-owner relationship for owners of brachycephalic dogs. *PLoS One*, 14(7), e0219918.

Packer R.M.A., O'Neill D.G., Fletcher F., Farnworth M.J. 2020. Come for the looks, stay for the personality? A mixed methods investigation of reacquisition and owner recommendation of Bulldogs, French Bulldogs and Pugs. *PLoS One*, 15(8), e0237276.

PDSA. 2018a. PDSA Animal Wellbeing (PAW) Report 2018. www.pdsa.org.uk/pawreport. (accessed 16 March 2020).

PDSA. 2018b. Flat faced dogs: Advice for owners. https://www.pdsa.org.uk/taking-care-of-your-pet/looking-after-your-pet/puppies-dogs/flat-faced-dogs-advice-for-owners. (accessed September 2, 2020).

PDSA. 2019. PDSA Animal Wellbeing (PAW) Report 2019. www.pdsa.org.uk/pawreport. (accessed 16 March 2020).

PDSA. 2020a. Running your own 'Which Pet' consultations. https://www.pdsa.org.uk/whichpet. (accessed March 16, 2020).

PDSA. 2020b. Reward-based training. https://www.pdsa.org.uk/taking-care-of-your-pet/looking-after-your-pet/puppies-dogs/reward-based-training. (accessed September 22, 2020).

Puppy Contract. 2020. www.puppycontract.org.uk. (accessed March 16, 2020).

Teng K.T., McGreevy P.D., Torbio J-A.L.M.L., and Dhand N.K. 2016. Trends in popularity of some morphological traits of purebred dogs in Australia. *Canine Genetics and Epidemiology*, 3(2). doi:10.1186/s40575-016-0032-2.

The Kennel Club. 2018. French Bulldogs overtake Labradors as UK's most popular dog breed. https://www.thekennelclub.org.uk/press-releases/2018/june/french-bulldogs-overtake-labradors-as-uks-most-popular-dog-breed/. (accessed September 2, 2020)

Trevors G., Muis K., Pekrun R., Sinatra G., and Winne P. 2016. Identity and epistemic emotions during knowledge revision: a potential account for the backfire effect. *Discourse Processes*. doi:10.1080/0163853X.2015.1136507.

Wensley S., Betton V., Martin N., and Tipton E. 2020. Advancing animal welfare and ethics in veterinary practice through a Pet Wellbeing Task Force, practice-based Champions and clinical audit. *Vet Record*. doi:10.1136/vr.105484.

Wensley S., Betton V., Gosschalk K., Hooker R., Main D.C.J., and Tipton E. 2021. Driving evidence-based improvements for the UK's "Stressed. Lonely. Overweight. Bored. Aggressive. Misunderstood…but loved" companion animals. *Vet Record*. doi:10.1002/vetr.7

Yeates J. 2013. *Animal Welfare in Veterinary Practice*. Wiley-Blackwell, Hoboken, NJ.

6 Nurses and the Brachycephalic Patient – Practical Considerations and the Role of Veterinary Nurses in Improving Brachycephalic Health

Kate Price
Swaffham Veterinary Centre

CONTENTS

Introduction...69
Brachycephalic Nurse Clinics..70
History Taking...71
Examination..71
    Eyes..71
Skin Folds...73
Breathing Assessment...74
Weight and Thermoregulation..76
Pre-purchase Consultations..77
Conclusions...80
References..80

INTRODUCTION

Veterinary nurses in small animal practice are seeing an increasing number of brachycephalic dogs presenting on a daily basis. A recent survey demonstrated the high prevalence of brachycephalic dogs in UK veterinary practice, with 79% of veterinary nurses reporting that they treat them at least once a day (Campbell 2020). This is perhaps unsurprising given the disease burden present in brachycephalic dogs, as outlined in Chapter 10–19 of this book, and the growing popularity of brachycephalic breeds, as outlined in Chapter 3. Veterinary nurses are well placed to drive improvements in animal welfare (Yeates 2014). Veterinary nurses can gather information and educate clients in an accessible format and are often seen as the bridge between veterinary surgeons and clients. There is clear motivation in the veterinary nursing community to improve the health and welfare of brachycephalic dogs. Although the normalisation of health problems in brachycephalic breeds is endemic in owners of these dogs (Packer, Hendricks, and Burn 2012) and may also be a problem with veterinary professionals who have become desensitised to what is abnormal for these breeds (O'Neill et al. 2018), 95% of veterinary nurses recently surveyed believed more needed to be done to prevent the normalisation of brachycephalic health problems (Campbell 2020). This chapter investigates how veterinary nurses can make a difference to brachycephalic health and welfare, by exploring activities veterinary nurses can take part in both at an individual dog level, e.g. nurse clinics, and at a wider population level, e.g. pre-purchase education (Figure 6.1).

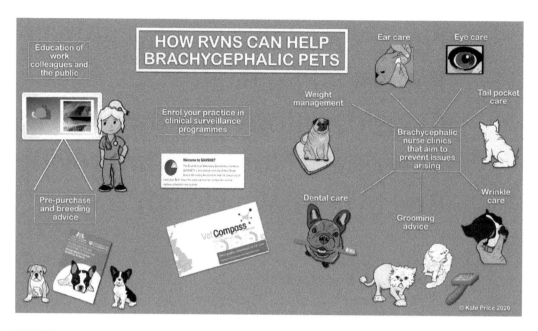

FIGURE 6.1   Infographic of potential ways veterinary nurses can help brachycephalic pets.

BRACHYCEPHALIC NURSE CLINICS

As a regulated body, veterinary nurses are not permitted to make a diagnosis under the Veterinary Surgeons Act (1966). However, veterinary nurses are allowed to discuss symptoms with owners and also discuss conditions that may cause these symptoms and recommend seeing a veterinary surgeon for further advice, diagnostics and potential treatment (Ackerman 2012). Many of the health issues that brachycephalic dogs experience are progressive and can be exacerbated by poor husbandry or poor owner understanding of the issues their dog faces. Veterinary nurses are therefore well placed to play a role in disease prevention, particularly when seeing dogs from a young age in puppy health checks.

Owners of brachycephalic are often unaware that their dog may have a progressing problem and only seek veterinary advice when the dogs are at a critical level of clinical severity. Indeed, studies have found that the more severe the clinical signs of a disease, the higher the likelihood the owner will report it as a problem (Packer et al. 2019). With the current high level of negative media attention around brachycephalic breeds (Usborne 2019), brachycephalic health and welfare is a sensitive subject, often exacerbated by biased beliefs in individual owners' belief about their own dog's health. Studies by Packer and colleagues (Packer et al. 2019, Packer, Hendricks, and Burn 2012) discuss how cognitive bias processes drive a deflection phenomenon, where owners of brachycephalic dogs may be aware of their chosen breed's health issues but refuse to accept that their own dog is affected, possibly because they find this psychologically uncomfortable. With brachycephalic dogs often presenting with multiple and complex health issues, trust from owners can be vital in obtaining a full history. Given that owners are often more likely to open up to nurses on health issues than with their vets (Yeates 2014), brachycephalic nurse clinics could be an ideal place to build up both trust with a client and glean more accurate information on a dog's health status. Having a thorough history is essential to improve the health, welfare and outcome of these animals whether they are in for surgery or seen in a nurse clinic, and allows for the implementation of individual care plans which will again improve outcomes for patients, but also job satisfaction for veterinary nurses (Lock 2011).

HISTORY TAKING

Because owners may normalise some of their dog's health issues (Packer et al. 2019, Packer, Hendricks, and Burn 2012), a good way of finding possible underlying health issues is to send out a pre-consultation questionnaire to owners covering all aspects of lifestyle. The questions included in these questionnaires should cover not only health but also husbandry (e.g. exercise regime, cleaning) that may uncover ongoing health issues that owners do not recognise as such or consider as part of their dog's breed. Taking a good history is not only important in brachycephalic nurse clinics but is especially important if veterinary nurses are admitting a brachycephalic animal for surgery, as a better understanding of their health status may reduce perioperative complications. Extra questions may be necessary to highlight any potential recovery issues such as regurgitation, especially in French Bulldogs (Fenner, Quinn, and Demetriou 2020, Costa et al. 2020). As such, having a bespoke set of questions for brachycephalic dogs on admission may be of high value and reflect other brachycephalic specific protocols that have become a normal part of veterinary practice (e.g. anaesthesia requirements).

EXAMINATION

Eyes

Brachycephalic breeds, due to the nature of their skull shape, tend to have much shallower orbits (eye sockets). This means that they can have overly protruding eyes (exophthalmos), incomplete blink (lagophthalmos) and a very large eyelid aperture (macroblepharon), as covered within the ophthalmology chapter in this book. These features, along with their foreshorted muzzle and in some individuals, pronounced nasal folds, predispose brachycephalic dogs to corneal ulcers (Packer, Hendricks, and Burn 2015). Their conformation makes them particularly vulnerable to direct ocular trauma and insufficient spreading of tears, both of which can lead to ulceration. A study by O'Neill and colleagues investigated the epidemiology of corneal ulcers in over 100,000 UK dogs and found that brachycephalic dogs were 11 times more likely to get corneal ulcers compared to crossbreed dogs of non-brachycephalic type (O'Neill et al. 2017). Within this, Pugs were 19 times more likely, and Bulldogs were 6 times more likely to be diagnosed with corneal ulcers than crossbreed dogs (O'Neill et al. 2017). In a study of French Bulldogs (O'Neill et al. 2018), ophthalmic disorders were the fifth most common group of disorders diagnosed in this breed.

Both brachycephalic cats and dogs are known to have reduced corneal sensitivity (Blocker and Van Der Woerdt 2001, Kafarnik, Fritsche, and Reese 2008, Barrett et al. 1991) which could in part explain why damage to the cornea often goes unnoticed by owners until the dog is clearly in pain. Consequently, nurses have an important role in educating owners on good preventative eye care and how to reduce the risk of trauma. Owners should be encouraged in nurse clinics to take a 'risk audit' of their dog's life, for example, to think about potential obstacles in their home at eye level, activities that could dry their dog's eyes, and if they own a cat, to consider their dog's interactions with them and whether they pose a risk to their eyes due to cat scratch injuries.

Within a brachycephalic nurse clinic, nurses can perform Schirmer tear tests to quantify tear production where an insufficient tear film is suspected. Care must be taken when handling small brachycephalic patients for this test to avoid damaging the eye and/or causing stress and dyspnoea to dogs with compromised respiratory systems. When handling the brachycephalic dog, less restraint is often better, to avoid dogs becoming distressed and panicking. Breed-specific considerations in the handling of these patients includes avoiding lifting their head up, as this can compromise their airways and subsequent breathing, instead, letting them sit with their heads in a normal relaxed forward-facing position. Gently supporting their heads with one hand under their chin, while the other is placed lightly behind their head can aid procedures to prevent patients from moving too much while avoiding distress from restraint (Figure 6.2).

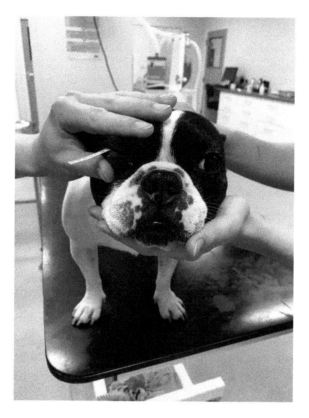

FIGURE 6.2   Schirmer tear test with gentle handling.

If dogs are found to have insufficient tear production, recommendations can then be made for chronic supplementation with lubrication. Husbandry advice should also be given regarding conditions that can additionally dry the eyes, where extra lubrication may be needed, for example in warm windy weather, or on journeys in cars with air conditioning. Veterinary nurses should always demonstrate how to apply the lubricant as this can avoid confusion by the owner and help to increase owner compliance. Indeed, previous studies have found that demonstrations were one of the communication types that owners felt were lacking in veterinary practices that contributed to non-compliance (Loftus 2012).

In addition to educating and advising owners on brachycephalic ocular health, the same principles should be applied when nursing inpatients. Some drug combinations can influence tear production, particularly some that are used in common anaesthetic premedications (Dodam, Branson, and Martin 1998). Anaesthetised patients have reduced tear production and thus require ocular lubrication during procedures (Bolzanni et al. 2020). This should not necessarily end once the dog has recovered; studies have found that dogs' tear production can be decreased for 24 hours, and as such it has been recommended that all patients have eyes lubricated every 2–4 hours, while in-patients (Chandler et al. 2013) have eyes lubricated for up to 36 hours after anaesthetic (Herring et al. 2000).

During nurse clinics, when examining the eyes, nurses should identify potential sources of irritation to the eyes. This includes the eyelids, eyelashes, nasal fold (if present) and hairs from the rest of the face. Medial canthal entropion is commonly seen in Pugs (Maini et al. 2019), where there is an abnormal inward rotation of the eyelid at the inner corner of the eye, such that the haired skin contacts the conjunctival and cornea. Many Pugs with medial canthal entropion will present with pigmentary keratitis (Maini et al. 2019, Labelle et al. 2013), a brown pigment classically in the four and eight o'clock positions in the eye, and if left untreated, can lead to severe visual impairment. During nurse clinics, owners can be talked through potential treatment options that they can then discuss with their veterinary surgeon, including lifelong topical treatment or more invasive surgery (e.g. a medial canthoplasty). In addition to eyelid abnormalities, large nasal folds can lead to hairs growing on the fold to rub against and irritate the cornea (nasal fold trichiasis), elevating the risk of corneal ulcers (Packer, Hendricks, and Burn 2015). In some cases, patients would benefit from the skin fold being surgically reduced in size or completely removed to reduce irritation; however, the thought of their dogs' looks changing dramatically will put some clients off the idea of this surgery despite its potential benefits. Clients whose dogs are experiencing recurrent ulcers because of a large nasal fold should be gently counselled over a number of visits, so they are aware of this as an option to markedly reduce their dog's risk of ulcers and are able to weigh up the costs and benefits of this procedure. Photos of the aesthetic outcome in healed patients who successfully underwent this procedure may help persuade owners who are worried about changes in their dog's appearance.

SKIN FOLDS

As dictated in their breed standards, brachycephalic breeds including Pugs and Bulldogs have additional skin folds on their faces. This excess skin is often not limited to their face, however, and skin folds can often be found at other locations including the vulva in female dogs and the tail base in screw-tailed breeds. Abnormal growth of the tail bone can result in a 'tail pocket' in screw-tailed breeds. Skin folds are problematic as they often rub against other areas of skin causing microtrauma (Jackson and Marsella 2012). Deep skin folds can also retain moisture and may lead to overgrowth of bacteria and yeast. These organisms feed on the skin secretions trapped in the fold, creating a perfect breeding ground for infections. These organisms also produce substances that cause further irritation to the skin. Without proper maintenance, skin folds can become infected and painful, a condition known as intertrigo, where the skin deep within the folds becomes pruritic and erythematous, often with a white greasy discharge which contains a large amount of bacteria (Paterson 2018) (Figure 6.3). Fluids such as tears, nasal discharge and saliva can collect in facial folds increasing the risk of infection, and urine may exacerbate problems in vulvar folds in female dogs. In addition to skin fold problems, some brachycephalic dogs are predisposed to allergic skin disease (O'Neill et al. 2018, 2019), which may cause chronic itchiness and make the skin more prone to infections.

Nurses should familiarise owners with the areas needing cleaning and demonstrate to owners how to clean and maintain the health of these areas. Nurses should emphasise and explain why it is important to keep skin folds dry and clean, and it is best time to start this educational process when brachycephalic dogs are presented as puppies. The younger the puppy when a cleaning regime is started, the more likely they are to tolerate it (Paterson 2018). Given infected skin can be extremely sore, dogs may be reluctant to allow it to be touched, and thus, keeping up a rigorous skin care routine will be of benefit to both dogs and their owners. Nurses should encourage owners to turn the cleaning regime into a positive experience using high value treats, and stopping while the dog is still calm and happy, before they show any signs of stress (Paterson 2018). The required frequency of cleaning will vary between dogs dependent on depth and size of folds and whether any concurrent skin conditions are present that may exacerbate the need to clean (e.g. allergies).

Some owners will use baby wipes to clean puppies' skin folds, but in the author's experience, they may not be sufficient and skin folds often need to be thoroughly washed with cotton wool balls soaked with dilute chlorhexidine at least once a week, remembering to thoroughly dry the area after. Chlorhexidine has good antibacterial properties, is non-irritant and has a long residual effect (Jackson and Marsella 2012). Sometimes, a protective emollient like petroleum jelly may be needed to reduce skin on skin abrasions (Paterson 2018). Discussing skin care at all nurse clinics is good practice to promote compliance and to encourage it to become part of daily life for owners.

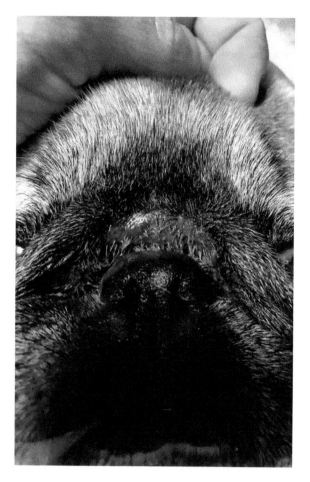

FIGURE 6.3   Infected nasal fold in a Pug.

BREATHING ASSESSMENT

Brachycephalic Obstructive Airway Disease (BOAS) is well documented in extreme brachycephalic dogs. Not all brachycephalic dogs will develop BOAS, but recent research found that 40%–50% of pet brachycephalic dogs (Bulldogs, French Bulldogs and Pugs) were clinically affected with BOAS (Liu et al. 2016). Brachycephalic dogs have been found to have a 3.5 times higher odds of being diagnosed with at least one upper respiratory tract disorder compared with crossbreeds (O'Neill et al. 2015). Despite this high prevalence of respiratory disease, over half (58%) of owners of BOAS affected dogs have been found to be unaware that their dog had a breathing problem (Packer, Hendricks, and Burn 2012). As such, veterinary nurses can play a central role in education of owners regarding the symptoms of BOAS, risk factors for BOAS and how to mitigate against them where possible. For those dogs that appear to struggle with their breathing, veterinary nurses can offer information on treatments available (including surgeries) and lifestyle modifications that may help their dogs both before and after surgery. BOAS clinical signs can vary according to the breed (Liu et al. 2016), and severity can vary between individuals, so early recognition of symptoms by nurses in brachycephalic clinics may help dogs that need treatment get earlier diagnosis and subsequent treatment before signs significantly progress.

During an examination, observe the dogs breathing while taking a history from the owner as this may be when the dog is at its most relaxed during a respiratory assessment. Listen for sounds

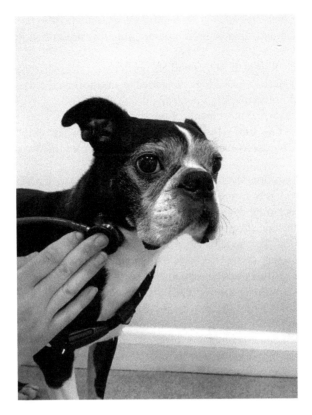

FIGURE 6.4   Stethoscope placement during pre- and post-exercise for laryngeal auscultation.

of stertor (lower pitched rattle) or stridor (higher pitched, like sawing wood) at rest, and then, listen using a stethoscope by placing to one side of the dog's neck but not underneath. By using a stethoscope, you will be able to hear whether or not there is laryngeal or pharyngeal stertor or stridor (Figure 6.4). Also observe for any respiratory effort at rest. In addition to a respiratory assessment at rest during a nurse clinic, an exercise assessment can be incorporated into this assessment, e.g. a 3-minute walk test (Riggs et al. 2019), as this is often a more effective way of detecting impaired airways. This is with the caveat that such challenges should never be undertaken if the animal is dyspnoeic pre-exercise, and if the dog shows any difficulties during the test, then it should be immediately stopped. Dogs are obligate nasal breathers, and brachycephalic dogs often have intranasal obstruction, so on exercise may be forced to switch to open mouth panting much faster than a normocephalic dog (Amis and Kurpershoek 1986, Oechtering et al. 2016). The idea of the exercise test in a nurse clinic is not to make any diagnosis of BOAS (which is the role of the veterinary surgeon), but to be able to talk through signs consistent with BOAS with the owner during the test and discuss possible lifestyle options to improve signs (e.g. weight loss) or make reference to surgical options available that may be offered by their veterinary surgeon following further assessment.

French Bulldogs and to a lesser extent Bulldogs and Pugs may suffer with gastrointestinal disorders concurrently with their airway disease, such as hiatal hernias, oesophageal dilation, chronic oesophagitis and reflux, the latter often being silent (Poncet et al. 2005, Kaye et al. 2018). After airway surgery, owners often report that their dogs eat better and put on weight. This has been the authors experience from working with brachycephalic dogs in clinics pre- and post-surgery. This may be due to less inflammation to the oesophageal lining from reduced reflux post-surgery. Finding out whether a dog is affected by reflux is important as it may also mean a careful care plan is needed both pre- and post-anaesthetic. Starvation pre-anaesthetic may need to be shorter to avoid build-up of stomach acid and suitable proton pump inhibitors prescribed pre- and postoperatively (Woodlands 2018).

Many brachycephalic dogs with BOAS will also suffer with sleep disorders. A 1987 study investigated the breathing patterns of Bulldogs compared to normocephalic dogs and found prolonged periods of oxygen desaturation to less than 90% during the REM (rapid eye movement) phase of sleep and observed paradoxical movements of the rib cage which are consistent with upper airway obstruction (Hendricks et al. 1987). Bulldogs were observed to wake up to 20 times a night. Once asleep, they also proved harder to wake up which made it harder for them to correct the desaturation of oxygen.

Despite this severe picture, sleep disorders have been little studied since this seminal work. However, signs that are likely to indicate sleep disorders have been reported in recent studies of brachycephalic dogs, including dogs developing postures to avoid upper airway obstruction while asleep. Such postures include sleeping while sitting up, resting their chins on an object or sleeping with their favourite toy in their mouths (Roedler, Pohl, and Oechtering 2013, Packer et al. 2019). Nurses need to make owners aware that these are not normal behaviours and not 'cute' or benign; infographics with these postures can be used to educate the public both in the surgery and on social media. These postures and behaviours usually signify severe nasal obstruction. Being unable to breathe through their noses while sleeping means affected dogs have adapted to breathing through their mouths by wedging them open to keep their airway patent. When lying with their heads flat on the ground, concurrent macroglossia and thickened palates get pushed upwards against the nasopharynx essentially narrowing this airway passage. If an owner reports that their dog is lazy and sleeps all day, it may well be due to fragmentation of sleep at night. In humans, excessive daytime sleepiness (hypersomnolence) is the result of sleep fragmentation related to recurrent central nervous system arousals in response to disordered breathing events (Caples, Gami, and Somers 2005).

WEIGHT AND THERMOREGULATION

Nurses are central to weight management in all companion animals in the veterinary clinic, but particularly in brachycephalic dogs where overweight and obesity can pose additional risks to their health. This role should ideally begin from puppyhood, with veterinary nurses educating owners on appropriate diet and body condition for their breed. Obesity in Pugs is a global issue (Liu et al. 2016, O'Neill et al. 2016, Mao et al. 2013). A study investigating the health of Pugs in primary veterinary care in England (2009–2015) found that overweight/obesity was the most prevalent disorder in this breed (O'Neill et al. 2016). Similarly in Bulldogs, obesity was found to be the third most prevalent disorder (O'Neill et al. 2019).

Weight is of high importance to respiratory health in brachycephalic patients. In both humans and dogs, obesity has direct negative effects on lung function (Jenkins and Moxham 1991, Bach et al. 2007, Manens et al. 2012), sleep (Caples, Gami, and Somers 2005) and quality of life (Yam et al. 2016). Obesity has been found to be a significant risk factor for BOAS (Liu et al. 2017, Packer et al. 2015). Oechtering (2010) describes the theory of the 'meat in the box' model, where excess soft tissues within the nose and the throat are confined within a hard box represented by the base of the skull and mandibles (Oechtering 2010). Excessive weight only acts to exacerbate this situation. Excess adipose tissue laid down within this 'box' puts extra pressure on the already restricted airways and worsens clinical signs. Excessive panting and stressful situations can lead to a vicious cycle of airway oedema and restriction caused by the need to work harder to cool down by moving air over the mucous membranes within the already restricted nose and throat (Davis, Cummings, and Payton 2017, Oechtering et al. 2016).

In addition to restricting the airway, overweight/obesity can exacerbate thermoregulatory capacities, already compromised in brachycephalic breeds. Thermoregulation occurs through a balance of heat gain versus heat dissipation; excessive fat also acts like a layer of insulation reducing the dogs' ability to lose heat (Byers et al. 2011, Davis, Cummings, and Payton 2017). Consequently, overweight brachycephalic dogs are more susceptible to hyperthermia during heat stress whether environmental or exercise induced. A recent study of the epidemiology of heat-related illness (HRI) in dogs identified brachycephalic skull shape as a risk factor, with 2.1 times the odds of being diagnosed with HRI than mesocephalic dogs (Hall, Carter, and O'Neill 2020b). Indeed, in a recent survey of owners of Bulldogs, French Bulldogs and Pugs, over one-third of owners reported that their dog had a problem with heat regulation (Packer et al. 2019). Ensuring that all owners of brachycephalic dogs are aware of this risk of heat to their dog is of high priority, particularly so for those that are overweight/obese, ensuring owners are aware that dogs should be kept indoors and well-cooled on warm as well as hot days. Educating owners on the risks of heat exposure has the potential to avoid preventable deaths, given the high mortality of HRI (Hall, Carter, and O'Neill 2020a).

Addressing a dog's weight with owners can be a sensitive issue. There is often an emotional connection between owner and dog when it comes to giving and restricting food (Linder 2014, White et al. 2011). This is further complicated by an owner's perceptions of their dog's weight sometimes being completely different to what the veterinary surgeon or veterinary nurse sees (White et al. 2011). Owners may be unaware of ideal weight for their dog and have skewed perceptions of what the ideal body condition is for their breed. Normalisation of obesity in brachycephalic breeds may be in part, due to the breed standards of brachycephalic breeds. Within the standards for Pugs, Bulldogs and French Bulldogs, descriptive words such as 'cobby', 'thick set' and 'thick neck' are described, which have been shown to increase the risk of developing BOAS (Liu et al. 2017, Packer et al. 2015). Nurses are often asked by owners what their dog's ideal weight should be. As Burkholder (2000) highlights, when advising on weight loss, there can be wide variation in size and frame between dogs even within breed, so ideal breed weights are often inaccurate. Weight ideally should be recorded alongside an assessment of body condition score (Burkholder 2000). For consistency in assessment, and to increase owner compliance, having the same veterinary nurse for each weight loss patient is a good idea in brachycephalic nursing clinics (Figure 6.5). By increasing an owner's understanding of the effect of excess weight on their dog's ability to breathe, exercise and thermoregulate, nurses can help brachycephalic dogs from a young age live healthier and happier lives.

PRE-PURCHASE CONSULTATIONS

In addition to helping individual brachycephalic dogs and their owners, veterinary nurses can help to influence the health and welfare of brachycephalic breeds at a wider level via the education of prospective owners. One such aspect of this is aiding in the choice of suitable breeds for individual households and educating the public regarding the health risks associated with brachycephalic breeds. There are over 200 registered pedigree breeds, and many 'designer' crossbreeds and mixed breeds that make up this population. What motivates households to purchase or rescue a particular breed is a complex subject influenced by many owner and dog related variables, but also subject to fashion and fads (Packer, Murphy, and Farnworth 2017, Herzog 2006).

The accessibility and quality (and indeed, whether owners seek out) pre-purchase information may affect choice of breed. In the 2019 PDSA PAW report (Peoples Dispensary for Sick Animals 2019), it was of concern that 21% of new owners surveyed conducted no pre-purchase research at all. For those that *do* conduct pre-purchase research, information is largely available via the Internet, but with online information comes misinformation that could lead to poor choices. It is the veterinary professions role to help owners make informed choices based on the best available evidence. This is with the aim of reducing health and welfare issues at both individual and population levels, and also avoiding financial strain for clients with unforeseen veterinary costs and the potential relinquishment of pets that clients were not prepared to care for.

FIGURE 6.5   Pug on weighing scales at a brachycephalic weight clinic.

A potential tool in the fight against poor purchasing decisions is the provision of pre-purchase consultations, to provide expert advice to owners on the responsibilities of owning a dog alongside counselling over breed choice. This concept is still relatively rare in veterinary medicine; however, in the 2019 PDSA PAW report (Peoples Dispensary for Sick Animals 2019), the number of people asking vets for pre-purchase advise rose slightly from 4% in 2015 to 7% in 2019. The proportion of these owners that initially desired or went on to purchase a brachycephalic dog is unknown; however, given the continual rise in popularity of these breeds, it is of concern that many owners purchase brachycephalic dogs despite information on their health becoming more widely accessible over the past decade. For example, campaigns on brachycephalic health have grown on social media (e.g. BVA's #Breedtobreathe campaign (British Veterinary Association 2020a)). This could indicate several issues, including, but not limited to, the following:

1. Some owners are not conducting any pre-purchase research and are thus not aware of the health problems in brachycephalic breeds.
2. Some owners have conducted research, are aware of potential health problems, but are willing to take the risks associated with owning a brachycephalic dog (or mitigate risks by taking out pet insurance, for example).

3. Owners have conducted research and are aware of some of the health issues associated with brachycephalic breeds, but have dissonant perceptions of breed health; for example, they downplay the risks of disease, see clinical signs as breed 'quirks', see disease as to be expected with that breed and/or disregard the welfare implications of these problems (Packer et al. 2019, Packer, Hendricks, and Burn 2012).

Which category or categories an owner falls into is likely to be important in the communication style taken with an owner; further discussion of communicating with clients is covered in Chapter 5 in this book. Overall, it is important to know the general reasons that 'pull' owners towards brachycephalic breeds, so if alternatives are suggested, they are in line with some of the owners key motivations for dog ownership. A recent study investigated the influence of breed type on pre-purchase motivation in both brachycephalic and non-brachycephalic owners of the top ten most registered breeds with The Kennel Club in 2014 (Packer, Murphy, and Farnworth 2017). This study found that appearance was a greater influence on breed choice than health of the breed for brachycephalic breed owners compared with non-brachycephalic breed owners, alongside size being suited to the owners' lifestyle and their chosen breed being perceived to be good with children. Similarly, in a Danish study of breed motivations, owners of French Bulldogs rated appearance and personality more highly than health (Sandøe et al. 2017) As such, deciphering whether an owner is entirely 'set' on the brachycephalic appearance (despite explaining the inherent health risks) before continuing a pre-purchase discussion is important; if they cannot be persuaded to purchase a non-brachycephalic breed with similar behavioural characteristics, then time may be better spent counselling them on how to buy the healthiest brachycephalic dog possible. Discuss what health tests the bitch and sire should have before breeding and what specific health issues are commonly seen in their chosen breed. In addition, discuss appropriate sources of puppies from higher welfare sources and promote the use of The Puppy Contract where possible (British Veterinary Association 2020b). Packer et al. (2017) found that brachycephalic owners were more likely to buy from puppy Internet selling sites (Packer, Murphy, and Farnworth 2017); however, this route runs the risk of inadvertently buying puppies via puppy farmers and puppy dealers and should thus be discouraged. Advise prospective owners to contact the breed clubs of their desired breed for a list of assured breeders that conduct appropriate health test (outlined in Chapter 8) and to join a waiting list for such a breeder.

For those that are more malleable in their breed choices, discussing other breeds that may fit their lifestyle, including characteristics such as suitability with children and the size of the breed, along with other important features of individual households may be fruitful in identifying other desirable breed options that have a lower risk of severe health problems. It is a fallacy that brachycephalic breeds do not need much exercise, but many owners value their perceived 'laziness' (Packer et al. 2020) and see it as a desirable 'lifestyle' feature, e.g. in a busy working home. However, this is not the case and a lack of exercise provision could exacerbate respiratory problems through obesity, and as such owners seeking out a brachycephalic breed based on this misunderstanding should be counselled.

A suitable pre-purchase information pack can be provided for all owners after the consultation for the prospective owners to take away. Resources included inside could include breed health information, health test requirements, information on The Puppy Contract (British Veterinary Association 2020b) and information on the importance of good lifetime pet insurance. Other methods of educating clients that may be thinking of getting a new dog are through educational waiting room displays (Figure 6.6), social media (e.g. practice Facebook accounts) and breed information leaflets provided within the practice.

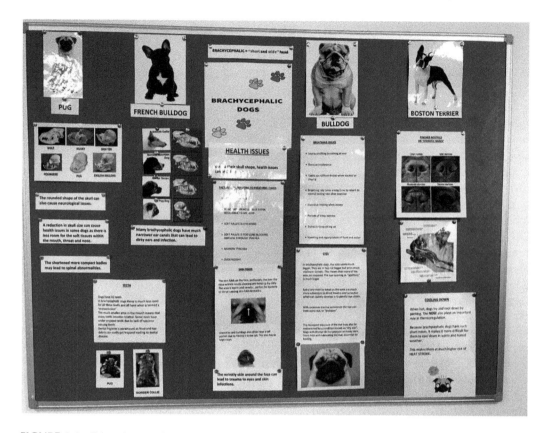

FIGURE 6.6　Educational waiting room displays about brachycephalic health.

CONCLUSIONS

Veterinary nurses are uniquely placed to make substantial positive improvements in the health and welfare of brachycephalic dogs through understanding, compassion and education, acting as a bridge between clients and veterinary surgeons. Through brachycephalic nurse clinics, a holistic approach can be taken to the multiple health issues that may arise in brachycephalic patients, with the aim of disease prevention where possible, by addressing potential health issues from a young age and promoting appropriate lifestyles and husbandry for health. Through pre-purchase consultations, veterinary nurses can also act to improve dog welfare at the population level, by educating potential buyers on the possible conformational and genetic risks that come with buying an animal with extreme brachycephalic features and providing counselling on the financial and emotional difficulties this could potentially cause. Veterinary nurses can promote lower-risk alternatives to brachycephalic breeds, and for owners of those who are insistent on buying a brachycephalic breed, direction can be provided on the best way to do this. Veterinary nurses are a potentially overlooked but important and passionate stakeholder group in the brachycephalic crisis whose potential may yet to be fully realised.

REFERENCES

Ackerman, N. 2012. *The Consulting Vet Nurse*. Chichester: Wiley.

Amis, T.C., and C. Kurpershoek. 1986. "Pattern of breathing in brachycephalic dogs." *American Journal of Veterinary Research* 47 (10):2200–2204.

Bach, J.F, E.A. Rozanski, D. Bedenice, D.L. Chan, L.M. Freeman, J.L.S. Lofgren, T.J. Oura, and A.M. Hoffman. 2007. "Association of expiratory airway dysfunction with marked obesity in healthy adult dogs." *American Journal of Veterinary Research* 68 (6):670–675.

Barrett, P.M., R.H. Scagliotti, R.E. Merideth, P. Jackson, and F. Alarcon. 1991. "Absolute corneal sensitivity and corneal trigeminal nerve anatomy in normal dogs." *Progress in Veterinary & Comparative Ophthalmology* 1 (4):245–254.

Blocker, T., and A. Van Der Woerdt. 2001. "A comparison of corneal sensitivity between brachycephalic and Domestic Short-haired cats." *Veterinary ophthalmology* 4 (2):127–130.

Bolzanni, H., A.P. Oriá, A.C.S. Raposo, and L. Sebbag. 2020. "Aqueous tear assessment in dogs: Impact of cephalic conformation, inter-test correlations, and test-retest repeatability." *Veterinary Ophthalmology* 23 (3):534–543.

British Veterinary Association. 2020a. "Breed to Breathe Campaign." accessed October. https://www.bva.co.uk/take-action/breed-to-breathe-campaign/.

British Veterinary Association. 2020b. "The Puppy Contract." accessed October. https://puppycontract.org.uk/.

Burkholder, W.J. 2000. "Use of body condition scores in clinical assessment of the provision of optimal nutrition." *Journal of the American Veterinary Medical Association* 217:650–654.

Byers, C.G., C.C. Wilson, M.B. Stephens, J. Goodie, F.E. Netting, and C. Olsen. 2011. "Obesity in dogs: Part 1: Exploring the causes and consequences of canine obesity." *Veterinary Medicine* 106 (4):184–192.

Campbell, H. 2020. "Nursing implications of over-breeding brachycephalic canines, focusing on British and French Bulldogs." *The Veterinary Nurse* 11 (5):235–239. doi:10.12968/vetn.2020.11.5.235.

Caples, S.M., A.S. Gami, and V.K. Somers. 2005. "Obstructive sleep apnea." *Annals of Internal Medicine* 142 (3):187–197.

Chandler, J.A., A. van der Woerdt, J.E. Prittie, and L. Chang. 2013. "Preliminary evaluation of tear production in dogs hospitalized in an intensive care unit." *Journal of Veterinary Emergency and Critical Care* 23 (3):274–279.

Costa, R.S., A.L. Abelson, J.C. Lindsey, and L.A. Wetmore. 2020. "Postoperative regurgitation and respiratory complications in brachycephalic dogs undergoing airway surgery before and after implementation of a standardized perianesthetic protocol." *Journal of the American Veterinary Medical Association* 256 (8):899–905.

Davis, M.S., S.L. Cummings, and M.E. Payton. 2017. "Effect of brachycephaly and body condition score on respiratory thermoregulation of healthy dogs." *Journal of the American Veterinary Medical Association* 251 (10):1160–1165. doi:10.2460/javma.251.10.1160.

Dodam, J.R., K.R. Branson, and D.D. Martin. 1998. "Effects of intramuscular sedative and opioid combinations on tear production in dogs." *Veterinary Ophthalmology* 1 (1):57–59.

Fenner, J.V.H., R.J. Quinn, and J.L. Demetriou. 2020. "Postoperative regurgitation in dogs after upper airway surgery to treat brachycephalic obstructive airway syndrome: 258 cases (2013-2017)." *Veterinary Surgery* 49 (1):53–60.

Hall, E.J., A.J. Carter, and D.G. O'Neill. 2020a. "Dogs don't die just in hot cars—Exertional heat-related illness (Heatstroke) is a greater threat to UK dogs." *Animals* 10:1324.

Hall, E.J., A.J. Carter, and D.G. O'Neill. 2020b. "Incidence and risk factors for heat-related illness (heatstroke) in UK dogs under primary veterinary care in 2016." *Scientific Reports* 10 (1):1–12.

Hendricks, J.C., L.R. Kline, R.J. Kovalski, J.A. O'Brien, A.R. Morrison, and A.I. Pack. 1987. "The English bulldog: A natural model of sleep-disordered breathing." *Journal of Applied Physiology* 63 (4):1344–1350.

Herring, I.P., J.P. Pickett, E.S. Champagne, and M. Marini. 2000. "Evaluation of aqueous tear production in dogs following general anesthesia." *Journal of the American Animal Hospital Association* 36 (5):427–430.

Herzog, H. 2006. "Forty-two thousand and one Dalmatians: Fads, social contagion, and dog breed popularity." *Society and Animals* 14 (4):383–397.

Jackson, H.A., and R. Marsella. 2012. *BSAVA Manual of Canine and Feline Dermatology.* Quedgeley: British Small Animal Veterinary Association.

Jenkins, S.C., and J. Moxham. 1991. "The effects of mild obesity on lung function." *Respiratory Medicine* 85 (4):309–311.

Kafarnik, C., J. Fritsche, and S. Reese. 2008. "Corneal innervation in mesocephalic and brachycephalic dogs and cats: Assessment using in vivo confocal microscopy." *Veterinary Ophthalmology* 11 (6):363–367.

Kaye, B.M., L. Rutherford, D.J. Perridge, and G.T. Haar. 2018. "Relationship between brachycephalic airway syndrome and gastrointestinal signs in three breeds of dog." *Journal of Small Animal Practice* 59 (11):670–673.

Labelle, A.L., C.B. Dresser, R.E. Hamor, M.C. Allender, and J.L. Disney. 2013. "Characteristics of, prevalence of, and risk factors for corneal pigmentation (pigmentary keratopathy) in Pugs." *Journal of the American Veterinary Medical Association* 243 (5):667–674.

Linder, D. 2014. "How to implement and manage a weight loss plan." *The Veterinary Nurse* 5:216–219.

Liu, N.-C., V.J. Adams, L. Kalmar, J.F. Ladlow, and D.R. Sargan. 2016. "Whole-body barometric plethysmography characterizes upper airway obstruction in 3 brachycephalic breeds of dogs." *Journal of Veterinary Internal Medicine* 30 (3):853–865.

Liu, N.-C., E.L. Troconis, L. Kalmar, D.J. Price, H.E. Wright, V.J. Adams, D.R. Sargan, and J.F. Ladlow. 2017. "Conformational risk factors of brachycephalic obstructive airway syndrome (BOAS) in pugs, French bulldogs, and bulldogs." *Plos One* 12 (8):e0181928. doi:10.1371/journal.pone.0181928.

Lock, K. 2011. "Reflections on designing and implementing a nursing care plan." *The Veterinary Nurse* 2 (5):272–277.

Loftus, L. 2012. "The non-compliant client." *Veterinary Nursing Journal* 27 (8):294–297.

Maini, S., R. Everson, C. Dawson, Y.M. Chang, C. Hartley, and R.F. Sanchez. 2019. "Pigmentary keratitis in pugs in the United Kingdom: Prevalence and associated features." *BMC Veterinary Research* 15 (1):384.

Manens, J., M. Bolognin, F. Bernaerts, M. Diez, N. Kirschvink, and C. Clercx. 2012. "Effects of obesity on lung function and airway reactivity in healthy dogs." *The Veterinary Journal* 193 (1):217–221. doi:10.1016/j.tvjl.2011.10.013.

Mao, J., Z. Xia, J. Chen, and J. Yu. 2013. "Prevalence and risk factors for canine obesity surveyed in veterinary practices in Beijing, China." *Preventive Veterinary Medicine* 112 (3–4):438–442.

O'Neill, D.G., A.M. Skipper, J. Kadhim, D.B. Church, D.C. Brodbelt, and R.M.A. Packer. 2019. "Disorders of Bulldogs under primary veterinary care in the UK in 2013." *PloS One* 14 (6):e0217928.

O'Neill, D.G., L. Baral, D.B. Church, D.C. Brodbelt, and R.M.A. Packer. 2018. "Demography and disorders of the French Bulldog population under primary veterinary care in the UK in 2013." *Canine Genetics and Epidemiology* 5 (1):3. doi:10.1186/s40575-018-0057-9.

O'Neill, D.G., E.C. Darwent, D.B. Church, and D.C. Brodbelt. 2016. "Demography and health of Pugs under primary veterinary care in England." *Canine Genetics and Epidemiology* 3 (1):5. doi:10.1186/s40575-016-0035-z.

O'Neill, D.G., C. Jackson, J.H. Guy, D.B. Church, P.D. McGreevy, P.C. Thomson, and D.C. Brodbelt. 2015. "Epidemiological associations between brachycephaly and upper respiratory tract disorders in dogs attending veterinary practices in England." *Canine Genetics and Epidemiology* 2 (1):10. doi:10.1186/s40575-015-0023-8.

O'Neill, D.G., M.M. Lee, D.C. Brodbelt, D.B. Church, and R.F. Sanchez. 2017. "Corneal ulcerative disease in dogs under primary veterinary care in England: Epidemiology and clinical management." *Canine Genetics and Epidemiology* 4 (1):5. doi:10.1186/s40575-017-0045-5.

Oechtering, G. 2010. "Brachycephalic syndrome-new information on an old congenital disease. " *Veterinary Focus* 20 (2):2–9.

Oechtering, G.U., S. Pohl, C. Schlueter, J.P. Lippert, M. Alef, I. Kiefer, E. Ludewig, and R. Schuenemann. 2016. "A novel approach to brachycephalic syndrome. 1. Evaluation of anatomical intranasal airway obstruction." *Veterinary Surgery* 45 (2):165–172.

Packer, R.M.A., D. Murphy, and M.J. Farnworth. 2017. "Purchasing popular purebreds: investigating the influence of breed-type on the pre-purchase motivations and behaviour of dog owners." *Animal Welfare* 26 (2):191–201. doi:10.7120/09627286.26.2.191.

Packer, R.M.A., A. Hendricks, and C.C. Burn. 2012. "Do dog owners perceive the clinical signs related to conformational inherited disorders as 'normal' for the breed? A potential constraint to improving canine welfare." *Animal Welfare* 21 (1):81.

Packer, R.M.A., A. Hendricks, and C.C. Burn. 2015. "Impact of facial conformation on canine health: Corneal ulceration." *Plos One* 10 (5):e0123827. doi:10.1371/journal.pone.0123827.

Packer, R.M.A., A. Hendricks, M.S. Tivers, and C.C. Burn. 2015. "Impact of facial conformation on canine health: Brachycephalic obstructive airway syndrome." *Plos One* 10 (10):e0137496. doi:10.1371/journal.pone.0137496.

Packer, R.M.A., D.G. O'Neill, F. Fletcher, and M.J. Farnworth. 2019. "Great expectations, inconvenient truths, and the paradoxes of the dog-owner relationship for owners of brachycephalic dogs." *Plos One* 14 (7):e0219918. doi:10.1371/journal.pone.0219918.

Packer, R.M.A., D.G. O'Neill, F. Fletcher, and M.J. Farnworth. 2020. "Come for the looks, stay for the personality? A mixed methods investigation of reacquisition and owner recommendation of Bulldogs, French Bulldogs and Pugs." *Plos one* 15 (8):e0237276.

Paterson, S. 2018. "Nursing intertrigo in the dog." *The Veterinary Nurse* 9 (8):402–408.

Peoples Dispensary for Sick Animals. 2019. "PDSA Animal Well-being (PAW) Report", accessed October. https://www.pdsa.org.uk/media/7420/2019-paw-report_downloadable.pdf.

Poncet, C.M., G.P. Dupre, V.G. Freiche, M.M. Estrada, Y.A. Poubanne, and B.M. Bouvy. 2005. "Prevalence of gastrointestinal tract lesions in 73 brachycephalic dogs with upper respiratory syndrome." *Journal of Small Animal Practice* 46 (6):273–279.

Riggs, J., N.-C. Liu, D.R. Sutton, D. Sargan, and J.F. Ladlow. 2019. "Validation of exercise testing and laryngeal auscultation for grading brachycephalic obstructive airway syndrome in pugs, French bulldogs, and English bulldogs by using whole-body barometric plethysmography." *Veterinary Surgery* 48 (4):488–496.

Roedler, F.S., S. Pohl, and G.U. Oechtering. 2013. "How does severe brachycephaly affect dog's lives? Results of a structured preoperative owner questionnaire." *The Veterinary Journal* 198 (3):606–610. doi:10.1016/j.tvjl.2013.09.009.

Sandøe, P., S.V. Kondrup, P.C. Bennett, B. Forkman, I. Meyer, H.F. Proschowsky, J.A. Serpell, and T.B. Lund. 2017. "Why do people buy dogs with potential welfare problems related to extreme conformation and inherited disease? A representative study of Danish owners of four small dog breeds." *PloS One* 12 (2):e0172091.

Usborne, S. 2019. "'This is a calamity': The surgeons keeping pugs and bulldogs alive". *The Guardian.* https://www.theguardian.com/lifeandstyle/2019/feb/27/this-is-a-calamity-the-surgeons-keeping-pugs-and-bulldogs-alive.

White, G.A., P. Hobson-West, K. Cobb, J. Craigon, R. Hammond, and K.M. Millar. 2011. "Canine obesity: Is there a difference between veterinarian and owner perception?" *Journal of Small Animal Practice* 52:622–626.

Woodlands, C. 2018. "Perioperative care of the brachycephalic patient and surgical management of brachycephalic obstructive airway syndrome." *The Veterinary Nurse* 9 (10):532–538.

Yam, P.S., C.F. Butowski, J.L. Chitty, G. Naughton, M.L. Wiseman-Orr, T. Parkin, and J. Reid. 2016. "Impact of canine overweight and obesity on health-related quality of life." *Preventive Veterinary Medicine* 127:64–69.

Yeates, J. 2014. "The role of the veterinary nurse in animal welfare." *Veterinary Nursing Journal* 29 (7):250–251. doi:10.1111/vnj.12160.

7 The Epidemiology of Brachycephalic Health – Understanding the Science and Exploring the Evidence on Demography, Disorder Frequency and Risk Factors

Dan G. O'Neill
The Royal Veterinary College London

CONTENTS

Introduction 85
Background 86
Epidemiological Principles 86
Predispositions and Protections to Disorders in Brachycephalic Companion Animals 88
Disorder Prevalence 92
Brachycephalic Demography and Popularity 94
Longevity and Mortality 98
Conclusions 101
References 101

INTRODUCTION

Until recently, beliefs on important companion animal health issues were generally moulded by expert opinion ('eminence-based veterinary medicine') from recognised experts or self-appointed commentators (Holmes and Cockcroft 2004). However, although often highly persuasive, experts opinion is now regarded as a weak source of evidence because the belief systems of individual experts are inevitably framed by their many personal and deeply held cognitive biases (Holmes and Ramey 2007). This partially explains why experts often disagree vehemently on specific topics. Consequently, 'evidence-based veterinary medicine' has gained increased traction since the 1990s, and we now rely increasingly on evidence generated from population-based studies for better decision-making (Huntley et al. 2016, Holmes and Ramey 2007). In this chapter, I welcome you into my exciting and entrancing world of epidemiology where you can explore the evidence we now have on brachycephalic topics such as disorder prevalence and predisposition, demography and longevity. This chapters aims to provide the enthusiastic reader with the tools and the data to appreciate a population-based view of brachycephalic health. Enjoy your trip.

BACKGROUND

Before we enter the world of brachycephalic data, it is important to grasp some core epidemiological principles. Epidemiology literally means "the study of what is upon the people" but is generally prefixed as 'veterinary epidemiology' when applied to animals. Since the time of Hippocrates over 2000 years ago, humankind has sought to understand the causation of disease using methods that counted disease occurrences and then extracted patterns from these data. The term 'epidemiology' was coined in the sixteenth century. Associations noted by Ignaz Semmelweiss between hygiene standards and maternal mortality due to childhood fever in Vienna in 1842, and by John Snow between water supplies and cholera outbreaks in London in 1854, helped to establish this new science of epidemiology as critical for understanding disease in man and later in animals (Paneth 2004). Veterinary epidemiology is now an established science within the core curriculum for veterinary medicine undergraduates and also offers specific post-graduate qualifications (London School of Hygiene and Tropical Medicine 2020).

Veterinary epidemiology offers structured scientific methods for systematic collection of data, hypothesis-driven research, use of analytical techniques, and implementation of results to inform actions and policy (Dohoo, Martin, and Stryhn 2009). In the veterinary companion animal sphere, these methods have classically been applied to epidemic diseases (i.e. new diseases or with sudden increases in case counts above normal expectations in that area) such as canine distemper in the 1920s (Bresalier and Worboys 2014) and parvovirus in dogs in the 1970s (Miranda and Thompson 2016). However, as immunisation and improved hygiene reduced the contribution of infectious diseases to the overall disease burden of companion animals, epidemiological understanding for the control of endemic diseases (i.e. diseases with a constant presence in a population within a geographic area) has become more important.

Although recognised for over a century (Skipper 2019), health issues associated with extreme body conformation have become increasingly prioritised as critical endemic diseases in companion animals over recent decades (McGreevy and Nicholas 1999). Given that breed is defined as "a group of animals *selected by man* to have a uniform appearance that distinguishes them from other members of the same species" (Hall and Bradley 1995, Langer 2018), this means that man has continued to 'gift' conformation-related disorder predispositions to animals born to breeds with extreme conformations. Breed conformation and breed popularity are not intrinsic features created by nature but are instead dictated by the whim of humans. It follows therefore that the levels of endemic diseases associated with these conformations and breeds in the overall animal population are also at the whim of humans to control (Galibert et al. 2011). This realisation has led to the development of specific breed and conformation health disciplines within companion animal epidemiology, with several large research programmes now specifically exploring these topics worldwide (VetCompass Australia 2020, VetCompass 2020, PETscan 2020). Using this framework of understanding, this chapter will focus specifically on epidemiological aspects related to brachycephaly, although the methods discussed here are equally applicable to other extreme conformations in companion animals (DogWellNet. 2020).

EPIDEMIOLOGICAL PRINCIPLES

The authors of many chapters of this book present epidemiological evidence to support their conclusions. A good understanding of the basic methods and terminology of epidemiology will assist the astute reader to best harness the powers of this wonderful science.

Epidemiological evidence comes at two levels of inference: descriptive and analytic. Descriptive epidemiology describes the world as it is, in relation to the population, location and time period under study. In the context of this chapter, for example, descriptive epidemiology will describe the popularity of specific breeds or the frequency of disease within these breeds. The frequency of disease is called *prevalence* and is reported as the proportion (or percentage) of study animals affected by disease at a point in time (point prevalence) or during a specified period of time (period prevalence). For example, a study using veterinary clinical data reported a 1-year period prevalence of 13.18% for overweight/obesity in Pugs in the UK (O'Neill, Darwent, et al. 2016). This is useful information to help prioritise this condition compared to other conditions in the breed; overweight/obesity was the most common disorder reported in Pugs in that study, providing some evidence that control of overweight/obesity is highly relevant to the health of this breed. However, if we wanted to learn whether Pugs were predisposed to overweight/obesity, we would need a study that compared the risk of overweight/obesity in Pugs to other breeds (Pegram et al. 2020). Equally, if we wanted to understand the overall welfare impact of overweight/obesity on Pugs, we would additionally need to take account of disorder severity and duration to get a fuller welfare picture (Summers et al. 2019). The key message for readers here is that each piece of epidemiological evidence is just part of a larger jigsaw; one piece of evidence on its own should not be assumed to answer broader welfare questions fully.

Whereas descriptive epidemiology describes 'what is', i.e. patterns of disease, humans are often more interested in 'why it is'. For this, we rely on analytic epidemiology to evaluate *risk factors*. Risk factors describe features of an animal or its environment where we can quantifiably report associations with disease likelihood (Pfeiffer 2010). Important risk factors for disease include sex, neutering and age, but in the context of this book, the most important risk factors are brachycephalic conformation and breed. Comparative results for risk factors are generally reported as odds ratios (OR) or risk ratios (RR). These compare the relative probability of a disorder (e.g. patellar luxation) between categories of the risk factor (French Bulldog versus crossbreds as categories within breed as a risk factor). For example, if breed is the risk factor for a study of patellar luxation in dogs, we can say that French Bulldogs have 5.4 times the odds (95% confidence interval [CI] 3.1–9.3, $P<0.001$) of patellar luxation compared to crossbred dogs (O'Neill, Meeson, et al. 2016). Odds and risk ratios are interpreted similarly; values above 1.0 suggest increased disease probability (risk), whereas values below 1.0 suggest reduced probability. If the 95% CI spans 1.0 (i.e. if 1.0 appears between the lower and upper bounds of the 95% CI), this suggests little or no evidence of difference between the categories. The wary reader should note that logical selection of the baseline category for these comparisons is critical. For example, a dystocia study (difficulty whelping) reported French Bulldogs as the most predisposed breed, with 15.9 times odds (95% CI 9.3–27.2, $P<0.001$) compared with crossbred bitches. However, if this comparison had been to another predisposed breed such as the Boston Terrier (OR 12.9, 95% CI 5.6–29.3, $P<0.001$), French Bulldogs would not have seemed at any increased risk (O'Neill, O'Sullivan, et al. 2017).

Due to financial and logistical constraints, epidemiological studies are generally based on subsets (samples) of animals rather than on the full population. Such studies therefore report estimated values that can be extrapolated from the sample to any wider population with similar characteristics to the sampled animals. These estimated values are accompanied by a 95% confidence interval (often abbreviated to 95% CI) that provides lower and upper limits between which the study is 95% confident the true value in the wider general population exists. The 95% CI therefore describes the uncertainty surrounding the estimated value; a wide 95% CI (i.e. a wide interval between the lower and upper limits) shows high uncertainty, whereas a narrow 95% CI shows higher precision (Pfeiffer 2010). The p-value is another useful statistic that helps to infer the strength of evidence for analytic test results. The older convention was to accept p-values under 0.05 as 'statistically significant'. However, perceptive readers are now advised to consider the exact p-value and to take a more holistic approach by also examining other aspects such as the size of the study, level of the effect, width of the confidence interval and the choice of comparator group before reaching conclusions (Jeffery 2015).

Epidemiological work on brachycephaly ultimately aims to identify key attributes and relationships that we can modify over time to improve welfare. Each new piece of epidemiological evidence should be integrated within our pre-existing overall body of scientific knowledge. We can then apply this new updated evidence base towards the creation of innovative solutions for key issues within the overall brachycephalic crisis. For example, based on epidemiological evidence that Pugs are predisposed to corneal ulceration (O'Neill, Darwent, et al. 2016, O'Neill, Lee, et al. 2017), owners can be alerted to this problem and the need for vigilant ocular care in the breed, e.g., via brachycephalic nursing clinics as explored in Chapter 6: nurses and the brachycephalic patient – practical considerations and the role of VNs in improving brachycephalic health. Additionally, veterinarians can increase their clinical vigilance during ophthalmological examinations and perhaps also ensure routine assessment for keratoconjunctivitis sicca (dry eye) in this breed, as explored in Chapter 12: Ophthalmology in practice for brachycephalic breeds.

Health issues of brachycephalic animals generally have very complex 'webs of causation' and therefore require deep epidemiological consideration to ensure robust results and interpretations. For example, brachycephalic dogs are 11.18 times more likely to have corneal ulceration than crossbred dogs (O'Neill, Lee, et al. 2017). However, older dogs are also more likely to show corneal ulceration; dogs aged 9–12 years have 2.74 times the odds of dogs aged under 3.0 years (O'Neill, Lee, et al. 2017). Given that average age often differs between breeds in the wider population because of changing popularity and typical lifespans, this suggests that direct comparison of corneal ulceration risk between breeds (univariable analysis) will fail to take this age effect into account and therefore give biased confounded results (Pfeiffer 2002). Younger breeds or groups of dogs will be confounded to show lowered risk because younger dogs are less likely to get corneal ulceration due to their age. Confounding (literally meaning to 'mix up' or 'confuse') describes mixing up effects from the risk factor of interest (e.g. breed) with some other factor (e.g. age) that is also associated with the disease outcome. Due to limited computing power at that time, many older epidemiological studies in veterinary medicine reported only univariable findings. These were likely to be heavily confounded and therefore should now be interpreted with extreme caution. Thankfully, modern powerful computing now enables researchers to carry out multivariable analyses that can account for the effects of many confounding variables (provided these have been collected). The cautious reader should seek out multivariable results wherever possible rather than accepting findings from univariable analyses that may be misleading.

PREDISPOSITIONS AND PROTECTIONS TO DISORDERS IN BRACHYCEPHALIC COMPANION ANIMALS

A key epidemiological question in relation to brachycephalic health asks which disorders are more common (i.e. predisposed) or less common (i.e. protected) in individual brachycephalic breeds or in brachycephalic animals overall. Although this might seem an easy question to answer at a conceptual level, it is not quite so simple at an epidemiological level. We need to consider carefully issues such as the case definitions for disorders (i.e. how to define some dogs as having the disorder and other as not) and to explore selection biases in how the animals were included in the study as well as considering how the data were analysed.

One approach to answering this predisposition question is to sift through the total body of published studies to date. This approach was applied in a recent book *Breed Predispositions to Disease in Dogs and Cats* that reported all disorders with evidence of predisposition in individual dog and cat breeds worldwide (Gough, Thomas, and O'Neill 2018). The quasi 'systematic review' findings described in this book showed us what science had reported so far on breed predispositions and perhaps also highlights some limitations for interpreting this evidence. Tables 7.1 and 7.2 show the counts of unique predisposed disorders in common brachycephalic dog and cat breeds, respectively, compared with some common non-brachycephalic breeds. These results make interesting reading because they do not appear, at first glance, to support a hypothesis that brachycephalic dog or cat breeds have more predispositions than their non-brachycephalic counterparts. However, it should be noted that few of the published studies included in that book began with a clearly defined plan on which breeds would be explored; instead, most just reported on the most common breeds that happened to be available at the time of the study. Consequently, popular breeds are more likely to be included in studies and therefore

TABLE 7.1

Counts of Individual Disorders with Evidence of Predisposition in Common Brachycephalic Breeds of Dogs, Compared with Some Common Mesocephalic and Dolichocephalic Breeds Worldwide (Gough, Thomas, and O'Neill 2018)

Breed	Skull-Shape Category	No. Disorder Predispositions
Boxer	Brachycephalic	76
Pug	Brachycephalic	41
English Bulldog	Brachycephalic	39
Shih-Tzu	Brachycephalic	39
Cavalier King Charles Spaniel	Brachycephalic	29
Pekingese	Brachycephalic	24
Boston Terrier	Brachycephalic	22
Lhasa Apso	Brachycephalic	21
Chihuahua	Brachycephalic	20
French Bulldog	Brachycephalic	18
Bull Mastiff	Brachycephalic	11
Dogue de Bordeaux	Brachycephalic	7
American Bulldog	Brachycephalic	3
King Charles Spaniel	Brachycephalic	0
German Shepherd Dog	Dolichocephalic	77
Dachshund	Dolichocephalic	47
Greyhound	Dolichocephalic	34
Whippet	Dolichocephalic	13
Labrador Retriever	Mesocephalic	70
Cocker Spaniel	Mesocephalic	63
Yorkshire Terrier	Mesocephalic	47
Staffordshire Bull Terrier	Mesocephalic	20

TABLE 7.2

Counts of Individual Disorders with Evidence of Predisposition in Common Brachycephalic Breeds of Cats, Compared with Some Common Mesocephalic and Dolichocephalic Breeds Worldwide (Gough, Thomas, and O'Neill 2018)

Breed	Skull-Shape Category	No. Disorder Predispositions
Persian	Brachycephalic	29
Burmese	Brachycephalic	24
Himalayan	Brachycephalic	15
Siamese	Non-brachycephalic	38
Abyssinian	Non-brachycephalic	19
Ragdoll	Non-brachycephalic	13

to show greater numbers of published predispositions, while rare breeds may be infrequently evaluated for predispositions. Absence of evidence is not evidence of absence, so we should interpret published predisposition counts cautiously as a measure of breed health. That said though, the evidence that is available for disorder predispositions within breeds still provides a useful resource to support health plans for these breeds (The Kennel Club 2020a). It is worth noting that predisposition evidence alone is insufficient to confirm the welfare importance of that disorder to breed health; evidence on the relative severity, duration and prevalence also need to be considered (Summers et al. 2019).

Studies designed *a priori* to specifically compare health between skull shape groups offer a more scientifically robust approach to explore overall brachycephalic health. This concept is being increasingly applied by the VetCompass Research Group in the UK (VetCompass 2020). VetCompass studies include all animals under first-opinion veterinary care at participating clinics and aim for results that are generalisable to the wider population of companion animals. Reliability of these results also benefits from having health information recorded by veterinarians contemporaneously at the time of these clinical events (O'Neill et al. 2014b).

There is substantial evidence that skull conformation is a risk factor for brachycephalic obstructive airway disease (BOAS) in brachycephalic dog breeds (Liu et al. 2017). However, by definition, BOAS cannot be a diagnosis in non-brachycephalic animals, and there is growing evidence that normalisation of BOAS in brachycephalic breeds leads to under-recognition and under-reporting of this disorder by both owners and veterinarians (Packer et al. 2015, 2019, 2020).

To overcome these issues, an early VetCompass study widened this research question to explore upper respiratory tract (URT) disorders in general. Random samples from three extreme brachycephalic breed types (Bulldog, French Bulldog and Pug) were compared to three other common small-to-medium-sized breed types (moderate brachycephalic: Yorkshire Terrier and non-brachycephalic: Border Terrier and West Highland White Terrier) (O'Neill, Jackson, et al. 2015). The prevalence of URT disorders among the study dogs overall was 15.8% (95% CI: 13.2–18.4), highlighting the relevance of URT disorders to the overall disorder burden of dogs. The prevalence of URT disorders was substantially higher in extreme brachycephalic dogs (22.0%, 95% CI: 18.0–26.0) than in the moderate and non-brachycephalic group (9.7%, 95% CI: 7.1–12.3) (P < 0.001). The prevalence of having at least one URT disorder varied significantly between the individual breed types: Bulldog 19.5%, French Bulldog 20.0%, Pug 26.5%, Yorkshire Terrier 13.0%, Border Terrier 9.0% and West Highland White Terrier 7.0% (*P* < 0.001). Following multivariable modelling to account for confounding (e.g. age differences between the dogs), the extreme brachycephalic breeds had 3.5 times (95% CI: 2.4–5.0, *P* < 0.001) the odds of diagnosis with at least one URT disorder compared to the moderate and non-brachycephalic group. Similarly after multivariable modelling, the English Bulldog (OR 4.0, 95% CI 2.1–7.9, *P* < 0.001), French Bulldog (OR 5.1, 95% CI 2.6–10.2, *P* < 0.001), Pug (OR 6.9, 95% CI 3.6–13.3, *P* < 0.001) and Yorkshire Terrier (OR 2.2, 95% CI 1.1–4.5, *P* = 0.026) had higher odds of having at least one URT disorder compared with the West Highland White Terrier. These results highlight the complexity of exploring health issues in brachycephalic dogs where we have many breeds that may have their own unique health patterns. It is also clear that there are also health issues in non-brachycephalic dogs that need to be considered before reaching overall conclusions. In short, this study confirmed a substantial predisposition to URT disorders in brachycephalic breeds but additionally showed that URT conditions are also common in non-brachycephalic dogs.

A more recent VetCompass study undertook a more holistic epidemiological exploration of predispositions and also protections to common disorders in brachycephalic dogs in the UK (O'Neill et al. 2020). Information on all disorders during 2016 was extracted on a random sample of 22,333 dogs from an overall population of 955,554 dogs attending 784 veterinary clinics. Breeds were categorised as either brachycephalic or non-brachycephalic. Multivariable analyses compared the odds between the skull shape categories for each of the 30 most common disorders across both categories. After accounting for confounding factors such as age, the odds for 10 of the 30 disorders differed between the brachycephalic and non-brachycephalic groups of dogs. This suggests that there are core health differences between brachycephalic dogs and the remaining dog population. However, review of the disorders that differed suggests that some of these differences may be unrelated to brachycephaly *per se* and that highly popular breeds in each group may have heavily influenced the results. For example, Cavalier King Charles Spaniels are highly predisposed to show heart murmurs (Mattin et al. 2015) and therefore their inclusion in the brachycephalic category may explain the high odds ratio (predisposition) for heart murmurs in this group.

Brachycephalic dogs showed predispositions in eight of the ten disorders that differed between the groups, and showed protection in the remaining two disorders (Table 7.3). This imbalance provides some evidence that the health of brachycephalic dogs in relation to common disorders is poorer than the health of non-brachycephalic dogs but also identifies that this deleterious effect is not universal across disorders. The predispositions identified here in brachycephalic dogs were supported by previous studies that also evidenced high prevalence and risk of disorders such

TABLE 7.3

Prevalence of the 30 Most Common Disorders Recorded in Brachycephalic (n = 4,169) and Non-brachycephalic (n = 18,079) Dog Types under UK Primary Veterinary Care from January 1, 2016 to December 31, 2016 at Practices Participating in the VetCompass Programme

Disorder	Brachycephalic No. (%)	Non-brachycephalic No. (%)	Odds Ratio	95% Confidence Interval	P-Value*
Corneal ulceration	100 (2.40)	72 (0.40)	8.40	5.21–13.56	**< 0.001**
Heart murmur	143 (3.43)	330 (1.83)	3.52	2.70–4.60	**< 0.001**
Umbilical hernia	91 (2.18)	117 (0.65)	3.16	1.94–5.18	**< 0.001**
Pododermatitis	71 (1.70)	230 (1.27)	1.66	1.20–2.28	**0.002**
Skin cyst	50 (1.20)	196 (1.08)	1.52	1.04–2.22	**0.029**
Patellar luxation	86 (2.06)	146 (0.81)	1.40	1.01–1.93	**0.038**
Otitis externa	303 (7.27)	1323 (7.32)	1.29	1.10–1.51	**0.002**
Retained deciduous tooth	88 (2.11)	137 (0.76)	1.30	0.85–2.01	0.221
Pyoderma	67 (1.61)	258 (1.43)	1.26	0.92–1.74	0.156
Anal sac impaction	249 (5.97)	822 (4.55)	1.24	1.03–1.50	**0.021**
Pruritus	81 (1.94)	282 (1.56)	1.22	0.90–1.67	0.203
Overgrown nail(s)	212 (5.09)	760 (4.20)	1.18	0.98–1.43	0.102
Wound	42 (1.01)	208 (1.15)	1.15	0.77–1.72	0.497
Disorder not diagnosed	20 (0.48)	161 (0.89)	1.09	0.55–2.16	0.805
Allergy	66 (1.58)	284 (1.57)	1.06	0.76–1.48	0.709
Diarrhoea	143 (3.43)	706 (3.91)	1.05	0.82–1.33	0.710
Gastroenteritis	64 (1.54)	233 (1.29)	1.05	0.74–1.51	0.778
Skin mass	57 (1.37)	406 (2.25)	1.01	0.73–1.39	0.972
Lameness	88 (2.11)	502 (2.78)	0.99	0.74–1.31	0.922
Flea infestation	101 (2.42)	356 (1.97)	0.98	0.73–1.31	0.878
Obesity	266 (6.38)	1311 (7.25)	0.96	0.81–1.14	0.657
Vomiting	131 (3.14)	546 (3.02)	0.96	0.74–1.24	0.748
Periodontal disease	485 (11.63)	2310 (12.78)	0.93	0.81–1.07	0.308
Aggression	86 (2.06)	414 (2.29)	0.91	0.67–1.22	0.511
Conjunctivitis	86 (2.06)	413 (2.28)	0.89	0.65–1.22	0.464
Foreign body	40 (0.96)	241 (1.33)	0.80	0.52–1.24	0.323
Osteoarthritis	39 (0.94)	483 (2.67)	0.79	0.53–1.16	0.230
Lipoma	17 (0.41)	303 (1.68)	0.59	0.34–1.01	0.056
Undesirable behaviour	42 (1.01)	291 (1.61)	0.52	0.34–0.81	**0.003**
Claw injury	31 (0.74)	278 (1.54)	0.45	0.29–0.70	**< 0.001**

The probability for each disorder in brachycephalic compared with non-brachycephalic dogs is reported using multivariable methods (mixed effects multivariable logistic regression modelling that included the *skull conformation, adult bodyweight category, bodyweight relative to breed/sex mean, age category, sex, neuter* and *insurance* with the *clinic* attended included as a random effect). *P*-values<0.05 in bold.

as corneal ulceration (O'Neill, Lee, et al. 2017, Packer, Hendricks, and Burn 2015) and patellar luxation (O'Neill, Meeson, et al. 2016) in brachycephalic breeds. However, the study also reported reduced odds for undesirable behaviours in brachycephalic dogs. Favourable disposition towards the behaviours of brachycephalic dogs may interact with the allure of their baby-like features to partly explain current human motivation for ownership of these breeds, despite well-reported health issues (Packer et al. 2019, 2020). Similar influences have been reported for appearance and behaviours on ownership decisions in cats (Plitman et al. 2019).

In summary, there is now strong evidence supporting associations between brachycephalic status in dog breeds and substantial health issues. However, the evidence also suggests that many of these predispositions may not be directly related to brachycephaly itself but to other health issues within popular examples of brachycephalic breeds. There is also evidence that some disorders such as undesirable behaviours are protected (i.e. reduced) in brachycephalic animals. That said, it is worth challenging the ecological validity of this conclusion about brachycephalic dogs overall; can we safely generalise this conclusion on brachycephalic animals overall as implying that each individual brachycephalic breed must therefore also have poorer health? To answer this question, we would need to apply the robust epidemiological methods described used above to a series of breed-specific studies. This approach has already started within VetCompass with the recent publication of a holistic predispositions and protections study on Staffordshire Bull Terriers (Pegram et al. 2020), and a series of studies on brachycephalic breeds are planned to follow over the coming years.

There is substantial published evidence that some key disorders are often considered as typical predispositions in brachycephalic dog breeds. Many of these studies will be cited and discussed in the relevant chapters of this book, but a summary of these disorders is shown in Table 7.4 along with some supporting evidence.

DISORDER PREVALENCE

The welfare relevance of disorders to breed health can be estimated from combined effects of their prevalence (how common they are), duration (how long animals are affected for) and severity (the degree of discomfort, pain or harm) (Collins et al. 2010, Summers et al. 2019). Of these three features, establishing good evidence on disorder prevalence in dogs has been identified as a first priority (Bateson 2010). Prevalence data can be derived from studies that focus on one disorder to high precision, e.g. corneal ulceration (O'Neill, Lee, et al. 2017), or from broader studies of multiple disorders in single breed (O'Neill, Darwent, et al. 2016) or across multiple breeds (Wiles et al. 2017).

A UK Kennel Club survey in 2014 collected data on 43,005 dogs across 187 breeds and used these to report the proportion of dogs within individual breeds with at least one disorder (Wiles et al. 2017). These values offered a proxy measure of overall breed health. Review of these univariable results for breeds with at least 200 responses suggests a trend towards higher proportions of dogs with at least one disorder from within common brachycephalic breeds (Table 7.5). The age information provided on these breeds also highlights the French Bulldog, Pug and English Bulldog as substantially younger than many other breeds. Advancing age is a recognised risk factor for many common disorders in dogs (Anderson et al. 2020, Pegram et al. 2019, O'Neill, Corah, et al. 2018, O'Neill, Lee, et al. 2017). This suggests that the proportion of dogs affected in these three younger breeds is confounded downwards by age and that a multivariable analysis taking account of age might correct these proportions upwards to make the trend towards higher disorder counts in these youthful brachycephalic breeds even steeper.

It is also possible to report the prevalence of all common disorders in key individual brachycephalic breeds. To date, the VetCompass Research Team have published breed prevalence results on five key brachycephalic breeds: French Bulldog (O'Neill, Baral, et al. 2018), Pug (O'Neill, Darwent, et al. 2016), English Bulldog (O'Neill, Skipper, et al. 2019), Chihuahua (O'Neill, Packer, et al. 2020) and Cavalier King Charles Spaniel (Summers et al. 2015). Table 7.6 creates a combined list of all disorders that were in the top ten for at least one of these breeds and shows the prevalence and rank

TABLE 7.4

Disorders Commonly Considered as Predisposed in Brachycephalic Dog Breeds, Along with Supporting Evidence

Disorder	Some Breeds Affected	Evidence
Brachycephalic obstructive airway syndrome (BOAS)	Pug, English Bulldog, French bulldog	(Liu et al. 2016, 2015, 2017, Fasanella et al. 2010)
Corneal exposure	Pekingese, Griffon Bruxellois, Pugs	(Ali and Mostafa 2019, Pe'er, Oron, and Ofri 2020)
Corneal ulceration	Pug, Boxer, Pekingese, Shih-Tzu, English Bulldog, Cavalier King Charles Spaniel, Lhasa Apso, French Bulldog, Chihuahua	(O'Neill, Lee, et al. 2017, Maini et al. 2019, OFA 2020)
Demodicosis	Pug, English Bulldog, French Bulldog, Dogue de Bordeaux, Boxer	(O'Neill, Turgoose, et al. 2020)
Dystocia	English Bulldog, French Bulldog, Boston Terrier, Pug, Chihuahua	(Bergstrom et al. 2006, Adams et al. 2010, O'Neill, O'Sullivan, et al. 2019, 2017, Dobak et al. 2018)
Elbow dysplasia	Pug, English Bulldog, French bulldog, Chihuahua	(OFA 2020, BVA 2020a, Sanchez Villamil et al. 2020)
Entropion and ectropion	Neopolitan Mastiff, English Bulldog, Shih-Tzu	(OFA 2020, Christmas 1992, Maini et al. 2019, Krecny et al. 2015)
Exaggerated neck thickness	Pug, English Bulldog, French bulldog	(Packer et al. 2015, Liu et al. 2017)
Hemivertebrae/kyphosis/ scoliosis	Pug, English Bulldog, French bulldog	(Ryan et al. 2017, Ryan et al. 2019, Bertram, ter Haar, and De Decker 2019, Wyatt et al. 2019)
Hip dysplasia	Dogue de Bordeaux, American bulldog, Pug, English Bulldog, French bulldog	(BVA 2020b)
Inverted or screw tails	Pug, English Bulldog, French bulldog, Boston Terrier	(Vasiadou and Papazoglou 2016)
Obesity	Pug, Shih Tzu	(O'Neill, Darwent, et al. 2016, Raffan et al. 2016, Courcier et al. 2010)
Patellar luxation	Chihuahua, American Bulldog, French Bulldog, Pug, Lhasa Apso, Cavalier King Charles Spaniel	(O'Neill, Meeson, et al. 2016, Nilsson, Zanders, and Malm 2018, Boge et al. 2019, OFA 2020)
Prolapse of the gland of the nictitating membrane	English Bulldog, French Bulldog	(Multari et al. 2015, Mazzucchelli et al. 2012)
Skin fold dermatitis	Pug, English Bulldog, French bulldog, Pekingese.	(Packer et al. 2019, Scott et al. 2001, Paterson 2017)
Stenotic nares	Pug, English Bulldog, French bulldog	(Wykes 1991, Hostnik et al. 2017, Liu et al. 2016, Fernández-Parra et al. 2019)

for any disorders that appeared in the 40 most common disorders overall. These data highlight some interesting messages. First, many of these common disorders in these breeds are not those that we typically associate with brachycephaly but instead are disorders that are typical of dogs in general (O'Neill et al. 2014a). This suggests that efforts to improve the welfare of brachyce-phalic breeds should extend beyond focusing just on brachycephalic predispositions and should also consider common disorders in dogs in general such as otitis externa, periodontal disease and anal sac disorders. Second, it is clear that the ranking of disorders by prevalence varies widely across the five brachycephalic breeds; many disorders highly ranked for one brachycephalic breed do not even appear within the top 40 for another. So, although each of these breeds may share a brachycephalic

TABLE 7.5

The Number, Proportion and Median Age of Dogs from Common Brachycephalic and Some Other Breeds with at Least One Disorder Reported in a UK Survey on Dogs Registered with The Kennel Club (Wiles et al. 2017)

Breed	Skull-Shape Category	No. Dogs	No. Dogs with at Least One Disorder	% dogs Affected	Median Age (years)
Boxer	Brachycephalic	724	391	54.01	4.85
Cavalier King Charles Spaniel	Brachycephalic	1244	607	48.79	5.25
English Bulldog	Brachycephalic	370	178	48.11	2.73
Shih-Tzu	Brachycephalic	351	139	39.60	3.92
Pug	Brachycephalic	555	216	38.92	2.67
French Bulldog	Brachycephalic	330	122	36.97	1.81
Lhasa Apso	Brachycephalic	470	121	25.74	3.57
German Shepherd Dog	Dolichocephalic	1410	540	38.30	4.47
Dachshund (miniature smooth-haired)	Dolichocephalic	296	98	33.11	3.42
Whippet	Dolichocephalic	707	147	20.79	4.29
Staffordshire Bull Terrier	Mesocephalic	797	295	37.01	4.76
Labrador Retriever	Mesocephalic	6938	2388	34.42	4.73
Cocker Spaniel	Mesocephalic	3723	977	26.24	3.99

conformation, disorder profiles show differing and unique signatures for each breed. The message here is that efforts to improve the health of brachycephalic dogs should be focused at the breed level (at the very least, and may need to be focused even more granularly than this) rather than on all brachycephalic dogs as a group. This is the approach currently taken by The Kennel Club's 'Breed Health and Conservation Plan' programme (The Kennel Club 2020a). Third, some disorders that are well-recognised as brachycephalic predispositions are both highly ranked and show worrying high prevalence values in some of these breeds. These disorders should be considered therefore as important candidates for breed-based health reforms. Examples from Table 7.6 include skin fold dermatitis in the English Bulldog, French Bulldog and Pug, prolapsed nictitating membrane in the English Bulldog and French Bulldog, and patellar luxation in the Chihuahua.

BRACHYCEPHALIC DEMOGRAPHY AND POPULARITY

It may seem amazing but despite extensive efforts expended on researching (Royal Veterinary College 2020, University of Cambridge 2020) and reforming (The Kennel Club 2020a) brachyce- phalic health in companion animals, there is still no universally agreed categorisation system of dogs breeds by skull shape. A recent VetCompass study aimed to redress this deficit by publishing a categorisation system that classifies UK dog breeds by skull shape: brachycephalic, mesocephalic, dolichocephalic and crossbreds (O'Neill et al. 2020). Based on this VetCompass system, the study reported that the general UK dog population comprised of 18.74% brachycephalic, 46.48% mesoce- phalic, 7.84% dolichocephalic and 26.94% crossbred types. With almost one-fifth of the overall UK dog population as brachycephalic, it is clear that welfare issues associated with this conformation will have a major welfare impact on dogs as a whole.

The VetCompass study identified that the brachycephalic group spanned 34 individual breeds (O'Neill et al. 2020). However, a relatively small subgroup of eight breeds accounted for over 90% of these dogs: Chihuahua (22.91%), Shih-Tzu (19.07%), Cavalier King Charles Spaniel (10.43%), Pug (9.91%), French Bulldog (9.55%), Lhasa Apso (7.46%), Boxer (5.88%) and English Bulldog (5.01%). This nugget of epidemiological information alone suggests that overall welfare impacts

TABLE 7.6

Prevalence of Disorders Featuring among the 40 Most Common Disorders for English Bulldog (O'Neill, Skipper, et al. 2019), Cavalier King Charles Spaniel (*CKCS) (Summers et al. 2015), Chihuahua (O'Neill, Packer, et al. 2020), French Bulldog (O'Neill, Baral et al. 2018) and Pug (O'Neill, Darwent et al. 2016)

Disorder	English Bulldog		CKCS*		Chihuahua		French Bulldog		Pug	
	Rank	%	Rank	%	Rank	%	Rank	%	Rank	%
Otitis externa	1	12.7	4	9.2	14	1.3	1	14.0	3	7.5
Pyoderma	2	8.8	26	2.1	30	0.5	8	2.7	24	1.7
Obesity	3	8.7			2	5.9			1	13.2
Skin fold dermatitis	4	7.8					5	3.0	12	3.2
Overgrown nail(s)	5	7.3			7	3.3	4	3.1	7	5.6
Prolapsed nictitating membrane	6	6.8					9	2.6		
Cryptorchidism	7	5.6			10	1.9	40	1.1	28	1.3
Conjunctivitis	8	5.4	5	7.5	20	0.8	3	3.2	21	1.9
Pododermatitis	9	5.4					10	2.5		
Alopecia	10	5.3			25	0.6	25	1.8	20	2.0
Diarrhoea	11	4.9	2	11.0	13	1.3	2	7.5	10	3.8
Vomiting	15	3.6			17	1.1	15	2.2	9	5.0
Brachycephalic airway obstruction syndrome (BOAS)	16	3.5					11	2.4	8	5.2
Aggression	24	2.4			5	4.2	13	2.3		
Periodontal disease	26	2.1	9	5.6	1	13.5	33	1.3	6	6.1
Upper respiratory tract infection	29	1.5			23	0.6	7	2.7		
Heart murmur	31	1.4	1	30.9	9	2.0	14	2.2		
Dental disease			3	9.5	24	0.6				
Anal sac infection			6	7.4	11	1.8				
Heart disease			7	7.3						
Corneal disorder			8	6.5			38	1.1	2	8.7
Mitral valve disorder			10	5.0						
Flea infestation			12	3.7	8	2.9			33	1.2
Anal sac impaction			13	3.6	4	4.9	6	2.9	5	6.5
Patellar luxation			17	3.3	6	4.0	18	2.1	34	1.2
Retained deciduous tooth					3	5.7			13	3.1
Ear disorder									4	7.4

from brachycephaly are driven by the unique health issues of a small subset of popular breeds and therefore highlights the value of focusing health reform efforts on key breeds (The Kennel Club 2020a).

This paper also highlighted some other key differences between brachycephalic and non-brachycephalic dogs with important messages for how we interpret epidemiological studies (O'Neill et al. 2020). Brachycephalic dogs were substantially younger on average than mesocephalic and dolichocephalic types (3.31 vs. 5.33 and 5.07 years, respectively, $P < 0.001$). Brachycephalic dogs were also substantially lighter on average than mesocephalic and dolichocephalic types (8.75 vs. 16.98 and 25.80 kg, respectively, $P < 0.001$). Given that age and bodyweight are commonly reported risk factors across a range of common disorders, the paper explored the impact of confounding by comparing results between univariable and multivariable analyses for 30 common disorders. The inference from univariable and multivariable analyses on disorder predisposition differed between brachycephalic and non-brachycephalic types for 11/30 (30.67%) disorders. These findings strongly reinforce the warning discussed above that older studies using only univariable analyses should be interpreted with caution due to high risks of confounding.

In line with our innate human tendency to always 'improve', humans soon began to adapt dogs after domestication for specific functions such as guarding, herding or pulling (Arman 2007). This resulted in varying conformations being selected across the 'types' of dogs that met these functions. With the advent of dog showing as a popular pastime in Victorian Britain, these 'types' were standardised to create uniformity within the types and diversity between the types. Breed standards were published to reduce acrimony over perceived injustices at dog shows and the concept of the modern dog breed was invented (Worboys, Strange, and Pemberton 2018).

Especially over the past century, dogs have moved from earlier working roles to fulfil companion or 'pet' roles instead. Consequently, aesthetic and social appeal now generally trumps working ability for owners deciding on which breed to own (Ghirlanda, Acerbi, and Herzog 2014, Ghirlanda et al. 2013). Given the subjectivity of aesthetic and social appeals, it is hardly surprising that breed popularity is therefore subject to large and often apparently whimsical fluctuations. In the UK, The Kennel Club have kept detailed records of breed registrations for over a century and, very generously, publish these data openly (The Kennel Club 2020b). A review of all registrations from 1908 to 2018 confirms that relative popularity of breeds has always varied widely across all breed; this phenomenon is not restricted to brachycephalic breeds.

Figure 7.1 highlights huge popularity of the Pekingese a century ago followed by a progressive decline to low current levels. The English Bulldog was similarly popular a century ago before

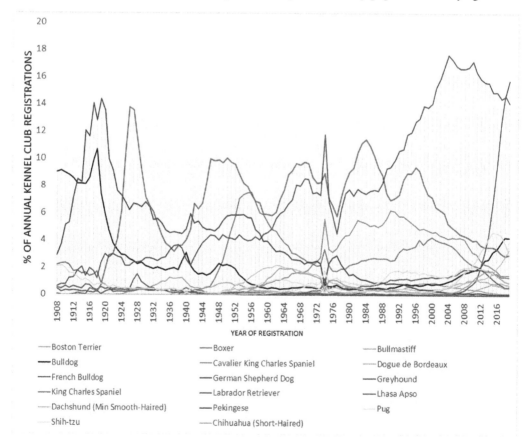

FIGURE 7.1   Proportional annual registrations with the UK Kennel Club (2008–2018) for common brachycephalic dog breeds as well as some common mesocephalic and dolichocephalic breeds (The Kennel Club 2020b).

also falling out of favour, but has shown a resurgence in popularity since the turn of the new century. Conversely, French Bulldogs were rare registrations until 15 years ago, but have since risen phenomenally to become one of the two most commonly registered breeds. Pug popularity has waxed and waned to lesser degrees over the past century before also rising substantially over the past two decades. However, similarly dramatic changing popularity has also affected many non-brachycephalic breeds. The German Shepherd Dog shows five peaks over the past century but has been in steep decline for the past 20 years. And even the Labrador Retriever that has been hugely popular in the UK for the past 30 years is now showing evidence of declining favour with UK owners. The conclusion here is that breeds have always come and gone in popularity. Even the current remarkable rising popularity of certain brachycephalic breeds is not unprecedented and is unlikely to be permanent; ownership preferences, and humans in general, have always been fickle.

About a third of UK dogs are estimated to be registered with The Kennel Club (The Kennel Club 2020b). For a more holistic view of UK dog ownership, data on the wider UK dog population are needed. The VetCompass Programme collects anonymised clinical record information on over eight million dogs from first-opinion veterinary practices in the UK (VetCompass 2020). Given that the majority of UK dogs are registered for veterinary care, VetCompass therefore offers a useful resource on general dog ownership in the UK. Evidence from published VetCompass studies on the Chihuahua, Pug, English Bulldog and French Bulldog shows similar patterns of recent rising popularity to The Kennel Club data, with the French Bulldog showing the steepest incline (Figure 7.2) (O'Neill, Darwent, et al. 2016, O'Neill, Baral, et al. 2018, O'Neill, Skipper, et al. 2019, O'Neill, Packer, et al. 2020). Rising popularity for many of these common brachycephalic breeds can at least partially explain the younger average age for brachycephalic dogs described above. High demand

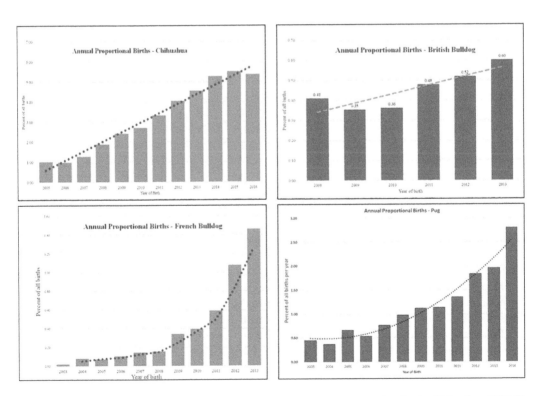

FIGURE 7.2 Proportional annual births within the VetCompass dog population for the Chihuahua (O'Neill, Packer et al. 2020), English Bulldog (O'Neill, Skipper et al. 2019), French Bulldog (O'Neill, Baral et al. 2018) and Pug (O'Neill, Darwent et al. 2016).

means that puppies of these breeds are entering the brachycephalic population at a higher rate than the older dogs are dying and therefore dragging the median population age downwards. We will revisit some impacts from this effect again in the longevity section. But for now, it is important to highlight that there is general agreement across stakeholders that sudden and dramatic increases in popularity often lead to adverse overall health and welfare outcomes for the breeds concerned and therefore efforts should be made to smooth out changes in population structures to dog breeds (BWG 2020). The UK Brachycephalic Working Group considers welfare issues associated with sudden population growth to include breed-related health issues, health deterioration as median ages of populations rise, improper sourcing of puppies (puppy farms and importation), unsuitable ownership profiles and rehoming/abandonment, and decreased genetic diversity. For these reasons, deeper understanding of breed population dynamics is more than just an academic exercise; it is a core criterion to help understand breed welfare at a population level.

LONGEVITY AND MORTALITY

Reliable and representative longevity (lifespan) and mortality (causes of death) statistics can provide useful insights into health and welfare variation of, and between, breeds and types of companion animals. However, safe interpretation of published statistics requires full consideration of potential biases that may be intrinsic to the data sources and methods used in these studies. The fundamental concept behind longevity studies is that average longevities for populations of dogs (e.g. a breed) are likely to be highly consistent despite the high uncertainty that exists on the projected lifespan of any individual dog. On this basis, longevity comparisons between breeds or types of dogs should offer a useful proxy measure of overall health and well-being, assuming that longer life is evidence of better health and welfare. Similarly, comparing mortality statistics (i.e. comparing the most common causes of death between breeds) should offer some special insights into the key life-limiting disorders of these breeds.

But could it really be as easy as this? In reality, safe comparison and interpretation of longevity and mortality is a highly challenging science and holds many caveats that are often ignored or not even recognised (Day and Reynolds 2018). Understanding the data source is critical when reviewing longevity and mortality statistics. The perfect study would hold full data on all animals of that type from the wider population over time from the moment of their births. However, this perfect data resource does not exist, and each current data resource holds its own unique selection biases. First-opinion veterinary clinical data exclude animals that are not under general veterinary care, while referral veterinary data are even more heavily biased towards the subset of first-opinion animals that are sick enough to warrant specialist veterinary care (O'Neill et al. 2014b). Data sourced from kennel club records are limited to pedigreed and registered individuals from recognised pure breeds (Lewis et al. 2018). Insurance databases are restricted to animals that are insured; insurance death claims are often capped at certain age limits and are restricted to certain policy types (Egenvall et al. 2009). However, if we bear these limitations in mind and compare results within data sources rather than trying to compare across sources, some useful comparative conclusions could still be reached, with the caveat that some other insidious biases still exist.

The perfect longevity study would apply a prospective cohort study design with many thousands of animals all followed from birth to death (Dohoo, Martin, and Stryhn 2009). There are many longstanding examples of cohort studies for humans that have cost millions of pounds (UKRI – UK Research and Innovation 2020). However, there are fewer examples of such cohort designs in companion animals (GenPup 2020, Dogslife 2020, Morris Animal Foundation 2020), and longevity results from these have been limited by relatively few deaths in the cohorts so far, as well as selection biases introduced from the high levels of drop-outs from the studies (Murray, Kinsman, et al. 2021, Pugh, Bronsvoort, et al. 2015, Ruple, Jones, et al. 2020). Consequently, currently published longevity studies for companion animals have mainly relied on retrospective cohort or cross-sectional study designs that are subject to biases including right censored data and changing popularity effects. Right censoring refers to cohort studies that do not follow all animals to death; data analysis before this point excludes these later deaths that would tend to have been older on average, and therefore, the longevity and mortality results will be biased towards a younger subset at death (Urfer 2011).

Changing popularity over time is another serious challenge to safe inference from current longevity studies. Extrapolation of results from the population under study (e.g. French Bulldogs in the UK) to the breed itself (e.g. French Bulldogs in general) assumes a steady state of breed popularity (i.e. neither rising nor falling) but, as Figures 7.1 and 7.2 show, this assumption is rarely true. If a breed comprises a constant relative proportion of all puppies born annually over a prolonged period (i.e. steady state), then the average ages at death for that breed across a short period could offer a reliable indicator of average breed longevity over a longer period of time. However, rising and falling breed popularity results in differing relative proportions of puppies entering the overall population each year and therefore differing age patterns of dogs available to die each year. For example, Figure 7.2 shows the popularity of French Bulldogs in the UK has increased sharply in recent years, rising from under 0.1% of all puppies born before 2005 to comprise over 1.0% of all puppies born after 2012 (O'Neill, Baral, et al. 2018). Given that the majority of French Bulldogs in 2013 are therefore likely to be young, it follows that younger dogs have increased opportunity to contribute death events during this period and therefore longevity results will be biased downwards. A converse effect applies for breeds with declining popularity such as German Shepherd Dogs (O'Neill, Coulson, et al. 2017) or West Highland White Terriers (O'Neill, Ballantyne, et al. 2019). The conclusion here is that the shrewd reader should interpret the longevity and mortality information provided in this chapter as describing the specific population (including its location and time period) under study but should avoid the temptation to rush into generalising these findings to all animals of that type.

Longevity results on some common brachycephalic dog breeds from four published studies are summarised in Table 7.7. Notwithstanding the caveats discussed above about over-interpreting these data, these results reinforce the recurring theme from this chapter that brachycephalic issues are better explored at a breed level than at an overall brachycephalic level. Longevity varies widely across brachycephalic breeds; breeds including the English Bulldog and the Dogue de Bordeaux show consistently short longevity, while other breeds including the Lhasa Apso and Shih Tzu show consistently longer lives. Common mesocephalic and dolichocephalic breeds appear to have a more consistent and longer lifespan.

A fuller exploration of mortality to report and compare the specific causes of death in brachycephalic breeds is outside the scope of this chapter. However, the curious reader can find useful information on mortality in dogs from the references at the end of this chapter (O'Neill et al. 2013, Fleming, Creevy, and Promislow 2011, Proschowsky, Rugbjerg, and Ersbøll 2003, Adams et al. 2010, Bonnett et al. 2005, O'Neill, Church, et al. 2015, Lewis et al. 2018, O' Neill, Darwent, et al. 2016, O'Neill, Baral, et al. 2018, O'Neill, Skipper, et al. 2019, Egenvall, Bonnett, Olson, et al. 2000, Egenvall, Bonnett, Shoukri, et al. 2000, Egenvall et al. 2005).

A final thought to consider; at the start of this section, I stated that longevity could offer a useful proxy metric to evaluate breed welfare and health *assuming that longer life is evidence of better health and welfare*. However, given that 86.4% of deaths in dogs involve euthanasia, there is merit in challenging the view that longevity and welfare are necessarily linearly related (O'Neill et al. 2013). Long life *de facto* may not equate to high animal welfare (Yeates 2010); we should consider the quality of the life lived as well as the quantity. In human medicine, attention is now moving to the concept of *healthspan* (healthy longevity) which describes the proportion of overall lifespan with a good quality of life (Waters 2011). Work is underway in companion animals to evaluate quality of life; this information could augment longevity information and help us to understand breed health more holistically (Mellor et al. 2020, Summers et al. 2019, Teng 2018, Teng et al. 2018).

TABLE 7.7

Summary Longevity Results for Common Brachycephalic Dog Breeds and Some Other Breeds Based on Agria Insurance Records in Sweden (Egenvall, Bonnett, Shoukri et al. 2000), Owner Surveys on The Kennel Club Registered Dogs in the UK (Lewis et al. 2018, Adams et al. 2010) and VetCompass primary care Veterinary Clinical Records in the UK (O'Neill et al. 2013)

Breed	Skull shape	Median Longevity (years)			
		Sweden – Agria insured (Egenvall, Bonnett, Shoukri, et al. 2000)	UK – The Kennel Club registered (Adams et al. 2010)	UK – VetCompass primary care veterinary (O'Neill et al. 2013)	UK – The Kennel Club registered (Lewis et al. 2018)
Boxer	Brachycephalic	7.5	10.3	10.0	8.8
English Bulldog	Brachycephalic		6.3	8.4	
Cavalier King Charles Spaniel	Brachycephalic	7.5	14.6	9.9	9.8
Chihuahua	Brachycephalic		12.4	7.1	
Dogue de Bordeaux	Brachycephalic		3.8	5.5	
King Charles Spaniel	Brachycephalic		10.0	12.0	
Lhasa Apso	Brachycephalic		14.3	13.0	
Mastiff	Brachycephalic		6.8	7.1	
Shih-Tzu	Brachycephalic		13.2	13.3	
Dachshund	Dolichocephalic		12.7		
German Shepherd Dog	Dolichocephalic	5.0		11.0	9.5
Greyhound	Dolichocephalic		9.1	10.8	
Miniature Dachshund	Dolichocephalic			13.5	
Whippet	Dolichocephalic				9.8
Cocker Spaniel	Mesocephalic		11.2	11.5	10.8
Labrador Retriever	Mesocephalic		12.3	12.5	11.5
Staffordshire Bull Terrier	Mesocephalic		12.8	10.7	10.8
Crossbreed		3.7		13.1	
All study dogs			11.3	12.0	

CONCLUSIONS

Epidemiology is the most powerful method yet devised to understand health issues of companion animals at a population level. However, great power brings great responsibility to execute and interpret these studies correctly. This chapter introduced the reader to some core epidemiological principles and caveats to start your epidemiological journey into understanding brachycephalic health. The chapter has, for the first time, merged and summarised relevant findings from many studies on prevalence, predisposition, demography and longevity in brachycephalic dogs. These results suggest that welfare reforms should be focused at the breed level. A recurring theme in this chapter is the complexity of issues behind the current brachycephalic crisis. Predisposed disorders in brachycephalic breeds are considered by the author as endemic. The impact from these health problems interacts with a demographic background of spectacularly changing popularity for some key brachycephalic dog breeds driven by human social effects. This tripartic interaction of health issues, breed population growth and social mores has created a 'perfect storm' leading to the current brachycephalic crisis. Resolving this crisis will require concerted efforts to combat the injurious effects from each of these three phenomena.

I hope that you have enjoyed this foray into the epidemiology of brachycephalic companion animals and are now fortified with a sufficiently robust epidemiological armoury to tackle these issues at a population level.

REFERENCES

Adams, V. J., K. M. Evans, J. Sampson, and J. L. N. Wood. 2010. "Methods and mortality results of a health survey of purebred dogs in the UK." *Journal of Small Animal Practice* 51 (10):512–524. doi:10.1111/j.1748–5827.2010.00974.x.

Ali, K.M., and A. A. Mostafa. 2019. "Clinical findings of traumatic proptosis in small-breed dogs and complications associated with globe replacement surgery." *Open Veterinary Journal* 9 (3):222–229.

Anderson, K. L., H. Zulch, D.G. O'Neill, R.L. Meeson, and L.M. Collins. 2020. "Risk factors for canine osteoarthritis and its predisposing arthropathies: A systematic review." *Frontiers in Veterinary Science* 7 (220). doi: 10.3389/fvets.2020.00220.

Arman, K. 2007. "A new direction for kennel club regulations and breed standards." *The Canadian Veterinary Journal [La Revue Veterinaire Canadienne]* 49:953–965.

Bateson, P. 2010. "Independent inquiry into dog breeding." University of Cambridge, accessed November 2. https://dogwellnet.com/files/file/308-independent-inquiry-into-dog-breeding-2010-patrick-bateson/.

Bergstrom, A., A. Nødtvedt, A. S. Lagerstedt, and A. Egenvall. 2006. "Incidence and breed predilection for dystocia and risk factors for cesarean section in a Swedish population of insured dogs." *Veterinary Surgery* 35 (8):786–791.

Bertram, S., G. ter Haar, and S. De Decker. 2019. "Congenital malformations of the lumbosacral vertebral column are common in neurologically normal French Bulldogs, English Bulldogs, and Pugs, with breed-specific differences." *Veterinary Radiology & Ultrasound* 60 (4):400–408. doi:10.1111/vru.12753.

Boge, G. S., E. R. Moldal, M. Dimopoulou, E. Skjerve, and A. Bergström. 2019. "Breed susceptibility for common surgically treated orthopaedic diseases in 12 dog breeds." *Acta Veterinaria Scandinavica* 61 (1):19. doi:10.1186/s13028-019-0454-4.

Bonnett, B. N., A. Egenvall, Å. Hedhammar, and P. Olson. 2005. "Mortality in over 350,000 insured Swedish dogs from 1995–2000: I. breed-, gender-, age- and cause-specific rates." *Acta Veterinaria Scandinavica* 46 (3):105–120.

Bresalier, M., and M. Worboys. 2014. "'Saving the lives of our dogs': the development of canine distemper vaccine in interwar Britain." *The British Journal for the History of Science* 47 (2):305–334. doi:10.1017/S0007087413000344.

BVA. 2020a. "Elbow Dysplasia Scheme for dogs." British Veterinary Association, accessed August 19. https://www.bva.co.uk/canine-health-schemes/elbow-scheme/.

BVA. 2020b. "Hip Dysplasia Scheme for Dogs." British Veterinary Association, accessed August 19. https://www.bva.co.uk/canine-health-schemes/hip-scheme/.

BWG. 2020. "The Brachycephalic Working Group." The Brachycephalic Working Group, accessed July 4. http://www.ukbwg.org.uk/.

Christmas, R. E. 1992. "Common ocular problems of Shih Tzu dogs." *The Canadian Veterinary Journal [La Revue Veterinaire Canadienne]* 33 (6):390–393.

Collins, L. M., L. Asher, J. F. Summers, G. Diesel, and P. D. McGreevy. 2010. "Welfare epidemiology as a tool to assess the welfare impact of inherited defects on the pedigree dog population." *Animal Welfare* 19:67–75.

Courcier, E. A., R. M. Thomson, D. J. Mellor, and P. S. Yam. 2010. "An epidemiological study of environmental factors associated with canine obesity." *Journal of Small Animal Practice* 51 (7):362–367. doi:10.1111/j.1748–5827.2010.00933.x.

Day, S. M. and R. J. Reynolds. 2018. "Survival, mortality, and life expectancy." In *Cerebral Palsy-Clinical and Therapeutic Aspects*, edited by I.J. Al-Zwaini, 45–64. IntechOpen.

Dobak, T. P., G. Voorhout, J. C. M. Vernooij, and S. A. E. B. Boroffka. 2018. "Computed tomographic pelvimetry in English bulldogs." *Theriogenology* 118:144–149 doi:10.1016/j.theriogenology.2018.05.025.

Dogslife. 2020. "Dogslife." Dogslife, accessed October 19. https://www.dogslife.ac.uk/.

DogWellNet. 2020. "International Working Group on Extremes of Conformation in Dogs (IWGECD)." IPFD DogWellNet, accessed July 25. https://dogwellnet.com/content/international-actions/extremes-of-conformation-brachycephalics/international-working-group-on-extremes-of-conformation-in-dogs-iwgecd-r695/.

Dohoo, I., W. Martin, and H. Stryhn. 2009. *Veterinary Epidemiologic Research*. 2nd ed. Charlottetown, Canada: VER Inc.

Egenvall, A., B. N. Bonnett, A. Hedhammar, and P. Olson. 2005. "Mortality in over 350,000 insured Swedish dogs from 1995–2000: II. breed-specific age and survival patterns and relative risk for causes of death." *Acta Veterinaria Scandinavica* 46 (3):121–36.

Egenvall, A., B. N. Bonnett, P. Olson, and Å. Hedhammar. 2000. "Gender, age, breed and distribution of morbidity and mortality in insured dogs in Sweden during 1995 and 1996." *Veterinary Record* 146 (8):519–525.

Egenvall, A., A. Nødtvedt, J. Penell, L. Gunnarsson, and B. N. Bonnett. 2009. "Insurance data for research in companion animals: Benefits and limitations." *Acta Veterinaria Scandinavica* 51:42. doi:10.1186/1751-0147-51-42.

Egenvall, A., B. N. Bonnett, M. Shoukri, P. Olson, Å. Hedhammar, and I. Dohoo. 2000. "Age pattern of mortality in eight breeds of insured dogs in Sweden." *Preventive Veterinary Medicine* 46 (1):1–14. doi:10.1016/s0167-5877(00)00135-5.

Fasanella, F. J., J. M. Shivley, J. L. Wardlaw, and S. Givaruangsawat. 2010. "Brachycephalic airway obstructive syndrome in dogs: 90 cases (1991–2008)." *Journal of the American Veterinary Medical Association* 237 (9):1048–1051. doi:10.2460/javma.237.9.1048.

Fernández-Parra, R., P. Pey, L. Zilberstein, and M. Malvè. 2019. "Use of computational fluid dynamics to compare upper airway pressures and airflow resistance in brachycephalic, mesocephalic, and dolichocephalic dogs." *The Veterinary Journal* 253:105392. doi:10.1016/j.tvjl.2019.105392.

Fleming, J. M., K. E. Creevy, and D. E. L. Promislow. 2011. "Mortality in North American dogs from 1984 to 2004: An investigation into age-, size-, and breed-related causes of death." *Journal of Veterinary Internal Medicine* 25 (2):187–198.

Galibert, F., P. Quignon, C. Hitte, and C. André. 2011. "Toward understanding dog evolutionary and domestication history." *Comptes Rendus Biologies* 334 (3):190–196. doi:10.1016/j.crvi.2010.12.011.

GenPup. 2020. "Generation Pup." Dogs Trust, accessed October 19. https://generationpup.ac.uk/.

Ghirlanda, S., A. Acerbi, and H. Herzog. 2014. "Dog movie stars and dog breed popularity: A case study in media influence on choice." *PLoS One* 9 (9):e106565. doi: 10.1371/journal.pone.0106565.

Ghirlanda, S., A. Acerbi, H. Herzog, and J. A. Serpell. 2013. "Fashion vs. Function in cultural evolution: The case of dog breed popularity." *PLoS One* 8 (9):1–6. doi: 10.1371/journal.pone.0074770.

Gough, A., A. Thomas, and D. O'Neill. 2018. *Breed Predispositions to Disease in Dogs and Cats*. 3rd ed. Chichester, West Sussex: Wiley-Blackwell.

Hall, S. J. G., and D. G. Bradley. 1995. "Conserving livestock breed biodiversity." *Trends in Ecology & Evolution* 10 (7):267–270. doi:10.1016/0169-5347(95)90005-5.

Holmes, M., and P. Cockcroft. 2004. "Evidence-based veterinary medicine 1. Why is it important and what skills are needed?" *In Practice* 26 (1):28–33.

Holmes, M. A., and D. W. Ramey. 2007. "An introduction to evidence-based veterinary medicine." *Veterinary Clinics of North America: Equine Practice* 23 (2):191–200. doi: 10.1016/j.cveq.2007.03.001.

Hostnik, E. T., B. A. Scansen, R. Zielinski, and S. N. Ghadiali. 2017. "Quantification of nasal airflow resistance in English bulldogs using computed tomography and computational fluid dynamics." *Veterinary Radiology & Ultrasound* 58 (5):542–551. doi:10.1111/vru.12531.

Huntley, S. J., R. S. Dean, A. Massey, and M. L. Brennan. 2016. "International evidence-based medicine survey of the veterinary profession: Information sources used by veterinarians." *PLoS One* 11 (7):e0159732. doi:10.1371/journal.pone.0159732.

Jeffery, N. 2015. "Liberating the (data) population from subjugation to the 5% (P-value)." *Journal of Small Animal Practice* 56 (8):483–484. doi:10.1111/jsap.12391.

Krecny, M., A. Tichy, J. Rushton, and B. Nell. 2015. "A retrospective survey of ocular abnormalities in pugs: 130 cases." *Journal of Small Animal Practice* 56 (2):96–102. doi:10.1111/jsap.12291.

Langer, G. 2018. "Possible mathematical definitions of the biological term "breed"." *Archives Animal Breeding* 61 (2):229–243. doi:10.5194/aab-61-229-2018.

Lewis, T. W., B. M. Wiles, A. M. Llewellyn-Zaidi, K. M. Evans, and D. G. O'Neill. 2018. "Longevity and mortality in Kennel Club registered dog breeds in the UK in 2014." *Canine Genetics and Epidemiology* 5 (1):10. doi:10.1186/s40575-018-0066-8.

Liu, N.-C., V. J. Adams, L. Kalmar, J. F. Ladlow, and D. R. Sargan. 2016. "Whole-body barometric plethysmography characterizes upper airway obstruction in 3 brachycephalic breeds of dogs." *Journal of Veterinary Internal Medicine* 30 (3):853–865. doi:10.1111/jvim.13933.

Liu, N.-C., D. R. Sargan, V. J. Adams, and J. F. Ladlow. 2015. "Characterisation of brachycephalic obstructive airway syndrome in french bulldogs using whole-body barometric plethysmography." *PLoS One* 10 (6):e0130741. doi:10.1371/journal.pone.0130741.

Liu, N.-C., E. L. Troconis, L. Kalmar, D. J. Price, H. E. Wright, V. J. Adams, D. R. Sargan, and J. F. Ladlow. 2017. "Conformational risk factors of brachycephalic obstructive airway syndrome (BOAS) in pugs, French bulldogs, and bulldogs." *PLoS One* 12 (8):e0181928. doi:10.1371/journal.pone.0181928.

London School of Hygiene and Tropical Medicine. 2020. "MSc Veterinary Epidemiology." LSHTM, accessed October 11. https://www.lshtm.ac.uk/study/courses/masters-degrees/veterinary-epidemiology.

Maini, S., R. Everson, C. Dawson, Y. M. Chang, C. Hartley, and R. F. Sanchez. 2019. "Pigmentary keratitis in pugs in the United Kingdom: Prevalence and associated features." *BMC Veterinary Research* 15 (1):384. doi:10.1186/s12917-019-2127-y.

Mattin, M. J., A. Boswood, D. B. Church, J. López-Alvarez, P. D. McGreevy, D. G. O'Neill, P. C. Thomson, and D. C. Brodbelt. 2015. "Prevalence of and risk factors for degenerative mitral valve disease in dogs attending primary-care veterinary practices in England." *Journal of Veterinary Internal Medicine* 29 (3):847–854. doi:10.1111/jvim.12591.

Mazzucchelli, S., M. D. Vaillant, F. Wéverberg, H. Arnold-Tavernier, N. Honegger, G. Payen, M. Vanore, L. Liscoet, O. Thomas, B. Clerc, and S. Chahory. 2012. "Retrospective study of 155 cases of prolapse of the nictitating membrane gland in dogs." *Veterinary Record* 170 (17):443. doi:10.1136/vr.100587.

McGreevy, P. D., and F. W. Nicholas. 1999. "Some practical solutions to welfare problems in dog breeding." *Animal Welfare* 8:329–341.

Mellor, D. J., N. J. Beausoleil, K. E. Littlewood, A. N. McLean, P. D. McGreevy, B. Jones, and C. Wilkins. 2020. "The 2020 five domains model: Including human–animal interactions in assessments of animal welfare." *Animals* 10 (10):1870.

Miranda, C., and G. Thompson. 2016. "Canine parvovirus: The worldwide occurrence of antigenic variants." *Journal of General Virology* 97 (9):2043–2057. doi:10.1099/jgv.0.000540.

Morris Animal Foundation. 2020. "Golden Retriever Lifetime Study." Morris Animal Foundation, accessed October 19. https://www.morrisanimalfoundation.org/golden-retriever-lifetime-study.

Multari, D., A. Perazzi, B. Contiero, G. De Mattia, and I. Iacopetti. 2015. "Pocket technique or pocket technique combined with modified orbital rim anchorage for the replacement of a prolapsed gland of the third eyelid in dogs: 353 dogs." *Veterinary Ophthalmology* 19:214.

Murray, J.K., Kinsman, R.H., Lord, M.S., Da Costa, R.E.P., Woodward, J.L., Owczarczak-Garstecka, S.C., Tasker, S., Knowles, T.G. and Casey, R.A. 2021. "'Generation Pup' – protocol for a longitudinal study of dog behaviour and health." *BMC Veterinary Research* 17(1): 1. doi:10.1186/s12917-020-02730-8.

Nilsson, K., S. Zanders, and S. Malm. 2018. "Heritability of patellar luxation in the Chihuahua and Bichon Frise breeds of dogs and effectiveness of a Swedish screening programme." *The Veterinary Journal* 234:136–141 doi:10.1016/j.tvjl.2018.01.010.

O'Neill, D. G., A. M. O'Sullivan, E. A. Manson, D. B. Church, A. K. Boag, P. D. McGreevy, and D. C. Brodbelt. 2017. "Canine dystocia in 50 UK first-opinion emergency-care veterinary practices: Prevalence and risk factors." *Veterinary Record* 181 (4). doi:10.1136/vr.104108.

O'Neill, D. G., E. Turgoose, D. B. Church, D. C. Brodbelt, and A. Hendricks. 2020. "Juvenile-onset and adult-onset demodicosis in dogs in the UK: Prevalence and breed associations." *Journal of Small Animal Practice* 61 (1):32–41. doi:10.1111/jsap.13126.

O'Neill, D. G., D. B. Church, P. D. McGreevy, P. C. Thomson, and D. C. Brodbelt. 2013. "Longevity and mortality of owned dogs in England." *The Veterinary Journal* 198 (3):638–643. doi:10.1016/j.tvjl.2013.09.020.

O'Neill, D.G., D. B. Church, P. D. McGreevy, P. C. Thomson, and D. C. Brodbelt. 2014a. "Prevalence of disorders recorded in dogs attending primary-care veterinary practices in England." *PLoS One* 9 (3):1–16. doi:10.1371/journal.pone.0090501.

O'Neill, D., D. Church, P. McGreevy, P. Thomson, and D. Brodbelt. 2014b. "Approaches to canine health surveillance." *Canine Genetics and Epidemiology* 1 (1):2.

O'Neill, D. G., D. B. Church, P. D. McGreevy, P. C. Thomson, and D. C. Brodbelt. 2015. "Longevity and mortality of cats attending primary care veterinary practices in England." *Journal of Feline Medicine and Surgery* 17 (2):125–133. doi:10.1177/1098612x14536176.

O'Neill, D. G., Z. F. Ballantyne, A. Hendricks, D. B. Church, D. C. Brodbelt, and C. Pegram. 2019. "West Highland White Terriers under primary veterinary care in the UK in 2016: demography, mortality and disorders." *Canine Genetics and Epidemiology* 6 (1):7. doi:10.1186/s40575-019-0075-2.

O'Neill, D. G., L. Baral, D. B. Church, D. C. Brodbelt, and R. M. A. Packer. 2018. "Demography and disorders of the French Bulldog population under primary veterinary care in the UK in 2013." *Canine Genetics and Epidemiology* 5 (1):3. doi:10.1186/s40575-018-0057-9.

O'Neill, D. G., C. H. Corah, D. B. Church, D. C. Brodbelt, and L. Rutherford. 2018. "Lipoma in dogs under primary veterinary care in the UK: Prevalence and breed associations." *Canine Genetics and Epidemiology* 5 (1):9. doi:10.1186/s40575-018-0065-9.

O'Neill, D. G., N. R. Coulson, D. B. Church, and D. C. Brodbelt. 2017. "Demography and disorders of German Shepherd Dogs under primary veterinary care in the UK." *Canine Genetics and Epidemiology* 4 (1):7. doi:10.1186/s40575-017-0046-4.

O'Neill, D. G., E. C. Darwent, D. B. Church, and D. C. Brodbelt. 2016. "Demography and health of Pugs under primary veterinary care in England." *Canine Genetics and Epidemiology* 3 (1):1–12. doi:10.1186/s40575-016-0035-z.

O'Neill, D. G., C. Jackson, J. H. Guy, D. B. Church, P. D. McGreevy, P. C. Thomson, and D. Brodbelt. 2015. "Epidemiological associations between brachycephaly and upper respiratory tract disorders in dogs attending veterinary practices in England." *Canine Genetics and Epidemiology* 2 (1):10.

O'Neill, D. G., R. L. Meeson, A. Sheridan, D. B. Church, and D. C. Brodbelt. 2016. "The epidemiology of patellar luxation in dogs attending primary-care veterinary practices in England." *Canine Genetics and Epidemiology* 3 (1):1–12. doi:10.1186/s40575-016-0034-0.

O'Neill, D. G., A. M. O'Sullivan, E. A. Manson, D. B. Church, P. D. McGreevy, A. K. Boag, and D. C. Brodbelt. 2019. "Canine dystocia in 50 UK first-opinion emergency care veterinary practices: Clinical management and outcomes." *Veterinary Record* 184:409. doi:10.1136/vr.104944.

O'Neill, D. G., R. M. A. Packer, M. Lobb, D. B. Church, D. C. Brodbelt, and C. Pegram. 2020. "Demography and commonly recorded clinical conditions of Chihuahuas under primary veterinary care in the UK in 2016." *BMC Veterinary Research* 16 (1):42. doi:10.1186/s12917-020-2258-1.

O'Neill, D. G., A. M. Skipper, J. Kadhim, D. B. Church, D. C. Brodbelt, and R. M. A. Packer. 2019. "Disorders of Bulldogs under primary veterinary care in the UK in 2013." *PLoS One* 14 (6):e0217928. doi:10.1371/journal.pone.0217928.

O'Neill, D. G., M. M. Lee, D. C. Brodbelt, D. B. Church, and R. F. Sanchez. 2017. "Corneal ulcerative disease in dogs under primary veterinary care in England: Epidemiology and clinical management." *Canine Genetics and Epidemiology* 4 (1):5. doi:10.1186/s40575-017-0045-5.

O'Neill, D. G., D. B. Church, P. D. McGreevy, P. C. Thomson, and D. C. Brodbelt. 2013. "Longevity and mortality of owned dogs in England." *The Veterinary Journal* 198:638–643.

O'Neill, D. G., C. Pegram, P. Crocker, D. C. Brodbelt, D. B. Church, and R. M. A. Packer. 2020. "Unravelling the health status of brachycephalic dogs in the UK using multivariable analysis." *Scientific Reports* 10 (1):17251. doi:10.1038/s41598-020-73088-y.

OFA. 2020. "Breed Statistics." Orthopedic Foundation for Animals, accessed August 19. https://www.ofa.org/diseases/breed-statistics.

Packer, R. M. A., A. Hendricks, and C. C. Burn. 2015. "Impact of facial conformation on canine health: Corneal ulceration." *PLoS One* 10 (5):1–16. doi:10.1371/journal.pone.0123827.

Packer, R. M. A., A. Hendricks, M. S. Tivers, and C. C. Burn. 2015. "Impact of facial conformation on canine health: Brachycephalic obstructive airway syndrome." *PLoS One* 10 (10):e0137496. doi:10.1371/journal.pone.0137496.

Packer, R. M. A., D. G. O'Neill, F. Fletcher, and M. J. Farnworth. 2019. "Great expectations, inconvenient truths, and the paradoxes of the dog-owner relationship for owners of brachycephalic dogs." *PLoS One* 14 (7):e0219918. doi:10.1371/journal.pone.0219918.

Packer, R. M. A., D. G. O'Neill, F. Fletcher, and M. J. Farnworth. 2020. "Come for the looks, stay for the personality? A mixed methods investigation of reacquisition and owner recommendation of Bulldogs, French Bulldogs and Pugs." *PLoS One* 15 (8):e0237276. doi:10.1371/journal.pone.0237276.

Paneth, N. 2004. "Assessing the contributions of john snow to epidemiology: 150 years after removal of the broad street pump handle." *Epidemiology* 15 (5):514.

Paterson, S. 2017. "Intertrigo in the dog: Aetiology, clinical signs and therapy." *Companion Animal* 22 (2):72–77. doi:10.12968/coan.2017.22.2.72.

Pe'er, O., L. Oron, and R. Ofri. 2020. "Prognostic indicators and outcome in dogs undergoing temporary tarsorrhaphy following traumatCic proptosis." *Veterinary Ophthalmology* 23 (2):245–251. doi:10.1111/vop.12713.

Pegram, C., D. G. O'Neill, D. B. Church, J. Hall, L. Owen, and D. C. Brodbelt. 2019. "Spaying and urinary incontinence in bitches under UK primary veterinary care: A case–control study. " *Journal of Small Animal Practice* 60 (7):395–403. doi:10.1111/jsap.13014.

Pegram, C., K. Wonham, D. C. Brodbelt, D. B. Church, and D. G. O'Neill. 2020. "Staffordshire Bull Terriers in the UK: Their disorder predispositions and protections." *Canine Medicine and Genetics* 7 (1):13. doi:10.1186/s40575-020-00092-w.

PETscan. 2020. "PETscan." Utrecht University, accessed March 31. https://www.uu.nl/en/organisation/veterinary-service-and-cooperation/patientcare-uvcu/the-companion-animals-genetics-expertise-centre/projects-and-services/petscan.

Pfeiffer, D. U. 2002. *Veterinary Epidemiology - An Introduction*. New York: John Wiley & Sons.

Pfeiffer, D. U. 2010. *Introduction to Veterinary Epidemiology*. Hoboken, NJ: Wiley-Blackwell.

Plitman, L., P. Černá, M. J. Farnworth, R. M. A. Packer, and D. A. Gunn-Moore. 2019. "Motivation of owners to purchase pedigree cats, with specific focus on the acquisition of brachycephalic cats." *Animals* 9 (7):394.

Proschowsky, H. F., H. Rugbjerg, and A. K. Ersbøll. 2003. "Mortality of purebred and mixed-breed dogs in Denmark." *Preventive Veterinary Medicine* 58 (1–2):63–74. doi:10.1016/s0167-5877(03)00010-2.

Pugh, C.A., Bronsvoort, B.M.d.C., Handel, I.G., Summers, K.M. and Clements, D.N. 2015. "Dogslife: a cohort study of Labrador retrievers in the UK." *Preventive Veterinary Medicine* 122(4): 426–435. doi:10.1016/j.prevetmed.2015.06.020.

Raffan, E., R. J. Dennis, C. J. O'Donovan, J. M. Becker, R. A. Scott, S. P. Smith, D. J. Withers, C. J. Wood, E. Conci, D. N. Clements, K. M. Summers, A. J. German, C. S. Mellersh, M. L. Arendt, V. P. Iyemere, E. Withers, J. Söder, S. Wernersson, G. Andersson, K. Lindblad-Toh, G. S. H. Yeo, and S. O'Rahilly. 2016. "A deletion in the canine pomc gene is associated with weight and appetite in obesity-prone labrador retriever dogs." *Cell Metabolism* 23 (5):893–900. doi:10.1016/j.cmet.2016.04.012.

Royal Veterinary College. 2020. "Brachycephalic Research Team." Royal Veterinary College, accessed February 27. https://www.rvc.ac.uk/research/focus/brachycephaly.

Ruple, A., Jones, M., Simpson, M. and Page, R. 2020. "The Golden Retriever Lifetime Study: Assessing factors associated with owner compliance after the first year of enrollment." *Journal of Veterinary Internal Medicine*. doi:10.1111/jvim.15921.

Ryan, R., R. Gutierrez-Quintana, G. ter Haar, and S. De Decker. 2017. "Prevalence of thoracic vertebral malformations in French Bulldogs, Pugs and English Bulldogs with and without associated neurological deficits." *The Veterinary Journal* 221:25–29 doi:10.1016/j.tvjl.2017.01.018.

Ryan, R., R. Gutierrez-Quintana, G. ter Haar, and S. De Decker. 2019. "Relationship between breed, hemivertebra subtype, and kyphosis in apparently neurologically normal French BulCldogs, English Bulldogs, and Pugs." *American Journal of Veterinary Research* 80 (2):189–194. doi:10.2460/ajvr.80.2.189.

Sanchez Villamil, C., A. S. J. Phillips, C. L. Pegram, D. G. O'Neill, and R. L. Meeson. 2020. "Impact of breed on canine humeral condylar fracture configuration, surgical management, and outcome." *Veterinary Surgery* 49 (4):639–647. doi:10.1111/vsu.13432.

Scott, D. W., W. H. Miller, C. E. Griffin, and G. H. Muller. 2001. *Muller & Kirk's Small Animal Dermatology*. 6th ed. Philadelphia, PA: Saunders.

Skipper, A. 2019. "The 'Dog Doctors' of Edwardian London: Elite canine veterinary care in the early twentieth century." *Social History of Medicine*. doi:10.1093/shm/hkz049.

Summers, J. F., D. G. O'Neill, D. Church, L. Collins, D. Sargan, and D. C. Brodbelt. 2019. "Health-related welfare prioritisation of canine disorders using electronic health records in primary care practice in the UK." *BMC Veterinary Research* 15 (1):163. doi:10.1186/s12917-019-1902-0.

Summers, J., D. O'Neill, D. Church, P. Thomson, P. McGreevy, and D. Brodbelt. 2015. "Prevalence of disorders recorded in Cavalier King Charles Spaniels attending primary-care veterinary practices in England." *Canine Genetics and Epidemiology* 2 (1):4.

Teng, K. T.-Y. 2018. "Epidemiology of Disorders Compromising the Welfare of Dogs and Cats." Sydney School of Veterinary Science, Faculty of Science, The University of Sydney.

Teng, K. T.-Y., B. Devleesschauwer, C. M. De Noordhout, P. Bennett, P. D. McGreevy, P.-Y. Chiu, J.-A. L. M. L. Toribio, and N. K. Dhand. 2018. "Welfare-Adjusted Life Years (WALY): A novel metric of animal welfare that combines the impacts of impaired welfare and abbreviated lifespan." *PLoS One* 13 (9):e0202580. doi:10.1371/journal.pone.0202580.

The Kennel Club. 2020a. "Breed health and conservation plans (BHCPs)." The Kennel Club Limited, accessed November 2. https://www.thekennelclub.org.uk/health/breed-health-and-conservation-plans/.

The Kennel Club. 2020b. "Breed registration statistics." The Kennel Club Limited, accessed October 17. https://www.thekennelclub.org.uk/media-centre/breed-registration-statistics/.

UKRI - UK Research and Innovation. 2020. "Cohort Directory." Medical Research Council, accessed October 19. https://mrc.ukri.org/research/facilities-and-resources-for-researchers/cohort-directory/.

University of Cambridge. 2020. "Cambridge BOAS Research Group". University of Cambridge, accessed January 3. https://www.vet.cam.ac.uk/boas.

Urfer, S. R. 2011. "Bias in canine lifespan estimates through right censored data." *Journal of Small Animal Practice* 52 (10):555. doi:10.1111/j.1748-5827.2011.01125.x.

Vasiadou, C., and L. G. Papazoglou. 2016. "Surgical management of screw tail and tail fold pyoderma in dogs." *Journal of the Hellenic Veterinary Medical Society* 67 (4):205–210.

VetCompass. 2020. "VetCompass Programme." RVC Electronic Media Unit, accessed February 11. http://www.rvc.ac.uk/VetCOMPASS/.

VetCompass Australia. 2020. "VetCompass Australia." The University of Sydney, accessed March 17. http://www.vetcompass.com.au/.

Waters, D. J. 2011. "Aging research 2011: Exploring the pet dog paradigm." *Institute of Laboratory Animal Research Journal* 52 (1):97–105.

Wiles, B. M., A. M. Llewellyn-Zaidi, K. M. Evans, D. G. O'Neill, and T. W. Lewis. 2017. "Large-scale survey to estimate the prevalence of disorders for 192 Kennel Club registered bCreeds." *Canine Genetics and Epidemiology* 4 (1):8. doi:10.1186/s40575-017-0047-3.

Worboys, M., Strange, J.-M. and Pemberton, N. 2018. *The invention of the modern dog : breed and blood in Victorian Britain.* 1st ed. Baltimore, Maryland, USA: Johns Hopkins University Press.

Wyatt, S. E., P. Lafuente, G. Ter Haar, R. M. A. Packer, H. Smith, and S. De Decker. 2019. "Gait analysis in French bulldogs with and without vertebral kyphosis." *The Veterinary Journal* 244:45–50 doi:10.1016/j.tvjl.2018.12.008.

Wykes, P. M. 1991. "Brachycephalic airway obstructive syndrome." *Problems in medicine* 3 (2):188.

Yeates, J. W. 2010. "Death is a welfare issue." *Journal of Agricultural and Environmental Ethics* 23 (3):229–241. doi:10.1007/s10806-009-9199-9.

8 The Genetics of Brachycephaly, Population Genetics and Current Health Testing for Brachycephalic Breeds

David Sargan
University of Cambridge

CONTENTS

Summary 107
Introduction 108
Genetics of Brachycephaly 109
Genetics of BOAS 113
Genetic Diversity in Brachycephalic Breeds and Likely Effects on Fitness 114
Maintaining and Restoring Genetic Diversity 117
Tests for BOAS 118
Conclusions 120
References 121

SUMMARY

Three breeds of extreme brachycephalic dog, the French bulldog, Pug and Bulldog have suddenly found themselves among the most numerous Kennel Club (KC)-registered breeds in the UK, with the French bulldog becoming the most popular breed registered in 2018. Ten years ago, numbers for these three breeds together were less than a quarter of those currently. Their non-pedigree pet populations have also grown quickly. A high proportion of dogs of these three breeds suffer problems caused by their conformation (brachycephalic syndrome), including difficulties with breathing, eye and eyelid conformation, skin fold dermatitis and dystocia, and they may also suffer from vertebral, hip and elbow problems.

Selection for extremes of skull shape and body conformation in combination with the waxing and waning popularity of these breeds has contributed to highly inbred populations both in the UK and elsewhere, with studies showing long regions under selective sweeps and highly reduced heterozygosity across the genome. The major brachycephalic breeds have very low effective population sizes and reduced major histocompatibility complex (MHC) diversity which may well reduce their ability to respond to infections and increase liability to autoimmune diseases. (The MHC is a set of polymorphic genes that encode proteins important in the immune response and in the recognition of self.)

Genetic studies have implicated eight of the selective sweep regions as being involved with conformation and skull shape of brachycephalic dogs, and a mutation in the coding or control regions of a single gene has been strongly implicated in producing that conformation for at least three of these. Not every selective sweep is present in every brachycephalic breed.

Dogs that are homozygous for these swept regions are not always affected by brachycephalic obstructive airway syndrome (BOAS). So, while the mutations in the regions may be primary to the presence of disease, they do not by themselves cause the disease. The loci associated with severe BOAS are now being isolated, and in the absence of a willingness to fundamentally change skull shape in these breeds, they may offer a more limited opportunity to reduce BOAS incidence by genetic means.

Genetic testing is not yet available for BOAS, but phenotypic tests based either on exercise performance or changes in respiration during exercise are now in use.

INTRODUCTION

Certain breeds of brachycephalic dogs such as the French Bulldog, Pug and Bulldog have become vastly popular over the last two decades. In part, this is almost certainly because humans are evolutionarily programmed to respond strongly to infant faces (reviewed Borgi and Cirulli, 2016), and these three breeds are highly brachycephalic. Consequently, many breeds have been selectively bred towards this pattern which is psychologically desirable to humans (Serpell, 2002, Ghirlanda et al., 2013). The brachycephalic breeds that are gaining in popularity are also considered convenient to own and keep – seen as needing minimal exercise, and therefore suited to the sedentary, elderly and infirm, and to those with little room, as well as to children. Three extreme brachycephalic breeds, the pug, bulldog and French bulldog are now among the most numerous KC-registered breeds in the UK, with the French bulldog briefly becoming most commonly registered breed in 2018. Twenty-five years ago, registrations of pugs and French bulldogs in the UK were in the low hundreds, while those of the bulldog have grown more slowly over a longer period, but were less than one thousand in the early 1980s. Even 10 years ago, total numbers of these three breeds were less than a quarter of those currently (The Kennel Club, 2018). Their non-pedigree pet populations have also grown, as evidenced by the age structure of these dogs in primary veterinary practice, although the recent nature of the number increase means that these breeds are probably still a good deal less popular in terms of total numbers than those breeds such as Labrador retrievers and the spaniel breeds that have had more sustained popularity and include older as well as younger dogs (O'Neill et al., 2013, 2016, 2018; Pet Food Manufacturers Association, 2018).

Brachycephaly has been defined in a number of ways including the use of indices comparing a variety of skull width and length measurements (Stockard et al., 1941; Koch et al., 2012; Evans and de Lahunta, 2016), the use of measurement of the craniofacial angle (Regodón et al., 1993), or variants and combinations of these measurements. Commonly, a skull width-to-length ratio of >0.81 has been used to define the condition (Koch et al., 2012; Evans and de Lahunta, 2016). But other subjective or context-based definitions have also been used, and furthermore, the skull shape is not static over an individual dog's lifetime, as it changes as the dog matures. Such measurements reflect a continuum of skull shapes, from moderate shortening of skull and muzzle in breeds like the St Bernard and Boxer, to extreme brachycephaly with muzzle lengths commonly below 1 cm and sometimes virtually no muzzle protrusion at all, as in some Pugs, Griffon Bruxellois and Japanese chins, as well as in some cats (Packer et al. 2015a,b; Farnworth et al., 2016; Ladlow, Liu and Sargan, unpublished data).

Brachycephalic syndrome consists of a group of diseases related to conformation which are more common in brachycephalic breeds than other dogs. They include BOAS and associated comorbidities such as exercise intolerance, heat intolerance, disturbed sleep, regurgitation of food and hiatus hernia but also morbidities related to other conformational exaggerations such as skeletal (hemivertebrae and similar abnormalities associated with screw tail and neurological problems causing pain, rear limb difficulties and/or incontinence; problems in birthing associated with narrow pelvic girdle; rear limb arthritis and other joint difficulties); ophthalmic (proptosis/exophthalmos, corneal damage, cherry eye, entropion and ectropion, etc.); dermatological (e.g. skin infections associated with excess skin); or dental (tooth overcrowding, undershot jaw). Deep conflict exists between those who perceive brachycephalic syndrome as a welfare crisis with many suffering animals and those convinced that most brachycephalic dogs are both happy and healthy. Evidence to establish the morbidity and prevalence of brachycephalic syndrome remains sparse but suggests 20%–40% of these dogs suffer moderate to severe morbidity, and these problems, if untreated, last for periods of years (Asher et al., 2009; Liu et al., 2015, 2016). The potential causes of morbidity listed above are reviewed elsewhere in this volume. I will only observe here that brachycephalic dogs live for 3–4 years less than other breeds of similar weight (see Figure 8.1, and O'Neill et al., 2013, 2016, 2018; Michell, 1999; Cassidy, 2007; Lewis et al., 2018).

BOAS and other aspects of brachycephalic syndrome are often stated to have become a more common and severe disease in recent decades than previously. Moderately extreme forms of brachycephaly already existed in or before the 1960s in the bulldog and some other breeds (Fox, 1963), but much hearsay suggests that average skull shapes in at least some brachycephalic breeds have shifted relentlessly towards more extreme forms over the last 80 or more years (Drake and Klingenberg, 2008). Is this really true? A study of the skulls that are representative of breeds over the last century from the collection of the Natural History Museum of Bern shows that skulls have become even shorter over time in the Bulldog, and some other brachycephalic breeds including Boxer, French Bulldog, Pug and Pekingese (Figure 8.2), and Bernese Mountain Dog (not shown in Figure 8.2) (Koch and Sturzenegger, 2015).

In this chapter, I will consider the genetics underlying brachycephaly and of BOAS, the genetic health of the key brachycephalic breeds and the effects of rapid expansion and subsequent reduction in population size on genetic diversity. I will also briefly describe some current and future likely means of health testing in brachycephalic breeds.

GENETICS OF BRACHYCEPHALY

Brachycephaly (skull and head shortening) in domestic dog breeds has been developed over a period of centuries. Popular breed histories suggest that this occurred initially by selection for reduction of

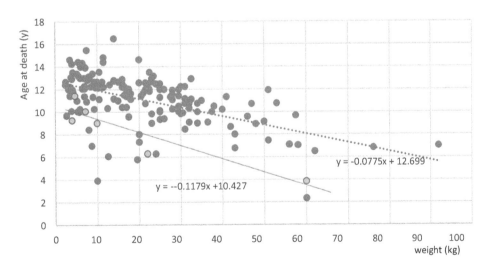

FIGURE 8.1 Extreme brachycephalic breeds on average live for 3–4 years less than expected from their weights. Extreme brachycephalic breeds (Bulldog, Dogue de Bordeaux, French Bulldog, Japanese Chin, Pekingese, Pugs) are shown in red and other breeds in blue. Median age at death are from UK Kennel Club breed surveys, 2004 and 2014, and were not available from these sources, from Michell (1999) and Cassidy (2007). Weights are mean of male and female weights from AKC breed descriptions and if these do not include a weight, from a variety of web-based breed club sources.

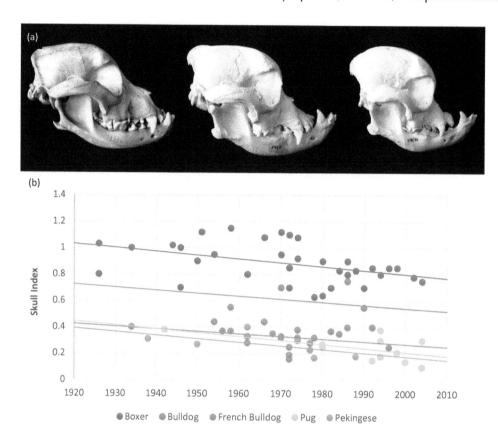

FIGURE 8.2 **Brachycephaly is becoming more exaggerated in a number of breeds. (a) Bulldog skulls in the collection of the Naturhistorisches Museum der Burgergemeinde, Bern**: Left: 1900–1920. Middle: Switzerland-Champion, born 1963, died 1971. Right: Born 1985, died 1993. (Marc Nussbaumer, licensed under a Creative Commons Attribution-Share Alike 4.0 International license.) (b) **Decreases in skull index in brachycephalic breeds** in the same collection. The skull index referred to here is that of Koch et al. (2012), in which the length of the facial skull is divided by the length of the cerebrum. Data here are adapted from Koch and Sturzenegger (2015). Linear trend lines are shown for each breed.

jaw and muzzle length to allow a stronger bite force in fighting dogs, and more recently, and especially in smaller breeds, there has been exaggeration of features that evoke those of baby animals and humans such as large eyes and flat face (Serpell, 2002). Popular brachycephalic dog breeds also show adaptations suited to domestic companion animal roles rather than hunting/working roles in external environments such as a well-developed cone-rich retinal *area centralis* rather than a visual streak (McGreevy et al., 2004) and behaviours including allo-grooming (social grooming between members of the same species) (McGreevy et al., 2013) that are signs of gregariousness in other canids. Compared with wolves or their cubs, the upward rotation of jaw and muzzle of brachycephalic dogs stems from a *de novo* cranial flexion in which the palate is tilted dorsally in brachycephalic and mesaticephalic breeds or ventrally in dolichocephalic and down-faced breeds (Nussbaumer, 1978; Drake, 2011). This flexion is never seen at any stage of wolf development. "Dogs are not paedomorphic wolves" (Drake, 2011) but have evolved along a specialist path to fit human needs.

Several studies have compared DNA sequence or single-nucleotide polymorphism (SNP) profiles of brachycephalic and mesaticephalic or dolichocephalic breeds (Bannasch et al., 2010; Boyko et al., 2010; Marchant et al., 2017; Schoenebeck et al., 2012; Schoenebeck and Ostrander, 2013, 2014).

These studies show that several loci (regions of the genome) carry mutations that segregate with skull shape across several breeds and that the actual mutation causing the change may be surrounded by a near homozygous tract (a selective sweep) varying from a few thousand base pairs in size up to several megabases. There are a number of loci which govern size and allopatric variation that are also involved in skull shape. Not all brachycephalic breeds carry all of these mutations. Homozygous loci associated with brachycephaly, and the associated genes and mutations, where known, are presented in Table 8.1.

The SMOC2 (SPARC-related modular calcium binding 2), DVL2 (dishevelled 2) and BMP3 (bone morphogenetic protein 3) are all genes with major effects on skeletal development. SMOC2 is expressed throughout the body on basement membranes and associated with fibronectin in the extracellular matrix, regulating cell matrix interactions. But during embryonic development, both SMOC2 and the closely related SMOC1 gene achieve high expression in the branchial arches in mid-gestation embryos, and SMOC2 mutations in mice and humans are associated with developmental dental defects. In mice, the SMOC1 gene is known to regulate BMP3 signalling and is associated with abnormal cranium, snout and maxilla morphology, while SMOC2 KO (knock out) mice have altered skull morphology including reduction in overall skull length and slightly increased maxilla width (Marchant et al., 2017; Mouse Genome Informatics, 2020). SMOC2 modulates signalling of

TABLE 8.1

Genetically Fixed, Non-variable Regions or Gene Locations Previously Associated with Brachycephaly Found in This Population of Pugs, French Bulldogs and English Bulldogs

Candidate Locus	Gene Implicated / Measurement Effected	Genetic Lesion	Homozygous Region Fixed in the Breed Population?			References
			Pug	French Bulldog	English Bulldog	
CFA1 (59.83M)	SMOC2/ allometric snout length, mandible length	Intronic insertion of 3'truncated LINE element[*]	Yes	Yes	Yes	Boyko et al. (2010) Marchant et al. (2017)
CFA5 (32.2M)	DVL2/Snout length	c.2044delC Leads to premature termination: (p.P684L fsX26)	No	Yes	Yes	Bannasch et al. (2010) Mansour et al. (2018)
CFA24 (23.4M)	Unknown /Zygomatic arch width		Yes	Yes	Yes	Boyko et al. (2010) Schoenebeck et al. (2012)
CFA30 (32.6M)	Unknown (KIF23? 33.2M) /palate width		Yes	Yes	Yes	Boyko et al. (2010)
CFA32 (5.2M)	BMP3/Total skull length, snout length, bone density/ early suture closure**	p.F425L	Yes	Yes	Yes	Schoenebeck et al. (2012)
CFAX (85.4M)	Skull length, mandible height		Yes	Yes	Yes	Boyko et al. (2010)
CFAX (101.7M)	(IGSF1?)/Allometric Palatal length and width; eye and zygomatic width		Yes	Yes	Yes	Boyko et al. (2010) Plassais et al. (2019)

The table follows the authors in using the term snout length. This equates closely to muzzle and viscerocranium length.

[*] This LINE element is in complete linkage disequilibrium with five SNP in non-brachycephalic breeds located in introns of SMOC gene, but none of these are in regions of high-sequence conservation across species.

[**] Observed in rats and humans.

BMP proteins in zebrafish embryos, while inhibition of SMOC2 using morpholinos causes cranio-facial reduction (Bloch-Zupan et al., 2011; Melvin et al, 2013; Mommaerts et al., 2014).

DVL2 is a member of a group of three genes encoding cytoplasmic scaffolding proteins that have multiple effects on skeletal development under the control of the WNT5A-ROR pathway (Gonzalez-Sancho et al., 2004; Nishita et al., 2010; Ho et al., 2012). Loss of function mutations in this pathway is associated with Robinow syndrome in humans, characterised by short-limbed dwarfism and ver-tebral abnormalities including hemivertebrae, and with faces characterised by a high forehead, short upturned nose and wide set eyes (reviewed in Patton and Afzal, 2002, see Person et al., 2010; White et al., 2016). In dogs, DVL2 is associated with the screw-tail phenotype and vertebral abnormalities. The mutation is fixed in bulldogs and French bulldogs, and is close to fixation in Boston terriers and present in a substantial proportion of pit bulls and Staffordshire terriers, and rarely in Shih Tzus as well as some mixed breed dogs (Mansour et al., 2018). These authors suggest that it may confer an especially broad muzzle. However, the mutation is not present in pugs, although this breed fre-quently does have vertebral malformation and is more prone to neurological sequelae (hind limb paresis) than the bulldog-related breeds (Bertram et al., 2019; Ryan et al., 2019). Just as in humans, it is probable that vertebral shape and its consequences and interaction with skull shape in dogs can have multiple genetic origins (Roifman et al., 2015).

BMP3 is a negative regulator of bone density in mice, and knock out increases the proportion and total amount of trabecular bone in this model (Daluiski et al., 2001; Bahamonde and Lyons, 2001). It is expressed in osteoblast lineage cells and suppresses osteoblastogenesis from bone mar-row stromal cells in a feedback loop (Kokabu et al., 2012). In a zebrafish model, Bmp3 antisense morpholinos cause reduced cartilage development, loss of jaw structures and reduced cranioskel-etal development (Schoenebeck et al., 2012). The inclusion of the BMP3 p.F425L mutation in the Wisdom Panel for SNP identification of dog breeds means that there is a great deal of information on the distribution of this mutation across breeds (Dreger et al., 2019). The allele is fixed or very close to fixed, as might be predicted, in the extreme brachycephalic breeds Bulldog, French bull-dog, Boston terrier and Pug, as well as Griffon Bruxellois, Pekingese and Japanese Chin, but it is also fixed in the Norfolk terrier (although having an allele frequency of only 25% in the Norwich terrier). The Scottish terrier, Sealyham terrier, smooth fox terrier and Staffordshire Bull terrier are also close to fixation for the mutant allele, although none of these latter terrier breeds show extreme brachycephaly. Furthermore, several breeds that have mesaticephalic faces combined with chondrodysplastic dwarfism, such as the Lancashire heeler and both corgi breeds, have substantial loads of the BMP3 mutation (as do several breeds with moderate levels of brachycephaly such as the Cavalier King Charles spaniel). Finally, there are some moderately brachycephalic dogs, mainly of larger breeds such as the Boxer and the Dogue de Bordeaux, that lack the BMP3 mutation. Clearly, the distribution of developmental anomalies attributable to BMP3 mutation is controlled by addi-tional elements in the genomic background of the breed (Ostrander et al., 2017).

Other genes have been an identified with important roles in canine head shape. Fondon and Garner (2007) have described the inclusion of short tandem repeats (STR) within exons of develop-mental transcription factors as one of the processes by which phenotypic change is driven quickly in dogs. These elements are highly variable in all species by addition or deletion of repeats. Strong selection by humans may have driven their rapid evolution in dogs where they are both more com-mon in exonic positions and more variable than in humans. Most prominently in the context of brachycephaly, a gene identified by Fondon and Garner as the runt-related transcription factor 2 gene (RUNX2), but related to a partial DNA sequence deposited by these authors in GenBank (AY308819.1) which in the current canine genome build (Canfam 3.0), shows considerably greater homology to RUNX1. The RUNX gene they identify contains such a repeat which encodes a run of between 18 and 20 glutamines followed by 12–17 alanines, depending on the allele present. Fondon and Garner showed that both the clinorhynchy (the degree of dorsoventral nose bend) and the mid-face length increase across 27 breeds as the proportion of glutamine to alanine increases in these repeats (Fondon and Garner, 2004, 2007). This is consistent with findings of a similar association in RUNX2 of human cleidocranial dysplasia patients (a condition characterised by abnormal or absent collar bones, as well as tooth abnormalities, prominent forehead, wideset eyes and a flat nose) (Ramesar et al., 1996). Other types of mutations in RUNX2 exist in dogs, which are more certainly matched to this gene and include several SNP changes predicted by SIFT (Ng and Henikoff, 2003) to cause harmful structural changes (ENSEMBL Genome Server, 2020; Kalmar, Liu, Ladlow and Sargan, unpublished). Similar SNP, insertion and deletion mutations in RUNX2 also cause human facial dysmorphias (e.g. Zhang et al., 2000), while RUNX1 expression in cartilage is also required for normal tooth, palatal and craniofacial development of mice (Charoenchaikorn et al., 2009; Sarper et al., 2018a, b; Yamashiro et al. 2002, 2004). Experiments in mice suggest a mechanism in which reductions in RUNX2 expression are associated with reduced palatal length and that longer alanine tracts reduce expression leading to reductions in osteo-progenitor cells (Amiel et al., 2004; Lou et al., 2009; Liu et al., 2011).

GENETICS OF BOAS

BOAS is not a new condition. In 1891, Dr Georg Muller's *Diseases of the Dog and Their Treatment*, page 102, states "In some breeds of dogs, such as Pugs and bulldogs the [nasal] passage is so narrow that a slight contraction can cause them to breathe through their mouth. The nasal sound is like a snore …". This passage occurs in a section on "Diseases of the Respiratory Organs Concentrating on a Variety of Causes of Dyspnoea". Stockard also reports the problem in his sublime 1941 study of the transmission and development of brachycephaly and other features of canine conformation.

BOAS has genetic risk factors including sex and breed, but also environmental ones including external temperature, age, body condition score and frequency of exercise (Liu et al., 2015, 2016, 2017; Wykes, 1991; Ladlow et al., 2018; Lilja-Maula et al., 2017, Packer et al., 2015b; Tarricone et al., 2019). Measuring the heritability of BOAS within a breed has been difficult until recently because of the difficulty of defining the presence and severity of the condition. Whole-body barometric plethysmography (WBBP) was first used in 2010 to examine sedated dogs with BOAS by Bernaerts et al., 2010, who noted significant differences in several airflow parameters between brachycephalic and control dogs. Recently, the Cambridge University BOAS group have developed analysis of whole-body barometric plethysmographic traces from fully awake dogs to allow trait quantitation in the three common extreme brachycephalic breeds and accurate comparison with exercise performance (Liu et al., 2015, 2016). It is clear that breed is a primary factor in the development of BOAS. Information from GWAS (genome-wide association studies) on 172–210 dogs of each of the bulldog, French bulldog and Pug breeds from the UK show and clinical populations has recently been used to look for genetic markers of BOAS. DNA from dogs classed as BOAS+ or BOAS− was analysed using Genome-wide Complex Trait Analysis (GCTA) software. This methodology allows calculation of within-breed narrow sense heritability estimates of 40%–60% for BOAS in these UK populations (Kalmar et al., in preparation). For details of GCTA, see Yang et al. (2011, 2013, 2016). The GWAS results suggest that multiple loci (8–11 loci per breed) are involved in determining presence and severity of BOAS, and that the majority of mutant alleles are specific to breed for the three common extreme brachycephalic breeds, the bulldog, French bulldog and pug (Kalmar et al., in preparation).

Laryngeal collapse is one component of brachycephalic syndrome for which an effect on presence or severity of BOAS has been noted in several (but not all) studies (Dupre et al., 2016; Riggs et al., 2019; Rutherford et al., 2017, but see Haimel and Dupre, 2015). Laryngeal collapse is also a common problem in Norwich terriers, although these dogs are not brachycephalic (Johnson et al., 2013). A mutant allele associated with laryngeal collapse in Norwich terriers has been identified in the ADAMTS3 gene (Marchant et al., 2019). This mutant allele is also present in Bulldogs where it appears to be fixed (that is, homozygous in all animals in the breed), and in French bulldogs, where it is not fixed but segregates in approximate Hardy–Weinberg equilibrium. But the mutation is not present in pugs. The mutant allele does not show association with laryngeal collapse or with BOAS severity in any of these three brachycephalic breeds (Marchant, Schoenebeck, Sargan and Mellersh, unpublished).

GENETIC DIVERSITY IN BRACHYCEPHALIC BREEDS AND LIKELY EFFECTS ON FITNESS

There are a large number of brachycephalic breeds, and similar facial shortening has occurred by different genetic routes across these breeds. A dog with some conformational similarities to the modern pug exists in Chinese artworks that are up to 2000 years old. Pugs are said to have been first imported from South East Asia by Marco Polo, but modern European pugs are now related closely to the Griffon Bruxellois and share only limited ancestry with other ancient East Asian companion dogs such as the Pekingese, Japanese Chin and Shih Tzu (Parker et al., 2017). However, all of these brachycephalic Asian breeds do represent very different canine lineages to Bulldogs, French bulldogs and Boston terriers, which are more closely related to each other. In both Asian and European lineage brachycephalic breed groupings, the BMP3 F425L mutation is found associated with a particular shared haplotype, but in each group, this haplotype differs from that shared in the other group (Parker et al., 2017). Each grouping also shares haplotype with a different population of those mesaticephalic breeds shown by Wisdom Panel data to carry the BMP3 mutation, but which are not brachycephalic (Dreger et al., 2019). This mutation in the Bulldog haplotype background seems to be shared mainly with terriers and perhaps has a UK origin, while that in the pug haplotype background is shared with spaniels and smaller schnauzer breeds, perhaps suggesting an origin in Asia and subsequent spread in continental Europe (Dreger et al., 2019).

This shows that there was much interbreeding of dog types to achieve the particular phenotype desired by any given owner up to the point that the studbook for the breed was closed. For the three key extreme brachycephalic breeds (bulldog, French bulldog and Pug), closure occurred after recognition by The Kennel Club well over a century ago, after which haplotype exchange became very limited, being reduced to off-studbook or specifically allowed matings. Furthermore, dogs have high fecundity, with a single litter producing more than the single male and single female needed to replace their parents in the next generation of a population of constant size. The ability of stud dogs to sire large numbers of litters, and of bitches to bear several large litters, means that the number of breeding individuals required to replace a given population with a new generation of the same size is far less than the population size. Breeding for both form and function occurs through a succession of human choices about which dogs most closely conform to the type required. The consequence is that many dogs are rejected for every dog chosen as a parent. As suggested by Table 8.1 tight selection for the brachycephalic type continues to the present day so that genetic variety in the populations has declined as many potential sires and dams are excluded from breeding.

Twenty-five years ago, both pugs and French bulldogs were numerically quite small breeds in the UK. In 1995, there were 505 registrations for Pugs and 244 for French bulldogs, and both breeds had long periods with their populations at roughly this level. Bulldogs, for which breed recognition occurred as early as 1873, had only 670 registrations in 1981, but started becoming more popular from then on. In each breed, the number of litters was very considerably below the number of available parents over a period of years. In any breed being selected for the show ring (or anything else), the total number of parents (over a period of years) is likely to be considerably below the number of litters, because the 'best' parents are used repeatedly. Hence, by 1980, each of these breeds will already have been quite highly inbred.

One way to quantify the diversity in a population is through measurement of the effective population size (designated N_e): the effective population size is the size of a population of randomly mating individuals that would lose genetic diversity (measured as heterozygosity) in the next generation at the same rate as the population being studied. For example, in 2006, the human population of Iceland (actually around 300,000, with an extremely well-documented population history dating back to 1100) was estimated using molecular techniques to have an N_e of about 5000 (Bataillon et al., 2006). Iceland's isolation had prevented much immigration after its initial settlement, and in any isolated population, alleles are continuously lost because only a half of each of the two parental genomes is found in a single child. Hence, if a couple have n children, the proportion of the parental genomes represented across all children is $(1-(1/2)^n)$. In addition, in real populations, a proportion of the population does not have children. Even in a perfectly mixing population, some diversity is therefore lost in each generation. In fact, in a population of constant size, heterozygosity loss across a whole population occurs at a rate of $1/(2N_e)$ per generation. Hybrid vigour is the extra vigour and healthiness seen when outcrosses are made and is thought to be caused at least in part by the loss of homozygosity in recessive alleles that have small disadvantageous effects. In the opposite case, as populations become inbred, some slightly disadvantageous alleles will be lost entirely, but others will become homozygous in more individuals. If N_e is small, the process happens in fewer generations. Because loss of hybrid vigour seems to occur in all naturally occurring species, zoos attempt to maintain N_e of at least 100 using worldwide inter-crossing of captive animals. An effective population size below 50 (which will cause an increase in inbreeding of more than 1.0% per generation) indicates the future of the breed may be at risk (Food and Agricultural Organisation, 2010). Inbreeding depression is not proven in dogs (for review, see Nicholas et al., 2016), but there is no reason to believe that dogs are protected from its ill effects when compared with other inbred species.

Using pedigree records dating from 1981 to 2014, Tom Lewis has estimated N_e for all KC breeds and has reported surprisingly low numbers – even for numerically large breeds (Lewis et al., 2015; Lewis 2015). For example, for the pedigree Labrador Retriever $N_e=82$ despite an average 36,919 registrations per year in the last 10 years, while for some breeds, it is well below 50.

There has been recent very rapid expansion of some brachycephalic breed populations in the UK. In particular that of the French bulldog (up about 150-fold in the last 25 years) and pug (up about 20-fold) but also that of the bulldog (up nearly tenfold). This rapid expansion has reduced selection pressures in recent generations a good deal, so that many more dogs will have contributed to the breeding pool to allow this population growth rate. Effective pedigree population sizes are relatively large when measured over the period 1980–2014 (Pug $N_e=134$, FB $N_e=132$, Bulldog $N_e=68$) (Lewis, 2015). Again using UK Kennel Club pedigree records, an earlier analysis of the Bulldog as well as nine mesaticephalic and dolichocephalic breeds covering the period 1970–2006 gave an N_e of 41 for the bulldog, so the effects of population expansion on reducing selection pressures are seen in the Bulldog also, albeit at a lower level (Calboli et al., 2016).

But molecular methods are less kind to two of the three brachycephalic breeds, highlighting the way that small founder groups and strong selection in closed populations over a much longer period have very seriously eroded their heterozygosity. The bulldog population in the USA is dominated by just four paternal Y chromosome haplotypes with a single one occurring in 93% of current males, while mitochondrial sequencing suggests only five separate maternal lines contribute to the current population. Diversity of dog leucocyte antigen (DLA) loci is also low, with only nine DLA class II haplotypes in all and 62% of the population sharing a single DLA class II haplotype (Pedersen et al., 2016). A survey with a senior author from the UK, but multinational authorship, of five bulldogs along with many other breeds showed a similarly skewed distribution of DLA haplotypes in those bulldogs, with one MHC class II haplotype present in three of these five. It is not explicit that these bulldogs were from the UK and differences in nomenclature make it impossible to ascertain that this was the same allotype as those surveyed by Pedersen (Kennedy et al., 2007; Pedersen et al., 2016). But MHC diversity is low in many canine breeds, compared with humans. The bulldog DLA homozygosity figures are not dissimilar to those seen in UK cocker spaniels in the pet population (KC-registered cocker spaniels do have a low N_e, at 49, despite their popularity in this country) (Aguirre-Hernandez et al., 2010). Diversity is also similar to that seen in many breeds in a study of 25 US breeds which shares authorship with the bulldog study (Angles et al., 2005). So it seems that even low levels of MHC diversity are fairly well tolerated in dog breeds in the developed world, although it is not clear whether this would be true in environments in which veterinary care and sterilised dogfood was less available.

In addition, genomic STR loci in the bulldog have rather small numbers of alleles, particularly when compared to human or wild animal populations. STR diversity figures for the bulldog are in the same range as those seen in flat-coated retrievers (N_e,=68, the same figure as the bulldog) (Pedersen et al., 2016; Aguirre-Hernandez and Sargan, unpublished). A large American genome-wide SNP diversity study shows the pug as having rather few haplotypes, extended linkage disequilibrium, long runs of homozygosity and a high number of autozygous individuals when compared with most other breeds (Boyko et al., 2010). However, the pug still shows more variation than the boxer and is comparable with breeds such as the Irish wolfhound and the Cavalier King Charles spaniel (Boyko et al., 2010), both of which have undergone well-documented very severe bottle-necking in their recent histories (due to the near elimination of giant dogs in the two world wars, and, for the Cavalier King Charles Spaniel, recent breed refoundation).

Another precise metric of genetic diversity is offered by observation of numbers of SNPs which can be derived from genome-wide association study (GWAS) data. A series of recent publications have used GWAS techniques to discover loci connected with particular canine diseases. Taking only those studies which have used the Illumina high-density canine array (which interrogates a standard group of SNP known to show variation in many dog breeds) and have used similar quality control steps (in particular, those studies that have set minimum minor allele frequency (MAF) in the collected data at MAF>0.05), it is broadly possible to simply compare the number of SNPs found as heterozygous in each breed to gain an idea of the breed's retained diversity. For 30 single breed GWA studies using a total of 25 different breeds, and each using more than 20 dogs, with the 173,662 SNP Illumina chip with similar QC, different breeds retained from 75,916 heterozygous SNP positions (in the Doberman) to 126,607 in the Boxer breed (Sargan, unpublished). As the model breed from which many of the SNPs on the chip were derived, the boxer would be expected to show a high level of heterozygosity in this assay and cannot be directly compared with other breeds). For the other breeds, these differences are not caused by the different numbers of dogs in each GWAS study. (For numbers of dogs versus numbers of SNP in analysis, $R^2 = -0.0143$, $p = 0.702$.) Compared to all of these breeds, our own unpublished figures show that the Pug (73,896 heterozygous SNP positions) and the Bulldog (75,465 heterozygous SNP positions) are the two least diverse breeds ever studied using this criterion, although the French bulldog (101,202 heterozygous SNP positions) is close to the average across all breeds in these studies (101,682 heterozygous SNP positions) (Kalmar and others, unpublished information). By these criteria, inbreeding had increased a long way in UK populations of the Pug and Bulldog by 2018. And although French bulldogs in our sample retain SNP heterozygosity, they show a higher level of identity by state (IBS) than the Bulldog and Pug (that is, each SNP is found embedded in the same set of haplotypes on each chromosome on which it occurs). This is what would be expected if the population has expanded rapidly from a recently genetically bottlenecked starting population through rapid expansion with minimal selection (Theunert et al., 2012) (see Figure 8.3).

There is now a danger that future reduced demand for those breeds with currently expanded populations means that only a small subgroup of each population will be needed for future breeding and therefore represented in later generations. Consequently, future generations will carry an even smaller proportion of the genetic variations that are important for breed health. This could raise new inherited problems previously unseen in the expanded breeds. These could include problems raised by continual selection for fashionable/extreme facial or conformational types, but also problems associated with single gene variants (mutations) that becomes widespread within any small group of highly related animals used for breeding. In the recessive case, this spread can happen without awareness by breeders. When an unaffected sire and a dam that both carry the recessive mutation are crossed, a proportion of the offspring will acquire and show the associated defect.

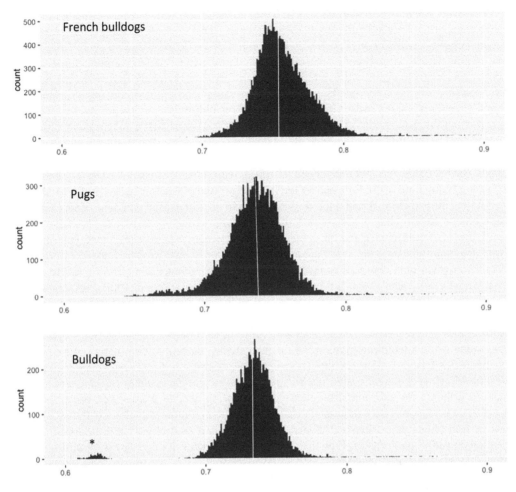

FIGURE 8.3   Identity by state in GWAS data from UK cohorts of the Extreme Brachycephalic Breeds in 2017. (a) French Bulldog. (b) Pugs. (c) Bulldogs. Pairwise comparisons of each member of the cohort at all variable SNP positions are shown. (* the individual showing least relation with the rest of the bulldog cohort was unusually small for a bulldog and seems likely to contain DNA from a second breed.)

MAINTAINING AND RESTORING GENETIC DIVERSITY

The term 'outcrossing' has been used in two different senses in dog breeding in recent years. It can mean inter-crossing of pure-bred dogs descended from two different lines, usually established by breeders who have been seeking to develop distinctive traits while using dogs that are both within a single KC-registered breed, or it can mean inter-crossing individuals from two different breeds in order to bring in a variation not previously present within the breed. Looking at the first of these senses, many breeders are attempting to address low N_e by deliberate outcrossing of lines within their breed. This can lead to high N_e values close to the actual population size and prevent further genetic diversity loss. For example, see the Sealyham and Dandie Dinmont terrier documents (Lewis and The Kennel Club, 2015). But it cannot undo losses of diversity that have already happened, because of the low rate at which errors occur in DNA replication and the fact that only a very small percentage of such errors are actively advantageous to health. Many hundreds of generations would be needed to rebuild genetic diversity once it has been lost.

A relatively rapid way to address this is by outcrossing to another breed, although results of such crosses are not always fully predictable either in terms of conformation or in the retention of other complex characteristics such as behavioural traits, which also have a personality component. In numerically large breeds, such crosses would have to be undertaken on a large scale for meaningful impact to occur within a few generations. Even when the desired allele can be identified by genetic testing rather than by a partially penetrant phenotype, introducing a variant across a whole breed is a lengthy task. An example of this is the introduction of a normal SLC2A9 gene from the German pointer into the Dalmatian breed allows removal of hyperuricosuria, hyperuricemia and associated gout (Safra et al., 2006; Bannasch et al., 2008). Such a cross was first made in 1973, followed by backcrossing, but because of resistance by breeders to accepting these crosses, the backcross was not recognised as falling within the breed standard for Dalmatians by The Kennel Club until 2010, followed by the American Kennel Club in 2011. As of now, the large majority of Dalmatians still retain the original Dalmatian mutant allele. In considering brachycephalic breeds, the problem is greater because of the large numbers of individuals in each breed, the relatively lower litter size and the complexity of the genetic basis for the phenotype. Bulldogs, French Bulldogs and Pugs have average litter sizes of four to just over five puppies (Borge et al., 2011). Even when only one gene is involved, expansion of a novel allele from a single cross to become a heterozygous variant throughout a constant population size of 1000 dogs with average litter size of four while maintaining the genetic diversity and type of the original population cannot be achieved in fewer than ten generations. When multiple genes on different chromosomes are involved, backcrossing from the F1 generation with the aim of placing genes in a single genetic background while attempting to retain all the new alleles relevant to elimination of a conformational defect is fraught with difficulty. Different chromosomes will segregate independently, and retention of all advantageous new loci will require a substantial number of inter-crosses to be made, to generate the substantial number of F1 offspring needed from which backcrosses can be made. Selection may rely on sets of genetic tests for each required allele (if these have been identified) or phenotypic means. Of course, breeders would naturally wish to impose their own selection on additional traits to allow them to produce the types of dogs they want throughout the process. If this additional selection is strong, much diversity will still be lost. This might include the newly introduced alleles. The process of spreading new alleles through the population could take longer if the pool of breeding dogs is reduced and selection by genetic testing for desired introduced alleles is not available.

TESTS FOR BOAS

Because BOAS is one of the major problems for brachycephalic dogs but not all brachycephalic dogs have BOAS, tests are being developed to allow removal from breeding of dogs likely to carry alleles that cause BOAS. Genetic tests offer the advantage that the result remains constant throughout life. In a disease of late onset like BOAS, a genetic test can allow removal of severely affected dogs from the breeding programme before they reach breeding age, even if they are not showing any signs of disease. Genetic tests are also very efficient because they allow direct measurement of all heritable components in the disorder without masking by environmental variables such as the dog being overweight, or unused to exercise or the day of testing being hot.

Although there are currently no genetic tests for BOAS, these are under development and will involve a substantial number of alleles (above ten alleles in some breeds) (Kalmar et al., unpublished observations). Hence, genetic estimated breeding values (based on the estimated contribution to BOAS of the set of alleles in the individual under test) will be used to plan breeding. The aim will be to remove only severely affected (BOAS Grade 3) dogs from breeding initially, while breeding Grade 2 moderately affected dogs only to BOAS–ve (Grades 0 and 1) dogs would continue. In the UK, this would mean removing from breeding about one-sixth of each generation if all dogs took the genetic test.

In the absence of genetic tests, other tests based on phenotype have recently come into use.

Skull shape: Evidence has been presented that BOAS and other aspects of the brachycephalic syndrome have strong relationship to muzzle length and in particular to the craniofacial ratio across many breeds (Packer et al., 2015a, b), although this conclusion has not been reproduced in some other studies (Liu et al., 2017). European Union guidance (Secretary General of the Council of Europe, 1987) has led to regulations in some member states that prevent breeding of any dog for which "it can reasonably be expected, on the basis of genotype, phenotype or state of health that breeding from it could have a detrimental effect on its health or welfare or the health or welfare of its offspring" (Government of the UK, 2018). These conformational studies led the Dutch government to ban the breeding of any dog whose muzzle is shorter than one-third of the length of the skull. The legislation has been implemented by the Dutch Kennel Club using a 'traffic light system'. No dog with a muzzle less than a third the length of the skull may be bred ('**red**'), effectively banning many or most Pugs from breeding, and also banning breeding of a large number of French bulldogs and Bulldogs. Additional limits are prescribed to breeding of dogs whose muzzle is between 30% and 50% in skull length ('**orange**'). For these animals, additional criteria are required to allow breeding including having open nostrils and not having proptotic eyes or excessive sclera visible. Unrestricted breeding is allowed for dogs with longer muzzles ('**green**'). For details, see Van Hagen (2019). This test is something of a catch-all in that not all dogs with muzzles in the red area suffer BOAS, although they may have other problems associated with the brachycephalic syndrome such as exposed and vulnerable eyes. Rigorous use of these tests will disqualify a large number of dogs in affected breeds from breeding and thus greatly reduce the genetic variation in the affected breeds unless measures such as outcrossing are used to counter this effect.

Exercise-based and exercise together with respiratory assessment-based tests. Three tests have been developed in different countries to assess directly the presence of BOAS by evaluating exercise performance or respiratory change under exercise. A 6-minute walk test originally developed to detect congestive heart failure in humans and then in dogs (Boddy et al., 2004) has been more recently been applied to examining exercise performance in brachycephalic bulldogs (Lilja Maula et al., 2017). Dogs are allowed to walk at their own pace and the distance covered is measured. Heart rate, respiratory rate, body temperature and respiratory noise are recorded, with respiratory noise, graded by a veterinarian using a scale modified from Liu et al. 2015, being used as a comparator measure of BOAS status. The test relies on measuring the distance walked in the 6 minutes. Using this test, there was a trend for Bulldogs graded as BOAS+ve to walk less far than bulldogs rated as BOAS−ve (494 ± 56 m, $n=8$ cf 543 ± 55 m, $n=15$; $p=0.059$), and the same group of BOAS−ve Bulldogs were able to walk less far than non-brachycephalic control dogs ($n=10$, $p=0.02$) (Lilja Maula et al., 2017).

A 1000 m walk test has also been used in Germany since 2009 to test for BOAS in Pugs, in Holland since 2014 to test for BOAS in Bulldogs and in Finland since 2017 to test for BOAS in eight different breeds (Bulldog, French Bulldog, Pug, Boston Terrier, King Charles Spaniel, Japanese Chin, Shih Tzu and Petit Brabancon) (Finnish Kennel Club 2017–2019). In the study of the 6-minute walk test of Lilja-Maula et al. (2017), when the same Bulldogs undertaking the 6-minute walk were asked to go on to complete 1000 m of walking within 12 minutes, seven Bulldogs were unable to complete the exercise: three from the BOAS+ve group and four from the BOAS−ve group. But in this case, the BOAS−ve group were significantly quicker than the BOAS+ve group ($p=0.046$), and there was a trend for the BOAS−ve group to normalise breathing, heart rate and body temperature faster, post exercise ($p=0.06$). Unfortunately, there was some overlap between the fastest BOAS+ve Bulldogs and the slowest BOAS−ve. Sensitivity and specificity of the test were not calculated, as sample size was small. Subsequently, both 6-minute and 1000 m walking tests have been applied to somewhat larger groups of French Bulldogs and Pugs by the Finnish group (Aromaa et al., 2019). This allowed them to show that for both breeds, the mean distance walked in 6 minutes was lower in BOAS+ve dogs (this was only a trend in French bulldogs ($p=0.066$) but was strongly established in pugs ($p=0.001$)). Similarly, the time taken to walk 1000 m and the recovery time after exercise were significantly higher in BOAS+ve than BOAS−ve dogs of both breeds ($p < 0.01$ for all comparisons). In this case, ROC curves were compiled to evaluate the tests. At optimal cut-off for both breeds, specificity is good: 96% for the French Bulldog and 90% for the pug, using the 1000 m test, which performs better than the 6-minute test for both breeds. But sensitivity is rather less good: 70% for the French Bulldog and 71% for the pug: to improve sensitivity to 80% reduces specificity to 64% in the French Bulldog and about 77% in the pug. This test is useful but when used in directing a breeding programme, its ability to detect and remove all cases will be bought at the cost of also removing 20%–35% of non-affected dogs. This would severely reduce the gene pool if the test was applied with the aim of removing all BOAS+ve dogs from breeding. In an attempt to increase sensitivity, the required time for completion of the 1000 m walk test by the Pugs and French bulldogs under the Finnish Kennel Club rules has now been reduced to 11 minutes. The effect of this is still being measured. It would be very interesting to know how the test would become more accurate if only Grade 3 dogs were compared with BOAS−ve dogs. This increases discrimination in the Cambridge test described below.

The Cambridge group has developed WBBP as an objective method to measure BOAS in dogs at rest (Liu et al., 2015, 2016). WBBP is not a practical test for whole populations of animals, but it reflects closely the results of an exercise tolerance-based respiratory functional test that actively seeks to place animals in mild respiratory deficit using a 3-minute trot at ±5 miles per hour (8 km per hour), in order to observe changes between resting and exercised respiration by auscultation (Liu et al., 2015, 2016). The relationship between WBBP and the Cambridge functional grading test has been developed for Pugs, French Bulldogs and Bulldogs. The 3-minute trot test was compared with a 3-minute walk test at nearly three miles per hour (4.8 km per hour), and with a simple auscultation test at rest, using novel cohorts different from those used originally to standardise the WBBP test (Riggs et al., 2019). While specificity in each of the tests was 100% (i.e. no dog was identified as BOAS+ve which on WBBP appeared as BOAS−ve), sensitivity was higher in the trot test than in the walk test, and both exercise based tests were more sensitive than auscultation of the resting animal alone; 93% (CI95 65%–100%) of dogs identified as BOAS+ve by WBBP were identified as BOAS+ve in the trot test, 70% (CI_{95} 52%–83%) in the walk test and 57% (CI_{95} 39%–73%) by resting auscultation. These findings have allowed the UK Kennel Club to introduce a testing programme, 'The Kennel Club and University of Cambridge Respiratory Function Grading Scheme' based on the clinical assessment of dogs before and after respiratory testing. Prior to performing auscultation in connection with this programme, veterinarians are specifically trained to grade the dogs based on respiratory sounds so that consistency can be achieved across the country. Dogs are graded into one of four BOAS grades (0, 1, 2 or 3) as specified by Liu et al. (2015) with Grade 0 considered to have no BOAS, Grade 1 mild respiratory signs detectable by auscultation but without clinically significant BOAS, Grade 2 dogs having clinically significant but moderate and not immediately life-threatening BOAS which should be monitored by the dog's veterinarian and Grade 3 having severe BOAS requiring treatment. It is suggested that Grade 3 dogs be withdrawn from breeding programmes and Grade 2 dogs should only be bred with Grade 0 or Grade 1 dogs. The expectation that this will lead to about 15%–20% of dogs in each breed being withdrawn from breeding programmes. Because the disease is genetically complex, it is likely that there will continue to be some dogs with BOAS born within the programme, but these should be greatly reduced.

CONCLUSIONS

The 2019 downturn in the numbers of brachycephalic dogs being registered in the UK may be welcomed as likely to reduce welfare problems in the pedigree dog population. But the cohort size of the three most popular breeds has been growing for many years. The numbers of brachycephalic dogs reaching the older age groups at which BOAS, and many of the other diseases in the brachycephalic syndrome reach their highest prevalence will still be increasing for a small number of years. We do yet know whether the recent fall seen in pedigree registrations of the most popular brachycephalic breeds reflects changes in the non-pedigree pet population, which (after many years in which the pedigree population was dominant in these breeds) appears more recently to have been growing in parallel with the boom in the pedigree one (for example, see Pet Food Manufacturers Association Data Report, 2018).

An understanding of the genetics of brachycephalic syndrome, and particularly of associated disorders such as BOAS, laryngeal collapse, vertebral malformations and IVDD, opens the possibility of reducing the prevalence of these diseases and the suffering that they cause affected animals in an efficient way. These tests are now in use or in development. They have many advantages but in a situation where populations of the breed may have peaked, care must be taken to ensure that the gene pool is not reduced dramatically in attempts to remove genes causing these particular diseases.

In the meantime, tests based on phenotype offer considerable power and are now being used in many countries. It is to be hoped that these tests will be taken up by large numbers of owners and can themselves reduce the problems dealt with in this chapter. The key to this lies in education of dog owners, and all those concerned with dog health need to concern themselves with this.

Most recently some countries have introduced regulations that may prevent any breeding of extreme brachycephalic breeds in their current form. It is not yet clear whether enforcement of these regulations will be effective.

REFERENCES

Aguirre-Hernandez, J., G. Polton, L. J. Kennedy, and D. R. Sargan. "Association between anal sac gland carcinoma and dog leukocyte antigen-dqb1 in the english cocker spaniel." [In eng]. *Tissue Antigens* 76, no. 6 (Dec 2010): 476–81.

Amiel J., D. Trochet, M. Clément-Ziza, A. Munnich, S. Lyonnet. "Polyalanine expansions in human." *Hum Mol Genet* 13, no. 2 (Oct 1 2004): R235–43.

Angles, J. M., L. J. Kennedy, and N. C. Pedersen. "Frequency and distribution of alleles of canine Mhc-Ii Dla-Dqb1, Dla-Dqa1 and Dla-Drb1 in 25 representative american kennel club breeds." [In eng]. *Tissue Antigens* 66, no. 3 (Sep 2005): 173–84.

Aromaa, M., L. Lilja-Maula, and M. M. Rajamäki. "Assessment of welfare and brachycephalic obstructive airway syndrome signs in young, breeding age french bulldogs and pugs, using owner questionnaire, physical examination and walk tests." *Anim. Welf.* 28, no. 3 (2019): 287–98.

Asher, L., G. Diesel, J. F. Summers, P. D. McGreevy, and L. M. Collins. "Inherited defects in pedigree dogs. Part 1: Disorders related to breed standards." [In eng]. *Vet J* 182, no. 3 (Dec 2009): 402–11.

Bahamonde, M. E., and K. M. Lyons. "Bmp3: To be or not to be a bmp." [In eng]. *J Bone Joint Surg Am* 83-A, no. Pt 1 (2001): S56–62.

Bannasch D., N. Safra, A. Young, N. Karmi, R.S. Schaible, G.V. Ling. "Mutations in the SLC2A9 gene cause hyperuricosuria and hyperuricemia in the dog." *PLoS Genet* 4, no. 11 (Nov 2008): e1000246.

Bannasch, D., A. Young, J. Myers, K. Truve, P. Dickinson, J. Gregg, R. Davis, et al. "Localization of canine brachycephaly using an across breed mapping approach." [In eng]. *PLoS One* 5, no. 3 (Mar 10 2010): e9632.

Bataillon, T., T. Mailund, S. Thorlacius, E. Steingrimsson, T. Rafnar, M. M. Halldorsson, V. Calian, and M. H. Schierup. "The effective size of the icelandic population and the prospects for ld mapping: Inference from unphased microsatellite markers." [In eng]. *Eur J Hum Genet* 14, no. 9 (Sep 2006): 1044–53.

Bernaerts, F., J. Talavera, J. Leemans, A. Hamaide, S. Claeys, N. Kirschvink, and C. Clercx. "Description of original endoscopic findings and respiratory functional assessment using barometric whole-body plethysmography in dogs suffering from brachycephalic airway obstruction syndrome." [In eng]. *Vet J* 183, no. 1 (Jan 2010): 95–102.

Bertram S., G. Ter Haar, S. De Decker. "Congenital malformations of the lumbosacral vertebral column are common in neurologically normal French Bulldogs, English Bulldogs, and Pugs, with breed-specific differences." *Vet Radiol Ultrasound* 60, no. 4 (Jul 2019): 400–408.

Bloch-Zupan, A., X. Jamet, C. Etard, V. Laugel, J. Muller, V. Geoffroy, J. P. Strauss, et al. "Homozygosity mapping and candidate prioritization identify mutations, missed by whole-exome sequencing, in Smoc2, causing major dental developmental defects." [In eng]. *Am J Hum Genet* 89, no. 6 (Dec 9 2011): 773–81.

Boddy, K. N., B. M. Roche, D. S. Schwartz, T. Nakayama, and R. L. Hamlin. "Evaluation of the six-minute walk test in dogs." [In eng]. *Am J Vet Res* 65, no. 3 (Mar 2004): 311–3.

Borge, K. S., R. Tonnessen, A. Nodtvedt, and A. Indrebo. "Litter size at birth in purebred dogs--a retrospective study of 224 breeds." [In eng]. *Theriogenology* 75, no. 5 (Mar 15 2011): 911–9.

Borgi, M., and F. Cirulli. "Pet face: Mechanisms underlying human-animal relationships." [In eng]. *Front Psychol* 7 (2016): 298.

Boyko, A. R., P. Quignon, L. Li, J. J. Schoenebeck, J. D. Degenhardt, K. E. Lohmueller, K. Zhao, et al. "A simple genetic architecture underlies morphological variation in dogs." [In eng]. *PLoS Biol* 8, no. 8 (Aug 10 2010): e1000451.

Calboli, F. C., J. Sampson, N. Fretwell, and D. J. Balding. "Population structure and inbreeding from pedigree analysis of purebred dogs." [In eng]. *Genetics* 179, no. 1 (May 2008): 593–601.

Cassidy, K. M. 2007 "Dog Longevity Webpage ". [cited 2016 11/11/2016]. Available from: http://users.pullman.com/lostriver/longhome.htm.

Charoenchaikorn, K., T. Yokomizo, D. P. Rice, T. Honjo, K. Matsuzaki, Y. Shintaku, Y. Imai, et al. "Runx1 is involved in the fusion of the primary and the secondary palatal shelves." [In eng]. *Dev Biol* 326, no. 2 (Feb 15 2009): 392–402.

Daluiski, A., T. Engstrand, M. E. Bahamonde, L. W. Gamer, E. Agius, S. L. Stevenson, K. Cox, V. Rosen, and K. M. Lyons. "Bone morphogenetic protein-3 is a negative regulator of bone density." [In eng]. *Nat Genet* 27, no. 1 (Jan 2001): 84–8.

Drake, A. G. "Dispelling dog dogma: An investigation of heterochrony in dogs using 3d geometric morphometric analysis of skull shape." [In eng]. *Evol Dev* 13, no. 2 (Mar-Apr 2011): 204–13.

Drake, A. G., and C. P. Klingenberg. "The pace of morphological change: Historical transformation of skull shape in st bernard dogs." [In eng]. *Proc Biol Sci* 275, no. 1630 (Jan 7 2008): 71–6.

Dreger, D. L., B. N. Hooser, A. M. Hughes, B. Ganesan, J. Donner, H. Anderson, L. Holtvoigt, and K. J. Ekenstedt. "True colors: Commercially-acquired morphological genotypes reveal hidden allele variation among dog breeds, informing both trait ancestry and breed potential." [In eng]. *PLoS One* 14, no. 10 (2019): e0223995.

Dupre, G., and D. Heidenreich. "Brachycephalic syndrome." [In eng]. *Vet Clin North Am Small Anim Pract* 46, no. 4 (Jul 2016): 691–707.

ENSEMBL_Genome_server_Canis_familiaris. "Runx2 Transcript Variants Table - Sift Analysis." http://www.ensembl.org/Canis_familiaris/Transcript/Variation_Transcript/Table?db=core;g=ENSCAFG000 00002008;r=12:13628539-13840785;t=ENSCAFT00000050131.

Evans, J., and de Lahunta A., *Miller's Anatomy of the Dog.* 4th ed.: Saunders/Elsevier, Philadelphia, PA, 2016.

FAO. "Breeding Strategies for Sustainable Management of Animal Genetic Resources." Food and Agricultural Organisation of the United Nations, http//:www.fao.org/3/i1103e/i1103e00.htm.

Farnworth, M. J., R. Chen, R. M. Packer, S. M. Caney, and D. A. Gunn-Moore. "Flat feline faces: Is brachycephaly associated with respiratory abnormalities in the domestic cat (Felis Catus)?" [In eng]. *PLoS One* 11, no. 8 (2016): e0161777.

Fondon 3rd, J. W., and H. R. Garner. "Detection of length-dependent effects of tandem repeat alleles by 3-D geometric decomposition of craniofacial variation." [In eng]. *Dev Genes Evol* 217, no. 1 (Jan 2007): 79–85.

Fondon 3rd, J. W., and H. R. Garner. "Molecular origins of rapid and continuous morphological evolution." [In eng]. *Proc Natl Acad Sci U S A* 101, no. 52 (Dec 28 2004): 18058–63.

Fox M. W. "Developmental Abnormalities of the Canine Skull." *Can J Comp Med Vet Sci* 27, no. 9 (Sep 1963): 219–22.

Ghirlanda, S., A. Acerbi, H. Herzog, and J. A. Serpell. "Fashion Vs. function in cultural evolution: The case of dog breed popularity." [In eng]. *PLoS One* 8, no. 9 (2013): e74770.

Gonzalez-Sancho, J. M., K. R. Brennan, L. A. Castelo-Soccio, and A. M. Brown. "Wnt proteins induce dishevelled phosphorylation via an lrp5/6- independent mechanism, irrespective of their ability to stabilize beta-catenin." [In eng]. *Mol Cell Biol* 24, no. 11 (Jun 2004): 4757–68.

Haimel, G., and G. Dupre. "Brachycephalic airway syndrome: A comparative study between pugs and french bulldogs." [In eng]. *J Small Anim Pract* 56, no. 12 (Dec 2015): 714–9.

Ho, H. Y., M. W. Susman, J. B. Bikoff, Y. K. Ryu, A. M. Jonas, L. Hu, R. Kuruvilla, and M. E. Greenberg. "Wnt5a-ror-dishevelled signaling constitutes a core developmental pathway that controls tissue morphogenesis." [In eng]. *Proc Natl Acad Sci U S A* 109, no. 11 (Mar 13 2012): 4044–51.

Johnson, L. R., P. D. Mayhew, M. A. Steffey, G. B. Hunt, A. H. Carr, and B. C. McKiernan. "Upper airway obstruction in norwich terriers: 16 Cases." [In eng]. *J Vet Intern Med* 27, no. 6 (Nov-Dec 2013): 1409–15.

Kennedy, L. J., A. Barnes, A. Short, J. J. Brown, S. Lester, J. Seddon, L. Fleeman, *et al.* "Canine dla diversity: 1. New alleles and haplotypes." [In eng]. *Tissue Antigens* 69, no. Suppl 1 (Apr 2007): 272–88.

Koch, D. A., and N. Sturzenegger. "Changes of the skull of brachycephalic dogs in the last 100 years]." [In ger]. *Schweiz Arch Tierheilkd* 157, no. 3 (Mar 2015): 161–3.

Koch, D., T. Wiestner, A. Balli, P. Montavon, E. Michel, G. Scharf, and S. Arnold. "Proposal for a new radiological index to determine skull conformation in the dog." [In eng]. *Schweiz Arch Tierheilkd* 154, no. 5 (May 2012): 217–20.

Kokabu, S., L. Gamer, K. Cox, J. Lowery, K. Tsuji, R. Raz, A. Economides, T. Katagiri, and V. Rosen. "Bmp3 suppresses osteoblast differentiation of bone marrow stromal cells via interaction with Acvr2b." [In eng]. *Mol Endocrinol* 26, no. 1 (Jan 2012): 87–94.

Ladlow, J., N. C. Liu, L. Kalmar, and D. Sargan. "Brachycephalic obstructive airway syndrome." [In eng]. *Vet Rec* 182, no. 13 (Mar 31 2018): 375–78.

Lewis T. W., B. M. Abhayaratne, S. C. Blott. "Trends in genetic diversity for all Kennel Club registered pedigree dog breeds." *Canine Genet Epidemiol* 2 (Sep 21 2015): 13.

Lewis, T. W., B. M. Wiles, A. M. Llewellyn-Zaidi, K. M. Evans, and D. G. O'Neill. "Longevity and mortality in kennel club registered dog breeds in the UK in 2014." [In eng]. *Canine Genet Epidemiol* 5 (2018): 10.

Lewis, T.W., and The_Kennel_Club. (2015) "Breed Population Analyses." Available from https:// www. thekennelclub/for-vets-and-researchers/kc-research-publication-and-health-data/breed-population-analyses/.

Lilja-Maula, L., A. K. Lappalainen, H. K. Hyytiainen, E. Kuusela, M. Kaimio, K. Schildt, S. Molsa, M. Morelius, and M. M. Rajamaki. "Comparison of submaximal exercise test results and severity of brachycephalic obstructive airway syndrome in english bulldogs." [In eng]. *Vet J* 219 (Jan 2017): 22–26.

Liu J. C., C. J. Lengner, T. Gaur, Y. Lou, S. Hussain, M.D. Jones, B. Borodic B, et al. "Runx2 protein expression utilizes the Runx2 P1 promoter to establish osteoprogenitor cell number for normal bone formation." *J Biol Chem* 286, no. 34 (Aug 26 2011): 30057–70. doi: 10.1074/jbc.M111.241505.

Liu, N. C., V. J. Adams, L. Kalmar, J. F. Ladlow, and D. R. Sargan. "Whole-body barometric plethysmography characterizes upper airway obstruction in 3 brachycephalic breeds of dogs." [In eng]. *J Vet Intern Med* 30, no. 3 (May 2016): 853–65.

Liu, N. C., D. R. Sargan, V. J. Adams, and J. F. Ladlow. "Characterisation of brachycephalic obstructive airway syndrome in French bulldogs using whole-body barometric plethysmography." [In eng]. *PLoS One* 10, no. 6 (2015): e0130741.

Liu, N. C., E. L. Troconis, L. Kalmar, D. J. Price, H. E. Wright, V. J. Adams, D. R. Sargan, and J. F. Ladlow. "Conformational risk factors of brachycephalic obstructive airway syndrome (boas) in pugs, french bulldogs, and bulldogs." [In eng]. *PLoS One* 12, no. 8 (2017): e0181928.

Lou Y., A. Javed, S. Hussain, J. Colby, D. Frederick, J. Pratap, R. Xie, et al. "A Runx2 threshold for the cleidocranial dysplasia phenotype." *Hum Mol Genet* 18, no. 3 (Feb 1 2009): 556–68. doi: 10.1093/hmg/ddn383.

Mansour, T. A., K. Lucot, S. E. Konopelski, P. J. Dickinson, B. K. Sturges, K. L. Vernau, S. Choi, et al. "Whole genome variant association across 100 dogs identifies a frame shift mutation in dishevelled 2 which contributes to Robinow-like syndrome in bulldogs and related screw tail dog breeds." [In eng]. *PLoS Genet* 14, no. 12 (Dec 2018): e1007850.

Marchant, T. W., E. Dietschi, U. Rytz, P. Schawalder, V. Jagannathan, S. Hadji Rasouliha, C. Gurtner, et al. "An adamts3 missense variant is associated with norwich terrier upper airway syndrome." [In eng]. *PLoS Genet* 15, no. 5 (May 2019): e1008102.

Marchant, T. W., E. J. Johnson, L. McTeir, C. I. Johnson, A. Gow, T. Liuti, D. Kuehn, et al. "Canine brachycephaly is associated with a retrotransposon-mediated missplicing of Smoc2." [In eng]. *Curr Biol* 27, no. 11 (Jun 5 2017): 1573–84.

McGreevy P., T. D. Grassi, A. M. Harman. A strong correlation exists between the distribution of retinal ganglion cells and nose length in the dog. *Brain Behav Evol* 63, no. 1 (2004): 13–22.

McGreevy P.D., D. Georgevsky, J. Carrasco, M. Valenzuela, D.L. Duffy, J.A. Serpell. Dog behavior co-varies with height, bodyweight and skull shape. *PLoS One* 8, no. 12 (Dec 16 2013): e80529.

Melvin, V. S., W. Feng, L. Hernandez-Lagunas, K. B. Artinger, and T. Williams. "A morpholino-based screen to identify novel genes involved in craniofacial morphogenesis." [In eng]. *Dev Dyn* 242, no. 7 (Jul 2013): 817–31.

Michell, A. R. " Longevity of british breeds of dog and its relationships with sex, size, cardiovascular variables and disease." *Vet Record* 145, no. 22 (1999): 625–29.

Mommaerts, H., C. V. Esguerra, U. Hartmann, F. P. Luyten, and P. Tylzanowski. "Smoc2 modulates embryonic myelopoiesis during zebrafish development." [In eng]. *Dev Dyn* 243, no. 11 (Nov 2014): 1375–90.

Mouse Genome Informatics (MGI). Phenotypes associated with Smocltm1b(EUCOMM)Wtsi/Smocltm1b(EUCOMM)Wtsi 2020. Available from: http://www. informatics.jax.org/diseasePortal/genoCluster/view/18208. Accessed 18.05.2020

Muller, G. "Diseases of the dog and their treatment." Translated by Dr A Glass. Chap. Diseases of the Respiratory Organs In *Diseases of the Dog and Their Treatment*, 102. Philadelphia, PA: W Horace Hoskins, 1891.

Ng, P. C., and S. Henikoff. "Sift: Predicting amino acid changes that affect protein function." [In eng]. *Nucleic Acids Res* 31, no. 13 (Jul 1 2003): 3812–4.

Nicholas, F. W., E. R. Arnott, and P. D. McGreevy. "Hybrid vigour in dogs?" [In eng]. *Vet J* 214 (Aug 2016): 77–83.

Nishita, M., S. Itsukushima, A. Nomachi, M. Endo, Z. Wang, D. Inaba, S. Qiao, et al. "Ror2/Frizzled complex mediates Wnt5a-Induced Ap-1 activation by regulating dishevelled polymerization." [In eng]. *Mol Cell Biol* 30, no. 14 (Jul 2010): 3610–9.

Nussbaumer, M. "Biometric analysis of the skull base in small and medium sized dogs (German)." *Z. Tierzüchtg. Züchtungsbiol.* 95 (1978): 1–14.

O'Neill D. G., D. B. Church, P. D. McGreevy, P. C. Thomson, D. C. Brodbelt. "Longevity and mortality of owned dogs in England." *Vet J* 198, no. 3 (2013): 638–43.

O'Neill, D. G., L. Baral, D. B. Church, D. C. Brodbelt, and R. M. A. Packer. "Demography and disorders of the french bulldog population under primary veterinary care in the UK in 2013." [In eng]. *Canine Genet Epidemiol* 5 (2018): 3.

O'Neill, D. G., E. C. Darwent, D. B. Church, and D. C. Brodbelt. "Demography and health of pugs under primary veterinary care in england." [In eng]. *Canine Genet Epidemiol* 3 (2016): 5.

O'Neill, D. G., A. M. Skipper, J. Kadhim, D. B. Church, D. C. Brodbelt, and R. M. A. Packer. "Disorders of bulldogs under primary veterinary care in the UK in 2013." [In eng]. *PLoS One* 14, no. 6 (2019): e0217928.

Ostrander, E. A., R. K. Wayne, A. H. Freedman, and B. W. Davis. "Demographic history, selection and functional diversity of the canine genome." [In eng]. *Nat Rev Genet* 18, no. 12 (Dec 2017): 705–20.

Packer, R. M., A. Hendricks, and C. C. Burn. "Impact of facial conformation on canine health: Corneal ulceration." [In eng]. *PLoS One* 10, no. 5 (2015a): e0123827.

Packer, R. M., A. Hendricks, M. S. Tivers, and C. C. Burn. "Impact of facial conformation on canine health: Brachycephalic obstructive airway syndrome." [In eng]. *PLoS One* 10, no. 10 (2015b): e0137496.

Parker, H. G., D. L. Dreger, M. Rimbault, B. W. Davis, A. B. Mullen, G. Carpintero-Ramirez, and E. A. Ostrander. "Genomic analyses reveal the influence of geographic origin, migration, and hybridization on modern dog breed development." [In eng]. *Cell Rep* 19, no. 4 (Apr 25 2017): 697–708.

Patton, M. A., and A. R. Afzal. "Robinow syndrome." [In eng]. *J Med Genet* 39, no. 5 (May 2002): 305–10.

Pedersen, N. C., A. S. Pooch, and H. Liu. "A genetic assessment of the english bulldog." [In eng]. *Canine Genet Epidemiol* 3 (2016): 6.

Person, A. D., S. Beiraghi, C. M. Sieben, S. Hermanson, A. N. Neumann, M. E. Robu, J. R. Schleiffarth, et al. "Wnt5a mutations in patients with autosomal dominant robinow syndrome." [In eng]. *Dev Dyn* 239, no. 1 (Jan 2010): 327–37.

Pet Food Manufacturers Association Pet Data Report 2018 web page https://www.pfma.org.uk/_assets/docs/annual-reports/PFMA-Pet-Data-Report-2018.pdf. Accessed 18.05.2020

Plassais J., J. Kim, B. W. Davis, D. M. Karyadi, A. N. Hogan, A. C. Harris, B. Decker, et al. "Whole genome sequencing of canids reveals genomic regions under selection and variants influencing morphology." *Nat Commun* 10, no. 1 (Apr 2 2019): 1489.

Ramesar, R.S., Greenberg, J., Martin, R., Goliath, R., Bardien, S., Mundlos, S., Beighton, P. "Mapping of the gene for cleidocranial dysplasia in the historical cape town (Arnold) Kindred and evidence for locus homogeneity." *J Med Genet.* 33, no. 6 (1996): 511–14.

Regodón S., J. M Vivo, A. Franco, M. T. Guillén, A. Robina. "Craniofacial angle in dolicho-, meso- and brachycephalic dogs: radiological determination and application." *Ann Anat* 175, no. 4 (Aug 1993): 361–63.

Riggs, J., N. C. Liu, D. R. Sutton, D. Sargan, and J. F. Ladlow. "Validation of exercise testing and laryngeal auscultation for grading brachycephalic obstructive airway syndrome in pugs, french bulldogs, and english bulldogs by using whole-body barometric plethysmography." [In eng]. *Vet Surg* 48, no. 4 (May 2019): 488–96.

Roifman M., H. Brunner, J. Lohr, J. Mazzeu, D. Chitayat. Autosomal Dominant Robinow Syndrome. 2015 Jan 8 [updated Oct 3 2019]. In: Adam M. P., H. H. Ardinger, R. A. Pagon, S. E. Wallace, L. J. H. Bean, K. Stephens, A. Amemiya, editors. *GeneReviews®* [Internet]. Seattle, WA: University of Washington, 1993–2020.

Rutherford, L., L. Beever, M. Bruce, and G. Ter Haar. "Assessment of computed tomography derived cricoid cartilage and tracheal dimensions to evaluate degree of cricoid narrowing in brachycephalic dogs." [In eng]. *Vet Radiol Ultrasound* 58, no. 6 (Nov 2017): 634–46.

Ryan R, R.Gutierrez-Quintana, G.T. Haar, S. De Decker. "Relationship between breed, hemivertebra subtype, and kyphosis in apparently neurologically normal French Bulldogs, English Bulldogs, and Pugs." *Am J Vet Res* 80, no. 2 (Feb 2019): 189–94.

Sarper, S. E., T. Inubushi, H. Kurosaka, H. Ono Minagi, K. I. Kuremoto, T. Sakai, I. Taniuchi, and T. Yamashiro. "Runx1-Stat3 signaling regulates the epithelial stem cells in continuously growing incisors." [In eng]. *Sci Rep* 8, no. 1 (Jul 19 2018a): 10906.

Sarper, S. E., H. Kurosaka, T. Inubushi, H. Ono Minagi, K. I. Kuremoto, T. Sakai, I. Taniuchi, and T. Yamashiro. "Runx1-Stat3-Tgfb3 signaling network regulating the anterior palatal development." [In eng]. *Sci Rep* 8, no. 1 (Jul 25 2018b): 11208.

Safra N., R. H. Schaible, D. L. Bannasch. "Linkage analysis with an interbreed backcross maps Dalmatian hyperuricosuria to CFA03." *Mamm Genome* 17, no. 4 (Apr 2006): 340–5.

Schoenebeck, J. J., S. A. Hutchinson, A. Byers, H. C. Beale, B. Carrington, D. L. Faden, M. Rimbault, et al. "Variation of Bmp3 contributes to dog breed skull diversity." [In eng]. *PLoS Genet* 8, no. 8 (2012): e1002849.

Schoenebeck, J. J., and E. A. Ostrander. "The genetics of canine skull shape variation." [In eng]. *Genetics* 193, no. 2 (Feb 2013): 317–25.

Schoenebeck, J. J., and E. A. Ostrander. "Insights into morphology and disease from the dog genome project." [In eng]. *Annu Rev Cell Dev Biol* 30 (2014): 535–60.

Secretary General of the Council of Europe. "European Convention for the Protection of Pet Animals, Article 5, 1987." https://treaties.un.org/doc/Publication/UNTS/Volume%201704/volume-1704-I-29470-English.pdf.

Serpell, J. A. "Anthropomorphism and anthropomorphic selection: Beyond the "Cute Response"." *Soc. Anim.* 10, no. 4 (2002): 437–54.

Stockard, C. R., Anderson, O.D., James, W. T. *He Genetic and Endocrinic Basis for Differences in Form and Behavior.* Wistar Institute of Anatomy and Biology, Philadelphia: The Wistar Institute of Anatomy and Biology, 1941.

Tarricone J., G. M., Hayes, A. Singh, G. Davis. "Development and validation of a brachycephalic risk (BRisk) score to predict the risk of complications in dogs presenting for surgical treatment of brachycephalic obstructive airway syndrome." *Vet Surg* 48, no. 7 (Oct 2019): 1253–61.

The_Kennel_Club. "Pedigree Breed Health Surveys." 2014. https://www.thekennelclub.org.uk/pedigreebreed healthsurvey.

The_Kennel_Club. "Purebred Breed Health Surveys." 2004 https://www.thekennelclub.org.uk/for-vets-and-researchers/purebred-breed-health-survey-2004/.

The Kennel Club. "Breed Registration Statistics." The Kennel Club, http://www.thekennelclub.org.uk/registration/breed-registration-statistics/.

Theunert, C., K. Tang, M. Lachmann, S. Hu, and M. Stoneking. "Inferring the history of population size change from genome-wide snp data." [In eng]. *Mol Biol Evol* 29, no. 12 (Dec 2012): 3653–67.

UK Government. "The Animal Welfare Act 2006, (Licensing of Activities Involving Animals) (England) Regulations 2018; Breeding Dogs." https://www.legislation.gov.uk/ukdsi/2018/9780111165485/pdfs/ukdsi_9780111165485_en.pdf.

van Hagen, M.A.E. "Breeding Short-Muzzled Dogs: Criteria for the Enforcement of Article 3.4. Of the Animal Keepers Decree (Besluit Houders Van Dieren)." Universiteit Utrecht, 2019.

White, J. J., J. F. Mazzeu, A. Hoischen, Y. Bayram, M. Withers, A. Gezdirici, V. Kimonis, et al. "Dvl3 alleles resulting in a -1 frameshift of the last exon mediate autosomal-dominant robinow syndrome." [In eng]. *Am J Hum Genet* 98, no. 3 (Mar 3 2016): 553–61.

Wykes, P. M. "Brachycephalic airway obstructive syndrome." [In eng]. *Probl Vet Med* 3, no. 2 (Jun 1991): 188–97.

Yamashiro, T., T. Aberg, D. Levanon, Y. Groner, and I. Thesleff. "Expression of runx1, -2 and -3 during tooth, palate and craniofacial bone development." [In eng]. *Mech Dev* 119, no. Suppl 1 (Dec 2002): S107–10.

Yamashiro, T., X. P. Wang, Z. Li, S. Oya, T. Aberg, T. Fukunaga, H. Kamioka, et al. "Possible roles of runx1 and sox9 in incipient intramembranous ossification." [In eng]. *J Bone Miner Res* 19, no. 10 (Oct 2004): 1671–7.

Yang, J., S. H. Lee, M. E. Goddard, and P. M. Visscher. "Gcta: A tool for genome-wide complex trait analysis." [In eng]. *Am J Hum Genet* 88, no. 1 (Jan 7 2011): 76–82.

Yang, J., S. H. Lee, M. E. Goddard, and P. M. Visscher. "Genome-wide complex trait analysis (Gcta): Methods, data analyses, and interpretations." [In eng]. *Methods Mol Biol* 1019 (2013): 215–36.

Yang, J., S. H. Lee, N. R. Wray, M. E. Goddard, and P. M. Visscher. "Gcta-Greml accounts for linkage disequilibrium when estimating genetic variance from genome-wide snps." [In eng]. *Proc Natl Acad Sci U S A* 113, no. 32 (Aug 9 2016): E4579–80.

Zhang, Y. W., N. Yasui, K. Ito, G. Huang, M. Fujii, J. Hanai, H. Nogami, et al. "A Runx2/Pebp2alpha a/Cbfa1 mutation displaying impaired transactivation and smad interaction in cleidocranial dysplasia." [In eng]. *Proc Natl Acad Sci U S A* 97, no. 19 (Sep 12 2000): 10549–54.

9 International and National Approaches to Brachycephalic Breed Health Reforms in Dogs

Brenda N. Bonnett
International Partnership for Dogs

Monique Megens
Plan B Veterinair

Dan G. O'Neill
The Royal Veterinary College

Åke Hedhammar
Swedish University of Agricultural Sciences

CONTENTS

Background and Introduction 127
Historical Perspective and Stakeholders 128
    Stakeholders in Showing and Breeding of Brachycephalic Dogs 134
    Veterinarians and the Veterinary Profession 136
Specific Actions and Efforts 137
    Education and Raising Awareness 138
    Commissions, *Working Groups, Statements* 138
        Sample Statements 139
        Proposed Actions 139
    Actions to Improve the Quality of the Dogs Supplied 140
    Actions Addressing Supply, Demand, Trade and Transport of Dogs 140
        Multi-national/European Efforts by Authorities 141
        The European Union (EU) – The European Parliament (EP) – The EU Council 142
    Regulatory Approaches – Breed-Specific 142
        National Example: The Netherlands 142
    International Multi-stakeholder 144
Conclusions 144
References 145

BACKGROUND AND INTRODUCTION

The issues affecting brachycephalic breed health and welfare are wide-ranging and complex for individual dogs, breeds, populations, dog owners, researchers, veterinarians and society in general. As demonstrated in Table 9.1 and below, although the timeline of attention to these issues spans at least the past six decades, many of the same challenges persist. Although scientific and clinical knowledge have increased markedly, reliable data on the true prevalence of severe health and welfare conditions within and across these breeds has been slow to accrue, or to be effectively utilised even when available (IPFD 2020c). Notwithstanding the limitations of current scientific knowledge, human behaviour is recognised increasingly as an integral and impactful component of the issue, sometimes related more to emotion than evidence. Consequently, expertise from the fields of communication and even psychology is urgently needed. At a societal level, public concern and even outrage about poor dog welfare, in general, and for brachycephalic breeds, in specific, have motivated regulators and government agencies, with issues arising from and influenced by long held and even historical attitudes. However, those in favour of promoting brachycephalic types of dogs/ breeds are also passionate and committed. Cultural, socio-economic and national differences complicate the situation across countries and groups (Bonnett 2020). Sourcing of dogs, breeding, trade, transport and marketing are factors involved in the interplay between animal welfare and the human desires to breed and own brachycephalic dogs. Unfortunately, the need to address public opinion may result in actions that fail to adequately take into consideration the complexity of the problem(s) or fully engage the range of parties needed for successful actions and outcomes (Bonnett 2019a, DogWellNet 2020b, IPFD 2020b). Beyond varied local challenges, brachycephalic health and welfare epitomise a dog problem requiring international efforts across multiple stakeholder groups.

A 2018 paper from Australia covers the breadth and depth of the brachycephalic issue, clinically and ethically. The authors conclude that "Veterinarians have a professional and moral obligation to prevent and minimise the negative health and welfare impacts of extreme morphology and inherited disorders, and they must address brachycephalic obstructive airway syndrome (BOAS) not only at the level of the patient, but also as a systemic welfare problem" (Fawcett et al. 2018). In this chapter, we focus mainly on how challenges are being addressed at the level of veterinary and cynological organisations. However, in terms of understanding the complexity from the perspective of a veterinarian in practice, it is worth noting that the best resolution for concurrent issues is not always clear. Competing interests and factors include the best interests of an individual dog overall or in relation to a specific health event; the owner's attachment, attitudes, wishes, needs and ability to recognise the issue and provide care; concerns for the breed overall; as well as the practical reality of the veterinarian as both a caregiver and a business person (Bonnett 2019a). The interplay between personal and collective responsibility is one that is a challenge for those from any stakeholder group, especially when there may not even be consensus on what responsible dog ownership entails (Bonnett 2019a, Westgarth et al. 2019, Bonnett 2019c).

This chapter will describe issues and actions across regions and countries, primarily Europe and North America, encompassing various stakeholder groups including kennel and breed clubs, veterinary organisations, research institutions and regulators. An historical perspective is included, as this informs efforts and challenges in many regions. The stage of development, degree and focus of interest on the brachycephalic issue are diverse, and there is a real need to combine forces to prevent redundant efforts and to maximise the sharing of resources. The case will be made that an even more collective approach to the problem could improve the situation for brachycephalic dogs in a timelier fashion. Work on the brachycephalic issue will – and should – impact and inform interest in other issues related to extreme characteristics in dogs.

HISTORICAL PERSPECTIVE AND STAKEHOLDERS

Due to their rapidly growing numbers and severe breed-specific health and welfare problems, brachycephalic breeds have increasingly come under a worldwide spotlight over recent years. The issues are diverse and include concerns associated with the sources and channels that supply the high and growing demand for these breeds. As stated above, and illustrated in Figure 9.1, almost everyone involved with brachycephalic breeds contributes in some way to these challenges and is impacted by the attitudes and actions of other stakeholders. This complex problem cannot be resolved by any one group acting alone. In fact, isolated and poorly coordinated efforts have sometimes only increased

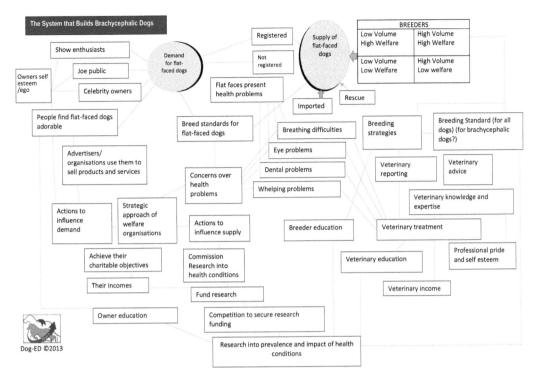

FIGURE 9.1 Graphic representation of 'The System That Builds Brachycephalic Dogs'.

the polarity across individuals and groups, or at least the most vocal or radical among them, without substantially improving the situation for the dogs themselves. To increase the success of local, regional and international efforts, an understanding of the separate and collective approaches is necessary, as is an understanding of the emotional and human side of the equation. In addition, as outlined in Table 9.1, examining the historical perspective not only helps to highlight past and current actions, but also creates a base from which to project forward to identify issues, barriers and actions that impact the likely success of future endeavours.

Table 9.1 presents a timeline for a selected list of multi-stakeholder attentions to brachycephalic health and welfare, with a critical commentary on the example or impact of these efforts. Some key takeaways from Table 9.1 include

- The realisation that aspects of the conformation of brachycephalic breeds contribute directly to reduced health and welfare for individuals of those breeds was being noticed and widely discussed as early as the 1960s and 1970s.
- Cynological organisations, including national kennel clubs, together with veterinarians and researchers identified a link with breed standards, insofar as some characteristics, presumably defined as 'ideal' and deemed as necessary or desirable in the show ring, also predisposed the dogs to health problems including dystocia, eye abnormalities and other issues.
- There have been repeated calls for, and initiatives undertaken, to re-evaluate and change breed standards to help reduce their negative impacts on health and welfare. However, barriers to the effectiveness of such efforts have included an inability of authorities (including cynological organisations) to compel the country of origin, as "the owners" of various breed standards, to make substantial changes, e.g. to limit extremes or to include more objective criteria; and a culture in the dog show community of not monitoring, evaluating

TABLE 9.1
Historical Timeline of International Focus on Brachycephalic Health and Welfare

Timeline/ Stakeholders	Topic	Comments
1960s Cynological organisations; Nordic Kennel Clubs; WSAVA (veterinary organisations), researchers	Early international recognition of exaggerations of anatomical characteristics as a risk factor for canine health and its relation to breed standards. A scientific conference in conjunction with the General Assembly of Federation Cynologic Internationale (FCI) in Stockholm, Sweden, **1964** as reported in Nord Vet. Medicine (Hansen 1964). **Actions: established a Committee** *to study the standards of every registered breed and to report on those which hinder physiological functions of organs or parts of the body.* The work of that group was reported in *Journal of Small Animal Practice* 1969 (Henricson 1969) and at the World Veterinary Congress, Paris 1967.	Multi-stakeholder approach initiated by veterinarians and geneticists involved in kennel clubs. Focus on extreme characteristics and health not restricted to brachycephalics due to the relatively low numbers within these breeds.
1970s Cynological organisations; kennel clubs (including the UK)	**Actions:** Extensive work to have changes in breed standards implemented was hampered by the concept that the country of origin 'owns' the text. Rather than robust changes, addenda to breed standards were introduced. When even that failed to create meaningful change, new programs for training of judges on how to interpret and apply breed standards were developed and enforced, e.g. in Sweden as the Breed-Specific Instructions (The Nordic Kennel Clubs 2018).	The culture, historical attitudes and beliefs in the dog show community seem to have limited promotion of more moderate conformation.
1980s–1990s Government/ regulatory; researchers; welfare groups	During the 1980s, Harold R. 'Harry' Spira, an Australian veterinarian and all breed judge toured extensively nationally and internationally with talks on breed standards and health, giving examples on all exaggerations of current concern.	Another example of an initiative by veterinarians with kennel club involvement.
	International attention to the issue as new research started to emerge. Developing actions in Europe included: *Actions: The European Convention for the Protection of Pet Animals of November 13, 1987, included general rules on breeding of pet animals and an additional resolution concerning breeding of dogs and cats;* Council of Europe, 1995a,b – Multilateral consultation of parties to the European Convention for the protection of pet animals (ETS 123) (Council of Europe 2020). Included were limits to the shortness of skull and nose, so that breathing difficulties and blockage of lacrimal ducts are avoided, as well as disposition to birth difficulties (e.g. Bulldogs, Japan Chin, King Charles Spaniel, Pug, Pekin Palace dog).	This was mostly driven by a European authority and welfare groups. However, evidence of implementation, monitoring, follow-up and specific outcomes is lacking.
	Research on upper airway obstruction in mainstream veterinary journals in Europe and North America (Aron and Crowe 1985, Wykes 1991, Hobson 1995). Introduction of the Brachycephalic Syndrome 1991 (Wykes 1991).	Raised awareness as a problem in individual dogs of these breeds more so than a breed population issue.
2000 Veterinary/ research	The concept of a Brachycephalic syndrome from late 1980s now reframed as brachycephalic obstructive airway syndrome (BOAS); research addressed issues beyond primarily respiratory and discussed surgical correction (Poncet et al. 2005, Pink et al. 2006, Torrez and Hunt 2006). Raised attention as a major welfare problem first in dogs but then also in cats.	In multiple countries, e.g. France, Ireland, Australia.
Humane/welfare	Campaigns to influence the public against brachycephalic breeds.	Raising awareness, extreme efforts about extreme conformation

(Continued)

TABLE 9.1 (*Continued*)

Historical Timeline of International Focus on Brachycephalic Health and Welfare

Timeline/ Stakeholders	Topic	Comments
2008 TV/media-prompted, widescale, multi-stakeholder and multi-national attention	Pedigree Dogs Exposed was originally broadcast on August 19, 2008 (BBC 2008). Brought breed health issues into the public domain, with graphic and provocative presentation. Initially, the impact was primarily in the UK and Europe; however, this and the subsequent actions (see below) stimulated participation by many other stakeholders and highlighted the need for international attention.	Although provocative, this event prompted kennel clubs to highlight ongoing work and create new initiatives. It also underlined the public's right to a say in dog welfare issues; stimulated actions by various authorities (see below).
2008–2014 Authorities; kennel clubs; veterinary organisations	Several commission reports and working groups; investigation into nature and extent of issues from public perspective; identification of responsibilities; promotion of research and collaborative efforts (Bateson 2010, Rooney and Sargan 2008, APGAW 2009).	Many still ongoing; outcomes and metrics? Mainly independent initiatives nationally versus internationally.
	Kennel clubs initiated or enhanced programs involving breeder and judge education and changes, e.g. Breed-Specific Instructions (The Nordic Kennel Clubs 2018) and Breed Watch (The Kennel Club 2020b).	Outcomes? Metrics? -------------------------------
Welfare/humane groups	Humane group focus on public education campaigns; e.g. RSPCA ("Bred for Looks, Born to Suffer" (Stilwell 2011).	
	Humane Society of the USA hosted the Purebred Paradox meeting, Washington, DC, USA (Allan 2009).	This was a multi-stakeholder meeting.
	Campaigners continue their work, e.g. PDE The Blog (Harrison 2020b).	Metrics? Impact? (Mills 2018)
	Multi-stakeholder efforts to promote responsible use of dogs in advertising and media, e.g. CRUFFA – targeting celebrities and animal industry advertisers (Harrison 2020a), have engaged veterinary organisations (BVA), welfare (Pets Trust) and others.	
National data collection on dog health	From the mid-1990s, Swedish insurance data on millions of dog-years-at-risk and breed-specific statistics have been published in the refereed literature and in breed profiles for breeders and breed clubs. The latter have been used in breed-specific breeding strategies, especially in Scandinavian countries (IPFD 2021d).	
	In the UK, VetCompass began collecting anonymised veterinary clinical records to measure canine health in 2010 and has published over 95 papers related to canine health up to mid 2021. VetCompass holds data on over 8 million UK dogs (VetCompass 2021). Similar approaches began later with PETscan in The Netherlands (PETscan 2021) and VetCompass Australia (2021).	
Veterinary organisations and conferences 2012 IDHW and early IPFD 2015	Increased attention to health AND welfare, and breeding. Proclamations and position statements. Identification of the extent of normalisation of health issues by owner AND vets; (WSAVA and FECAVA 2017, WSAVA 2018, FVE and FECAVA 2018, BVA 2020a, b).	Even with increased efforts at collaboration, there is also increasing polarity within and across stakeholder groups.

(Continued)

TABLE 9.1 (*Continued*)

Historical Timeline of International Focus on Brachycephalic Health and Welfare

Timeline/ Stakeholders	Topic	Comments
Veterinary organisations and conferences	Calls for data on prevalence and severity of the condition within the population; claims on both sides of the issue, in the absence of data, fuelled the debate.	Still a difference of opinion about extent and severity within breeds.
2012 IDHW and early IPFD 2015	In summer 2015, more than 500 Swedish veterinarians forwarded a petition to the Swedish board of agriculture with the intention "to bring about collaboration between clinically active veterinarians, the Swedish Kennel Club (SKK) and the Swedish Board of Agriculture for the right of snub nosed dogs to breathe normally without an operation" (Rodin and Rundkvist 2020). They had six suggestions for how to come to terms with the problems. The Brachycephalic Working Group was formed in the UK in 2015 (Brachycephalic Working Group. 2019).	Followed by a petition from UK veterinarians. These efforts did spawn initiatives. However, the way in which they were initiated also promoted distrust and polarisation. Challenges continue.
2015-Current IPFD DogWellNet.com	Curated compilation of information, articles, research, videos … multi-stakeholder development. Archived historical by year (from 2015) (DogWellNet 2020e, c). Highlights the plethora of efforts and information from many stakeholder groups (see Table 9.2).	These links are dynamic; material will continue to be updated.
2012, 2015, 2017, 2019 IPFD International Dog Health Workshops (IDHWs)	**International multi-stakeholder meetings with Exaggeration as one specific theme.** Exaggeration as a specific theme continued with evolution of focus within IDHWs (DogWellNet 2020f). **2012** Raise awareness. **2015** Expand awareness and discuss need to determine the nature of the issues; what is needed to define the problem. **2017** No further discussion of '*if*' there is a problem – Multi-stakeholder, international consensus that there are major health and welfare issues in brachys … now a focus on what to do about it; sharing approaches etc.: many current approaches to education not having a major impact … need for Human Behaviour Change (HBC) approach (O'Neill, Keijser, et al. 2017). **2019** Need to change communication strategies, encourage HBC, coordinate and consolidate strategies; need for international working group (see below) (Pegram et al. 2020).	IDHWs were initially a Swedish Kennel Club initiative to bring stakeholders together, that introduced the concept of an international platform, which in turn resulted in the International Partnership for Dogs (IPFD) and ongoing IDHWs.
Paradox	During a period of repeated public education campaigns to dissuade owners, influence breeders, vets, etc., there was concurrent massive upsurge in numbers within affected breeds, especially French Bulldogs. Whereas concerted campaigns against purebred dogs, in general, with increase in demand for 'rescues' have been successful.	New research on human attitudes to dog health – health not the primary driver in pet acquisition for many (Sandøe et al. 2017, Bonnett 2019d, b). And this has resulted in new and ongoing welfare and health issues related to supply, etc.

(Continued)

or compelling judges to avoid rewarding dogs with extreme characteristics. Presumably, there are those among the dog fancy who do not believe that the extreme conformations described in some breed standards have much influence on the health of the general breed population. However, in contrast, there is a general consensus among many other groups, e.g. veterinarians and welfare groups, that statements such as 'the nose should be as short

TABLE 9.1 (*Continued*)

Historical Timeline of International Focus on Brachycephalic Health and Welfare

Timeline/ Stakeholders	Topic	Comments
2020 and beyond	The International Collaborative on Extreme Conformation in Dogs was formed in 2020, based on recommended actions at the fourth IDHW.	This group will collate resources for use by new and existing local and regional multi-stakeholder groups.
	There are many national efforts (Brachycephalic Working Group, 2019); however, recent regulatory efforts in the Netherlands and elsewhere (Bonnett 2019a, IPFD 2020b) have highlighted the challenges that arise when limited rather than inclusive actions are initiated without due consideration of the complexity of factors and players involved in the health and welfare of brachycephalic dogs.	
	Clearly, the public continues to embrace extremes, memes of perceived 'funny' (but not truly funny) dogs through ignorance, or denial? Is it 'Don't know or don't care'? Time will tell.	

as possible' from the American standard for the English Bulldog (American Kennel Club 2020) are unacceptable. Specific efforts related to standards and judges are discussed below.

- Differences in attitudes on diverse topics across stakeholder groups complicate collaborative efforts, regardless of the level of available statistical or scientific evidence on the topic.
- There continue to be calls for further actions, at all levels and across stakeholder groups.
- There has been a substantial and increasing body of both qualitative and quantitative scientific evidence, and media coverage on brachycephalic issues. Recently, there has been further elucidation of the nature, extent and complexity of the human side of this issue as well as individual dog and breed population impacts (Packer, Murphy, and Farnworth 2017, Packer, Hendricks, and Burn 2012, Packer et al. 2019, Sandøe et al. 2017, Bonnett 2019d); and various commissions and regulatory interventions have allocated responsibility and proposed actions (BWG 2020). There is a now sustained momentum for change, although often without clear outcomes for the dogs themselves.
- Perhaps the biggest challenge is the diversity of stakeholders and the complex interactions and inputs that come together under the banner of resolving the issues of brachycephalic health and welfare. Even collaborative efforts have failed to holistically address or coordinate all the issues, and no sole regulatory body or stakeholder group can mandate change. The lack of monitoring, metrics and outcomes assessment has perhaps contributed to the slow progress seen on issues of brachycephalic health and welfare, and limit our ability to determine the most effective strategies.
- The recognition of the need for efforts to be coordinated across stakeholders helped drive the creation of the International Partnership for Dogs (IPFD) and their platform DogWellNet.com (International Partnership for Dogs 2020), and has fuelled the ongoing IPFD International Dog Health Workshops (IDHWs) (Pegram et al. 2020, O'Neill, Keijser, et al. 2017). A wealth of information on the diverse initiatives is now being catalogued on DogWellNet.com, an information hub that is valuable for those initiating efforts or wanting to see what is being/has been done, by whom and where.
- Given that, even now, certain stakeholders/groups are yet to reach agreement on e.g. the true scale of the problem, need for action, severity and numbers affected, then it remains to be seen how well new international resources and actions will facilitate meaningful progress.

Table 9.1 follows a chronological perspective; Table 9.2 lists types of actions and challenges associated with them. A few more issues are worth noting that apply across all efforts.

- Social media impacts hugely on all aspects of the brachycephalic issue. While many groups try to educate the public on challenges in the breed or promote responsibility in advertising and the media (British Veterinary Association 2020, Harrison 2020a), memes of 'funny' brachycephalics continue to go viral online. While there are efforts to engage in meaningful dialog and collaboration, the attacks within and across stakeholder groups often become personal and vituperative.
- Despite a scarcity of population data and substantial differences across regions, there is a tendency to generalise concerns to 'all brachycephalics' or 'all Pugs'. However, a more key 'population' issue is that, e.g., the Pugs seen by judges in the show rings may be quite different from those seen in daily veterinary practice; perhaps, these show dogs reflect the 'best of the best' and the veterinary patients reflect the 'worst of the worst', leading to different stakeholders holding different beliefs and therefore making contrasting claims that the majority of dogs are either 'very healthy' or 'all sick'.
- The world in general is learning that propounding facts is not the most effective way to change attitudes for all groups of people (Bonnett 2020). This underlines the need for new approaches to communication and education, especially where there is evidence that repeated attempts at suggesting consumers avoid brachycephalics have been concurrent with a massive upsurge in popularity (The Kennel Club 2020a, Bonnett 2015). Understanding the drivers of both adverse and preferred behaviour must be sought (Holland 2019).
- Nothing exists in a vacuum and focus/priorities may change over time. For example, in countries where dog breeding comes under the supervision of agriculture departments, the emphasis and support for brachycephalic efforts may wax and wane as alternative urgent issues arise in other species. A similar situation may arise relative to the changing prioritisations of welfare campaigners or humane groups.
- Unfortunately, but not atypical of many health and welfare situations, many – if not most – initiatives by various stakeholder groups have lacked clear outcome definitions or monitoring strategies that could provide the metrics by which to measure success. Recommendations and programs have been instituted without the structure or enforcement needed for meaningful realisation. The lack of bodies with a mandate to enforce change across stakeholders, internationally or even nationally, certainly impacts the possibilities. When regulatory bodies have become engaged, it may be without a full understanding of the complexity of the issue that consequently limits activity to narrow, prescriptive acts (DogWellNet 2020b).

Building on material in Tables 9.1 and 9.2, the following section provides further comments from the perspective of specific stakeholder groups.

Stakeholders in Showing and Breeding of Brachycephalic Dogs

Although definitions may be blurred, there is a general differentiation between breeders who are involved in the show or pedigree world and associated with kennel and breed clubs versus other more casual or commercial breeders. The former group is the highly visible one, and consequently, the major responsibility for problems regarding the trends towards extremes in various breeds is often placed on show breeders and organisations. Comparisons between historical versions of these extreme breeds and today's dogs are often used to 'confirm' a drift towards increasingly extreme characteristics and the potential negative influence of breed standards; these topics are mentioned in Tables 9.1 and 9.2, as is the role of show judges. However, it is increasingly apparent that the proportion of dogs, especially in certain breeds and countries, being registered with major national

TABLE 9.2

Actions Addressing Brachycephalic Issues in Dogs: Benefits and Challenges

Types of Actions	Benefits	Challenges
Educate prospective owners	Raise awareness	Prioritisation of health may not be a priority; especially for some breeds; looks, trends may come first. Impulse buying. Integrated with issues of supply and demand.
Educate breeders, veterinarians, judges	Increased awareness	Moving from awareness to effective behaviour change is a slow process; organisational position statements need to be combined with increased individual responsibility and accountability.
Influence advertisers, marketing	Wide potential influence	Seems to be a slow sell; companies get value from pandering to popularity vs. for addressing health; risk alienating purebred breed aficionados; definitely a 'double-edged sword'.
Influence celebrities, media	Wide influence potentially	Tried, but sporadically? How to achieve leverage? Difficult to measure impact (Mills 2018).
Veterinary actions – practitioners	Stop normalising	Addressing ethical responsibility and owner sensitivities; vets need tools to help promote behaviour change.
Veterinary actions – organisations	Collective actions	Relatively easy to make proclamations vs. sustained collaborative actions or implementing substantive programs; limited by resources and mandate; balancing ethical and business aspects.
Regulatory – limit sales/ movement	Potentially limit availability … but producers find another way; and yet industry and other sectors have a vested interest in maintaining pet populations.	Challenges of harmonising international regulations with national differences. Lack of definitions of breeders; parameters that distinguish quality, etc. In both Europe and North America, the profitability of commercial breeding, online trade, e.g., is extremely high; may motivate criminals, but also tax authorities. Stopping trafficking requires law enforcement agencies. Some increased interest in consumer protection, reducing disease/zoonoses risk.
Regulatory, e.g. regulate breeding, e.g. the Netherlands	Better health and welfare of breeding animals and puppies	Likely to be ineffective if targeted at just kennel club-associated breeders as majority of dogs come from commercial and other breeders; within-country regulations may be negated by international realities. Regulations must be well-constructed, harmonised with other efforts, and relevant to the complexity of the brachycephalic health and welfare issues; enforceable; with metrics and follow-up. May be needed when other stakeholders are not making necessary changes.
Change breed standards	Recognises issues	Implementation? Enforcement? Just words or, again, behaviour change (e.g. judges)? How to make conformation that fits with health and welfare a priority?
Specific actions – fitness testing, etc.	Identify best breeding stock	Actions that are only voluntary likely to have limited impact (The Kennel Club 2020d); must be associated with, e.g., regulatory/registration rules. Some standardisation of protocols, while allowing some flexibility? Multiple strategies; various levels/lack of evidence/consensus on the measurements and actions that are most accurate at defining the 'best' or 'healthiest' dogs. Desire for one protocol for determining individual dog health and potential use as a breeding animal (The Kennel Club 2020d).

pedigree organisations is shrinking and therefore reflects a reducing direct relationship between these registries and the wider dog population (see below). Sources other than the classic pedigree dog breeders are now producing the majority of dogs for the companion pet market.

Scandinavia has been very proactive in terms of dog health and welfare, before and since 2008. Based on the awareness raised already in the 1960s (Hansen 1964, Henricson 1969), conferences for show judges were arranged by the Swedish Kennel Club in 1997 and 2007, and again in 2016

(Lindholm 2016), focusing on detrimental effects of exaggerated anatomical features. A video on breathing assessment was produced (prior to 2008 and updated in 2012) (Milligan 2018, Swedish Kennel Club 2012). Spearheaded by all-breed judge Göran Bodegård and contributed to by other judges and cynological organisations, in consultation with breed clubs, accessing data from a survey done by the Swedish Veterinary Association and using statistics available from widely subscribed dog health insurance, this work culminated in the first publication of Breed-Specific Instructions (BSI) for judges by Bodegard, Sporre-Willis and Hedhammar, 2009 (DogWellNet 2020a). These guidelines have been updated in subsequent years and adapted by several other countries (The Nordic Kennel Clubs 2018, DogWellNet 2020e). The BSI is a multi-stakeholder national initiative that became international and now goes hand in hand with the UK initiative – Breed Watch (The Kennel Club 2020b).

Despite their intrinsic shared ethos to improve dog health and welfare, there is great diversity in attitudes and actions on the brachycephalic issues across cynological organisations, kennel and breed clubs, regionally, nationally and internationally. For many years, some have been proactively tackling the challenges in health and welfare of brachycephalics and other breeds, whereas others have only recently acknowledged the issues and still others have even yet to prioritise these concerns at all. For example, the kennel clubs in Sweden, Finland and Norway have separately, and together, been very proactive on healthy breeding in general, as well as specifically addressing issues in brachycephalic dogs for many years (Hansen 1964, The Nordic Kennel Clubs 2018). The Kennel Club in the UK has established and supported numerous initiatives including its Breed Health and Conservation Plans project (The Kennel Club. 2021). The Canadian Kennel Club has recently established a specific committee on brachycephalic health and joined the IPFD International Collaborative on Extreme Conformation in Dogs (Pegram et al. 2020, DogWellNet 2020b). As the core challenges for brachycephalics are similar globally, the benefits from increasing international cooperation are obvious, in terms of building on and sharing existing resources and expertise and reducing redundant efforts. Those involved in the show world tend to be extremely invested, personally and emotionally, with their breeds, and it is clear that understanding the human and communication needs are key. It is encouraging that many breed and kennel clubs are participating actively in international efforts, e.g. the IDHWs, accessing resources on DogWellNet.com and promoting new initiatives (Pegram et al. 2020, O'Neill, Keijser, et al. 2017).

The Federation Cynologic Internationale has published an informative package on healthy breeding and dog welfare that includes a report on brachycephalic obstructive airway syndrome as well as data on brachycephalic breed registrations and health test results (FCI Scientific Commission 2020).

Veterinarians and the Veterinary Profession

In daily practice, companion-animal veterinarians are increasingly presented with brachycephalic dogs suffering from a variety of breed-related health and welfare problems (O'Neill et al. 2020). The rapidly increasing popularity and demand for brachycephalic dogs has led to an escalation in populations of these dogs (IPFD 2020c, O'Neill et al. 2020). As mentioned in the Australian paper (Fawcett et al. 2018), the ethical conflicts experienced by veterinarians and the extreme devotion of many brachycephalic owners to their dogs mean that this complex problem has many aspects that cannot be resolved by science alone.

The veterinary profession, as an advocate of animal health and welfare, recognises its responsibilities to help raise awareness and to ensure breed-related health and welfare problems related to brachycephaly are not normalised or minimised by being considered simply 'typical for the breed' (FVE and FECAVA 2018, FECAVA and FVE 2018, BVA 2020c 2018).

However, when clinicians are confronted with so many brachycephalic dogs with signs of compromised health, e.g. snoring and worse, even veterinarians may tend to normalise these signs, especially when faced with the prospect of upsetting clients by telling them their dogs have innate issues of health and welfare (Ravetz 2017). This again highlights the complexity of the issue. Sometimes, veterinary organisations have made position statements or used an approach of confrontational petitions to publicise their concerns (Bonnett 2015, WSAVA and FECAVA 2017, BVA 2020b). These may have been undertaken without an adequate understanding of, or sensitivity to, the complexity of the issues or with ascribing blame to others without knowing enough about what efforts these other stakeholders have undertaken. For example, as described above, many cynological organisations have been actively addressing breed-specific health for many years, and therefore, the awareness and actions on these issues have increased. However, the proportion of dogs bred by members of a kennel club are often only a minority (less than 20% in many countries, much less than that in some, and over 60% in only a few, e.g. Nordic countries) (The Kennel Club 2020d). This reality means that expecting kennel clubs to improve the brachycephalic situation on their own, e.g. through breeding regulations or changing breed standards, is unreasonable. As many different stakeholders are involved in the challenges, on both the supply and demand sides of the situation, it is crucial for all stakeholders to join forces.

The Federation of European Companion Animal Veterinary Associations (FECAVA), as well as the Federation of Veterinarians of Europe (FVE), have recognised this complexity and taken steps to highlight the roles and responsibilities of many stakeholders as well as the need for actions along several axes including supporting healthy breeding practices, consumer education, as well as issues of supply and demand/trade and transport (FECAVA 2018). Many veterinarians and organisations are participating in other international initiatives, as well (see more below).

The 2018 FVE and FECAVA joint position paper (FVE and FECAVA 2018) (endorsed later by the World Small Animal Veterinary Association (WSAVA) (2018)) states that the breeding of certain types of dogs such as brachycephalic dogs is clearly driven by consumer demand. Therefore, the most sustainable change is likely to happen by raising awareness and educating the public to avoid buying dogs of breeds with extreme hereditary traits that are likely to lead to animal health and welfare issues. Most consumers however may not recognise unhealthy attributes or be able to judge whether they have bought a healthy dog or a 'defective product' (Packer, Hendricks, and Burn 2012, Sandøe et al. 2017, Holland 2019). Over 90% of owners of brachycephalic dogs in the UK say they would get another dog of the same breed, in spite of extensive health problems (Packer et al. 2020). Buying a dog is often an emotional and impulsive rather than a rational decision, which makes the buyer less critical, and health is not prioritised. Simple information and education may not lead to the needed behaviour change of consumers (DogWellNet 2020e). It is also clear that public demand for specific breeds is heavily influenced by celebrities, media and social trends (Ghirlanda et al. 2013). The major veterinary organisations in the UK are part of the Brachycephalic Working Group that has released a strapline for wide dissemination: "Stop and think before buying a flat-faced dog" (BWG 2021). Veterinary organisations have partnered with the Campaign for the Responsible Use of Flat-Faced Animals (CRUFFA) initiative which targets industry, advertising and other media, pushing for a decrease in the use of brachycephalic images (Harrison 2020a). Many educational initiatives have been organised by veterinary groups within various countries (DogWellNet 2020a). More recently, veterinary associations have started to focus on what individual veterinarians can do to address brachycephalic issues and provide practical tools (BVA 2020a).

SPECIFIC ACTIONS AND EFFORTS

As mentioned above and in Table 9.1, the IPFD is facilitating collaborative actions by a wide array of stakeholders and is working to increase the power of initiatives through coordinated, informed regional actions (IPFD 2020b). At a recent IPFD International Dog Health Workshop (IDHW), for example, in addition to a specific theme on extremes of conformation, another stream highlighted new work on improving the health and welfare of dogs from commercial breeders and efforts to limit illegal trade (Pegram et al. 2020). The involvement of decision-makers from many of the organisations and stakeholders actively working on these issues means recommended actions can be taken in a coordinated way (Pegram et al. 2020, DogWellNet 2019). A specific outcome from the fourth IDHW was the initiation of the International Working Group on Extremes of Conformation in Dogs (IWGECD) with International Collaborative on Extreme Conformation in Dogs (ICECDogs) with a mission to provide resources and facilitate regional and international efforts covering all extremes of conformation in dogs (DogWellNet 2020g, Pegram et al. 2020).

The last decade has seen an increasing research focus on brachycephalics, reporting not only on the clinical problems of individual dogs, but also on the collection of epidemiological data to further define issues at a population level (Royal Veterinary College 2020). Epidemiological research programmes such as VetCompass in the UK (VetCompass 2020) and Australia (VetCompass Australia 2020) share anonymised veterinary clinical records to report on the health status of animals under general veterinary care. In the UK, VetCompass shares records on over eight million dogs from over 30% of UK veterinary clinics and has used these to support over 95 peer-reviewed publications (up to mid-2021) including key brachycephalic breeds (O'Neill et al. 2018, O'Neill, Skipper, et al. 2019, O'Neill et al. 2016) and conditions (O'Neill et al. 2020, O'Neill, O'Sullivan, et al. 2019, 2017, O'Neill, Lee, et al. 2017, O'Neill et al. 2015) related to brachycephaly. Qualitative research has also examined owner attitudes, communication strategies and the need for human behaviour change, not merely 'education' (Sandøe et al. 2017, Packer, Murphy, and Farnworth 2017, Packer, Hendricks, and Burn 2012, Packer et al. 2019, 2020). This holistic scientific approach towards brachycephalic evidence has been promoted and supported by the IDHWs. Details of the specific research findings are covered in other chapters in this text.

This following section gives examples of the many actions that have been taken by various stakeholders at national, multi-national and truly coordinated, international levels to (a) curb the demand for brachycephalic dogs by educating the public and raising awareness and (b) by improving the quality of the dogs supplied through encouragement of better breeding practices that put the health and welfare of the dogs first.

EDUCATION AND RAISING AWARENESS

Although a plethora of efforts have been undertaken to raise awareness and educate consumers, metrics on their success are generally lacking. This is not to imply that they have individually or collectively been without impact. There is no doubt that awareness of the brachycephalic issue has increased markedly in the first decade of the twenty-first century, as exemplified by the material in Tables 9.1 and 9.2. However, if the main outcome desired was either a decrease in popularity of these breeds or a marked improvement in their health and welfare overall, it would be challenging to claim much success on either. Possible reasons for these limitations have already been discussed above.

Examples of awareness campaigns include huge promotional campaigns with powerful titles from welfare organisations, e.g. *The Purebred Paradox* from the Human Society of the United States (Allan 2009) and the *Bred for Looks Born to Suffer*, a Royal Society for the Prevention of Cruelty to Animals campaign from 2011 (Stilwell 2011). Kennel clubs have organised many events to educate and support specific target groups, e.g. breeders or judges, in the UK, Scandinavia and elsewhere (DogWellNet 2020a, d). Various blog sites have used Facebook and other platforms to share their messages, e.g. Pedigree Dogs Exposed (Harrison 2020b). Veterinary organisations have also had this as a priority, including FECAVA (FECAVA 2018) and the British Veterinary Association, who has in 2020 updated resources for individual veterinarians (BVA 2020a, b).

COMMISSIONS, WORKING GROUPS, STATEMENTS...

A series of national initiatives were undertaken in the UK, especially following the 2008 documentary Pedigree Dogs Exposed (BBC 2008), including an independent inquiry report (Bateson 2010), an RSPCA funded report (Rooney 2009) and a government report (APGAW 2009). In

2010, the Advisory Council on the Welfare Issues of Dog Breeding was formed in the UK with funding from a range of animal welfare bodies to provide independent, expert advice and make recommendations on methods and priorities for improving the welfare issues of dog breeding (Crispin 2011). Particular regard was given to surveillance, research and development, breeding strategies, legislation and regulation, and to education and publicity as identified priorities. Although the group is now dissolved, a number of important reports were produced (Advisory Council on the Welfare Issues of Dog Breeding 2014, 2012) and many of the expert members continue to be involved in other initiatives related to improving dog breeding and welfare.

The Brachycephalic Working Group (BWG) is an active, multiple stakeholder effort within the UK, and their specific activities cover most of the topics mentioned in this article (BWG 2020). The framework document of the UK BWG addresses a series of targetted areas including supply and demand, breed-specific strategies, breed standards, show judging, brachycephalic health assessment, marketing and advertising, owner education and actions within the veterinary and research areas (Brachycephalic Working Group 2019). Buy-in from a wide range of stakeholders, with actions being undertaken by each, has resulted from societal pressure related to previous media campaigns and provides evidence that progress can be made, even in the absence of specific metrics for success and if on a slower timeframe than might be ideal. The UK BWG has been particularly effective in achieving consensus statements signed off by the diverse membership that provide a unified 'ground truth' to support directed actions on topics such as discouraging purchase of flat-faced dogs, obesity, using brachycephalic imagery in advertising, brachycephalic obstructive airway syndrome and the impacts of sudden population changes (BWG 2021). The collaborative, collective and coordinated approach of the UK BWG may provide a template for activities in other countries and for international efforts. Members of the BWG have been actively participating in international efforts with the IPFD and the IDHWs.

The Nordic Kennel Union (NKU), a multinational group, has had several initiatives addressing problems of extremes of conformation. In 2017, a working group appointed by the NKU, composed of veterinarians, geneticists and breed club representatives, presented "Statements and Proposals Regarding Respiratory Health in Brachycephalic Dogs" (NKU 2017) which included the following.

Sample Statements

- The prevalence of health problems related to BOAS is a serious and complex welfare problem.
- This calls for collaboration between multiple stakeholders and, because of the diverse manifestations of BOAS, various different strategies will be needed.
- Dogs clinically affected by BOAS or operated for BOAS should never be bred from.
- To decrease the prevalence of BOAS, selection for less exaggerated anatomical features is needed.

Proposed Actions

- Increase education of show judges and breeders, as well as puppy buyers.
- Develop and implement methods to examine breeding animals regarding respiratory function and temperature regulation and to promote central registration of dogs diagnosed and operated for BOAS.
- Work internationally towards breed standards with non-exaggerated text that consider the importance of health and to produce a guideline to how clubs are to organise and execute lectures/seminars.

Although the increased focus on BOAS is crucial, there has been a tendency to focus on the respiratory issues, sometimes to the cost of underplaying the broader problems of extreme conformation in these breeds (O'Neill et al. 2020). This has spawned legislation focused on, e.g., increasing the length of muzzle without due consideration of the complex of problems, including abnormal spines and other problems (IPFD 2020b).

ACTIONS TO IMPROVE THE QUALITY OF THE DOGS SUPPLIED

Already in 1967, WSAVA appointed a Committee to consider Breed Standards in relation to the Health and Welfare of Dogs who stated: "The meeting at the World Veterinary Congress in Paris unanimously declares that concern for the health and welfare of dogs demands that breed standards should not include requirements and recommendations that hinder physiological function of organs and parts of the body" (Anonymous 1969).

In North America, the American Veterinary Medical Association (AVMA) has described the challenges in bringing together various stakeholders to agree on a policy for breeding (Burns 2017). Although most groups agreed in principle, concerns were raised by, e.g., the American Kennel Club, that the wording could support legislation against specific breeds. The policy as sent to the AVMA House of Delegates differed somewhat in language from the original welfare committee proposal (Anonymous 2016) and was passed in 2017. The AVMA policy aligns with the existing Canadian Veterinary Medical Association policy (CVMA 2019) which addresses both breed- and health-specific issues. The CVMA also publishes the Code of Practice for Canadian Kennel Operations (CVMA 2018).

More recently, a program called Canine Care Certified at Purdue University in the USA has been developed and led by Prof. Candace Croney in collaboration with commercial dog breeders (Purdue University College of Veterinary Medicine 2020). It aims to provide a nationwide certification program that establishes rigorous standards for breeders. It is an example of defining and encouraging best practices, with the use of specific guidelines and oversight and using metrics to define and evaluate outcomes. Guidelines are based on peer-reviewed research studies including physical aspects of facilities to protect welfare and promote socialisation (Stella et al. 2019, Mugenda, Shreyer, and Croney 2019, Stella et al. 2018). This program was initiated due to interest from commercial breeders who committed to better health and welfare for breeding dogs and puppies, and who are embracing best practices in their breeding and care programs. Although the program is not specifically directed towards brachycephalic breeds, these breeds comprise an important proportion of the dog populations and this program offers a template for what can be accomplished.

ACTIONS ADDRESSING SUPPLY, DEMAND, TRADE AND TRANSPORT OF DOGS

As mentioned above, the supply of certain breeds of dogs by local and show breeders cannot meet demand in many countries. For example, in the UK, The Kennel Club registers around 250,000 dogs annually which contributes about 30% of the estimated 800,000 new puppies needed each year in the UK (The Kennel Club 2020c). Only a few countries have registries for all dogs, e.g. the Danish Dog Register (Danish Dog Register 2020). However, in many countries, the majority of dogs, especially of popular breeds, both with and without pedigrees (many are so-called purebred 'look a likes'), are not recorded within national registries, and the availability of national population data is limited. These dogs are often produced by casual or unscrupulous breeders or puppy farms (Cushing 2018). Because the economic value of a brachycephalic dog is often high and the production costs can be kept relatively low, often at the expense of welfare, this has resulted in a booming industry in Europe. The European Parliament has stated that illegal puppy trade is not only an animal welfare issue but believed to be the third most profitable organised crime within the EU after narcotics and weapons trafficking (Eurogroup for Animals 2020a, DogWellNet 2020g). Thus, an international multi-stakeholder approach to issues of supply and demand is crucial. At the time of writing this chapter, we are in the midst of the COVID-19 pandemic, and global demand for dogs has led to unprecedented prices being charged for dogs, increased supply from questionable sources and concerns over the longer-term welfare consequences. This has clearly demonstrated that the extensive attempts to educate the public about responsible acquisition and ownership of dogs are not uppermost in the minds of the public (IPFD 2020a).

Huge numbers of puppies are legally and illegally traded across borders. The umbrella organisations the Federation of Veterinarians of Europe (FVE), the Union of European Veterinary Practitioners (UEVP) and the Federation of Companion Animal Veterinary Associations (FECAVA) joined forces in Europe in 2016 to adopt the position paper 'Working towards responsible dog trade' (FVE, UEVP, and FECAVA 2016). The paper, among others, emphasises that online purchasing (classified ads, social networks and websites) is now a major channel for buying and selling dogs and is difficult to regulate. The scale of online advertising by dog breeders and sellers/dealers has significantly increased within the EU in recent years. In 2017 FVE, UEVP and FECAVA, together with the international animal welfare organisation Four Paws International, expressed their concerns about the increasing problems arising from the unregulated trade in animals on the Internet in a letter to the European Commission (FVE 2017). Together, they strongly called for regulation of dog sales on the internet. In October 2018, the European Commission adopted a recommendation for a coordinated control plan for official controls on online sales of dogs and cats to support the EU countries (European Commission 2018).

In 2018, Four Paws International launched the 'Thanks eBay!' campaign aimed at the global market leader in the classified ad sites industry to better regulate adverts (Four Paws 2018b). One of the actions was a letter signed by many stakeholders to the CEO and President of eBay in the USA calling on eBay, Inc. to make pet trade safer on all its classified ad sites worldwide (Four Paws 2018a).

In November 2018, the Austrian Presidency of the Council of the EU, Eurogroup for Animals and Four Paws International held a conference entitled "The Illegal Online Puppy Trade: towards a safer EU for animals and people online" (Eurogroup for Animals 2018a). In April 2020, an expert workshop on illegal trade was organised by the Croatian Council of the EU and Eurogroup for Animals. The report from this workshop was published in 2020: Illegal Pet Trade: Game Over (Eurogroup for Animals 2020b).

In spite of an increasing number of efforts at defining problems and finding solutions, enforcement tends to lag and there are few clear metrics of the achievements of reports, legislation and regulation. Time will tell whether legislation like Lucy's Law in the UK, aimed at restricting third-party sales, will achieve the desired endpoints (Gov.UK 2020), whereas the summer of 2020 saw a horrendous example of the failure of existing regulations about transport of dogs in Canada (Seglins and Thomas 2020).

Multi-national/European Efforts by Authorities

Examples in this section describe efforts by various authorities in Europe to address some of the issues described above. Many of these efforts cannot be considered as truly regulatory because they are not accompanied by effective enforceable strategies. Many are really efforts to raise awareness without measurable actions. However, they do highlight that political attention is being paid to the issue in response to concerns from various stakeholders.

When the Lisbon Treaty came into force in 2009, it amended the 'Treaty on the Functioning of the European Union' (TFEU) (Thomson Reuters Practical Law 2020) and introduced the recognition that animals are sentient beings. Although, at one time, it was also the intention to include a framework 'Animal Welfare Law', the previous European Commission under Juncker (2014–2019) did not proceed with this and, therefore, no harmonised animal welfare law is in place nor foreseen in the EU in the near future. Currently, the EU legislation lays down minimum standards, and national governments may adopt more stringent rules provided they are compatible. These minimum standards focus mainly on farm animals.

Instead of legislation, the European Commission under Juncker initiated the Animal Welfare Platform, with the motto "Everyone is responsible" (European Commission 2020b). Although companion animals were included, no priority was given to dogs and cats, and no official subgroup was formed. Therefore, on the initiative of the Dutch Ministry of Agriculture, Nature and Food Quality, a voluntary initiative working group was formed concerning 'Health and welfare of pets (dogs) in trade' (Arnts 2019, Eurogroup for Animals 2018b). Leaflets and infograms for buying dogs and selling dogs online were developed and endorsed by the Animal Welfare Platform in June 2020 (FVE 2020). No specific focus is given to brachycephalic dogs.

In March 2016, the European Parliament and the Council adopted the Regulation on Transmissible Animal Diseases (known as the "Animal Health Law") (European Commission 2020a, 2016). It is a single, comprehensive animal health law to replace all the previous complicated animal health rules (some 400 individual acts!) and will be applicable from 2021. The principle "Prevention is better than cure" is at its heart. the Animal Health Law applies to terrestrial and aquatic animals, and therefore also includes dogs. Although the focus is very much on farming, the Animal Health Law requires all commercial sellers and breeders, transporters and assembly centres of dogs to register their establishments with the competent authority. Although this is a good step forward, the legislation does not define what is exactly a commercial breeder and enforcement may be challenging.

The European Commission states the following on its website: "This Regulation does not provide rules on animal welfare, although it recognises that animal health and welfare are linked and it requires, for the first time universally, that animal welfare is taken into account when considering the impacts of diseases and measures to combat diseases" (Europa.ec 2016).

The European Union (EU) – The European Parliament (EP) – The EU Council

In 2017, a report on Animal Welfare in the European Union commissioned by the EP states: "a general animal welfare law and specific laws on several species are needed" (Broom 2017). In 2018, to inform the Members of European Parliament (MEPs) on breed-related health and welfare issues due to breeding for extreme conformation in dogs and cats, the EU Dog & Cat Alliance, the FVE and FECAVA jointly organised an event in the EP hosted by MEPs Petras Auštrevičius and Marlene Mizzi. The conference 'Health before looks – Breeding for extremes in dogs and cats' was well attended by MEPs as well as NGOs and cynological organisations. The health and welfare of brachycephalic breeds were highlighted, and the message was clear (Johnson 2018). An infographic was made for the event and some national members of FVE and FECAVA have translated it in to their national language and use it for education purposes (FECAVA 2020).

In December 2019, the Council of the EU adopted conclusions and recommendations on the animal welfare policies in the EU (Council of the European Union 2019). The conclusions highlight the importance of animal welfare as an integral part of sustainable animal production. They recognise the need to further update the current legislation. The EU recommends including dogs and cats within the context of economic activity.

REGULATORY APPROACHES – BREED-SPECIFIC

There are examples of numerous pieces of dog legislation in almost every country that address spay neuter, licensing and other dog welfare issues that are non-breed-specific (CFSG 2020). In many countries, health and welfare issues concerning the breeding of dogs are included in the legislation but often in general wording; for example, 'It is forbidden to breed dogs in a way that affects the health and welfare of the bitch and male dog as well as the offspring negatively'. However, specific criteria to enforce legislation concerning brachycephalic dogs were adopted in the Netherlands in 2019 which is worth exploring in more detail (see below) (Bonnett 2019a).

National Example: The Netherlands

On the first of July 2014, the Dutch Ministry of Agriculture, Nature and Food Quality amended the legislation concerning keepers of animals (Besluit Houders van Dieren) to improve animal welfare in commercial breeding (Overheid.nl 2014). In Article 3.4 of this legislation, it states that it is forbidden to breed companion animals in a way that influences the animal health and welfare in a negative way for the mature animals as well as the offspring. This includes breeding of dogs with conformational features that cause health and welfare issues. However, Article 3.4 did not lead to its intended improvement in health and welfare. The legislation failed to provide the necessary tools to the inspectors, resulting in making enforcement impossible. Therefore, the Dutch Ministry of Agriculture, Nature and Food Quality gave the Faculty of Veterinary Medicine at Utrecht University the task to draft criteria to enforce Article 3.4 targetted specifically at breeding of short-muzzled dogs. The official report was published in January 2019 (van Hagen 2019).

The health criteria focused mainly on brachycephalic obstructive airway syndrome (BOAS) and brachycephalic ocular syndrome (BOS). All criteria are labelled using a traffic light system: green, orange or red. For example, open nostrils are green, mild stenosis is orange and severe stenosis is red. Orange means currently accepted, except if more than one criterion is orange. Red means breaching the standard; therefore, future breeding of this animal is forbidden. Another criterion, for example, is the craniofacial ratio (CFR). The CFR green standard is more than 0.5, and orange is between 0.3 and 0.5.

There are virtually no dogs of some breeds, e.g. the pedigree Pug, currently in the Netherlands with a CFR that is even close to reaching up to 0.3. In consequence, the Dutch Pug Club has decided to immediately stop breeding Pugs. However, the minister does not want to ban breeds and therefore temporarily allows breeding if one parent dog does not meet the muzzle length criterion (CFR smaller or equal to 0.3) but meets all the other five criteria. The other parent dog must meet all the six criteria including muzzle length (CFR larger than 0.3).

The Dutch Kennel Club agrees, in general, with the report and supports the government's desire to enforce stricter rules for breeding short-muzzled dogs. However, it does not agree with using a CFR of less than 0.3 as a single criterion to ban breeding. Instead, the Dutch Kennel Club proposes that there should be an exception for dogs with a CFR of less than 0.3 that meet all other criteria and pass the Respiratory Function Grading Scheme developed by Cambridge University, in order to allow improvement of the breed in the future and to work towards an acceptable muzzle length. The Dutch Kennel Club also recommends adding additional criteria including, e.g. body condition score, with obese dogs excluded from breeding. Although the recommendations in the Utrecht University report came from veterinary experts and specialists, they unfortunately did not adequately include inputs from all stakeholders or consider all aspects of the complex brachycephalic issue that had the potential to limit the success of the program (van Hagen 2019, DogWellNet 2020b).

In the spring of 2020, the enhanced regulations demanded by the Dutch Government were enacted by the Dutch Kennel Club. This sparked outrage and concern from cynological organisations and dog breeders but conversely was met with approval from some veterinary and welfare groups. There are now heightened concerns that this reflects the beginning of a sequence of efforts to eliminate pure-bred/pedigreed dogs. These controversies and concerns, including the dangers of essentially unilaterally enacted legislation that does not consider the broadest consequences, are presented by IPFD in an evolving article following the situation (DogWellNet 2020b). Within that article are examples of comments on the legislation and the response of the Dutch Kennel Club from various cynological organisations, welfare and veterinary groups; they range from congratulatory, to considered to confrontational, highlighting a widening gap across stakeholders in terms of their understanding and views on the challenges and possible solutions. Subsequently in 2020, a comprehensive and balanced "Finnish Investigation: Problems and Means of Intervention in the Breeding of Dogs" was published, at the same time as a report "Finnish report: an investigation would curb problems with dog breeding through monitoring criteria and ethical delegation". The investigation discusses the use of the CFR, but not as unilaterally as in the Dutch legislation, but instead as part of a broader evaluation of an animal's condition (IPFD 2020b). No doubt more reports and investigations will come, and more countries will consider regulatory approaches, with the potential for both benefits and unintended consequences. As this publication goes to press, these situations are continuing to evolve, including a legal action against breed clubs, the Nowegian Kennel Club, and specific breeders (IPFD 2021).

As mentioned in several sections above, the IPFD is helping to stimulate and coordinate efforts, and to collate and distribute information on the issues and actions that are relevant for addressing health and welfare issues of extreme conformation. Many of the references throughout this chapter link to the IPFD platform, DogWellNet.com which is continually expanding and updated. Extensive information and links are available covering health programs, work of kennel and breed clubs, regulatory efforts, work by welfare and veterinary groups and more; resources are continually updated as this is a dynamic area of interest and concern (DogWellNet 2020a, c). Beyond this platform, the IPFD will continue to work via the IDHWs and the international working groups, and in engaging all stakeholders.

The fourth IDHW, co-organised by IPFD and The Kennel Club, took place in Windsor (UK) in June 2019. As mentioned above, it was decided to create an International Collaborative on Extreme Conformation in Dogs (DogWellNet 2020b). This collaborative aims to offer a platform from which national and international working groups and stakeholders join forces to address issues of extreme conformation that negatively influence the health, well-being, and welfare of dogs. The initial focus was on brachycephaly but the current plan is to cover all extreme conformations in dogs from the outset. The collaborative does not intend to harmonise working groups, nor will it prescribe stakeholder strategies. Instead, it offers a forum for collaboration and sharing of information and tools to enhance the work of new and existing efforts, nationally and internationally.

CONCLUSIONS

This chapter does not intend to provide an exhaustive or complete review, but rather, examples of actions and approaches across stakeholder groups, internationally. It is evident that a wide array of stakeholders has undertaken many specific actions, locally, nationally and regionally. These include many discussion papers, official and unofficial statements of problems and issues, needs and proposed actions. Strategies fall under categories including education/raising awareness, especially directed at consumers; addressing sources of puppies; and defining aspects of best practices for breeding. Statements, position papers and petitions have been written by the veterinary community, researchers, welfare groups, cynological organisations and regulatory bodies. However, despite these, there are still very few programs or actions that are designed to enforce called-for changes, or that have identified clear outcomes and defined how these will be monitored and evaluated. Notwithstanding, there has been progress, at least in terms of awareness and a marked increase in activities by various groups.

Emerging from all this activity are firm and consistent statements on the nature, extent and severity of the problem and a consensus – among many stakeholders – that the current state of health and welfare within certain brachycephalic breeds is unacceptably poor and must be addressed. Notwithstanding, there is passionate resistance and disagreement from other groups. Although there is still a scarcity of data on the true population levels of occurrence of individual problems in these brachycephalic breeds, the disparity of opinions globally may stem more from differences in attitudes than evidence. There is increasing consensus in Europe, the UK and Scandinavia (BWG 2020); the extent of recognition and action is apparently less in North America, although new initiatives are underway in both Canada and the USA, and many breed clubs are exploring BOAS programs. Other stakeholders are addressing responsible breeding, issues of commercial breeding and exploring challenges of online trade. However, as stated above, some pedigree breeders are certainly not in agreement, and brachycephalic breeds are still sought after by the public. Many stakeholders in other parts of the world may be mainly focused on other issues of welfare, including stray dogs and zoonotic problems. However, as groups and regions begin to expand their work on issues of extreme conformation, it is important that they have access to the history and resources developed by those who have gone ahead.

It remains to be seen to what extent individual and collaborative efforts can be effective in the face of the diverse and complex factors at play around the world. One thing is certain: despite how long it has taken to reach the point where there is widespread concern for brachycephalic health and welfare, global media and communication ensure that there will only be increased attention in the future. The question remains as to what the future will hold for individual brachycephalic dogs, the breeds and their many owners, whose devotion to them remains.

REFERENCES

Advisory Council on the Welfare Issues of Dog Breeding. 2012. "Advisory council on the welfare issues of dog breeding—standard for breeding dogs." accessed November 2. http://www.dogbreedhealth.com/wp-content/uploads/2012/05/DAC-Breeding-Standard-Final-10-04-12.pdf.

Advisory Council on the Welfare Issues of Dog Breeding. 2014. "Summary of progress since the Bateson report of 2010—a report of the Advisory Council on the Welfare Issues of Dog Breeding." International Partnership for Dogs, accessed November 2. https://dogwellnet.com/files/file/307-advisory-council-final-report-welfare-issues-of-dog-breeding-2014/.

Allan, C. 2009. "The Purebred Paradox." Humane Society of the United States (HSUS), accessed July 25. https://www.humanesociety.org/sites/default/files/docs/purebred%20paradox%20story%20%28004%29.pdf.

American Kennel Club. 2020. "Bulldog - Breed Standard." The American Kennel Club, accessed March 17. https://www.akc.org/dog-breeds/bulldog/.

Anonymous. 1969. "Report of the World Small Animal Veterinary Association Committee appointed to consider Breed Standards in relation to Health and Welfare of Dogs." *Journal of Small Animal Practice* 10:135–141.

Anonymous. 2016. "Board sends HOD a policy on responsible pet breeding Proposal addresses inherited disorders in/ companion animals." American Veterinary Medical Association (AVMA), accessed July 25. https://www.avma.org/javma-news/2017-01-01/board-sends-hod-policy-responsible-pet-breeding.

APGAW. 2009. A healthier future for pedigree dogs. London—The Associate Parliamentary Group for Animal Welfare.

Arnts, Léon. 2019. "Health & Welfare of Pets (dogs) in trade—Voluntary Initiative Group." EU Platform on Animal Welfare, accessed March 30. https://ec.europa.eu/food/sites/food/files/animals/docs/aw_platform_20190617_pres-07.pdf.

Aron, D.N., and D.T. Crowe. 1985. "Upper airway obstruction. General principles and selected conditions in the dog and cat." *The Veterinary Clinics of North America. Small Animal Practice* 15 (5):891–917.

Bateson, P. 2010. "Independent inquiry into dog breeding." University of Cambridge, accessed November 2. https://dogwellnet.com/files/file/308-independent-inquiry-into-dog-breeding-2010-patrick-bateson/.

BBC. 2008. "Pedigree Dogs Exposed." accessed August 6. http://www.bbc.co.uk/pressoffice/pressreleases/stories/2008/08_august/19/dogs.shtml.

Bonnett, B. 2019a. "Brachycephalic dogs in The Netherlands." IPFD - DogWellNet, accessed July 24. https://dogwellnet.com/blogs/entry/135-brachycephalic-dogs-in-the-netherlands/.

Bonnett, B. 2019b. "French Bulldog Health Seminar October 2018." IPFD DogWellNet, accessed August 6. https://dogwellnet.com/blogs/entry/117-french-bulldog-health-seminar-october-2018/.

Bonnett, B. 2019c. "Owners' perception of 'responsible dog ownership'." IPFD DogWellNet, accessed October 20. https://dogwellnet.com/blogs/entry/141-owners-perception-of-responsible-dog-ownership/.

Bonnett, B. 2019d. "Why do people choose the dogs they do - and what is the impact on dog health and welfare?". IPFD DogWellNet, accessed August 6. https://dogwellnet.com/blogs/entry/123-why-do-people-choose-the-dogs-they-do-and-what-is-the-impact-on-dog-health-and-welfare/.

Bonnett, B. 2015. "The Brachycephalic Issue—1. Sweden Fall 2015." IPFD DogWellNet, accessed July 25. https://dogwellnet.com/content/hot-topics/brachycephalics/the-brachycephalic-issue-1-sweden-fall-2015-r309/.

Bonnett, B. 2020. "Metrics, Process and CULTURE! Impacts on discussions on health and welfare of dogs." IPFD DogWellNet, accessed July 25. https://dogwellnet.com/blogs/entry/166-metrics-process-and-culture-impacts-on-discussions-on-health-and-welfare-of-dogs/.

Brachycephalic Working Group. 2019. "Brachycephalic Working Group Framework for a partnership approach to improving brachycephalic dog health and welfare—2019." The Brachycephalic Working Group, accessed April 17. http://www.ukbwg.org.uk/wp-content/uploads/2019/01/190118-UK-Brachycephalic-Working-Group-BWG-Framework-for-a-partnership-approach.pdf.

British Veterinary Association. 2020. "Advertising guidelines - Pets in advertising—A social concern." British Veterinary Association, accessed January 7. https://www.bva.co.uk/resources-support/ethical-guidance/advertising-guidelines-pets-in-advertising-a-social-concern/.

Broom, D.M. 2017. Animal Welfare in the European Union—Study for the PETI Committee. European Parliament - Directorate General for Internal Policies.

Burns, K. 2017. "AVMA passes policy on responsible pet breeding." American Veterinary Medical Association, accessed July 18. https://www.avma.org/News/JAVMANews/Pages/170301c.aspx.

BVA. 2018. "BVA Policy position on extreme conformation." British Veterinary Association, accessed September 24. https://www.bva.co.uk/uploadedFiles/Content/News,_campaigns_and_policies/Policies/Ethics_and_welfare/BVA%20Position%20on%20Extreme%20Conformation%20Full-%20Amended.pdf.

BVA. 2020a. "All animals should be bred for health over looks." BVA, accessed July 25. https://www.bva.co.uk/take-action/breed-to-breathe-campaign/.

BVA. 2020b. "Brachycephalic dogs." British Veterinary Association, accessed July 25. https://www.bva.co.uk/take-action/our-policies/brachycephalic-dogs/.

BVA. 2020c. "Reporting conformational changes and caesareans to the Kennel Club guidance." British Veterinary Association, accessed January 6. https://www.bva.co.uk/resources-support/ethical-guidance/reporting-conformational-changes-and-caesareans-to-the-kennel-club-guidance/.

BWG. 2020. "The Brachycephalic Working Group." The Brachycephalic Working Group, accessed July 4. http://www.ukbwg.org.uk/.

CFSG. 2020. "Legislation and Guidance." Canine and Feline Sector group (CFSG), accessed July 25. http://www.cfsg.org.uk/_layouts/15/start.aspx#/SitePages/Legislation%20and%20Guidance.aspx.

Council of Europe. 2020. "Details of Treaty No.125—European Convention for the Protection of Pet Animals." Council of Europe, accessed August 6. https://www.coe.int/en/web/conventions/full-list/-/conventions/treaty/125.

Council of the European Union. 2019. "Council conclusions on animal welfare - an integral part of sustainable animal production." Council of the European Union, accessed July 26. https://www.consilium.europa.eu/media/41863/st14975-en19.pdf.

Crispin, S. 2011. "The advisory council on the welfare issues of dog breeding." *The Veterinary Journal* 189 (2):129–131. doi:10.1016/j.tvjl.2011.06.008.

Cushing, M. 2018. "The looming dog shortage." Today's Veterinary Business, accessed July 25. https://todaysveterinarybusiness.com/looming-dog-shortage/.

CVMA. 2018. "A Code of Practice for Canadian Kennel Operations." The Canadian Veterinary Medical Association (CVMA) accessed July 25. https://www.canadianveterinarians.net/documents/Code-of-Practice-for-Canadian-Kennel-Operations.

CVMA. 2019. "Dog Breeding – Position Statement." Canadian Veterinary Medical Association (CVMA), accessed July 25. https://www.canadianveterinarians.net/documents/dog-breeding.

Danish Dog Register. 2020. "Danish Dog Register." The Dog Register, accessed October 20. https://www.hunderegister.dk/Home.

DogWellNet. 2020a. "Brachycephalics." IPFD DogWellNet, accessed July 26. https://dogwellnet.com/content/hot-topics/brachycephalics/.

DogWellNet. 2020b. "Challenges for Pedigree Dogs—Regulatory Enforcement of Brachycephalic Dogs in the Netherlands." IPFD DogWellNet, accessed July 25. https://dogwellnet.com/content/international-actions/extremes-of-conformation-brachycephalics/challenges-for-pedigree-dogs-regulatory-enforcement-of-brachycephalic-dogs-in-the-netherlands-r686/.

DogWellNet. 2020c. "Extremes of Conformation | Brachycephalics." IPFD DogWellNet, accessed July 26. https://dogwellnet.com/content/international-actions/extremes-of-conformation-brachycephalics/.

DogWellNet. 2020d. "The French Kennel Club BREATH Test Protocol for Brachycephalic Breeds." IPFD DogWellNet, accessed July 26. https://dogwellnet.com/content/hot-topics/brachycephalics/the-french-kennel-club-breath-test-protocol-for-brachycephalic-breeds-r688/.

DogWellNet. 2020e. "Hot Topics - Brachycephalics." IPFD - DogWellNet, accessed August 6. https://dogwellnet.com/content/international-actions/extremes-of-conformation-brachycephalics/the-brachycephalic-issue-archives/hot-topics-brachycephalics-r563/.

DogWellNet. 2020f. "IPFD International Dog Health Workshops." IPFD DogWellNet, accessed August 6. https://dogwellnet.com/content/ipfd-international-dog-health-workshops/.

DogWellNet. 2020g. "Report Illegal Pet Trade—Game Over." IPFD DogWellNet, accessed July 25. https://dogwellnet.com/content/welfare-legislation/dog-specific-legislation/transport-of-dogs-pets/report-illegal-pet-trade-game-over-r691/.

DogWellNet. 2019. "4th IDHW Theme Outcomes." IPFD DogWellNet, accessed July 25. https://dogwellnet.com/content/ipfd-international-dog-health-workshops/ipfd-international-dog-health-workshop-4/4th-idhw-post-meeting-resources/4th-idhw-theme-outcomes-r658/.

DogWellNet. 2020a. "Brachycephalic Resources—External Links." IPFD DogWellNet, accessed July 25. https://dogwellnet.com/content/hot-topics/brachycephalics/brachycephalic-resources-external-links-r326/.

DogWellNet. 2020b. "International Working Group on Extremes of Conformation in Dogs (IWGECD)." IPFD DogWellNet, accessed July 25. https://dogwellnet.com/content/international-actions/extremes-of-conformation-brachycephalics/international-working-group-on-extremes-of-conformation-in-dogs-iwgecd-r695/.

Eurogroup for Animals. 2018a. "Conference; "The Illegal Online Puppy Trade—towards a safer EU for animals and people online" ". Eurogroup for Animals, accessed July 26. https://www.euro-groupforanimals. rg/news/booming-online-puppy-trade-illegal-fraudulent-and-heartbreaking-dogs-and-buyers.

Eurogroup for Animals. 2018b. "EU Animal Welfare Platform – Pet subgroup on dog trade." Eurogroup for Animals, accessed October 20. https://www.eurogroupforanimals.org/news/eu-animal-welfare-plat-form-pet-subgroup-dog-trade.

Eurogroup for Animals. 2020a. "Commissioner Kyriakides welcomes new 'Illegal Pet Trade—Game Over Report'." Eurogroup for Animals, accessed July 25. https://www.eurogroupforanimals.org/news/commissioner-kyriakides-welcomes-new-illegal-pet-trade-game-over-report.

Eurogroup for Animals. 2020b. The Illegal Pet Trade—GAME OVER.

Europa.ec. 2016. "General Q&A—New EU Regulation on transmissible animal diseases ("Animal Health Law")." Europa.ec, accessed July 26. https://ec.europa.eu/food/sites/food/files/animals/docs/ah_law_regulation-proposal_qanda.pdf.

European Commission. 2016. General Q&A - New EU Regulation on transmissible animal diseases ("Animal Health Law"). European Commission.

European Commission. 2018. "Recommendation of 16.8.2018 on a coordinated control plan for the official controls on online sales of dogs and cats." European Commission accessed July 25. https://ec.europa.eu/food/sites/food/files/animals/docs/aw_other_euccp_recommend_2018-5488-f1_987143.pdf.

European Commission. 2020a. "Animal Health Law." European Commission, accessed March 30. https://ec.europa.eu/food/animals/health/regulation_en.

European Commission. 2020b. "EU Platform on Animal Welfare - "Everyone is responsible"." European Commission, accessed March 30. https://webgate.ec.europa.eu/awp/.

Fawcett, A., V. Barrs, M. Awad, G. Child, L. Brunel, E. Mooney, F. Martinez-Taboada, B. McDonald, and P. McGreevy. 2018. "Consequences and management of canine brachycephaly in veterinary practice—perspectives from Australian veterinarians and veterinary specialists." *Animals* 9 (1):3.

FCI Scientific Commission. 2020. "Healthy Breeding & Dog Welfare—informative package. Brachycephalic Breeds and Brachycephalic Obstructive Airway Syndrome (BOAS) Report, Strategy and Recommendations". Federation Cynologique Internationale, accessed October 28. http://www.fci.be/en/Healthy-Breeding-Dog-Welfare-informative-package-3573.html.

FECAVA. 2018. " 'Extreme breeding' of companion animals—Raising public awareness is key." FECAVA, accessed July 25. https://www.fecava.org/news-and-events/press-releases/extreme-breeding-of-com-panion-animals-raising-public-awareness-is-key/.

FECAVA. 2020. "FECAVA Downloads on Healthy Breeding." FECAVA, accessed July 26. https://www.fecava.org/policies-actions/healthy-breeding-3/.

FECAVA and FVE. 2018. "Breeding For Extreme Conformations—What Is The Problem?". FECAVA & FVE, accessed November 2. https://www.fecava.org/wp-content/uploads/2019/03/FLYER_Extreme-breeding_RVau21_06_18_BAT.pdf.

FVE., and FECAVA. 2018. "Breeding healthy dogs: The effect of selective breeding on the health and welfare of dogs." FVE & FECAVA, accessed May 8, 2021. https://www.fecava.org/wp-content/uploads/2019/03/2018_06_Extreme_breeding_adopted.pdf.

Four Paws. 2018a. "FOUR PAWS petition letter to eBay." Four Paws, accessed July 26. https://media.4-paws.org/2/5/1/7/2517fd230c36a9e41faf5f07d1a7debc8e67dbf1/20181212_Letter%20with%20Co-Signers_signed%20by%20Heli%20Dungler.pdf.

Four Paws. 2018b. "#THANKSEBAY!". FOUR PAWS International, accessed March 30. https://www.four-paws.org/campaigns-topics/campaigns/thanksebay.

FVE. 2020. "What to check before buying a dog, especially online?" Federation of Veterinarians in Europe (FVE), accessed July 26. https://www.fve.org/what-to-check-before-buying-online-a-puppy-dog/.

FVE, UEVP, and FECAVA. 2016. "Working towards responsible dog trade - The position of the veterinary profession in Europe– 'Profit should never take priority over animal health and welfare'." FVE & UEVP & FECAVA, accessed November 2. https://www.fecava.org/wp-content/uploads/2019/03/2016_Working-towards-responsible-dog-trade-final-draft.pdf.

FVE, UEVP, FECAVA and Four Paws International. 2017. "Letter to the European Commission." FVE, UEVP, FECAVA and Four Paws International, accessed July 25. https://www.fve.org/cms/wp-content/uploads/033_InternetTradeCA.pdf.

FVE, and FECAVA. 2018. "Breeding healthy dogs—The effect of selective breeding on the health and welfare of dogs." FVE & FECAVA, accessed November 2. https://www.fecava.org/wp-content/uploads/2019/03/2018_06_Extreme_breeding_adopted.pdf.

FVE, UEVP, and FECAVA. 2016. "Working towards responsible dog trade - The position of the veterinary profession in Europe—'Profit should never take priority over animal health and welfare'." FVE & UEVP & FECAVA, accessed November 2. https://www.fecava.org/wp-content/uploads/2019/03/2016_ Working-towards-responsible-dog-trade-final-draft.pdf.

Ghirlanda, S., A. Acerbi, H. Herzog, and J.A. Serpell. 2013. "Fashion vs. Function in cultural evolution—The case of dog breed popularity." *PLoS One* 8 (9):1–6. doi:10.1371/journal.pone.0074770.

Gov.UK. 2020. "Lucy's Law spells the beginning of the end for puppy farming." Crown, accessed July 26. https://www.gov.uk/government/news/lucys-law-spells-the-beginning-of-the-end-for-puppy-farming.

Hansen, H.J. 1964. "The body constitution of dogs and its importance for the occurrence of disease." *Nordisk Veterinaermedicin* 16:977–987.

Harrison, J. 2020a. "CRUFFA—Campaign for the Responsible Use of Flat-Faced Animals." Facebook, accessed July 25. https://www.facebook.com/groups/cruffa/.

Harrison, J. 2020b. "Pedigree Dogs Exposed - The Blog." Pedigree Dogs Exposed, accessed July 25. http://pedigreedogsexposed.blogspot.com/.

Henricson, B. 1969. "Breed defects in Dogs." *Svensk Veterinärtidskrift* 21 (12):317–320.

Hobson, H.P. 1995. "Brachycephalic syndrome." Seminars in veterinary medicine and surgery (small animal) (USA).

Holland, K.E. 2019. "Acquiring a pet dog—a review of factors affecting the decision-making of prospective dog owners." *Animals* 9 (4):124.

International Partnership for Dogs. 2020. "DogWellNet." International Partnership for Dogs, accessed March 17. https://dogwellnet.com.

IPFD. 2020a. "'Pandemic Puppies' and COVID-19–How to Navigate This Complex Issue." International Partnership for Dogs, accessed October 28. https://dogwellnet.com/content/education/education-of-consumers/responsible-dog-ownership/pandemic-puppies-and-covid-19-how-to-navigate-this-complex-issue-r690/.

IPFD. 2020b. "Think Globally, Act Locally - Promoting Open Dialogue and Collective Actions." International Partnership for Dogs, accessed October 28. https://dogwellnet.com/content/international-actions/think-globally-act-locally/think-globally-act-locally-promoting-open-dialogue-and-collective-actions-r704/.

IPFD. 2020c. "Get a GRIHP! on French Bulldogs." International Partnership for Dogs, accessed October 28. https://dogwellnet.com/content/health-and-breeding/breeds/breed-specific-health-reports/get-a-grihp-on-french-bulldogs-r700/.

IPFD. 2021. "Norwegian Lawsuit on Dog Breeds and Breeding - The "First" But Not the Last?" International Partnership for Dogs, accessed May 05 2021. https://dogwellnet.com/content/international-actions/think-globally-act-locally/norwegian-lawsuit-on-dog-breeds-and-breeding-the-first-but-not-the-last-r738/

IPFD. 2021d. "Breeds with Swedish Insurance Data." International Partnership for Dogs, accessed May 6. https://dogwellnet.com/breeds/additional-breed-resources/breeds-with-swedish-insurance-data-rl11/.

Johnson, B. 2018. "Health before looks - Collaborative action is urgently needed to stop the practice of extreme breeding in dogs and cats." Parliament Magazine, accessed November 2. https://www.theparliament-magazine.eu/news/article/health-before-looks.

Lindholm, A. 2016. "THE SKK CONFERENCE ON BRACHYCEPHALIC DOGS." IPFD DogWellNet, accessed July 25. https://dogwellnet.com/content/hot-topics/brachycephalics/the-skk-conference-on-brachycephalic-dogs-r367/.

Milligan, A. 2018. "Making assessments of dogs' respiration - BSI." IPFD DogWellNet, accessed July 25. https://dogwellnet.com/media/media/5-making-assessments-of-dogs-respiration-bsi/.

Mills, G. 2018. "Help stop 'pugmania', brachy group tells Disney." *Veterinary Record* 182 (23):649. doi:10.1136/vr.k2518.

Mugenda, L., T. Shreyer, and C. Croney. 2019. "Refining canine welfare assessment in kennels—Evaluating the reliability of Field Instantaneous Dog Observation (FIDO) scoring." *Applied Animal Behaviour Science* 221:104874. doi:10.1016/j.applanim.2019.104874.

NKU. 2017. "Statements and proposals regarding respiratory health in brachycephalic dogs." Nordic Kennel Union, accessed July 25. https://www.skk.se/globalassets/nku-en/documents/brachyreport.pdf.

O'Neill, D.G., A.M. O'Sullivan, E.A. Manson, D.B. Church, A.K. Boag, P.D. McGreevy, and D.C. Brodbelt. 2017. "Canine dystocia in 50 UK first-opinion emergency-care veterinary practices—prevalence and risk factors." *Veterinary Record* 181 (4). doi:10.1136/vr.104108.

O'Neill, D.G., E. Turgoose, D.B. Church, D.C. Brodbelt, and A. Hendricks. 2020. "Juvenile-onset and adult-onset demodicosis in dogs in the UK—prevalence and breed associations." *Journal of Small Animal Practice* 61 (1):32–41. doi:10.1111/jsap.13126.

O'Neill, D.G., L. Baral, D.B. Church, D.C. Brodbelt, and R.M.A. Packer. 2018. "Demography and disorders of the French Bulldog population under primary veterinary care in the UK in 2013." *Canine Genetics and Epidemiology* 5 (1):3. doi:10.1186/s40575-018-0057-9.

O'Neill, D.G., E.C. Darwent, D.B. Church, and D.C. Brodbelt. 2016. "Demography and health of Pugs under primary veterinary care in England." *Canine Genetics and Epidemiology* 3 (1):1–12. doi:10.1186/s40575-016-0035-z.

O'Neill, D.G., C. Jackson, J.H. Guy, D.B. Church, P.D. McGreevy, P.C. Thomson, and D.C. Brodbelt. 2015. "Epidemiological associations between brachycephaly and upper respiratory tract disorders in dogs attending veterinary practices in England." *Canine Genetics and Epidemiology* 2 (1):10.

O'Neill, D.G., S.F.A. Keijser, Å. Hedhammar, C. Kisko, G. Leroy, A. Llewellyn-Zaidi, S. Malm, P.N. Olson, R.M.A. Packer, J.F. Rousselot, I.J. Seath, J.W. Stull, and B.N. Bonnett. 2017. "Moving from information and collaboration to action—report from the 3rd International Dog Health Workshop, Paris in April 2017." *Canine Genetics and Epidemiology* 4 (1):16. doi:10.1186/s40575-017-0054-4.

O'Neill, D.G., A.M. O'Sullivan, E.A. Manson, D.B. Church, P.D. McGreevy, A.K. Boag, and D.C. Brodbelt. 2019. "Canine dystocia in 50 UK first-opinion emergency care veterinary practices—clinical management and outcomes." *Veterinary Record* 184:409. doi:10.1136/vr.104944.

O'Neill, D.G., A.M. Skipper, J. Kadhim, D.B. Church, D.C. Brodbelt, and R.M.A. Packer. 2019. "Disorders of Bulldogs under primary veterinary care in the UK in 2013." *PLoS One* 14 (6):e0217928. doi:10.1371/journal.pone.0217928.

O'Neill, D.G., M.M. Lee, D.C. Brodbelt, D.B. Church, and R.F. Sanchez. 2017. "Corneal ulcerative disease in dogs under primary veterinary care in England—epidemiology and clinical management." *Canine Genetics and Epidemiology* 4 (1):5. doi:10.1186/s40575-017-0045-5.

O'Neill, D.G., C. Pegram, P. Crocker, D.C. Brodbelt, D.B. Church, and R.M.A. Packer. 2020. "Unravelling the health status of brachycephalic dogs in the UK using multivariable analysis." *Scientific Reports* 10 (1):17251. doi:10.1038/s41598-020-73088-y.

Overheid.nl. 2014. "Decree of 5 June 2014, containing rules with regard to animal keepers (Animal Keepers Decree) / Besluit van 5 juni 2014, houdende regels met betrekking tot houders van dieren (Besluit houders van dieren)." Overheid.nl - Staatsblad van het Koninkrijk der Nederlanden, accessed July 26. https://zoek.officielebekendmakingen.nl/stb-2014-210.html.

Packer, R.M.A., A. Hendricks, and C.C. Burn. 2012. "Do dog owners perceive the clinical signs related to conformational inherited disorders as 'normal' for the breed? A potential constraint to improving canine welfare." *Animal Welfare* 21 (Supplement 1):81–93. doi:10.7120/096272812x13345905673809.

Packer, R.M.A., D. Murphy, and M.J. Farnworth. 2017. "Purchasing popular purebreds—investigating the influence of breed-type on the pre-purchase motivations and behaviour of dog owners." *Animal Welfare* 26:191–201. doi:10.7120/09627286.26.2.191.

Packer, R.M.A., D.G. O'Neill, F. Fletcher, and M.J. Farnworth. 2019. "Great expectations, inconvenient truths, and the paradoxes of the dog-owner relationship for owners of brachycephalic dogs." *PLoS One* 14 (7):e0219918. doi:10.1371/journal.pone.0219918.

Packer, R.M.A., D.G. O'Neill, F. Fletcher, and M.J. Farnworth. 2020. "Come for the looks, stay for the personality? A mixed methods investigation of reacquisition and owner recommendation of Bulldogs, French Bulldogs and Pugs." *PLoS One* 15 (8):e0237276. doi:10.1371/journal.pone.0237276.

Pegram, C.L., B.N. Bonnett, H. Skarp, G. Arnott, H. James, Å. Hedhammar, G. Leroy, A. Llewellyn-Zaidi, I.J. Seath, and D.G. O'Neill. 2020. "Moving from information and collaboration to action—report from the 4th international dog health workshop, Windsor in May 2019." *Canine Medicine and Genetics* 7 (1):4. doi:10.1186/s40575-020-00083-x.

PETscan (2021) *PETscan*, available: https://www.uu.nl/en/organisation/veterinary-service-and-cooperation/patientcare-uvcu/the-companion-animals-genetics-expertise-centre/projects-and-services/petscan [accessed March 20].

Pink, J.J., R.S. Doyle, J.M.L. Hughes, E. Tobin, and C.R. Bellenger. 2006. "Laryngeal collapse in seven brachycephalic puppies." *Journal of Small Animal Practice* 47 (3):131–135. doi:10.1111/j.1748-5827.2006.00056.x.

Poncet, C.M., G.P. Dupre, V.G. Freiche, M.M. Estrada, Y.A. Poubanne, and B.M. Bouvy. 2005. "Prevalence of gastrointestinal tract lesions in 73 brachycephalic dogs with upper respiratory syndrome." *Journal of Small Animal Practice* 46 (6):273–279. doi:10.1111/j.1748-5827.2005.tb00320.x.

Purdue University College of Veterinary Medicine. 2020. "Canine Care Certified." Purdue University College of Veterinary Medicine, accessed March 20. https://www.purdue.edu/vet/ccc/.

Ravetz, G. 2017. "Stop normalising suffering—vets speaking out about brachys." British Veterinary Association, accessed April 8. https://www.bva.co.uk/news-and-blog/blog-article/stop-normalising-suffering-vets-speaking-out-about-brachys/.

Rodin, T., and M. Rundkvist. 2020. "Pug Dog Passion—Awareness if the first step to improvement." Pug Dog Passion, accessed August 6. https://pugdogpassion.com/.

Rooney, N., and D. Sargan. 2008. "Pedigree dog breeding in the UK—a major welfare concern?". Royal Society for the Prevention of Cruelty to Animals, accessed November 2. https://dogwellnet.com/files/file/311-pedigree-dog-breeding-in-the-uk-a-major-welfare-concern/.

Rooney, N.J. 2009. "The welfare of pedigree dogs—cause for concern." *Journal of Veterinary Behavior—Clinical Applications and Research* 4 (5):180–186. doi:10.1016/j.jveb.2009.06.002.

Royal Veterinary College. 2020. "Brachycephalic Research Team." Royal Veterinary College, accessed February 27. https://www.rvc.ac.uk/research/focus/brachycephaly.

Sandøe, P., S.V. Kondrup, P.C. Bennett, B. Forkman, I. Meyer, H.F. Proschowsky, J.A. Serpell, and T.B. Lund. 2017. "Why do people buy dogs with potential welfare problems related to extreme conformation and inherited disease? A representative study of Danish owners of four small dog breeds." *PLoS One* 12 (2):e0172091. doi:10.1371/journal.pone.0172091.

Seglins, D., and C. Thomas. 2020. "Officials probe arrival of 500 puppies, 38 of them dead, aboard flight from Ukraine - Canadians' taste for exotic breeds fuelling global black market, say animal welfare advocates." CBC News, accessed July 26. https://www.cbc.ca/news/canada/ukraine-flight-puppies-1.5620691.

Stella, J., M. Hurt, A. Bauer, P. Gomes, A. Ruple, A. Beck, and C. Croney. 2018. "Does flooring substrate impact kennel and dog cleanliness in commercial breeding facilities?" *Animals* 8 (4):59.

Stella, J., T. Shreyer, J. Ha, and C. Croney. 2019. "Improving canine welfare in commercial breeding (CB) operations—Evaluating rehoming candidates." *Applied Animal Behaviour Science* 220:104861. doi:10.1016/j.applanim.2019.104861.

Stilwell, V. 2011. "RSPCA—Bred for looks, Born to suffer." Veterinary Times, accessed July 25. https://www.vettimes.co.uk/news/puppies-bred-for-looks-born-to-suffer-warns-rspca/.

Swedish Kennel Club. 2012. "Special Breed-specific Instructions for Judges - BSI - Making assessments of dogs' respiration." Swedish Kennel Club, accessed May 8, 2021. https://www.youtube.com/watch?v=kQ_3f4bLkME&ab_channel=SvenskaKennelklubben.

The Kennel Club. 2020a. "Breed registration statistics." The Kennel Club Limited, accessed October 17. https://www.thekennelclub.org.uk/media-centre/breed-registration-statistics/.

The Kennel Club. 2020b. "Breed Watch." The Kennel Club, accessed November 2. https://www.thekennelclub.org.uk/services/public/breed/watch/Default.aspx.

The Kennel Club. 2020c. "Facts and figures." The Kennel Club, accessed March 30. https://www.thekennelclub.org.uk/our-resources/media-centre/facts-and-figures.

The Kennel Club. 2020d. "The Kennel Club and University of Cambridge Respiratory Function Grading Scheme." The Kennel Club Limited, accessed November 2. https://www.thekennelclub.org.uk/health-and-dog-care/health/getting-started-with-health-testing-and-screening/respiratory-function-grading-scheme/.

The Kennel Club. 2021. "Breed health and conservation plans (BHCPs)." The Kennel Club Limited, accessed February 21. https://www.thekennelclub.org.uk/health/breed-health-and-conservation-plans/.

The Nordic Kennel Clubs. 2018. "Breed Specific Instructions (BSI) regarding exaggerations in pedigree dogs." Nordisk Kennel Union, accessed November 2. https://www.skk.se/globalassets/dokument/utstallning/breed-special-specific-instructions—bsi-a8.pdf.

Thomson Reuters Practical Law. 2020. "Treaty on the Functioning of the European Union (TFEU)." Thomson Reuters, accessed March 30. https://uk.practicallaw.thomsonreuters.com/PLCCoreDocument/ViewDocument.html?comp=pluk&DocumentGuid=Ia840c93e65fc11e598dc8b09b4f043e0&ViewType=FullText&HasDraftingNotes=False&ResearchReportViewMode=False&SessionScopeIsValid=True&IsCourtWireDocument=False&IsSuperPrivateDocument=False&IsPrivateDocument=False&ClientMatter=Cobalt.Website.Platform.Web.UserData.ClientMatter&AuthenticationStrength=0&IsMedLitStubDocument=False&IsOutOfPlanDocumentViewClicked=False&TransitionType=Default&ContextData=%28sc.Default%29&BillingContextData=%28sc.Default%29.

Torrez, C.V., and G.B. Hunt. 2006. "Results of surgical correction of abnormalities associated with brachycephalic airway obstruction syndrome in dogs in Australia." *Journal of Small Animal Practice* 47 (3):150–154. doi:10.1111/j.1748-5827.2006.00059.x.

van Hagen, M.A.E. 2019. "Breeding Short-Muzzled Dogs." Utrecht University - Department of Animals in Science and Society and the Expertise Centre Genetics of Companion Animals. Commissioned by the Ministry of Agriculture, Nature and Food Quality., accessed November 2. https://dogwellnet.com/applications/core/interface/file/attachment.php?id=4056,_https://dogwellnet.com/applications/core/interface/file/attachment.php?id=4288,_https://dogwellnet.com/applications/core/interface/file/attachment.php?id=4289.

VetCompass. 2021. "VetCompass Programme." RVC Electronic Media Unit, accessed March 20, 2021. http://www.rvc.ac.uk/VetCOMPASS/.

VetCompass Australia. 2021. "VetCompass Australia. " The University of Sydney, accessed March 20, 2021. http://www.vetcompass.com.au/.

Westgarth, C., R.M. Christley, G. Marvin, and E. Perkins. 2019. "The responsible dog owner—The construction of responsibility." *Anthrozoös* 32 (5):631–646. doi:10.1080/08927936.2019.1645506.

WSAVA. 2018. "WSAVA Endorses FVE/FECAVA Position Paper on Healthy Breeding." WSAVA, accessed July 25. https://wsava.org/wp-content/uploads/2020/01/WSAVA-Endorses-FVE-FECAVA-Position-Paper-on-Healthy-Breeding.pdf.

WSAVA and FECAVA. 2017. "Vets must 'dare to speak out'—Urgent action on brachycephalic dogs called for during panel discussion at FECAVA/WSAVA/DSAVA Congress in Copenhagen." WSAVA/FECAVA, accessed November 2. https://wsava.org/wp-content/uploads/2020/01/Urgent-action-on-brachycephalic-dogs-called-for-during-panel-discussion-at-Congress-in-Copenhagen.pdf.

Wykes, P.M. 1991. "Brachycephalic airway obstructive syndrome." *Problems in Veterinary Medicine* 3 (2):188.

Part 2

Clinical Viewpoints

10 Brachycephalic Obstructive Airway Syndrome (BOAS) – Clinical Assessment and Decision-Making

Jane Ladlow and Dr Nai-Chieh Liu
University of Cambridge

CONTENTS

Introduction.. 155
Disease Characteristics 158
Lesion Sites for BOAS... 160
Stenotic Nostrils.. 160
Nasal Turbinate Hypertrophy... 161
Hyperplastic Palate 163
Macroglossia.. 163
   Nasopharyngeal Restriction and Collapse 163
Tracheal Hypoplasia 163
Laryngeal Collapse 164
Everted Laryngeal Saccules 164
Tonsillar Hypertrophy 165
Gastrointestinal Signs 165
Disease Recognition, History and Clinical Signs 165
   Disease Recognition.. 165
BOAS: Diagnosis.. 166
History... 166
Clinical Examination 166
Clinical Signs.. 166
Respiratory Function Testing 168
Imaging 168
Treatment Decision-Making 171
Conclusion 172
Appendix... 172
   BOAS Functional Grading 173
References... 174

INTRODUCTION

Brachycephalic obstructive airway syndrome (BOAS) encompasses a variety of clinical conditions that arise from structural and functional upper airway obstruction (Harvey 1989). This syndrome is reported in both dogs and cats (Farnworth et al. 2016; Corgozinho et al. 2012). It may also occur in some other species such as the rabbit and horse that are being selected towards extreme conformation to show brachycephalic morphology, although breathing issues in these species have not been reported yet.

In dogs, the combination of effects from these respiratory conditions can manifest as noisy breathing, difficulty in breathing, exercise intolerance, gastrointestinal reflux, heat intolerance, cyanosis, collapse and sleep disorders (Roedler, Pohl, and Oechtering 2013). VetCompass has been used to show that the extreme brachycephalic breeds died younger than equivalent sized dogs, with a higher proportion of deaths due to upper respiratory disease (and additionally that certain brachycephalic breeds were at greater risk of upper respiratory disorders than non-brachycephalic breeds (O'Neill et al. 2015). In cats, the syndrome is recognised by increased upper airway noise, snoring and decreased activity levels (Farnworth et al. 2016).

Brachycephalic dogs and cats have, by definition, a short and wide skull and muzzle (Evans 1993) (Figure 10.1). This is hypothesised to result from premature closure of the sutures at the base of the skull leading to constriction of the basioccipital and basisphenoid bones with a reduction in the basicranial axis (Evans 1993). Although the skull is shortened, the soft tissues are not always proportionately reduced, particularly the soft palate, tongue and nasal turbinates (Grand and Bureau 2011; Jones, Stanley, and Nelson 2019; Schuenemann and Oechtering 2014) (Figure 10.2). There may also be constriction of the bones at the base of the skull, resulting in narrowing of the nasopharynx and pharynx (Heidenreich et al. 2016). In dogs affected with BOAS, the increased soft tissue in the pharynx and nasal cavity combined with narrowed airway passages results in restricted airflow. In certain breeds, cartilage abnormalities occur consistent with a generalised chondrodysplasia (Marchant et al. 2017; Bannasch et al. 2010). Consequently, hypoplasic trachea occurs in affected Bulldogs and French Bulldogs and, perhaps separately, laryngeal collapse in affected Pugs. These tissue disproportions (trachea, palate, tongue volume, larynx) seem breed specific and to differ between the breeds, which is logical when one considers the very different types of skull morphology seen across breeds which are all labelled as brachycephalic.

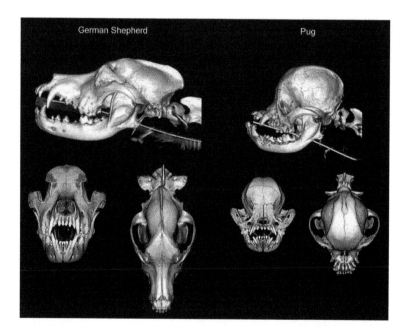

FIGURE 10.1   Computed tomographic 3D rendering images show the morphological comparisons of skull dimensions between a mesaticephalic dog (German shepherd) and a brachycephalic dog (Pug). (Reprinted with permission from Liu (2016) Figure 1-1.)

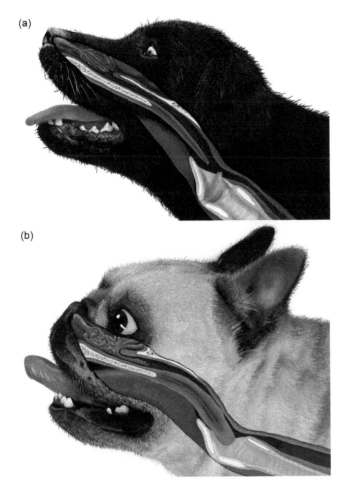

FIGURE 10.2   Comparison of the upper airway anatomy of a Labrador Retriever (A; mesaticephalic breed) and a BOAS-affected French Bulldog (B; brachycephalic breed). (Adapted with permission from Liu (2016) Figure 1-4.)

Brachycephaly has been defined in various ways, with a skull index (width×100/length) ratio of >80 perhaps the easiest to understand (Evans 1993). According to this definition, many breeds can be defined confidently as brachycephalic, including the Pug, French Bulldog, Bulldog, Boston Terrier, Cavalier King Charles Spaniel, Pekingese, Japanese Chin, Dogue de Bordeaux, Shih-Tzu, Boxer, Pomeranian, Griffon Bruxellois and Affenpinscher. It should be noted that there is variation among these breeds in both skull morphology and BOAS status.

Recent work by the Royal Veterinary College (RVC) suggests that the craniofacial ratio (CFR), which is muzzle length divided by cranial length, is associated with BOAS status (Packer et al. 2015). Although this intuitively explains why Pugs may be more affected by BOAS than Boxers, some breeds such as the Griffon Bruxellois and Affenpinscher within the RVC's data were less affected than would be expected from the CFR.

BOAS occurs predominantly in the more extreme brachycephalic breeds, including the Pug, French Bulldog and Bulldog. Despite the authors having spent the last 10 years at the Cambridge BOAS Research Group (Ladlow, Liu, and Sargan 2012) studying airway function in the extreme brachycephalic breeds, the true prevalence of BOAS within these breeds is not yet clear. Although some veterinary professionals consider that all brachycephalic dogs must, by definition, be affected to some extent, the authors have seen some dogs of extreme brachycephalic breeds that have no upper airway noise and that show normal exercise tolerance. In the study population (excluding clinical cases) of our Cambridge BOAS Research Group, we have classified approximately 40% of Bulldogs, 45% of French Bulldogs and 65% of Pugs with clinically significant disease (Liu et al. 2016). Clinically significant was defined to describe disease that negatively impacted on quality of life.

Other breeds reported with BOAS include the Boston Terrier, Pekingese, Shih-Tzu, Cavalier King Charles Spaniel and Japanese Chin (Packer et al. 2015). The prevalence and severity of BOAS in these breeds is not yet clear but seems to be less than that of the Pug, French Bulldog and Bulldog.

DISEASE CHARACTERISTICS

BOAS is a progressive disease over a lifetime, with affected dogs generally showing early signs of disease between 1 and 2 years of age, with an increase in disease prevalence with age, particularly up to the age of 4 years (Liu et al. 2016). Occasionally, severely affected puppies can present as early as 6 months of age.

A small proportion of Bulldog puppies will present before 4 months of age with respiratory difficulties, predominately from lower airway disease; these puppies usually have a severely hypoplastic trachea on imaging. Respiratory difficulty is secondary to regurgitation and aspiration pneumonia. With supportive care (oxygen therapy, antibiotic support and anti-inflammatories), some of these dogs will recover with clinical signs resolving as the trachea increases in size as the dogs mature (Clarke, Holt, and King 2011).

Considering a sex predisposition, a study of effects from conformation reported that male Bulldogs and French Bulldogs were predisposed (Liu et al. 2017), presumably because these breeds have a muscular 'bull' head with the males having more 'meat in the box' or muscle to fit or squeeze into the confined pharyngeal and nasopharyngeal area. Conversely in Pugs, more females than males are affected, which may be due to the softer cartilage that seems present in this breed, and the effect of weight on these structures. The trachea in Pugs has a natural C-shape cross-sectionally with a relatively wide dorsal tracheal membrane.

Other conformational factors have also been identified that increase the risk of BOAS. The most important of these are stenotic nostrils – in all brachycephalic breeds but particularly in the French Bulldog (Liu et al. 2016; Liu, Troconis, et al. 2017; Packer et al. 2015). Other breeds that are commonly affected by stenotic nostrils include the Shih-Tzu, Japanese Chin and Pekingese. In addition to being narrowed/stenosed, the nostrils of many extreme brachycephalic dogs are relatively immobile. In most non-brachycephalic dogs, the nostrils naturally move laterally on inspiration, particularly under exertion. However, this ability to move the nostrils has been lost in some of the extreme brachycephalics, particularly those with moderate or severe stenosis which probably compounds airway resistance at the nostril.

In the Bulldog and French Bulldog, increasing skull width is also identified as a conformational risk factor for BOAS (Liu, Troconis, et al. 2017). In Bulldogs, the thick neck is an additional risk factor for BOAS, with a neck:chest girth of 0.71 or above shown to increase the risk of BOAS in male Bulldogs (Liu, Troconis, et al. 2017). In French Bulldogs, having a longer neck decreases the risk of BOAS and being generally less 'cobby' or compact with a longer muzzle is likely to improve the airway. In Pugs, having eyes that are set relatively closer together is protective against BOAS (Liu, Troconis, et al. 2017). The external conformational factors (Table 10.1) account for about 40%–50% of the BOAS variance in breeds which suggests that breeding towards less extreme versions of the conformations described above could be expected to substantially reduce the occurrence of BOAS (Liu, Troconis, et al. 2017) (Figure 10.3).

In additional to breed-associated (i.e. inherited) conformations, acquired disease may also exacerbate BOAS severity in dogs. The most important of these is obesity, particularly in Pugs (Liu, Troconis, et al. 2017). Differing impacts from obesity on airway resistance between the extreme brachycephalic breeds may be partially explained by their differing tracheal shapes. Pugs tend to lay down excess fat around their necks and chest, which may compress their C-shaped trachea that

TABLE 10.1

Conformational Risk Factors

	Pug	**French Bulldog**	**Bulldog**
Definite	Nostril grade Body condition score Gender Skull index (head width/length) Eye width ratio	Nostril grade Neck/girth ratio Neck length ratio Gender	Neck/girth ratio Body condition score Skull index (head width/length)
Likely	Neck length ratio (neck length/ body length)	Craniofacial ratio	Body condition score Gender Nostril grade

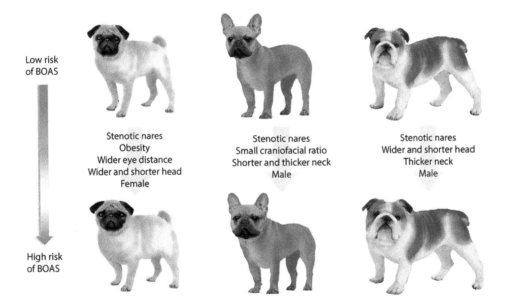

Low risk
of BOAS

Stenotic nares
Obesity
Wider eye distance
Wider and shorter head
Female

Stenotic nares
Small craniofacial ratio
Shorter and thicker neck
Male

Stenotic nares
Wider and shorter head
Thicker neck
Male

High risk
of BOAS

FIGURE 10.3   Conformational risk factors of BOAS in the Pug, French Bulldog, and Bulldog. (Reprinted with permission from Cambridge BOAS Research Group (Ladlow, Liu, and Sargan 2012).)

has a wide dorsal membrane. However, Bulldogs and French Bulldogs usually have an O-shaped trachea, often with the cartilage rings overlapping slightly. This conformation, without a compressible tracheal membrane, may explain why obesity has a less marked effect on respiratory function in these breeds. Excessive body condition score in any brachycephalic dog, however, is likely to exacerbate airway obstruction and will decrease tidal volume (Manens et al. 2012), but in Pugs, each incremental increase in BCS is linked to an increase BOAS index (airway obstruction severity score) of 6.4% (Liu, Troconis, et al. 2017). Reducing an obese dog's weight is very likely to improve respiratory function, although this has yet to be confirmed conclusively in clinical BOAS cases.

Other diseases that may worsen airway obstruction include those causing generalised muscular weakness, such as hyperadrenocorticism or myasthenia gravis, or space-occupying airway lesions such as laryngeal masses or granulomas, nasal tumours, rhinitis or unerupted tooth roots that reduce the airspace in the nasal cavity. Diseases that impair lower airway function may put increased stress on the upper airway; these include pneumonia (often related to regurgitation and aspiration in severe BOAS cases), lung masses and lung lobe torsion (especially in the Pug (Holmes et al. 2018).

LESION SITES FOR BOAS

In BOAS, primary conformational abnormalities restrict airflow, which in turn leads to a vicious cycle of secondary changes because significant negative airway pressures are created as the body tries to overcome the increased resistance to flow. The increased negative pressure during inspiration results in mucosal inflammation and laryngeal distortion. Palatine muscles hypertrophy in an attempt to maintain airway patency.

The elongated soft palate can trap the epiglottis of the larynx, making it difficult to switch from nasal breathing to open mouth breathing. In addition, as the dog struggles to breathe, the soft palate and surrounding tissues can become quite swollen and oedematous, thereby further impeding the airflow.

The veterinary literature offers little conclusive evidence to distinguish which specific anatomical structures are primarily responsible for the obstruction, which are secondary lesions and how these vary in effect between the different breeds. The primary lesion sites linked to BOAS include stenotic nares, a narrowed nasopharynx, an oversized soft palate, hypoplastic trachea and bronchial collapse, oversized tongue, and hypertrophic and aberrant turbinate crowding in the nasal cavity. These lesions can cause secondary areas of collapse and inflammation, notably laryngeal collapse, laryngeal saccule eversion and tonsillar hypertrophy (Table 10.2). Increased intra-thoracic pressures may result in pulmonary hypertension in addition to the gastrointestinal signs of reflux and hiatal hernia with the accompanying risk of aspiration pneumonia.

The early literature describing BOAS was by Harvey (1989). He reported a mixed cohort of brachycephalic dogs affected by a long soft palate (100%), stenotic nostrils (51%), laryngeal collapse (31%) including laryngeal saccule eversion and hypertrophic tonsils (9%). Fasanella and colleagues (2010) reported a case series of 90 dogs where 94% had an elongated soft palate, 77% stenotic nares, 66% everted laryngeal saccules and 56% everted tonsils (Fasanella et al. 2010).

However, Fasanella and Harvey may have underestimated the nasal component to BOAS because they worked from radiographs and direct observation of the upper airway lesions (Fasanella et al. 2010; Harvey 1989). Using the advanced imaging techniques that are now available, Oechtering in 2007 reported treating hypertrophic and malformed turbinates within BOAS in the extreme brachycephalic breeds using laser turbinectomy (Oechtering, Oechtering, and Noeller 2007).

STENOTIC NOSTRILS

Brachycephalic dogs have narrower or more stenotic nostrils compared to the longer nosed mesocephalic dogs (Packer et al. 2015). There are varying degrees of stenosis in different breeds. Nasal grading systems in Bulldogs, French Bulldogs and Pugs score dogs from 0 (open), 1 (mild), 2 (moderate) to 3 (severe stenosis). These grading systems are breed specific (Liu, Troconis, et al. 2017) (Figure 10.4).

Previously, stenotic nares were thought to be of paramount importance, and early treatment was advised to open the nostrils and prevent development of secondary airway obstructive lesions. Recent

---

TABLE 10.2

Lesion Sites for Brachycephalic Obstructive Airway Syndrome

Primary	Secondary
• Stenotic nostrils	• Thickened soft palate
• Elongated and thickened soft palate	• Nasopharyngeal collapse
• Nasal turbinate hypertrophy and malposition	• Tonsillar eversion and hypertrophy
• Nasopharyngeal restriction	• Laryngeal collapse
• Macroglossia	• Bronchial collapse
• Tracheal hypoplasia	• Gastrointestinal reflux with oesophagitis
• Laryngeal hypoplasia	• Hiatal hernia

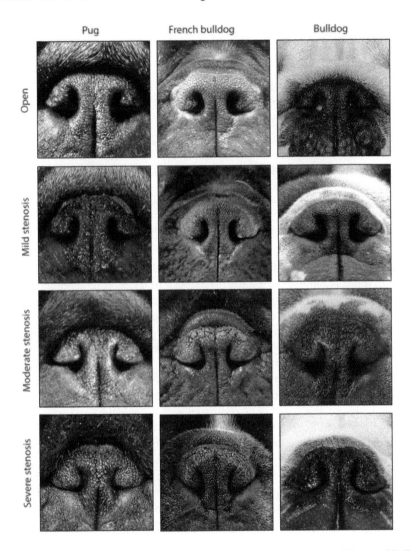

FIGURE 10.4   Example of different degrees of nostril stenosis in Pugs, French Bulldogs and Bulldogs. Open nostrils: nostrils are wide open; mildly stenotic nostrils: slightly narrowed nostrils where the lateral nostril walls do not touch the septum; moderately stenotic nostrils: the lateral nostril walls touch the septum at the dorsal part of the nostrils, and the nostrils are only open at the bottom; severely stenotic nostrils: nostrils are almost closed. The dog switches to oral breathing from nasal breathing with stress or very gentle exercise such as playing. (Reprinted with permission from Liu et al. (2017) Figure 1.)

work from the Cambridge BOAS Research Group shows that the nostrils in the French Bulldog are definitely of particular importance but, as yet, this nostril stenosis had not been linked with an increase in obstructive lesions at any other site (Liu et al. 2016). It is also worth noting that we have also seen mature dogs with no BOAS-related respiratory signs yet with severely stenotic nostrils.

NASAL TURBINATE HYPERTROPHY

The nasal turbinates in brachycephalic dogs are compacted inside an extremely short muzzle and have fewer branches compared to mesocephalics. This leads to a decreased nasal mucosal surface area (Figure 10.5). Oechtering and colleagues (2007) used CT to report that nasal obstruction from rostrally positioned ventral nasal conchae and nasopharyngeal obstruction from caudal aberrant

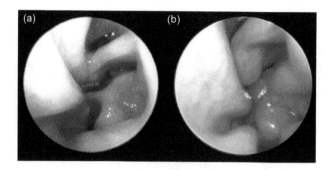

FIGURE 10.5 Examples of rhinoscopic images of nasal turbinates. (a) The nasal turbinate of a Shih-Tzu dog. The meatuses are patent with only one mucosal contact point between the plica dorsalis and the dorsal swell body; (b) the nasal turbinates of a French Bulldog. The meatuses are obstructed by the aberrant and hypertrophic nasal turbinates.

turbinates originating from the middle or ventral nasal conchae were seen in 43% of brachycephalic dogs (11, Pugs, 7 French Bulldogs, 5 English Bulldogs). In addition to abnormal positioning, the conchae had reduced branching compared to a normocephalic dog (German shepherd dog). These brachycephalic dogs, particularly the Pugs, frequently also had marked septal deviation and stenotic nares. These findings were mirrored on rhinoscopic examination where rostrally positioned conchae and enlarged plica recta resulted in turbinate overcrowding and nasal airflow obstruction (Schuenemann and Oechtering 2014).

The originating cause of the turbinate abnormalities is not yet clear. Walter and colleagues (2008) described histologic studies of brachycephalic conchae that reveal both thickened hyperaemic mucosa and increased thickened bone compared with normocephalic dogs – consistent with a mixed hypertrophy (Walter, Seeger, and Oechtering 2008). However, the thickened mucosa may be a primary lesion in brachycephalic dogs or could be secondary to chronic rhinitis. Puppies, as young as 5 months of age, were recognised with thickened mucosa in the study.

In an attempt to measure the turbinate crowding objectively, Schuenemann and Oechtering (2014) reported increased mucosal contact points between the ventral nasal concha and surrounding mucosal surfaces on rhinoscopic examination in the brachycephalic dogs compared with normocephalic dogs (Schuenemann and Oechtering 2014). These contact points were assessed using rigid rhinoscopy and therefore assess just the rostral rather than the whole nasal cavity. On average, French Bulldogs had three mucosal contact points/nasal cavity, and Pugs had 1.7 mucosal contact points/nasal cavity, whereas any contact between intranasal structures was uncommon in normocephalic dogs, with 76% having no mucosal contact. Rhinomanometry offers another technique to assess nasal cavity obstruction. Wiestner and colleagues (2007) used rhinomanometry to demonstrate that the nasal cavity of Bulldogs affected with BOAS had six times greater nasal airflow resistance than beagles (Wiestner et al. 2007).

The caudal aberrant turbinate tissue was also reported by Ginn who identified aberrant turbinates in the nasopharynx in 21% of 53 brachycephalic dogs undergoing rhinoscopy, including in 82% of Pugs (Ginn et al. 2008). On a note of caution though, Grosso and colleagues (2015) reported aberrant caudal turbinates in Bulldogs with only mild signs of BOAS, and the authors concluded that the presence of aberrant turbinates is often subclinical in this breed (Grosso, Ter Haar, and Boroffka 2015).

The decreased turbinate branching and clubbing, especially of the rostral ventral nasal conchae, with increased contact points, disrupts laminar flow and almost certainly compromises thermoregulation. This is difficult to measure in dogs though and thus has not been quantified. However, heat intolerance is reported as a component of BOAS on owner questionnaires (Roedler, Pohl, and Oechtering 2013). Studies on heat stroke in dogs have shown that brachycephalic breeds, particularly the Bulldog and French Bulldog, are at increased risk (Hall, Carter, and O'Neill 2020; Bruchim et al. 2006).

HYPERPLASTIC PALATE

An elongated palate is often cited as the most contributory lesion in BOAS. However, in a soft palate study comparing mildly and severely affected BOAS cases, Grand and Bureau (2011) found significantly increased thickness of the soft palate in the severely affected cases but not significant difference in the length of the soft palate, which was long in both affected and non-affected brachycephalic dogs (Grand and Bureau 2011). We have certainly seen long thin soft palates in both affected and minimally affected extreme brachycephalic breeds which raises the question of whether the soft palate thickness is inflammatory in nature due to restricted airflow further forward or whether the thickness is secondary to denervation and vibratory damage or whether it can be a primary lesion in some dogs.

Histopathology of the palate in extreme brachycephalic breeds has revealed reduced muscle mass, with myofibril atrophy and hyaline degeneration, increased stromal tissue with an amplified myxoid matrix in the lamina propria and increased salivary glandular tissue (Crosse et al. 2015; Pichetto et al. 2011). Interestingly, these changes are also seen in Grade 1 (mildly affected) dogs and as yet no association has been found between BOAS severity and histopathological changes.

In human patients with obstructive sleep apnoea, uvular histopathological samples demonstrated nerve degeneration that was linked to low-frequency vibration causing a mechanical stretch trauma with subsequent muscle atrophy (Shah et al. 2018). This is a self-perpetuating cycle of respiratory obstruction as the progressive weakness of the palatine muscles decreases the patency of the upper airway, leading to nasopharyngeal collapse. A recent Japanese paper that examined the palates of BOAS affected dogs also found hyaline degeneration, atrophy and hypertrophy of the muscle with decreased peripheral nerve supply to the palates which is consistent with the changes seen in humans with sleep apnoea (Arai et al. 2016).

In the caseloads of extreme breeds where the authors decide to operate, the soft palate seems to be the most important obstructive lesion site in the Bulldog.

MACROGLOSSIA

The tongue volume relative to body weight and ratio of skull length/width is higher in the extreme brachycephalics compared with mesocephalic dogs when examined with CT (Jones, Stanley, and Nelson 2019). This increase in tongue volume is probably a primary cause of obstruction and results in approximately 60% decreased oropharyngeal and nasopharyngeal airspace. Tongues of the brachycephalic dogs were also denser than the control dogs, which may reduce flexibility.

NASOPHARYNGEAL RESTRICTION AND COLLAPSE

The thickened soft palate and relative macroglossia may contribute to the nasopharyngeal constriction seen in some brachycephalic dogs (Kim et al. 2019). A study using fluoroscopy in awake dogs showed that dynamic nasopharyngeal collapse was more commonly seen in brachycephalic dogs rather than mesocephalic dogs (Rubin et al. 2015).

In addition to the soft tissues of the palate and tongue, the nasopharyngeal dimensions mediolaterally appear compressed in some of the brachycephalic dogs, and this may fit with the abnormal development of the base of the skull that contributes to the typical Bulldog head conformation (Stockard 1941).

TRACHEAL HYPOPLASIA

Tracheal hypoplasia is usually defined as tracheal lumen diameter (TD) to thoracic inlet diameter (TI) based on radiographic examination, although there are also CT assessment methods available (Kaye et al. 2015). In most brachycephalic dogs, a TD:TI ratio of <0.16 is considered hypoplastic, whereas in the Bulldogs, it is <0.12. Tracheal hypoplasia occurrence is higher in the screw-tailed brachycephalic dogs compared to non-screw-tailed brachycephalics (Komsta et al. 2019). A recent Polish study reported a much lower proportion of Bulldogs with tracheal hypoplasia (12.9%) compared with some of the older papers at 55% (Coyne and Fingland 1992), which is encouraging (Komsta et al. 2019). This study also reported that Pugs and French Bulldogs were more commonly affected by tracheal hypoplasia than the Bulldogs, although these differences may vary between countries.

Although tracheal hypoplasia is often cited as a major causative factor of BOAS, early work by Harvey and Fink suggested it did not affect prognosis after upper airway surgery (Harvey and Fink 1982) and that tracheal hypoplasia alone can be subclinical in the Bulldog (Coyne and Fingland 1992). Whether it affects the severity of BOAS when other upper airway obstructive lesions are present is not yet known.

LARYNGEAL COLLAPSE

Although laryngeal collapse is generally regarded as a secondary event in BOAS, the Pug can show laryngeal stridor and collapse with few other obvious airway obstructive sites. In these Pugs, the laryngeal cartilage seems very soft and may reflect a primary chondromalacia in this breed. The cases of laryngeal collapse that we see in Bulldogs and French Bulldogs typically reflect secondary change following increased airway resistance, increased negative intra-glottic luminal pressure and increased air velocity causing displacement of the rostral laryngeal structures medially with loss of the supporting function of the laryngeal cartilage.

Laryngeal collapse is classified into three grades (Leonard 1960) (Figure 10.6):

- Grade 1: Everted laryngeal ventricles (seen in 50% of BOAS cases)
- Grade 2: Deviation of the cuneiform cartilage medially
- Grade 3: Medial collapse of the cuneiform and corniculate cartilage of the arytenoid obstructing the airway

EVERTED LARYNGEAL SACCULES

Healthy laryngeal saccules (or ventricles) sit in the larynx, between the vestibular and vocal folds. The saccule lies medial to the thyroid cartilage and lateral to the vocal and vestibular folds.

In the first stage of laryngeal collapse, the laryngeal saccules evert under the action of negative pressure because the saccules represent the tissue of least resistance in the wall of the larynx. It was originally thought the saccules became oedematous initially but eventually fibrosed to create a ventral obstruction to the larynx. However, using histopathology, Cantatore and colleagues identified no fibrotic saccules but instead only oedema and a chronic lymphoplasmacytic

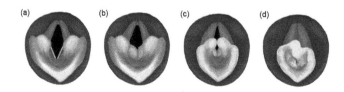

FIGURE 10.6 Grading of laryngeal collapse in brachycephalic dogs. (a) Normal larynx; (b) Grade 1 laryngeal collapse, with the feature of everted and oedematous laryngeal ventricles; (c) Grade 2 laryngeal collapse, in addition to everted laryngeal ventricles, the cuneiform processes collapse medially; and (d) Grade 3 laryngeal collapse, both the cuneiform processes and cornicular processes collapse medially and lead to obstruction of the rima glottidis.

inflammation (Cantatore et al. 2012). The histological features were unchanged between saccules resected months apart, suggesting the process is non-reversible: it is thought that, once a saccule is everted and swollen, the compression at its base interferes with lymphatic return, preventing spontaneous resolution.

The authors find that most of our BOAS-affected Pugs have everted laryngeal saccules, while this lesion is restricted to mainly the more severely BOAS-affected French Bulldogs or Bulldogs. This may reflect the more elliptically shaped and smaller rima glottidis in the Pug (Rutherford et al. 2017) or may be due to the seemingly softer cartilage which occurs in this breed.

Once a dog starts to have Grade 2 or 3 secondary collapse, the larynx is often the most difficult part of BOAS to treat. Laryngeal collapse is a negative prognostic indicator for a good outcome post-surgery (Liu, Oechtering, et al. 2017).

TONSILLAR HYPERTROPHY

Inflammation of the tonsils (amygdalitis) was diagnosed in 15/73 (20.5%) dogs, predominately French Bulldogs, with BOAS (Poncet et al. 2005). Partial tonsillectomy may decrease pharyngeal pressure and was included in the surgical procedure that we recently published that reported improved prognosis compared to the traditional technique without tonsillectomy (Liu, Oechtering, et al. 2017).

GASTROINTESTINAL SIGNS

Increased gastro-oesophageal reflux in BOAS cases can trigger an oesophagitis which manifests as regurgitation and vomiting (Lecoindre and Richard 2004).

Brachycephalic dogs also have increased incidence of hiatal hernia (Freiche and Poncet 2007). Excessive ingestion of air during dyspnoeic episodes can result in gastric tympany and distention with increased gastrin secretion.

Poncet and colleagues (2005) reported 98% cases (based on 73 dogs, predominately French Bulldogs) had chronic gastritis, 90% had lymphoplasmacytic duodenitis, and there was a positive correlation between severity of respiratory and gastro-intestinal signs (Poncet et al. 2005). GI signs resolved in 80% of these cases after BOAS surgery, either with or without GI medication, although immediate postoperative vomiting and regurgitation were reduced in dogs given 0.7 mg/kg omeprazole, 0.2 mg/kg cisapride and 1 g sucralfate. Poncet and colleagues (2005) also believed that concurrent medical management of GI tract pathology may have helped reduce the progression of the respiratory disease (laryngeal collapse).

DISEASE RECOGNITION, HISTORY AND CLINICAL SIGNS

DISEASE RECOGNITION

A major challenge for BOAS is that it is routinely under-recognised by both owners and veterinary professionals. A questionnaire-based study on disease recognition revealed that 60% of owners did not perceive their dogs to have airway obstruction even though concurrent clinical examination and history were compatible with the disease (Packer, Hendricks, and Burn 2012). Respiratory noises that would be alarming in a longer nosed dog are tolerated as 'normal for the breed' in extreme brachycephalics (O'Neill et al. 2013, 2016, 2019). Even for veterinary professionals, it can be challenging to decide which dogs are clinically affected. If upper airway noise is an indication of obstruction, then the majority of the extreme brachycephalics have some degree of BOAS. In most studies, however, the proportion of brachycephalic dogs reported with BOAS in general practice is about 15%–25% which suggests that only the severely affected dogs are being formally recorded as BOAS clinical cases (O'Neill et al. 2013, 2016, 2018, 2019).

BOAS: DIAGNOSIS

- History
- Clinical signs at rest and during exercise
- Oral examination under sedation/GA
- Head/thoracic CT or radiographs
- Endoscopic examination of the nasopharynx +/− oesophagus
- (Plethysmography: flow-volume respiratory traces)

HISTORY

A full history contributes to diagnosis and assessment of the full impact of BOAS on a dog's quality of life. Relevant questions include

- Exercise intolerance or reluctance
- Collapse or obvious cyanosis on exercise or excitement
- Eating disorders including choking on eating or difficulties eating
- Regurgitation
- Any sleeping disorders, including frequent 'choking' arousals when sleeping, sleeping with a raised head or in abnormal positions (sitting up).

These clinical signs are often more severe during the summer, and thus, questioning owners about signs at different times of the year is sensible.

CLINICAL EXAMINATION

Upper airway auscultation is an important part of the clinical examination, but the challenge is deciding whether all upper respiratory noise is clinically relevant. This is a difficult question to answer. While upper airway noise indicates some respiratory obstruction, many animals seem to tolerate some mild degree of respiratory noise while still living a seemingly acceptable quality of life, albeit with some snoring or stertorous breathing when exercising or stressed.

This inability to clearly define BOAS, along with clinical signs that are often described as 'normal' for the breed, makes determining the prevalence of BOAS within breeds very difficult. Although anatomical lesions are often used to define disease, not all dogs with stenotic nares necessarily show BOAS, as Shar Pei owners are often quick to point out.

CLINICAL SIGNS

- Noisy breathing
- Exercise and heat intolerance
- Hyperthermia
- Cyanosis and collapse
- Regurgitation and vomiting
- Flatulence.

During assessment for BOAS, the authors initially perform a (low stress) full clinical examination. We then observe respiratory effort and auscultate the upper airway directly, holding the stethoscope gently against the **SIDE** of the pharynx and larynx.

We are specifically listening for two noises:

- Stertor – a lower pitched vibratory sound (usually pharyngeal or nasopharynx)
- Stridor – a higher pitched – sawing sound often heard directly over the larynx.

However, interpretation can be complex; many affected French Bulldogs show nasal stertor; we have seen Pugs with nasal stridor and Bulldogs with both nasal and pharyngeal stertors. Some BOAS-affected dogs have no nasal noise because they have minimal nasal airflow, but the lack of nasal stertor masks their more compromised airways compared to dogs with obvious nasal airflow noise. Other signs of obstructed nasal airflow include flaring of the skin just behind the nasal planum, lack of nasal wing mobility and excessive panting.

The clinical picture is further obscured in some dogs, particularly Pugs, that appear to have reasonable respiratory function until stressed or exercised, whereupon marked dyspnoea becomes evident. Although this effect may be written off as 'excitement', we have yet to see a healthy Collie or Labrador Retriever show signs of respiratory obstruction after a short trot.

The authors have introduced an exercise stress test into our BOAS examination to identify dogs that sound reasonable when completely calm but have obvious marked airway noise when stressed or exercised. We use a trot test, which is designed to stress the upper airway and provides more effective grading of upper airway obstruction than a walk test based on results from whole body plethysmography in 44 dogs (93% agreement of disease severity) (Riggs et al. 2019). This is a slow trot (3–4 mph) for 3 minutes with the intention of keeping the dogs moving and increasing the oxygen demand and thus exertion of the upper airway. Some dogs, particularly young French Bulldogs or Pugs, will trot naturally at a faster pace.

A clinical examination that includes an exercise test (Riggs et al. 2019) is a more sensitive method of detecting stertor and stridor (86% versus 77%). Dogs with a flaccid larynx (e.g. Pugs) may only present with stridor after exercise when the intraluminal negative pressure increases.

In our research at Cambridge, we use a functional grading system for BOAS that includes the exercise test (Liu et al. 2015). BOAS is a disease of graduating severity, and airway noise is a good way to differentiate these grades. We grade airway noise as not present, mild (only audible with a stethoscope), moderate (heard without a stethoscope) and severe (easily audible without a stethoscope and constant with additional signs of dyspnoea). Respiratory effort and cyanosis are included in this scheme (appendix).

- Grade 0: No signs of airway obstruction
- Grade 1: Mild BOAS and are considered not to have clinically relevant disease
- Grade 2: Moderate disease
- Grade 3: Severely affected, usually requiring intervention.

While this grading system is not perfect because exercise tolerance can be affected by the ambient temperature and also the animal's fitness levels, it does expose those animals that are teetering on the edge of obstruction when calm but are firmly tipped into 'affected' once stressed. Exercise tests also reveal clinical signs in those dogs that only have respiratory noise when mouth breathing (often the long but not particularly thickened soft palates) after switching from nasal breathing in the consult room. We find about 40% of dogs that show few clinical signs in the consult room will be classified as affected after the exercise tolerance test.

This grading system was formally launched in February 2019 as 'The Kennel Club/ University of Cambridge Functional Grading Scheme' (RFG Scheme) with the objective of grading both breeding dogs and pet dogs to gain more information of disease prevalence, guide owners about whether treatment is required and avoid breeding from the severely affected dogs (The_Kennel_Club 2019). General practitioners are being trained as assessors, and the RFG Scheme can be used also to audit clinical cases in practice and assess response to treatment.

Other BOAS grading systems have been described, including assessing how far the dogs can walk and whether they can walk 1000 m in 6 minutes (Villedieu, Rutherford and Ter Haar 2019) or the time taken to walk 1000 m (Lilja-Maula et al. 2017).

In summary, BOAS diagnosis will be improved by inclusion of any exercise test into the examination because BOAS is a functional disease.

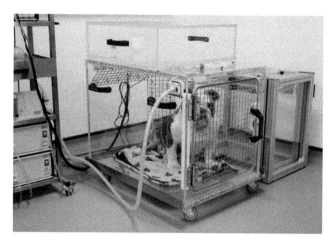

FIGURE 10.7   Whole-body barometric plethysmography (WBBP) chamber with a Bulldog being tested.

RESPIRATORY FUNCTION TESTING

In human respiratory medicine, in addition to clinical examination, respiratory function can be tested with whole body plethysmography in conjunction with forced spirometry. An objective measure of respiratory function in brachycephalic dogs would help to determine which dogs require treatment, response to such treatment and to guide breeding towards healthier individuals. Unfortunately, our clinical patients are not compliant with requests to breathe out into a tube with maximum force or breath-hold before inspiration. Instead, the technique that we use for the extreme brachycephalic dogs is whole-body barometric plethysmography (WBBP) where dogs are placed in a sealed chamber with biased airflow moving across it (Figure 10.7). Changes in the air pressure within the chamber that result from the humidification and expansion of air on breathing are detected by a pressure transducer (Liu et al. 2015).

Over the last 10 years, we have assessed over 2000 dogs with WBBP. Approximately 10%–20% of dogs do not provide usable test results due to persistent panting, barking or movement (Liu 2016). Exploration of three key markers of respiratory obstruction (expiratory time/inspiratory time, minute volume/body weight and peak expiratory airflow/peak inspiratory airflow) enables us to use quadratic component analysis (QDA) to characterise WBBP flow traces (Figure 10.8). This test for respiratory obstruction is 93% specific and 90% sensitive compared to clinical functional assessment in the Pug, French Bulldog and Bulldog (Liu et al. 2015, 2016; Liu 2016).

This WBBP test has been further developed to generate a BOAS index to describe relative BOAS severity: 0% is non-affected, and 100% are our most severely affected individuals (Liu et al. 2015, 2016; Liu 2016). The BOAS index now offers an objective measurement that evaluates risk factors for BOAS and the effectiveness of our surgical treatments, and takes account of the whole upper airway rather than just individual sections. Although WBBP is a useful and non-invasive method to assess respiratory function, it requires specialist equipment and operator training. The functional grading system is a relatively swift and straightforward technique of assessing airway function in general practice.

IMAGING

What imaging should we be using for BOAS diagnosis? This unfortunately largely depends, as often in veterinary medicine, on the owner's budget. Direct laryngoscopy is sufficient to evaluate the larynx and tonsils and gives useful information on soft palate length and thickness. A CT scan of the head will reveal turbinate morphology, nasopharyngeal dimensions, soft palate thickness and cricoid and tracheal size (Figure 10.9). Rhinoscopy is a complementary method of assessing

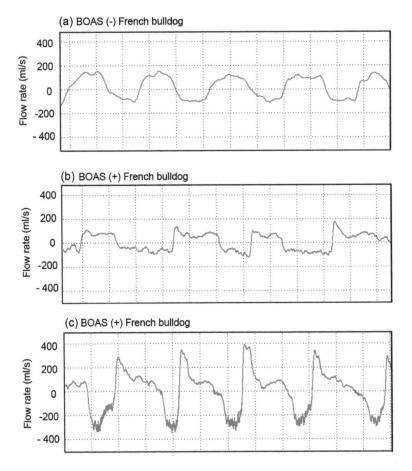

FIGURE 10.8 Examples of air flow traces obtained from the WBBP. (a) Air flow trace of a French Bulldog without BOAS; (b) air flow trace of a French Bulldog demonstrating a fixed-type upper airway obstruction; and (c) air flow trace of a French Bulldog demonstrating a dynamic-type upper airway obstruction. (Reprinted from Ladlow (2018) Figure 4 with permission.)

turbinate crowding, with a rigid scope used for the rostral nasal cavity combined with a flexible scope retroflexed behind the soft palate to assess caudal aberrant turbinates, trachea and bronchi and oesophageal lesions. CT of the thorax enables assessment for hypoplastic trachea, bronchial collapse, aspiration pneumonia and oesophageal diverticulum and may detect hiatal hernia (De Lorenzi, Bertoncello, and Drigo 2009). Thoracic cavity radiography allows the tracheal diameter to be gauged and exclusion of aspiration pneumonia if CT is not possible.

It is reasonable to treat without imaging if finances are very limited, although nasal or nasopharyngeal obstruction or aspiration pneumonia may have been missed in cases that do not respond successfully to conventional surgery.

Both mildly and severely affected dogs may have aberrant turbinates, stenotic nares and long soft palates, although the severely affected dogs are more likely to have thickened soft palates. This demonstrates the dangers from interpreting lesion sites on imaging independently to clinical signs. We also need to remember that static images may be affected by the presence of an endotracheal tube (Liu et al. 2018), phase of respiration, head position and degree of mouth opening, and thus need to be interpreted with caution.

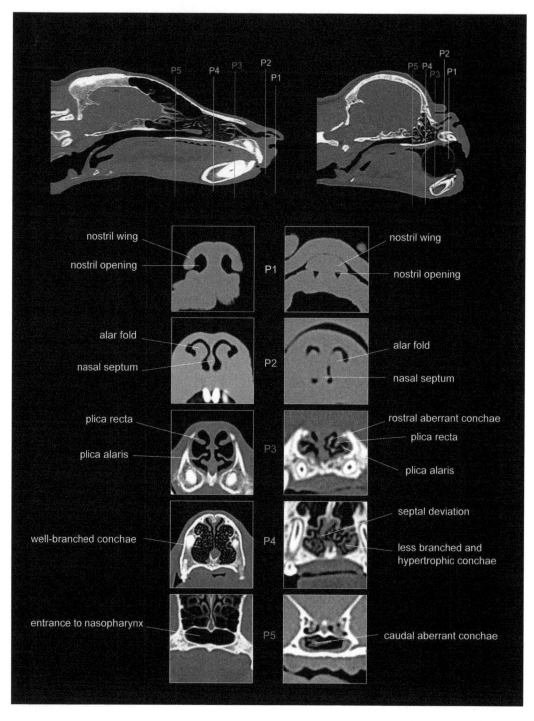

FIGURE 10.9   Computed tomographic images demonstrate the anatomical abnormalities of the nasal cavity in a BOAS-affected dog. A German shepherd dog is used as reference (left column); a Pug is used as a representative of BOAS-affected brachycephalic dogs. (Reprinted from Liu (2016) Figure 1-5 with permission.)

TREATMENT DECISION-MAKING

Many owners may be cautious or reluctant about commencing BOAS treatment in extreme brachycephalics, particularly as therapeutic surgery, even in experienced hands, carries the risk of complications including a low risk of death. Because many owners and primary practice veterinary surgeons do not recognise the early clinical signs as evidence of marked BOAS disease, this may explain why we see so many severely affected dogs as referrals that could have been treated with less risk earlier in the disease process. Many dogs are best treated when clinical signs first present at a younger age because BOAS is a progressive disease: earlier treatment can reduce development of secondary changes, particularly laryngeal collapse, and therefore offer a better long-term prognosis (Liu, Oechtering, et al. 2017).

Preventative surgery, however, is not yet proven as an effective option in the extreme brachycephalics. As yet, only the Shih-Tzu is reported to benefit from early nostril surgery (Huck, Stanley, and Hauptman 2008). For several reasons, the authors do not advise pre-emptive nostril surgery when performing other elective surgeries. There is no evidence as yet in the Bulldog, French Bulldog and Pug that early opening of the nostrils will prevent BOAS but such surgery can delay subsequent presentation for clinical BOAS because these owners may perceive that their pet has 'already had the surgery'. Lastly, we are trying to encourage breeding for more open nostrils. If many dogs undergo nasal surgery as puppies, it becomes much harder to monitor the changing incidence of stenotic nostrils over time.

Having thoroughly clinically assessed a brachycephalic dog, the decision on BOAS intervention in affected dogs should consider the following factors:

- Age
- Disease presentation
- Body condition/comorbidity
- Exercise requirement
- Temperament.

The RFG Scheme can also be used to guide treatment (The_Kennel_Club 2019; Liu et al. 2015):

- **Grade 0** dogs are not surgical candidates (even if their nostrils are narrowed) as we should only operate to treat clinical disease, not anatomical structures.
- **Grade 1** dogs rarely require surgery. Some mildly BOAS-affected Pugs that have epilepsy seem to show improved seizure control after airway surgery. We are not certain of the relationship between seizures and collapse, but it is possible transient hypoxia may trigger seizures. Mildly affected French Bulldogs that regurgitate regularly may show improvement of gastrointestinal signs after airway surgery.
- **Grade 2** dogs that are young and active generally benefit from surgery, particularly French Bulldogs with regurgitation. Older and more sedentary Grade 2 dogs may be benefit more from initial weight loss if overweight or may be monitored if comfortable in a non-stressful home environment.
- **Grade 3** dogs usually require surgery and marked weight loss for morbidly obese Pugs.

It is important that owners are counselled that surgery improves, rather than cures, most dogs and thus long-term BOAS monitoring is important. Reassessing dogs 6–8 weeks after surgery allows audit of surgical results and also provides a baseline for life-long monitoring.

It is also sensible to perform respiratory functional grading in clinically unaffected dogs because BOAS is a progressive disease. Introducing respiratory monitoring when the dogs are young is a gentle way to inform owners about a disease that may occur later and alert them about important clinical signs to monitor for.

All breeding animals should undergo respiratory assessment prior to breeding. Our current advice is that Grade 3 dogs should not be used for breeding for both their own welfare and the welfare of future progeny. Grade 2 dogs can still be used because their exclusion would reduce drastically the genetic pool but should be bred with Grade 1 or 0 dogs. Grade 2 dogs may be removed from the breeding programme in the longer term.

While this advice is not guaranteed to produce unaffected offspring because BOAS is a polygenic complex disease, it is a sensible way forward until genetic testing becomes available.

CONCLUSION

Early recognition of the clinical signs of BOAS by owners, breeders and veterinarians will allow appropriate treatment and decrease the number of diseased dogs that are bred, thus reducing the lifelong suffering that is experienced by too many of the extreme brachycephalic breeds.

At present, not enough is known about the breathing of many less popular brachycephalic breeds and more research is required to determine how many dogs in breeds other than the Bulldog, French Bulldog and Pugs are affected.

In breeds using health schemes to improve the breeding stock, close monitoring is needed to ensure that the desired objectives are being met and the breeds are indeed becoming healthier. The public also have a large part in determining the fate of these appealing breeds by only purchasing puppies from assured breeders who have invested in health testing of their dogs and not accepting snorting and snuffly dogs as 'normal for the breed'.

APPENDIX

The Kennel Club/University of Cambridge Respiratory Function Grading Scheme
     Respiratory Function Assessment Form

Name of dog:	Breed: male/female
Kennel Club registration number (if applicable):	Microchip number (if applicable):
Name of owner:	

Age:	Weight: _____kg Body condition score (BCS): _____/9	Colour:

Physical examination: * please note pre- and post- exercise tolerance test, if different.
Respiratory patterns * :   □ Normal   □ Irregular breathing   □ Inspiratory effort
Nostrils: □ Open   □ Mild stenosis   □ Moderate stenosis   □ Severe stenosis
Stertors (low pitch noise) * : □ Not audible   □ Mild   □ Moderate   □ Severe
Stridors (high pitch noise) * : □ Not audible   □ Mild   □ Moderate   □ Severe
Inspiratory effort * : □ Not present   □ Mild   □ Moderate   □ Severe
Dyspnoea/cyanosis/syncope * : □ No □ Yes _____
Heart/lung auscultation: □ Normal □ Abnormal _____

BOAS Functional Grading: □ Grade 0   □ Grade 1   □ Grade 2   □ Grade 3

Whole-body barometric plethysmography (WBBP) BOAS index: _____%

The above dog shows the physical characteristics and underwent the procedures as marked. The above report and its results are not a guarantee against any hereditary or acquired condition that may develop in the future.

Additional comments:
   □ Monitoring needed (under 2 years old / moderate respiratory signs / severe respiratory signs / obese)

Signature of Veterinary Surgeon   Date

_____   _____

BOAS FUNCTIONAL GRADING

Grade 0 – Free of respiratory signs; annual health check is suggested if the dog is under 2 years old.

Grade 1 – Mild respiratory signs of BOAS but does not affect exercise tolerance. Annual health check is suggested if the dog is under 2 years old.

Grade 2 – Moderate respiratory signs of BOAS. The dog has a clinically relevant disease and requires management, including weight loss and/or surgical intervention.

Grade 3 – Severe respiratory signs of BOAS. The dog should have a thorough veterinary examination with surgical intervention.

		Respiratory Noise[a]	**Inspiratory Effort**[b]	**Dyspnoea/Cyanosis/Syncope**[c]
Grade 0	Pre-ETT	Not audible	Not present	Not present
	Post-ETT	Not audible	Not present	Not present
Grade 1	Pre-ETT	Not audible to mild, or moderate intermittent nasal stertor when sniffing [d]	Not present	Not present
	Post-ETT	Mild or moderate intermittent nasal stertor when sniffing [d]	Not present to mild	Not present
Grade 2	Pre-ETT	Mild to moderate	Mild to moderate	Not present
	Post-ETT	Moderate to severe	Moderate to severe	Mild dyspnoea; cyanosis or syncope not present
Grade 3	Pre-ETT	Moderate to severe	Moderate to severe	Moderate to severe dyspnoea; may or may not present cyanosis. Inability to exercise.
	Post-ETT	Severe	Severe	Severe dyspnoea; may or may not present cyanosis or syncope.

The clinical grading was based on respiratory signs before (pre-ETT) and immediately after a 3-minute exercise tolerance test (post-ETT) with trotting speed of approximately 4–5 miles per hour performed by the study investigators or the dog owner. **Presentation of at least one sign in the highest grade determines the final grading result.**

[a] *Respiratory noise (stertor and/or stridor) was diagnosed by pharyngolaryngeal auscultation. Mild:* only audible under auscultation; moderate: intermittent audible noise that can be heard without stethoscope; severe: constant audible noise that can be heard without stethoscope.

[b] *An abnormal respiratory cycle characterised by evidence of increased effort to inhale the air in with the use of diaphragm and/or accessory muscles of respiration and/or nasal flaring with an increase in breathing rate. Mild:* regular breathing patterns with minimal use of diaphragm; moderate: evidence of use of diaphragm and accessary muscles of respiration; severe: marked movement of diaphragm and accessary muscles of respiration.

[c] *Dogs that have had episodes of syncope and /or cyanosis as documented by owner's report are classified into Grade III without ETT. Mild dyspnoea:* presents sign of discomfort; moderate dyspnoea: irregular breathing, signs of discomfort; severe dyspnoea: irregular breathing with signs of breathing discomfort and difficulty in breathing.

[d] Dogs with moderate intermittent nasal stertor when sniffing have similar BOAS index (objective respiratory function) to dogs only with mild respiratory noise; therefore, these dogs are considered Grade 1.

REFERENCES

Arai, K., M. Kobayashi, Y. Harada, Y. Hara, M. Michishita, K. Ohkusu-Tsukada, and K. Takahashi. 2016. "Histopathologic and immunohistochemical features of soft palate muscles and nerves in dogs with an elongated soft palate." *American Journal of Veterinary Research* 77 (1):77–83. doi:10.2460/ajvr.77.1.77.

Bannasch, D., A. Young, J. Myers, K. Truve, P. Dickinson, J. Gregg, R. Davis, et al. 2010. "Localization of canine brachycephaly using an across breed mapping approach." *PLoS One* 5 (3):e9632. doi:10.1371/journal.pone.0009632.

Bruchim, Y., E. Klement, J. Saragusty, E. Finkeilstein, P. Kass, and I. Aroch. 2006. "Heat stroke in dogs—A retrospective study of 54 cases (1999–2004) and analysis of risk factors for death." *Journal of Veterinary Internal Medicine* 20 (1):38–46. doi:10.1892/0891–6640(2006)20[38:HSIDAR]2.0.CO;2.

Cantatore, M., M. Gobbetti, S. Romussi, G. Brambilla, C. Giudice, V. Grieco, D. Stefanello, C. Giudice, and V. Grieco. 2012. "Medium term endoscopic assessment of the surgical outcome following laryngeal saccule resection in brachycephalic dogs." *Veterinary Record* 170 (20):518. doi:10.1136/vr.100289.

Clarke, D. L., D. E. Holt, and L. G. King. 2011. "Partial resolution of hypoplastic trachea in six english bulldog puppies with bronchopneumonia." *Journal of the American Animal Hospital Association* 47 (5):329–35. doi:10.5326/JAAHA-MS-5596.

Corgozinho, K. B., A. N. Pereira, S. C. dos Santos Cunha, C. B. Damico, A. M. R. Ferreira, and H. J. M. de Souza. 2012. "Recurrent pulmonary edema secondary to elongated soft palate in a cat." *Journal of Feline Medicine and Surgery* 14 (6):417–19. doi:10.1177/1098612X12442024.

Coyne, B. E., and R. B. Fingland. 1992. "Hypoplasia of the trachea in dogs—103 Cases (1974–1990)." *Journal of the American Veterinary Medicine Association* 201 (5):768–72. http://www.ncbi.nlm.nih.gov/pubmed/1399783.

Crosse, K. R., J. P. Bray, G. Orbell, and C. A. Preston. 2015. "Histological evaluation of the soft palate in dogs affected by brachycephalic obstructive airway syndrome." *New Zealand Veterinary Journal* 63 (6):319–25. doi:10.1080/00480169.2015.1061464.

Evans, H E. 1993. "The Skeleton." In *Miller's Anotomy of the Dog*, 3rd ed., 122–218. Philadelphia, PA: Saunders.

Farnworth, M. J., R. Chen, R. M. A. Packer, S. M. A. Caney, and D. A. Gunn-Moore. 2016. "Flat feline faces—Is brachycephaly associated with respiratory abnormalities in the domestic cat (Felis Catus)?" *PLoS One* 11 (8):e0161777. doi:10.1371/journal.pone.0161777.

Fasanella, F. J., J. M. Shivley, J. L. Wardlaw, and S. Givaruangsawat. 2010. "Brachycephalic airway obstructive syndrome in dogs—90 Cases (1991–2008)." *Journal of the American Veterinary Medical Association* 237 (9):1048–51. doi:10.2460/javma.237.9.1048.

Freiche, V., and C. Poncet. 2007. "Upper airway and gastro-intestinal syndrome in brachycephalic dogs." *Veterinary Focus* 17 (2):4–10.

Ginn, J. A., M. S. A. Kumar, B. C. McKiernan, and B. E. Powers. 2008. "Nasopharyngeal turbinates in brachycephalic dogs and cats." *Journal of the American Animal Hospital Association* 44 (5):243–49. http://www.jaaha.org/content/44/5/243.short.

Grand, J.-G. R., and S. Bureau. 2011. "Structural characteristics of the soft palate and meatus nasopharyngeus in brachycephalic and non-brachycephalic dogs analysed by CT." *Journal of Small Animal Practice* 52 (5):232–39. http://doi.wiley.com/10.1111/j.1748-5827.2011.01047.x.

Grosso, F. V., G. Ter Haar, and S. A. E. B. Boroffka. 2015. "Gender, weight, and age effects on prevalence of caudal aberrant nasal turbinates in clinically healthy english bulldogs—A computed tomographic study and classification." *Veterinary Radiology and Ultrasound* 56 (5):486–93. doi:10.1111/vru.12249.

Hall, E. J., A. J. Carter, and D. G. O'Neill. 2020. "Incidence and risk factors for heat-related illness (Heatstroke) in UK dogs under primary veterinary care in 2016." *Scientific Reports* 10:9128. doi:10.1038/s41598-020-66015-8.

Harvey, C. E. 1989. "Inherited and congenital airway conditions." *Journal of Small Animal Practice* 30 (3):184–87. http://doi.wiley.com/10.1111/j.1748-5827.1989.tb01531.x.

Harvey, C. E., and E. A. Fink. 1982. "Tracheal diameter—Analysis of radiographic measurements in brachycephalic and nonbrachycephalic dogs." *Journal of the American Animal Hospital Association* 18:570–76.

Heidenreich, D., G. Gradner, S. Kneissl, and G. Dupré. 2016. "Nasopharyngeal dimensions from computed tomography of pugs and french bulldogs with brachycephalic airway syndrome." *Veterinary Surgery* 45 (1):83–90. Wiley/Blackwell. doi:10.1111/vsu.12418.

Holmes, A. C., M. Tivers, K. Humm, and S. Adamantos. 2018. "Lung lobe torsion in adult and juvenile pugs." *Veterinary Record Case Reports* 6 (3):e000655. doi:10.1136/vetreccr-2018-000655.

Huck, J. L., B. J. Stanley, and J. G. Hauptman. 2008. "Technique and outcome of nares amputation (trader's technique) in immature shih tzus." *Journal of the American Animal Hospital Association* 44 (2):82–85.

Jones, B. A., B. J. Stanley, and N. C. Nelson. 2019. "The impact of tongue dimension on air volume in brachycephalic dogs." *Veterinary Surgery* 49 (3):512–20. doi:10.1111/vsu.13302.

Kaye, B. M., S. A. Boroffka, A. N. Haagsman, and G. Ter Haar. 2015. "Computed tomographic, radiographic, and endoscopic tracheal dimensions in english bulldogs with grade 1 clinical signs of brachycephalic airway syndrome." *Veterinary Radiology and Ultrasound* 56 (6):609–16. doi:10.1111/vru.12277.

Kim, Y. J., N. Lee, J. Yu, H. Lee, G. An, S. Bang, J. Chang, and D. Chang. 2019. "Three-dimensional volumetric magnetic resonance imaging (MRI) analysis of the soft palate and nasopharynx in brachycephalic and non-brachycephalic dog breeds." *Journal of Veterinary Medical Science* 81 (1):113–19. doi:10.1292/jvms.17-0711.

Komsta, R., Z. Osiński, P. D biak, P. Twardowski, and B. Lisiak. 2019. "Prevalence of Pectus Excavatum (PE), Pectus Carinatum (PC), Tracheal Hypoplasia, thoracic spine deformities and lateral heart displacement in thoracic radiographs of screw-tailed brachycephalic dogs." *PLoS One*. doi:10.1371/journal.pone.0223642.

Ladlow, J. F., N-C. Liu, and D. R. Sargan. 2012. "Cambridge BOAS Research Group." *University of Cambridge.* https://www.vet.cam.ac.uk/boas.

Ladlow, J.L 2018. "Brachycephalic obstructive airway syndrome - an update." *Companion.* March 4–13.

Lecoindre, P., and S. Richard. 2004. "Digestive disorders associated with the chronic obstructive respiratory syndrome of brachycephalic dogs—30 Cases (1999–2001)." *Revue de Médecine Vétérinaire* 155 (3):141–46.

Leonard, H. C. 1960. Collapse of the larynx and adjacent structures in the dog. *Journal of the American Veterinary Medical Association.* 137. http://www.ncbi.nlm.nih.gov/pubmed/14415784.

Lilja-Maula, L., A. K. Lappalainen, H. K. Hyytiäinen, E. Kuusela, M. Kaimio, K. Schildt, S. Mölsä, M. Morelius, and M. M. Rajamäki. 2017. "Comparison of submaximal exercise test results and severity of brachycephalic obstructive airway syndrome in english bulldogs." *Veterinary Journal* 219:22–26. doi:10.1016/j.tvjl.2016.11.019.

Liu, N.-C. 2016. "Novel diagnostic test for brachycephalic obstructive airway syndrome (BOAS) using whole-body barometric plethysmography (WBBP)." University of Cambridge.

Liu, N- C., V. J. Adams, L. Kalmar, J. F. Ladlow, and D. R. Sargan. 2016. "Whole-body barometric plethysmography characterizes upper airway obstruction in 3 brachycephalic breeds of dogs." *Journal of Veterinary Internal Medicine* 30 (3):853–65. doi:10.1111/jvim.13933.

Liu, N- C., G. U. Oechtering, V. J. Adams, L. Kalmar, D. R. Sargan, and J. F. Ladlow. 2017. "Outcomes and prognostic factors of surgical treatments for brachycephalic obstructive airway syndrome in 3 breeds." *Veterinary Surgery* 46 (2):271–80. doi:10.1111/vsu.12608.

Liu, N- C., D. R. Sargan, V. J. Adams, and J. F. Ladlow. 2015. "Characterisation of brachycephalic obstructive airway syndrome in french bulldogs using whole-body barometric plethysmography." *PLoS One* 10 (6):e0130741. doi:10.1371/journal.pone.0130741.

Liu, N- C., E. L. Troconis, L. Kalmar, D. J. Price, H. E. Wright, V. J. Adams, D. R. Sargan, and J. F. Ladlow. 2017. "Conformational risk factors of brachycephalic obstructive airway syndrome (BOAS) in pugs, french bulldogs, and bulldogs." *PLoS One* 12 (8):1–24. doi:10.1371/journal.pone.0181928.

Liu, N.-C., E. L. Troconis, M. McMillan, M.-A. Genain, L. Kalmar, D. J. Price, D. R. Sargan, and J. F. Ladlow. 2018. "Endotracheal tube placement during computed tomography of brachycephalic dogs alters upper airway dimensional measurements." *Veterinary Radiology & Ultrasound* 59 (3):289–304. Wiley/Blackwell. doi:10.1111/vru.12590.

Lorenzi, D. D., D. Bertoncello, and M. Drigo. 2009. "Bronchial abnormalities found in a consecutive series of 40 brachycephalic dogs." *Journal of the American Veterinary Medical Association.* doi:10.2460/javma.235.7.835.

Manens, J., M. Bolognin, F. Bernaerts, M. Diez, N. Kirschvink, and C. Clercx. 2012. "Effects of obesity on lung function and airway reactivity in healthy dogs." *Veterinary Journal* 193 (1):217–21. http://linkinghub.elsevier.com/retrieve/pii/S1090023311003984.

Marchant, T. W., E. J. Johnson, L. McTeir, C. I. Johnson, A. Gow, T. Liuti, D. Kuehn, et al. 2017. "Canine brachycephaly is associated with a retrotransposon-mediated missplicing of SMOC2." *Current Biology* 27 (11):1573–84. doi:10.1016/j.cub.2017.04.057.

O'Neill, D. G., D. B. Church, P. D. McGreevy, P. C. Thomson, and D. C. Brodbelt. 2013. "Longevity and mortality of owned dogs in England." *Veterinary Journal* 198 (3):638–43. doi:10.1016/j.tvjl.2013.09.020.

O'Neill, D. G., E. C. Darwent, D. B. Church, and D. C. Brodbelt. 2016. "Demography and health of pugs under primary veterinary care in England." *Canine Genetics and Epidemiology* 3:5. doi:10.1186/s40575-016-0035-z.

O'Neill, D. G., C. Jackson, J. H. Guy, D. B. Church, P. D. McGreevy, P. C. Thomson, and D. C. Brodbelt. 2015. "Epidemiological associations between brachycephaly and upper respiratory tract disorders in dogs attending veterinary practices in England." *Canine Genetics and Epidemiology* 2:10. doi:10.1186/s40575-015-0023-8.

O'Neill, D. G., A. M. Skipper, J. Kadhim, D. B. Church, D. C. Brodbelt, and R. M. A. Packer. 2019. "Disorders of bulldogs under primary veterinary care in the UK in 2013." *PLoS One* 14 (6):e0217928. doi:10.1371/journal.pone.0217928.

O'Neill, D. G., L. Baral, D. B. Church, D. C. Brodbelt, and R. M. A. Packer. 2018. "Demography and disorders of the french bulldog population under primary veterinary care in the UK in 2013." *Canine Genetics and Epidemiology*. doi:10.1186/s40575-018-0057-9.

Oechtering, T. H., G. U. Oechtering, and C. Noeller. 2007. "Computed tomographic imaging of the nose in brachycephalic dog breeds." *Tierärztliche Praxis Kleintiere* 35 (K):177–87. doi:10.1136/vr.i3215.

Packer, R. M. A., A. Hendricks, and C. C. Burn. 2012. "Do dog owners perceive the clinical signs related to conformational inherited disorders as 'Normal' for the breed? A potential constraint to improving canine welfare." *Animal Welfare* 21 (1):81–93. doi:10.7120/096272812X13345905673809.

Packer, R. M. A., A. Hendricks, M. S. Tivers, and C. C. Burn. 2015. "Impact of facial conformation on canine health—Brachycephalic obstructive airway syndrome." *PLoS One* 10 (10):e0137496. doi:10.1371/journal.pone.0137496.

Pichetto, M., S. Arrighi, P. Roccabianca, and S. Romussi. 2011. "The anatomy of the dog soft palate. II. Histological evaluation of the caudal soft palate in brachycephalic breeds with grade i brachycephalic airway obstructive syndrome." *Anatomical Record* 294 (7):1267–72. http://onlinelibrary.wiley.com/doi/10.1002/ar.21417/full.

Poncet, C. M., G. P. Dupre, V. G. Freiche, M. M. Estrada, Y. A. Poubanne, B. M. Bouvy, G. P. Dupré, et al. 2005. "Prevalence of gastrointestinal tract lesions in 73 brachycephalic dogs with upper respiratory syndrome." *Journal of Small Animal Practice* 46:273–79. doi:10.1111/j.1748–5827.2005.tb00320.x.

Riggs, J., N.-C. Liu, D. R. Sutton, D. R. Sargan, and J. F. Ladlow. 2019. "Validation of exercise testing and laryngeal auscultation for grading brachycephalic obstructive airway syndrome in pugs, french bulldogs, and english bulldogs by using whole-body barometric plethysmography." *Veterinary Surgery* 48 (4):488–96. doi:10.1111/vsu.13159.

Roedler, F. S., S. Pohl, and G. U. Oechtering. 2013. "How does severe brachycephaly affect dog's lives? results of a structured preoperative owner questionnaire." *Veterinary Journal* 198 (3):606–10. doi:10.1016/j.tvjl.2013.09.009.

Rubin, J. A., D. E. Holt, J. A. Reetz, and D. L. Clarke. 2015. "Signalment, clinical presentation, concurrent diseases, and diagnostic findings in 28 dogs with dynamic pharyngeal collapse (2008–2013)." *Journal of Veterinary Internal Medicine* 29 (3):815–21. doi:10.1111/jvim.12598.

Rutherford, L., L. Beever, M. Bruce, and G. Ter Haar. 2017. "Assessment of computed tomography derived cricoid cartilage and tracheal dimensions to evaluate degree of cricoid narrowing in brachycephalic dogs." *Veterinary Radiology and Ultrasound* 58 (6):634–46. doi:10.1111/vru.12526.

Schuenemann, R., and G. U. Oechtering. 2014. "Inside the brachycephalic nose—Intranasal mucosal contact points." *Journal of the American Animal Hospital Association* 50 (3):149–58. doi:10.5326/JAAHA-MS-5991.

Shah, F., T. Holmlund, E. L. Jäghagen, D. Berggren, K. Franklin, S. Forsgren, and P. Stål. 2018. "Axon and schwann cell degeneration in nerves of upper airway relates to pharyngeal dysfunction in snorers and patients with sleep apnea." *Chest* 154 (5):1091–98. doi:10.1016/j.chest.2018.06.017.

Stockard, C. R. 1941. The Genetic and Endocrinic Basis for Differences in Form and Behavior. The Wistar Institute of Anatomy and Biology.

The_Kennel_Club. 2019. "The Kennel Club and University of Cambridge Respiratory Function Grading Scheme." https://www.thekennelclub.org.uk/health/for-breeders/respiratory-function-grading-scheme/.

Villedieu, E., L. Rutherford, and G. Ter Haar. 2019. "Brachycephalic obstructive airway surgery outcome assessment using the 6-minute walk test—A pilot study." *Journal of Small Animal Practice* 60 (2):132–35. doi:10.1111/jsap.12942.

Walter, A., J. Seeger, and G. U. Oechtering. 2008. "Dolichocephalic versus brachycephalic conchae nasales - a microscopic anatomical analysis in dogs." *Hungarian Veterinary Journal* 130:123–24.

Wiestner, T. S., D. A. Koch, N. Nad, A. Balli, M. Roos, R. Weilenmann, E. Michel, and S. Arnold. 2007. "Evalualtion of the repeatability of rhinomanometry and its use in assessing transnasal resistance and pressure in dogs." *American Journal of Veterinary Research* 68 (2):178–84. doi:10.2460/ajvr.68.2.178.

11 Brachycephalic Obstructive Airway Syndrome – Surgical Management and Postoperative Management

Mickey Tivers and Elizabeth Leece
Paragon Veterinary Referrals

CONTENTS

Introduction 178
Medical Management of BOAS 178
   Gastrointestinal Abnormalities in Brachycephalic Dogs 178
Pre-surgical Considerations 179
   Anaesthetic Considerations 179
Surgery 179
   Surgical Positioning and Other Considerations 179
   Soft Palate 180
   Nostrils/Nares 182
   Tonsils 184
   Larynx/Laryngeal Collapse 184
      Everted Laryngeal Saccules/Grade I Laryngeal Collapse 184
      Advanced (Grades II and III) Laryngeal Collapse 185
   Aberrant Nasal Turbinates 186
Postoperative Care and Complications 187
   Complications 187
   Postoperative Care 188
      Recovery from Anaesthetic 188
      Postoperative Airway Swelling 188
      Postoperative Haemorrhage 189
      Oxygen Supplementation 190
      Avoiding Stress 190
      Regurgitation and Vomiting 190
      Aspiration Pneumonia 190
   Temporary Tracheostomy 191
Strategies for Surgery 192
Prognosis/Long-Term Outcome 193
   Prognostic Indicators 195
Conclusions 196
References 197

INTRODUCTION

Surgical management of brachycephalic obstructive airway syndrome (BOAS) is performed to reduce obstruction of the airway in order to improve the dog's function and quality of life. This reduction in airway obstruction should improve clinical signs and potentially prevent or ameliorate further deterioration in function in the future. Surgical correction, with shortening of the soft palate and widening of the nostrils, has been performed for over 70 years (Farquharson and Smith 1942, Trader 1949). More recently, our understanding of BOAS has improved enormously, particularly with advanced imaging providing more information on the less obvious components, such as intranasal obstruction (Oechtering 2010, Oechtering, Pohl, Schlueter, and Schuenemann 2016). Concurrently, due to the rise in popularity of these breeds, the veterinary profession has also been faced with a dramatic increase in the number of affected dogs, and subjectively an increased severity of disease, related to more severe conformational abnormalities (Oechtering 2010). Together, these factors have prompted the development of new techniques and the refinement of existing treatments in order to improve surgical effectiveness and hence outcomes. Most dogs treated for BOAS will have a procedure to reduce the size of their soft palate and widen their nostrils. Other procedures may be used based on surgeon preference and more complex procedures are typically reserved for dogs with very severe clinical signs or failure to respond to initial treatment. This chapter will explore up-to-date knowledge on surgical methods to treat BOAS in dogs.

MEDICAL MANAGEMENT OF BOAS

Owners should be counselled prior to surgery about appropriate husbandry and non-surgical management. This primarily involves weight loss as many dogs with BOAS are overweight, and this can exacerbate their condition (considered in greater depth in Chapter 18) (Liu et al. 2016, Packer et al. 2015). The author (MT) will emphasise to owners that one of the best things that they can do for their dog is to keep it at a lean weight long term. Dogs should be walked with a harness, and exercising during hot weather and stressful situations should be avoided.

Gastrointestinal Abnormalities in Brachycephalic Dogs

A significant proportion of dogs with BOAS have concurrent gastro-oesophageal issues. Regurgitation is the most common and obvious sign, but dogs can also suffer hypersalivation, gagging, retching and vomiting. This is thought to be secondary to BOAS due to the effect of negative airway pressure causing or exacerbating oesophagitis, gastro-oesophageal reflux and/or hiatal hernia. These problems seem to be particularly prevalent in French Bulldogs (Reeve et al. 2017), which may suggest that the breed has a tendency for pre-existing disease, which is exacerbated by BOAS. In one study of 73 dogs with BOAS, 54 (74.0%) had moderate to severe gastro-oesophageal signs including hypersalivation, regurgitation and vomiting (Poncet et al. 2005). On endoscopy, 71 of the dogs (97.3%) had evidence of gastro-oesophageal disease including oesophageal deviation, oesophagitis, gastro-oesophageal reflux, hiatal hernia and gastritis (Poncet et al. 2005). Another publication reported the results of barium swallow studies performed under fluoroscopic guidance in 30 brachycephalic dogs with signs of BOAS and concurrent chronic regurgitation (Reeve et al. 2017). A high rate of abnormalities was detected with delayed oesophageal transit in 27 dogs (90.0%), gastro-oesophageal reflux in 23 (76.7%) and hiatal hernia in 15 (50.0%). Surgical treatment of BOAS has been associated with a significant improvement in these gastro-oesophageal signs (Poncet et al. 2006). Dogs with significant gastro-oesophageal signs should be treated for this presurgery. For dogs with severe regurgitation, it can be beneficial to treat with omeprazole at a dose of 1 mg/kg twice daily per os for 1–2 weeks prior to surgery. Hopefully, this will reduce the risk of postoperative regurgitation, oesophagitis and aspiration pneumonia.

Some dogs are presented in respiratory distress. Emergency medical treatment consists of sedation, corticosteroids to reduce any airway swelling and oxygen treatment. Most dogs can be readily stabilised in this manner, and placement of a temporary tracheostomy is reserved for very severely affected animals. Very young brachycephalic puppies can be seen with acute respiratory distress, and this is often more likely to be due to aspiration pneumonia rather than primarily due to BOAS.

PRE-SURGICAL CONSIDERATIONS

BOAS surgery is typically an elective procedure, although rarely, dogs may be treated as emergencies. The surgeon should plan to assess the airway and perform corrective surgery under the same anaesthetic, to avoid the need to recover an animal with a compromised airway from anaesthesia unnecessarily. For this reason, any additional investigations, such as radiography, computed tomography (CT) or rhinoscopy, should be performed under the same anaesthetic (or alternatively at a later date).

Pre-operative stress should be avoided, and surgery is typically scheduled for early in the day so that the dog can be monitored post-surgery. Dogs should be placed in a large as cage or kennel as possible to reduce stress, and large 'walk-in' kennels are preferred.

Anaesthetic Considerations

For premedication, sedative drugs such as acepromazine or alpha-2 agonists in conjunction with an opioid may alleviate stress pre-surgery. However, these drugs can also cause relaxation of the smooth muscle of the nasopharynx and oropharynx, risking further airway obstruction and so low doses are recommended and dogs should be closely monitored once premedicated (Downing and Gibson 2018). Intravenous anaesthetic induction agents such as propofol and alfaxalone are recommended for dogs with BOAS for a smooth induction of anaesthesia, rapid endotracheal (ET) intubation and airway control. ET tube placement can be challenging due to the elongated soft palate, relatively oversized tongue and narrow 'gape' in some animals, limiting the view and access to the larynx. A laryngoscope and good light source are essential. A second laryngoscope or 'tongue depressor' is invaluable for lifting the soft palate to reveal the airway. In addition, a range of ET tubes should be available, as many brachycephalic dogs require a smaller tube than expected. Palpation of the diameter of the trachea prior to anaesthesia can help to estimate the size of tube likely to be required. Anaesthesia of brachycephalic dogs is considered in depth in Chapter 11.

SURGERY

Surgical Positioning and Other Considerations

For intra-oral surgery, the dog should be positioned in sternal recumbency, with the maxilla suspended to allow access (Figure 11.1). The mandible can also be secured to the operating table or a mouth gag placed to maintain an open mouth. The ET tube is typically fastened to the lower jaw with tie or tape. The use of magnifying loupes can be invaluable to facilitate intra-oral surgery. For nostril surgery, the maxilla is released, and the dog's head is repositioned on the table or on a sandbag with the mouth closed.

Many surgeons administer corticosteroids perioperatively to reduce postoperative swelling (Poncet et al. 2006, Riecks, Birchard, and Stephens 2007, White 2012, Davidson et al. 2001). Non-steroidal anti-inflammatory drugs (NSAIDs) can be given as an alternative, although the authors avoid NSAIDs in dogs with significant gastro-oesophageal signs.

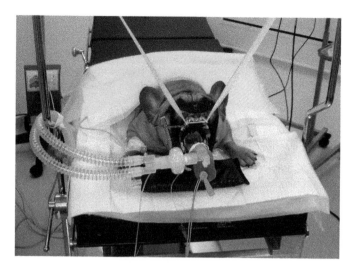

FIGURE 11.1   French Bulldog positioned for intra-oral surgery. Note the endotracheal tube is secured to the lower jaw, and the mouth is held open with a gag.

Due to the prevalence of gastro-oesophageal issues in brachycephalic dogs and the risk of post-operative regurgitation, it is sensible to consider perioperative treatment with omeprazole (Poncet et al. 2006). The authors will pre-emptively treat dogs with omeprazole at a dose of 1mg/kg twice daily, given prior to surgery and continued postoperatively. Hopefully, this will reduce the frequency and/or severity of any postoperative regurgitation and oesophagitis, although this is unproven as yet. For dogs with significant signs of regurgitation, omeprazole can be given for a few weeks pre-operatively (see above, the 'Medical Management of BOAS' section). Perioperative use of metoclopramide and cisapride have also been suggested (Poncet et al. 2006). Maropitant can also be given to reduce post-operative nausea that could contribute to complications,

Soft Palate

Virtually all dogs that are surgically treated for BOAS will have palate surgery, with 86.3%–100% reported in the literature (Torrez and Hunt 2006, Poncet et al. 2006, Riecks, Birchard, and Stephens 2007, Fasanella et al. 2010, Ree et al. 2016, Lindsay et al. 2020). Staphylectomy is shortening of the soft palate to prevent it from obstructing the rima glottidis. The traditional method for staphylectomy involves sharp excision of the caudal part of the soft palate, typically with scissors, and then apposition of the oral and nasal mucosa with synthetic, absorbable sutures (Riecks, Birchard, and Stephens 2007, Davidson et al. 2001, Poncet et al. 2006) (Figure 11.2). A wide variety of alternative methods of resection have been described including, but not limited to, the following devices: monopolar electrocautery, CO_2 laser, bipolar vessel sealing device, diode laser, harmonic scalpel and air plasma device (Riecks, Birchard, and Stephens 2007, Davidson et al. 2001, Brdecka et al. 2008, Dunie-Merigot, Bouvy, and Poncet 2010, Cook, Moses, and Mackie 2015, Michelsen 2011, Tamburro et al. 2019, Kirsch et al. 2019). Several studies have directly compared some of these techniques, and all have been shown to be associated with a good clinical outcome. The main advantages proposed for these devices are their speed and reduced haemorrhage, although there are specific health and safety requirements for the use of lasers. It has been shown that staphylectomy with a CO_2 laser is faster compared with the 'cut and sew' technique, but there is no apparent difference in outcome or complications (Worth et al. 2018, Davidson et al. 2001, Riecks, Birchard, and Stephens 2007). Ideally, soft palate surgery should be carried out with minimal tissue trauma and bleeding in order to limit postoperative swelling. Some surgeons prefer to avoid the use of 'energy'

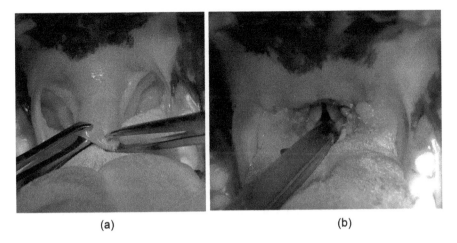

(a)   (b)

FIGURE 11.2   Intraoperative photographs of the oropharynx of a French Bulldog undergoing routine staphylectomy. (a) The caudal edge of the soft tissue has been grasped with Allis tissue forceps and pulled rostrally.
Note the long palate, obscuring the larynx and the enlarged tonsils. (b) The caudal portion of the soft palate
has been excised to the level of the rostral tonsil and sutured. Note that the larynx is now clearly visible. In this
dog, a tonsillectomy has also been performed.

devices as thermal energy may cause more tissue damage and consequent scarring and this is also
the author's (MT) preference (Ter Haar and Oechtering 2018). The ideal method of resection is yet
to be determined scientifically.

These techniques shorten the soft palate, but this does not specifically address the increased
thickness, which also obstructs the nasopharynx. A folded flap palatoplasty has been recommended
as an alternative technique, allowing both shortening and thinning of the soft palate and reducing nasopharyngeal obstruction (Findji and Dupre 2008, Haimel and Dupre 2015) (Figure 11.3).
Surgery involves incising a trapezoid shape in the mucosa of the ventral soft palate (Findji and
Dupre 2008). The mucosa and the underlying muscles are excised with monopolar electrocautery,
leaving the nasopharyngeal mucosa intact. The soft palate is then folded into the defect and sutured
using synthetic absorbable suture material. The rationale for this technique makes sound clinical
sense, as reducing the thickness of the soft palate should improve the airway further. Folded flap
palatoplasty is a more complicated technique than traditional 'cut and sew' staphylectomy, but it
has not been reported to have any increased risk of complications, although one study included a

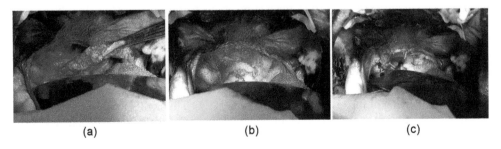

(a)   (b)   (c)

FIGURE 11.3   Intraoperative photographs of the oropharynx of a French Bulldog undergoing folded flap
palatoplasty (courtesy of Jane Ladlow). (a) The cut flap is grasped by forceps prior to complete excision. Note
that the muscle has been dissected out along with the overlying oropharyngeal mucosa. (b) The flap has been
excised leaving the defect in the soft palate prior to folding the caudal edge of the palate rostrally. (c) The flap
has been sutured. Note that the larynx is now clearly visible.

dog that had necrosis of their soft palate and another reported two dogs with dehiscence (Findji and Dupre 2008, Haimel and Dupre 2015, Stordalen et al. 2020). Although the technique is well described and associated with a good outcome, there are no studies directly comparing the technique with conventional staphylectomy.

Traditionally, recommendations have been to shorten the soft palate to the caudal border of the tonsils or just rostral to the tip of the epiglottis (Poncet et al. 2006, Riecks, Birchard, and Stephens 2007, Davidson et al. 2001). There was concern that making the palate shorter than this could result in nasal reflux of food and water leading to rhinitis (Brdecka et al. 2008, Findji and Dupre 2008, Sarran et al. 2018, Bright and Wheaton 1983). However, there has been a move towards more aggressive shortening of the soft palate, and it has been recommended to shorten the palate to the level of the mid- to rostral tonsil (Brdecka et al. 2008, Dunie-Merigot, Bouvy, and Poncet 2010, Findji and Dupre 2008). This seems to be well tolerated and importantly studies have not found complications of nasopharyngeal reflux and rhinitis with more rostral resections (Brdecka et al. 2008, Dunie-Merigot, Bouvy, and Poncet 2010, Haimel and Dupre 2015, Findji and Dupre 2008, Sarran et al. 2018, Lindsay et al. 2020). One study measured the area of greatest nasopharyngeal obstruction caused by the soft palate on pre-operative CT scans in French Bulldogs with BOAS (Sarran et al. 2018). The greatest obstruction was typically located relatively rostrally on the soft palate, with a mean distance of 9.4 mm from the hamular processes. Staphylectomy was performed at the level of maximal occlusion based on the CT findings, which was more rostral than more standard positioning. No dogs experienced complications relating to nasopharyngeal reflux or rhinitis (Sarran et al. 2018).

Intuitively removing as much soft palate as feasible should result in the greatest improvement in the airway. There is also a risk that insufficient resection of the soft palate could result in persistent clinical signs, requiring revision surgery (Sarran et al. 2018). However, the ideal amount of soft palate that should be removed has not been determined and may be different for each dog. The author routinely shortens the soft palate to the level of the rostral tonsils.

Nostrils/Nares

A variety of different surgical techniques are described to enlarge stenotic nares. Although this is a relatively minor procedure, it results in a good clinical improvement and is very well tolerated. Studies report that 25%–100% of dogs treated for BOAS have their nostrils widened (Torrez and Hunt 2006, Poncet et al. 2006, Riecks, Birchard, and Stephens 2007, Fasanella et al. 2010, Dunie-Merigot, Bouvy, and Poncet 2010, Findji and Dupre 2008, Haimel and Dupre 2015, Ree et al. 2016, Lindsay et al. 2020, Senn et al. 2011, Brdecka et al. 2008).

The original procedure, known as the 'Trader technique', involves simple excision of the ventral portion of the wing of the nostril (ala nasi) with a scalpel blade (Trader 1949). Haemorrhage is controlled by direct pressure and topical application of xylometazoline or adrenaline. The resulting wound is left to heal by second intention. The procedure is very effective at enlarging the nostril but fell out of favour due to concerns over the cosmetic appearance and preference for other techniques. Nevertheless, a study showed very good functional and cosmetic results in 13 Shih Tzu puppies with stenotic nares (Huck, Stanley, and Hauptman 2008).

Alarplasty (or rhinoplasty) techniques involve resection of a section of tissue from the wing of the nostril and suturing of the resulting defect (Figure 11.4). This will widen the nostril by reducing the thickness of the ala nasi and by lifting the axial border laterally. These techniques are commonly performed as they allow direct healing, while providing adequate enlargement of the nares and a good cosmetic appearance. A wide variety of techniques have been described, with the most common involving removal of a vertical, lateral or horizontal wedge of tissue from the dorsolateral cartilage (Dupre and Heidenreich 2016, Schmiedt 2018, Riecks, Birchard, and Stephens 2007). The incision is made through the epithelium and mucosa and should be deep into the underlying

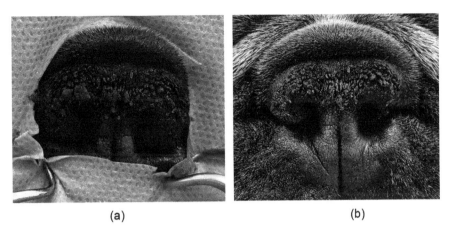

(a) (b)

FIGURE 11.4 Intraoperative photographs of the nasal planum of a French Bulldog undergoing wedge resection alarplasty. (a) The nares are very stenotic. (b) A vertical wedge of tissue has been excised from the wing of the nostril, and the defect has been sutured. Note the increased aperture of the nares.

cartilage to ensure that adequate tissue is removed. The author (MT) prefers a scalpel blade to a laser or electrocautery as the latter techniques have been associated with increased damage to the surrounding tissues, prolonged wound healing and scarring (Schmiedt 2018, Ter Haar and Oechtering 2018). Alternative techniques described include the use of a skin biopsy punch to remove a 'plug' of tissue from the centre of the alar fold (Trostel and Frankel 2010) and removal of a section of the dorsal nasal planum (dorsal offset rhinoplasty) (Dickerson et al. 2020). For all alarplasty techniques, the resulting defects are closed with synthetic absorbable suture material, and this controls haemorrhage. Absorbable suture material is used to avoid the need to remove the sutures, which will require sedation in most dogs. Alapexy is another technique that has been recommended for dogs with flaccid cartilage or when other techniques have failed (Ellison 2004). It involves fixing the wing of the nostril to the adjacent skin via the suturing of corresponding incisions, thus abducting it laterally.

These techniques address the more superficial stenosis but do not address any narrowing deeper within the nasal vestibulum. Ala vestibuloplasty has been recommended to correct this narrowing and involves removing a large portion of the alar fold (plica alaris) in order to open up the vestibule (Oechtering 2010, Dupre and Heidenreich 2016, Oechtering, Pohl, Schlueter, and Schuenemann 2016, Pohl, Roedler, and Oechtering 2016, Sarran et al. 2018). Trader's alarplasty has been combined with vestibuloplasty into a modified rhinoplasty technique, in order to maximise enlargement of the nares and address both the external and the internal narrowing (Liu et al. 2017) (Figure 11.5). This technique does not seem to be associated with increased complications, although healing takes 2–3 weeks. A modified horizontal wedge resection and a partial vestibulectomy has also been described (Ter Haar and Oechtering 2018).

There is no evidence to support the use of any specific technique for widening the nostrils in dogs with BOAS. We can presume that increasing the size of the nostrils as much as possible should improve air flow in a corresponding manner and that cosmetic appearance should not be a consideration. The author (MT) has had owners expressing concern about their dog's postoperative appearance before surgery. However, this has not been an issue following surgery. More positively, owners have commented on an apparent improvement in their dog's olfactory abilities. There is limited information on how wide nostrils should be but, in a similar way to soft palate procedures, there is a trend to removal of increasing amounts of tissue in order to maximise the clinical result.

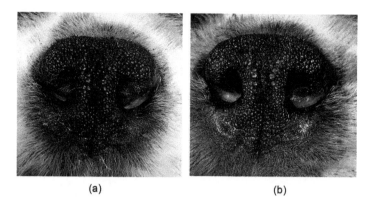

(a)   (b)

FIGURE 11.5   Intraoperative photographs of the nasal planum of a Cavalier King Charles Spaniel undergoing ala vestibuloplasty combined with a Trader's alarplasty. (a) Appearance of the nostrils prior to surgery. Note, this dog had previously had a wedge alarplasty, but there is still obstruction of the nasal vestibulum by the alar fold. (b) The completed surgery. Note that this has addressed the narrowing within the nasal vestibulum.

Tonsils

In some dogs with BOAS, the tonsils are everted from their crypts and appear hyperplastic and mobile. There is limited information on the benefit of removal of tonsils in affected dogs, with many studies not performing tonsillectomy or only very rarely (Poncet et al. 2006, Riecks, Birchard, and Stephens 2007, Torrez and Hunt 2006, Tarricone et al. 2019, Worth et al. 2018). Other studies have performed tonsillectomy routinely, where needed, in 42.7%–63.2% of dogs and even in 100% (Pohl, Roedler, and Oechtering 2016, Fasanella et al. 2010, Fenner, Quinn, and Demetriou 2020, Lindsay et al. 2020).

The author manipulates the tonsils in all dogs to see if they are likely to be drawn into the airway during inspiration and will perform tonsillectomy if this is the case. Tonsillectomy can be performed by sharp excision with electrocautery for haemostasis and suturing of the tonsillar crypt (Fenner, Quinn, and Demetriou 2020, Lindsay et al. 2020, Anderson 2018). A bipolar vessel sealing device or a harmonic scalpel can also be used (Cook, Moses, and Mackie 2015, Fenner, Quinn, and Demetriou 2020, Liu et al. 2017, Anderson 2018). Tonsillotomy is a sub-total or partial tonsillectomy, whereby only the protruding portion of the tonsil is excised. This has also been reported for the management of enlarged tonsils associated with BOAS although specific data on how the technique differs from tonsillectomy is lacking (Oechtering, Pohl, Schlueter, and Schuenemann 2016, Pohl, Roedler, and Oechtering 2016, Liu et al. 2017).

Larynx/Laryngeal Collapse

Laryngeal collapse is a secondary problem and, as described in Chapter 9, it can be graded out of three in terms of severity (Leonard 1960, Monnet and Tobias 2018).

Everted Laryngeal Saccules/Grade I Laryngeal Collapse

Eversion of the laryngeal saccules is typically referred to as mild or Grade I/III laryngeal collapse and is not uncommon in dogs with BOAS, causing some degree of obstruction of the airway (Leonard 1960, Monnet and Tobias 2018). It can be treated via surgical resection of the everted saccules (known as sacculectomy or ventriculectomy). The saccules are readily removed by grasping them with forceps and sharp excision with Metzenbaum Scissors, or similar (Riecks, Birchard, and Stephens 2007, Pink et al. 2006, Sarran et al. 2018, Lindsay et al. 2020). It has been recommended to remove the saccules with magnification in order to facilitate the procedure, and this has been described endoscopically with a diode laser (Oechtering, Pohl, Schlueter, and Schuenemann 2016, Pohl, Roedler, and Oechtering 2016). Sacculectomy can be facilitated by temporary removal of the ET tube, which otherwise can get in the way (Riecks, Birchard, and Stephens 2007, Sarran et al. 2018, Lindsay et al. 2020, Hughes et al. 2018). This may be more important in smaller breeds such as Pugs, as the saccules are smaller, and the access is more restricted.

There is some controversy over sacculectomy in dogs with BOAS, and the specific benefit of the technique is unclear. Several studies have reported the routine use of sacculectomy as part of BOAS surgery, with good results (Fasanella et al. 2010, Riecks, Birchard, and Stephens 2007, De Lorenzi, Bertoncello, and Drigo 2009, Torrez and Hunt 2006, Ree et al. 2016, Fenner, Quinn, and Demetriou 2020, Lindsay et al. 2020). The author (MT) believes that everted saccules contribute to airway obstruction and routinely performs sacculectomy when the saccules are causing obstruction of the ventral rima glottidis.

However, the role of sacculectomy has been questioned as it has been associated with an increased rate of complications (Hughes et al. 2018). It has also been suggested that everted saccules can resolve following correction of the other components of BOAS (e.g. alarplasty and staphylectomy) (Riecks, Birchard, and Stephens 2007, Poncet et al. 2006). However, in a prospective study, ten dogs were treated with unilateral sacculectomy, in addition to alarplasty and staphylectomy, and the remaining everted saccules did not resolve suggesting that spontaneous resolution is unlikely (Cantatore et al. 2012). However, concern over complications and uncertainty over efficacy means that many surgeons do not perform sacculectomy and several studies have not routinely removed everted saccules (Poncet et al. 2006, Dunie-Merigot, Bouvy, and Poncet 2010, Haimel and Dupre 2015).

Advanced (Grades II and III) Laryngeal Collapse

Historically, it was generally believed that dogs with more advanced collapse will be more severely affected, with a greater risk of complications and a poorer prognosis (White 2012, Pink et al. 2006, Harvey 1982). Indeed, dogs with severe clinical signs relating to laryngeal collapse can be very challenging to manage. However, laryngeal collapse is very complex, with a variety of different laryngeal abnormalities and variations between different breeds, including size and rigidity (Ter Haar and Oechtering 2018, Rutherford et al. 2017, Oechtering 2010, Caccamo et al. 2014, Dupre and Heidenreich 2016). Importantly, the grading scheme is very subjective and probably is not good at differentiating the clinical severity of an individual's condition when the larynx is taken in isolation. In addition, there is some discrepancy with how laryngeal collapse is graded, which complicates direct comparison between studies. The authors have treated dogs with apparently severe laryngeal collapse which are coping relatively well. It is important to assess the individual's anatomical abnormalities in concert with their clinical signs and treat them accordingly.

For most dogs, treatment consists of correcting the primary abnormalities via staphylectomy and alarplasty, along with husbandry changes such as weight loss (Poncet et al. 2006, Monnet and Tobias 2018, White 2012, Torrez and Hunt 2006, Haimel and Dupre 2015). By correcting the primary problems, this should improve the airway and therefore prevent or delay further progression of laryngeal collapse (and other secondary changes). Several studies have reported a good outcome for dogs with laryngeal collapse treated in this way, and the rate of 'life-threatening' collapse may be quite low (White 2012, Torrez and Hunt 2006, Haimel and Dupre 2015).

Those animals that are severely affected or fail to improve following initial surgery may benefit from more specific treatment of their laryngeal collapse. Surgical management strategies for dogs with severe laryngeal collapse are limited but include partial laryngectomy, cuneiformectomy, arytenoid lateralisation and permanent tracheostomy.

Partial laryngectomy has been used for management of laryngeal collapse, to enlarge the rima glottidis. However, it is associated with a very high rate of complications, aspiration pneumonia and a 50% mortality postoperatively (Harvey 1983). It is therefore not recommended.

Cuneiformectomy, the removal of one or both of the cuneiform process of the arytenoid, has been suggested as a treatment for laryngeal collapse (Pohl, Roedler, and Oechtering 2016, Liu et al. 2017). The procedure widens the rima glottidis and prevents dynamic obstruction. This technique requires further assessment, but removal of the obstructing cuneiform process is a logical treatment and has not been associated with major complications (Liu et al. 2017).

Arytenoid lateralisation (tie back) has also been reported for the management of severe laryngeal collapse (White 2012, Fasanella et al. 2010, Ree et al. 2016). One study reported arytenoid lateralisation in 12 dogs with Grade II or III laryngeal collapse that had continued, severe upper respiratory tract obstruction following routine surgery (White 2012). Ten dogs (83.3%) responded well to the treatment, with confirmed enlargement of the rima glottidis postsurgery and improved respiratory function. However, two dogs were euthanased postoperatively. The owners of the ten surviving dogs reported a marked improvement at follow-up (median 3.5 years), although all dogs had mild ongoing signs. The success of surgery depends on the rigidity of the laryngeal cartilages so it may not be effective in animals with chondromalacia or loss of integrity of the larynx (White 2012, Dupre and Heidenreich 2016). Arytenoid lateralisation in brachycephalic dogs is technically challenging and should ideally be performed by experienced surgeons. It should be reserved for dogs with severe laryngeal collapse that have continued, severe signs of upper respiratory tract obstruction following routine surgery.

Permanent tracheostomy is an alternative for the management of severe laryngeal collapse. However, the authors consider that this should be considered a salvage procedure, and it is reserved for dogs that fail standard surgery and tie-back or for those that are tracheostomy dependent (Gobbetti et al. 2018). The procedure is technically challenging, and for all breeds, it is associated with significant morbidity with 50%–61% dogs suffering major complications (Occhipinti and Hauptman 2014, Grimes et al. 2019). Dogs are at risk of acute airway obstruction and death, due to the stoma being blocked by mucous or adjacent skin folds. There is also a risk of stenosis of the stoma, and this can be complicated by concurrent weakening and collapse of the tracheal cartilage. Brachycephalic dogs may be at increased risk of complications due to their conformation, with thick, wide necks, excessive skin folds and narrow or hypoplastic tracheas. One study showed a significantly greater requirement for revision surgery in brachycephalic dogs, 53% compared to 26% (Grimes et al. 2019). In another study, just including dogs with BOAS and severe laryngeal collapse, 12/15 dogs (80%) suffered a major complication (Gobbetti et al. 2018). Complications included severe dyspnoea secondary to mucous obstruction (46.7%), stenosis of the stoma (13.3%) and obstruction due to skin folds (20.0%). Eight dogs (53%) died or were put to sleep, and four dogs (27%) required revision of their stoma. Nevertheless, five dogs had a good quality of life, with long-term survival exceeding 5 years. Dogs require intensive postoperative monitoring and nursing. Importantly, there is also a requirement for ongoing care by the owner, with cleaning of the stoma and monitoring for complications. For some owners, this is considered demanding, and this may influence their decision to treat (Gobbetti et al. 2018).

Aberrant Nasal Turbinates

One of the drawbacks of traditional surgical management of BOAS is that it does not address the intranasal disease, despite the fact that significant obstruction may be present. Affected dogs have abnormal growth of their turbinates and crowding of the turbinates within the nasal cavity, causing obstruction (Oechtering, Pohl, Schlueter, Lippert, et al. 2016, Ginn et al. 2008, Schuenemann and Oechtering 2014b). Recent studies have described a novel procedure to treat aberrant nasal turbinates: laser-assisted turbinectomy (LATE) (Oechtering, Pohl, Schlueter, and Schuenemann 2016, Schuenemann, Pohl, and Oechtering 2017, Liu et al. 2019). This uses rhinoscopy to allow the removal of the aberrant turbinates with a diode laser (Figure 11.6). LATE is a very specialised procedure requiring considerable expertise and equipment (Oechtering, Pohl, Schlueter, Lippert, et al. 2016, Oechtering, Pohl, Schlueter, and Schuenemann 2016, Liu et al. 2019, Schuenemann, Pohl, and

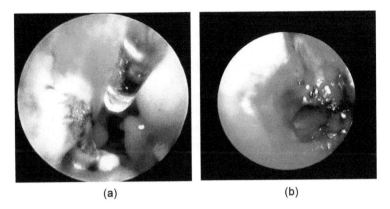

(a) (b)

FIGURE 11.6 Rhinoscopic images of the nasal cavity of a French bulldog undergoing LATE b (courtesy of Jane Ladlow). (a) A diode laser is being used to remove the aberrant ventral nasal choncha from the nasal cavity. (b) View of the nasopharynx following removal of the ventral nasal choncha.

Oechtering 2017). However, the technique is very effective in improving nasal airflow and is well tolerated, with a low rate of complications in experienced hands (Oechtering, Pohl, Schlueter, and Schuenemann 2016, Liu et al. 2019, Schuenemann, Pohl, and Oechtering 2017).

One study described 254 dogs treated with traditional multi-level BOAS surgery and LATE as part of the same procedure and reported mortality rate of 2.4%, with no dogs requiring a tracheostomy (Oechtering, Pohl, Schlueter, and Schuenemann 2016). Another study, from the same group, reported the owner assessed outcome of multilevel airway surgery combined with LATE in 62 dogs (Pohl, Roedler, and Oechtering 2016). All owners reported improvement in all categories (breathing, exercise and heat tolerance, eating, sleeping, general welfare and ability to smell), with a marked improvement in breathing in 72% and a marked improvement in general welfare in 73%. Importantly, the frequency of life-threatening events were decreased in 90% of dogs. While these studies performed LATE in all dogs with BOAS, an alternative is to perform LATE in dogs that do not respond well to conventional surgery (Liu et al. 2019). One study reported 57 dogs with BOAS treated with conventional surgery (Liu et al. 2019). Based on objective assessment of airway obstruction with whole-body barometric plethysmography (WBBP) and clinical assessment with BOAS functional grading, 50.9% of these dogs were considered candidates for LATE. Short-term complications included transient reverse sneezing in 60% of dogs for 4–7 days after LATE and intermittent nasal noise in 40%.

Interestingly, it has been shown that nasal conchae can regrow following LATE, with regrowth in 96% of French Bulldogs and 65% of Pugs (Schuenemann and Oechtering 2014a). This is important as conchae fulfil an important function in thermoregulation and some regrowth is considered beneficial (Schuenemann and Oechtering 2014a). However, despite this regrowth there remain significantly fewer mucosal contact points, and therefore, it is generally not associated with re-obstruction of air flow. Hence, revision surgery is only indicated in animals with recurrent clinical signs related to obstruction. In a study, 15.8% of dogs required revision surgery with LATE to remove the regrown conchae 6 months after the first procedure (Oechtering, Pohl, Schlueter, and Schuenemann 2016).

POSTOPERATIVE CARE AND COMPLICATIONS

COMPLICATIONS

Although BOAS surgery aims to improve the airway, in the short-term postoperative swelling and inflammation can cause further obstruction. This can lead to worsening of respiratory signs and dyspnoea, which can be further exacerbated by stress and aspiration pneumonia secondary to regurgitation or vomiting (Haimel and Dupre 2015, Torrez and Hunt 2006, Poncet et al. 2006, Fenner, Quinn, and Demetriou 2020, Hughes et al. 2018, Ree et al. 2016, Senn et al. 2011, Tarricone et al. 2019, Lindsay et al. 2020). Other reported complications include coughing, nasal discharge, haemorrhage and dehiscence of alarplasty or folded flap palatoplasty (Poncet et al. 2006, Fasanella et al. 2010, Haimel and Dupre 2015, Tamburro et al. 2019, Torrez and Hunt 2006). Post-surgical complication rates of between 6.5% and 41.7% have been reported (Poncet et al. 2006, Tarricone et al. 2019, Hughes et al. 2018, Lindsay et al. 2020, Ree et al. 2016, Riecks, Birchard, and Stephens 2007, Fasanella et al. 2010, Senn et al. 2011, Kirsch et al. 2019). There is considerable variation in how complications are reported with some studies including all complications and others only including major ones or specific types.

Mortality is low, although dogs can die as a result of upper respiratory tract obstruction due to postoperative swelling and/or aspiration pneumonia. Dogs can also have a vasovagal reaction following regurgitation or vomiting that can lead to respiratory arrest. Overall mortality for surgical correction of BOAS varies with many studies reporting a rate of 0%–3.7% (Tarricone et al. 2019, Torrez and Hunt 2006, Poncet et al. 2006, Oechtering, Pohl, Schlueter, and Schuenemann 2016, Worth et al. 2018, Fenner, Quinn, and Demetriou 2020, Findji and Dupre 2008, Riecks, Birchard, and Stephens 2007, Liu et al. 2019, Hughes et al. 2018, Lindsay et al. 2020). However, higher rates of 5.6%–10.0% have also been reported (Senn et al. 2011, Haimel and Dupre 2015, Ree et al. 2016, Kirsch et al. 2019).

POSTOPERATIVE CARE

Postoperative care is extremely important following airway surgery for BOAS, and many dogs will require strict observation and intensive nursing. Dogs should be monitored closely for complications, in particular dyspnoea related to airway swelling, to allow early intervention and treatment. Analgesia is important after BOAS surgery to minimise postoperative stress and worsening dynamic airway collapse. Postoperative analgesia should be provided based on each individual patient instead of prescriptive opioid administration, which may result in inappropriate panting and increased regurgitation (Bini et al. 2018). The authors avoid routine use of NSAIDs in dogs with significant gastro-oesophageal signs. Active warming may be required in hypothermic dogs, but this should be discontinued if they start to pant as overheating may be detrimental.

Recovery from Anaesthetic

Due to their compromised airway dogs are most at risk during recovery from anaesthetic.

Extubation should not occur while the dog is still sedated to reduce the risk of airway obstruction and is postponed for as long as possible and until they can lift their heads, swallow and protect their own airway. Dogs should be positioned in sternal recumbency with the head elevated to reduce risk of regurgitation and aspiration of fluid. They should be closely monitored following extubation with regular assessment of their respiratory rate and effort. Any signs of respiratory compromise should be addressed immediately (see below). An intravenous catheter should be maintained postoperatively in case there is a need for sedation or anaesthesia. In addition, equipment for suction and re-intubation, including induction agent, should be readily available. If an animal does experience respiratory obstruction following extubation, then they should be anaesthetised and re-intubated if necessary (see below).

Postoperative Airway Swelling

Postoperative swelling can cause further obstruction and lead to dyspnoea. Postoperative dyspnoea is reported as a complication in 1.6%–23.4% of dogs (Haimel and Dupre 2015, Fenner, Quinn, and Demetriou 2020, Tarricone et al. 2019, Torrez and Hunt 2006, Riecks, Birchard, and Stephens 2007, Hughes et al. 2018, Senn et al. 2011, Lindsay et al. 2020). A study of the use of LATE in 20 dogs

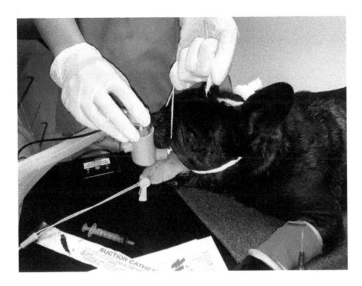

FIGURE 11.7   Nebulisation with adrenaline in a French Bulldog with marked airway swelling and dyspnoea following surgery for BOAS.

reported postoperative dyspnoea in 6 dogs (30%) (Liu et al. 2019). However, this study included dogs that had significant ongoing airway obstruction following conventional surgery and therefore an increase rate of complications may have been expected.

Dogs should be monitored closely and prompt action taken when airway swelling is suspected. Corticosteroids are recommended (e.g. dexamethasone at a dose of 0.1–0.2 mg/kg intravenously) to try to reduce swelling (Torrez and Hunt 2006, Lindsay et al. 2020). The authors also recommend nebulised adrenaline, which can be very effective (Ellis and Leece 2017) (Figure 11.7). Dogs should also be kept sedated to avoid stress than can exacerbate the swelling (see below). Supplemental oxygen should be provided (see below) (Lindsay et al. 2020, Tarricone et al. 2019). It is important to address any dyspnoea and swelling in a timely fashion to try to prevent further deterioration.

Dogs that do not respond to treatment may need to be re-anaesthetised and re-intubated (Lindsay et al. 2020). Surgeons should be proactive and consider re-intubation promptly, when dogs have postoperative dyspnoea. A period of intubation under anaesthesia will typically allow time for other treatments to work, for the swelling to improve and break the cycle of stress and dyspnoea. Dogs should then be slowly recovered from the anaesthetic and kept sedated and calm (see above and below). In the authors' experience, this process is successful in most animals with postoperative swelling. In a study, 8.9% of dogs (22/248) required repeat anaesthesia and re-intubation due to postoperative dyspnoea (Lindsay et al. 2020). Dogs were anaesthetised for a mean of 19 hours (range 1–72 hours). Half (50%; 11/22) of these dogs were successfully managed with re-intubation alone. The overall survival for dogs with significant dyspnoea was 84.4% in this study. Dogs that have severe swelling and that cannot be successfully extubated or require repeat episodes of intubation require a temporary tracheostomy (see below) (Lindsay et al. 2020, Tarricone et al. 2019).

Postoperative Haemorrhage

Postoperative haemorrhage is uncommon in the authors' experience but can occur following staphylectomy or alarplasty. Haemorrhage from the soft palate can cause upper respiratory tract obstruction and contribute to postoperative dyspnoea. Dogs with significant haemorrhage should be anaesthetised, and the surgical site is revised as needed to stop the bleeding. Haemorrhage from the nares may be more common with Trader's technique and also following LATE, and this can be controlled by vasoconstriction with topical xylometazoline or adrenaline.

Oxygen Supplementation

Supplemental oxygen is beneficial in the recovery period. Due to the upper respiratory obstruction in these dogs, oxygen administration directly into the trachea has significant advantages (Mann et al. 1992). It has been shown that placement of a catheter percutaneously into the trachea can be used to deliver oxygen and is effective at increasing the fraction of inspired oxygen and increasing arterial oxygen partial pressure (Mann et al. 1992). This is a useful technique in an emergency. An alternative is the use of a nasotracheal tube (Senn et al. 2011, Poncet et al. 2006, Haimel and Dupre 2015). This is a simple technique that involves the placement of a naso-oesophageal feeding tube through the nose and into the trachea. This can be placed following surgical correction of BOAS and then used to provide supplemental oxygen in the postoperative period. Nasotracheal tubes have been shown to decrease postoperative respiratory distress but do not seem to affect the incidence of complications or mortality (Senn et al. 2011). The use of continuous positive airway pressure (CPAP) masks has been investigated in brachycephalic dogs in the recovery period, after being demonstrated to improve arterial oxygenation, and reduce carbon dioxide levels and respiratory rates in sedated dogs (Rondelli et al. 2016, Staffieri et al. 2014).

Avoiding Stress

Stress during the postoperative period can lead to significant upper respiratory tract compromise and dogs must be kept calm and cool. Sedation is recommended for anxious dogs or those that become stressed (Haimel and Dupre 2015, Downing and Gibson 2018). BOAS patients should be placed in a kennel so they can be monitored constantly. However, small kennels can cause stress for some dogs and in the authors' experience it can be helpful to move these dogs to a visible 'walk-in' kennel before sedation is required. More importantly, a quiet, cool environment can encourage a smooth recovery.

For sedation, the authors prefer the use of medetomidine at a dose of 0.5 μg/kg intravenously as needed, although acepromazine 5 μg/kg can also be useful. If the animal becomes heavily sedated, the jaws can be held open and the tongue protracted to aid airflow.

Regurgitation and Vomiting

Regurgitation and/or vomiting is reported in 4.7%–24% of dogs following BOAS surgery (Poncet et al. 2006, Torrez and Hunt 2006, Haimel and Dupre 2015, Ree et al. 2016, Lindsay et al. 2020). A recent study looked at risk factors for regurgitation after BOAS surgery and found that it occurred in 34.5% (89/258) of dogs within 24 hours (Fenner, Quinn, and Demetriou 2020). Younger age and a history of regurgitation were significantly associated with increased odds of postoperative regurgitation. However, breed and stage of laryngeal collapse were not.

Significant regurgitation or vomiting can predispose to aspiration pneumonia, although a recent study with a high rate of regurgitation (34.5%) had a very low rate of aspiration pneumonia (0.8%) (Fenner, Quinn, and Demetriou 2020). Dogs can also suffer vasovagal syncope following an episode, and this can result in respiratory arrest and death. As mentioned above, animals can be treated with omeprazole and/or metoclopramide pre-surgically and these should be continued post-surgery. One study suggested that medical treatment of gastro-oesophageal abnormalities post-surgery improved success and reduced complications (Poncet et al. 2006). Animals that are nauseous should be treated with maropitant at a dose of 1 mg/kg once daily. Withholding food and water for 12–24 hours is suggested by some surgeons (Lindsay et al. 2020).

For dogs with excessive mucoid secretions that may compromise the airway, a mucolytic such as nebulised acetylcysteine, diluted to 100 mg/mL, can be used with care.

Aspiration Pneumonia

Bulldogs and French Bulldogs have been shown to be at increased risk of developing aspiration pneumonia, compared to the overall dog population (Darcy, Humm, and Ter Haar 2018). Dogs undergoing surgical treatment of BOAS are at risk of aspiration pneumonia both pre- and post-surgery, particularly those with significant regurgitation or vomiting. Many studies do not include information on the frequency of pre-operative aspiration pneumonia, but it has been identified in 1.6%–8.3% of dogs prior to surgery (Kirsch et al. 2019, Lindsay et al. 2020, Ree et al. 2016). Aspiration pneumonia has been reported following BOAS surgery in 0.8%–5.5% of dogs (Haimel and Dupre 2015, Ree et al. 2016, Lindsay et al. 2020, Kirsch et al. 2019, Fenner, Quinn, and Demetriou 2020). Conversely, several studies did not identify any aspiration pneumonia in their populations (Poncet et al. 2006, Cook, Moses, and Mackie 2015, Riecks, Birchard, and Stephens 2007, Tamburro et al. 2019). Postoperative aspiration pneumonia has been significantly associated with temporary tracheostomy and death in several studies (Ree et al. 2016, Kirsch et al. 2019, Worth et al. 2018). Animals should be monitored for evidence of aspiration pneumonia and treated with appropriate antibiotics and nebulisation and coupage as appropriate.

Temporary Tracheostomy

Dogs with severe postoperative swelling and dyspnoea may require a temporary tracheostomy. Fortunately, the proportion of dogs requiring temporary tracheostomy is relatively low, although this varies considerably between studies with a range of 0%–10.9% (Stordalen et al. 2020, Oechtering, Pohl, Schlueter, and Schuenemann 2016, Findji and Dupre 2008, Tamburro et al. 2019, Lindsay et al. 2020, Fenner, Quinn, and Demetriou 2020, Cook, Moses, and Mackie 2015, Riecks, Birchard, and Stephens 2007, Poncet et al. 2006, Tarricone et al. 2019, Torrez and Hunt 2006, Haimel and Dupre 2015, Ree et al. 2016, Worth et al. 2018). A study looked at specific risk factors for temporary tracheostomy following BOAS surgery and found that 12/198 dogs (6.1%) required a temporary tracheostomy and increasing age was a significant risk factor (Worth et al. 2018). Temporary tracheostomy has also been significantly associated with aspiration pneumonia, but it is unclear whether this is cause or effect (Ree et al. 2016, Kirsch et al. 2019).

Dogs that are dyspnoeic post-surgery due to swelling and that do not respond to medical treatment and repeat anaesthetic and intubation should be treated with a temporary tracheostomy (Figure 11.8). The authors recommend early intervention, once it is apparent that more conservative treatment is

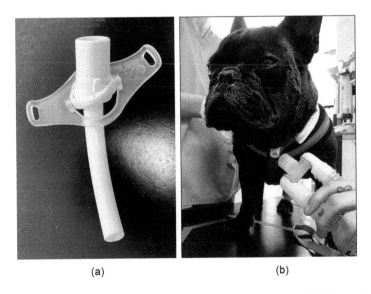

(a)   (b)

FIGURE 11.8   Tracheostomy tube. (a) A 5-mm silicone tracheostomy tube suitable for use in dogs. Note that the tube is uncuffed and has an inner sleeve that can be removed for cleaning. (b) Tracheostomy tube in place in a French Bulldog with marked airway swelling and dyspnoea following surgery for BOAS. The dog is being oxygenated prior to removal of the tube.

not working. Dogs with temporary tracheostomy must be closely monitored as complications can be life-threatening. Complications can include obstruction of the tube with mucous secretion, dislodgement of the tube, swelling around the stoma and aspiration pneumonia (Nicholson and Baines 2012, Bird et al. 2018, Stordalen et al. 2020). For dogs with excessive mucoid secretions causing obstruction of the tube, nebulised acetylcysteine (as mentioned above) may be useful.

It has been suggested that brachycephalic dogs are more at risk of complications due to their conformation with short, thick necks, relatively narrow or hypoplastic trachea and excessive skin folds and English Bulldogs were at significantly increased risk of dislodgement of the tube in one study (Nicholson and Baines 2012). A study looked at the outcome for temporary tracheostomy in 42 dogs treated for BOAS (Stordalen et al. 2020); 83.3% of the dogs experienced a major complication, including obstruction of the tube in 76.2% and dislodgement in 38.1%. Minor complications included cough (47.6%), nasal discharge (23.8%), inflammation around the stoma (11.9%) and subcutaneous emphysema (9.5%).

Silicone stents have been reported as an alternative to traditional tracheostomy tubes in order to reduce the rate of complications (Trinterud, Nelissen, and White 2014). Indeed, this study found a low rate of complications, with no obstruction, early dislodgement or stomal swelling. However, 60% of dogs that had the stent in situ for 5 days or more had dislodgement of the stent due to granulation tissue formation, suggesting that they are best used for short periods of time. Another modification suggested to reduce complications is the use of an 'H-shaped' tracheostomy and a Penrose drain 'sling' dorsal to the trachea to keep the stoma ventral (Bird et al. 2018). This study included 15/21 dogs with BOAS and 100% of the tubes were successfully managed. Complications included tube dislodgement in 6/21 dogs (28.6%) and obstruction due to discharge in one dog (4.8%). Tube replacement after dislodgement was subjectively much easier.

Although associated with a high rate of complications, temporary tracheostomy typically has a very good outcome (Nicholson and Baines 2012, Trinterud, Nelissen, and White 2014, Bird et al. 2018). Studies have reported survival rates of 83.3%–100% for dogs with temporary tracheostomy after BOAS surgery (Torrez and Hunt 2006, Riecks, Birchard, and Stephens 2007, Poncet et al. 2006, Findji and Dupre 2008, Dunie-Merigot, Bouvy, and Poncet 2010, Worth et al. 2018, Liu et al. 2019, Fenner, Quinn, and Demetriou 2020, Lindsay et al. 2020, Stordalen et al. 2020). However, some dogs remain tracheostomy dependent and require a permanent tracheostomy (see above). Less favourable survival has been reported with 0% in one study and 33.3% in another (Ree et al. 2016, Kirsch et al. 2019).

STRATEGIES FOR SURGERY

BOAS surgery needs to be tailored to the individual. It is worth noting that surgical intervention should be combined with appropriate husbandry changes including weight loss and exercise management. The first line of treatment for most dogs is alarplasty (+/– vestibuloplasty) and staphylectomy (or folding flap palatoplasty) with sacculectomy and tonsillectomy as required and based on surgeon preference. This will have a positive effect on most dogs and is generally associated with a good outcome (Torrez and Hunt 2006, Poncet et al. 2006, Riecks, Birchard, and Stephens 2007).

This approach has been described as a 'multilevel surgery', with the levels referring to the different areas of the upper respiratory tract (e.g. nostrils, oropharynx and larynx) (Liu et al. 2017, Oechtering, Pohl, Schlueter, and Schuenemann 2016). Although it does not address Grade II or III laryngeal collapse, this is still sufficient in most dogs (see the section above). In addition, it does not address any intranasal obstruction and treatment of this with LATE has been suggested in combination with traditional surgery either simultaneously in all dogs or in a staged manner in those with ongoing signs (Oechtering, Pohl, Schlueter, and Schuenemann 2016, Liu et al. 2019). A modified 'multilevel surgery' has been advocated in order to try to maximise the surgical result by refining the improvement in the functional airway (Liu et al. 2017). It involves a modified rhinoplasty (combined Trader's alarplasty and nasal vestibuloplasty), a folded flap palatoplasty, sacculectomy, tonsillectomy and partial cuneiformectomy (for Grades II and III laryngeal collapse) as needed. This protocol has been associated with a statistically significantly better outcome compared with traditional surgery, with no increase in complications (Liu et al. 2017). However, it is unclear whether all dogs would benefit from the modified surgery or whether it should be reserved for more severely affected individuals. In practice, a wide variation exists in the combination of surgeries performed and the specific techniques used.

The timing of surgery is considered important with early treatment recommended for clinically affected animals. This should improve their function and quality of life (see below). Importantly, early treatment should reduce the development and progression of secondary abnormalities such as laryngeal collapse, which has been reported in puppies as young as 4.5–6 months of age (see the previous section) (Pink et al. 2006).

Recently, there has been a move towards combining BOAS surgery with routine neutering in young brachycephalic dogs. This makes excellent sense in dogs already showing clinical signs. However, surgery can be associated with significant morbidity and, although overall mortality is very low, there must be a balance between potential improvement in quality of life and the risks of these procedures. It is questionable whether surgery should be used prophylactically in young brachycephalic dogs without clinical signs, as they may not be at risk of developing BOAS. A provocative test, such as the 3-minute exercise test, can be used to guide treatment in brachycephalic dogs (Riggs et al. 2019), with respiratory assessment discussed in detail in Chapter 2.

Dogs should be reassessed after surgery to ensure that they have responded appropriately. Assessment is typically limited to physical examination and owner feedback. This is subjective and complicated by the fact that owners are poor at judging the severity of clinical signs in dogs with BOAS and potentially biased to the outcome (Packer, Hendricks, and Burn 2012, Liu et al. 2015). A clinical grading scheme has been published and used in several studies and can be used to subjectively assess treatment response (Haimel and Dupre 2015, Sarran et al. 2018). WBBP offers an objective measure of BOAS and can be used to assess outcome and hence identify dogs that would benefit from additional intervention, although its availability is limited (Liu et al. 2019, Liu et al. 2017). A provocative exercise test has been suggested as a useful way of assessing response to surgery in practice (Riggs et al. 2019, Villedieu, Rutherford, and Ter Haar 2019). A 3-minute exercise test has been validated against WBBP, giving a BOAS functional grade, which can be used to guide treatment (see Chapter 2) (Riggs et al. 2019, Liu et al. 2019). The authors routinely perform the 3-minute exercise test before and at approximately 4–6 weeks following BOAS surgery in order to assess response to surgery.

Dogs that do not respond well to surgery may need additional intervention. Consideration should be given to whether initial surgery was adequate or might need revision (e.g. removing more soft palate) (Sarran et al. 2018). Dogs with significant laryngeal collapse or intranasal obstruction may benefit from further specific treatment of these abnormalities (see the previous sections).

PROGNOSIS/LONG-TERM OUTCOME

The success of surgical interventions is variable. Most dogs will have a positive response to surgery, but many do not become 'normal', with ongoing clinical sings and limitations on function (Torrez and Hunt 2006, Findji and Dupre 2008, Pohl, Roedler, and Oechtering 2016, Riecks, Birchard, and Stephens 2007, De Lorenzi, Bertoncello, and Drigo 2009). Studies have shown subjective improvement in 86.5%–100% of dogs but with only a marked improvement or resolution of signs in 34.6%–78.9% (see Table 11.1) (De Lorenzi, Bertoncello, and Drigo 2009, Dunie-Merigot, Bouvy, and Poncet 2010, Haimel and Dupre 2015, Poncet et al. 2006, Riecks, Birchard, and Stephens 2007, Torrez and Hunt 2006). Another study reported the owner assessed outcome of multilevel airway surgery combined with LATE (Pohl, Roedler, and Oechtering 2016). The clinical signs improved markedly after surgery, with life-threatening events reduced by 90%. However, despite the marked improvement, dogs did not become normal and clinical signs persisted.

TABLE 11.1

Summary of Subjective, Owner-Assessed Long-Term Outcome for Dogs Treated Surgically for Brachycephalic Obstructive Airway Syndrome (BOAS)

Study, number of dogs and follow-up time	Worsened	Stable	Some improvement	Good improvement
Torrez 2006 46 dogs 19-77 months	N/A	Poor No improvement 10.9%	Good Some improvement 32.6%	Excellent Marked improvement 56.5%
Poncet 2006 51 dogs 6-23 months (median 11 months)	Worse Worsening of one of more clinical signs 7.8%	Poor Little or no improvement 3.9%	Good Some remaining signs but minor in nature 21.6%	Excellent Resolution of all signs 66.7%
Riecks 2007 34 dogs >1 year	Poor Severity of clinical signs increased after surgery 2.9%	Fair No improvement in clinical signs 2.9%	Good Improvement in clinical signs with some limits on physical activity 47.1%	Excellent Marked improvement in clinical signs with no restriction on physical activity 47.1%
De Lorenzi 2009 36 dogs ≥6 months	Poor Clinical situation worsened or did not improve after surgery 11.1%	Fair	Moderate Good improvement of the clinical situation with some limitations in physical activity after surgery 30.6%	Adequate Excellent improvement in both the clinical situation and physical activity after surgery 58.3%
Dunie-Merigot 2010 57 dogs 6 months	Poor Worsening of the clinical signs 0%	Stabilisation of clinical signs 0%	Good Improvement of the clinical signs with residual signs and restriction of activity 21.1%	Excellent Marked improvement of the clinical signs with few or no residual clinical signs and no restriction of activity 78.9%
Haimel 2015 52 dogs Mean 24.3 months (SD +/-14.4)	Worse Worsening of signs 1.9%	No improvement No improvement 11.5%	Good Marked improvement 51.9%	Very good No current respiratory tract signs 34.6%

The results of additional studies, with different styles of assessment, are included in the text.

Although there are several studies reporting the subjective long-term outcome following surgery, a lack of consistency between scoring systems makes direct comparison challenging. This is compounded by variable population demographics between studies (e.g. breed distribution and dog age), the number of abnormalities present, the severity of the abnormalities and different procedures used by different surgeons. Many studies investigate one aspect of the disease when it is not possible to standardise the other components and their treatment, making it hard to draw conclusions about the specific efficacy of a given procedure. Importantly, most studies use a subjective owner assessment of outcome, which is prone to a variety of bias. As mentioned in Chapter 1, owners are not good at identifying clinical signs, making assessment of outcome in this manner potentially unreliable (Packer, Hendricks, and Burn 2012, Liu et al. 2015).

WBBP can provide an objective assessment of response to BOAS surgery. One study reported the results of multilevel surgery in 50 dogs (Liu et al. 2017). Surgery had a positive effect with a reduction in the median BOAS index from 75.8% pre-surgery to 63.2% post-surgery at a mean follow-up of 42 days. However, based on the BOAS index, only 32% of dogs had a satisfactory outcome (although this was relatively short term). Another study performed modified multilevel surgery in 57 dogs, and the median BOAS index decreased from 80.7% to 56.9% at 2–6 months post-surgery, with 50.9% of dogs having a BOAS index of >50%, suggesting the need for further intervention (Liu et al. 2019). Twenty of these dogs underwent LATE and the BOAS index significantly decreased at 2–6 months, from 66.7% to 42.3%, with only 20% of the dogs having a BOAS index of >50%. Overall, 55.6% of owners were very satisfied with the outcome, and 44.4% were satisfied with the outcome.

Taken together, this outcome data shows that although dogs improve with BOAS surgery, based on subjective and objective assessment there is a significant rate of ongoing clinical signs and functional impairment. It seems likely that this is due to a combination of a variety of factors, including disease severity in some individuals where it may not be possible to achieve a good outcome, the ongoing effect of untreated areas of obstruction (such as aberrant turbinates and tracheal hypoplasia) and possibly insufficient surgical correction in some individuals. Normal respiratory function in all dogs is therefore not a realistic expectation.

PROGNOSTIC INDICATORS

Several studies have looked at risk factors or prognostic indicators for short- and long-term outcomes, with conflicting findings. This is perhaps to be expected as most studies are retrospective and include a relatively small number of dogs for statistical analysis (even the larger studies). There are also considerable differences in the apparent populations of dogs, with variation in disease severity, treatments and complications reported. Importantly, studies use different outcome measures, such as frequency of specific complications and subjective or objective outcomes. Although it might be reasonable to assume that risk factors for complications and mortality would be the same as those for long-term outcome and improvement in functional airway improvement, this may not be the case. Overall, the authors do believe that the more severely affected dogs are more likely to suffer complications or have a poor outcome, but this has not been consistently demonstrated in the literature.

Age has been implicated as an important prognostic indicator. However, findings are inconsistent. Younger age has been significantly associated with a poor outcome in terms or BOAS index as assessed on WBBP (Liu et al. 2017) and with an increased risk of postoperative regurgitation (Fenner, Quinn, and Demetriou 2020). However, studies have also found no association between age and owner assessed outcome (Riecks, Birchard, and Stephens 2007) and also significant associations between increased age and the need for tracheostomy and overall complication rate (Lindsay et al. 2020, Worth et al. 2018). These seem conflicting with younger dogs and older dogs at increased risk in different studies. It seems likely that this dichotomy is primarily related to population differences, with variation in disease severity and hence dogs presenting at different ages and with varying clinical signs. For example, some practices may see dogs with particularly extreme conformation and severe airway obstruction, particularly if they are a tertiary referral centre. These animals are likely to be referred relatively early in their disease process and may be at increased risk of a poor outcome compared to any older animals in the population that may not be as severely affected. Other practices may see a greater proportion of younger dogs with less severe conformation and milder signs, referred for early intervention and then older animals that were not referred earlier and have gone on to develop more severe signs due to secondary changes and are therefore at greater risk.

Many studies have failed to find an association between significant laryngeal collapse and complications and/or outcome (Haimel and Dupre 2015, Torrez and Hunt 2006, Fenner, Quinn, and Demetriou 2020, De Lorenzi, Bertoncello, and Drigo 2009). Conversely, recent studies found that laryngeal collapse was a significant negative prognostic indicator for postsurgical outcome in terms of BOAS index (Liu et al. 2017) and for a negative outcome in terms of complications (Tarricone et al. 2019). This may relate to differences in the assessment of laryngeal collapse and the frequency of severe collapse in the different study populations (see previous section). The authors believe that dogs with severe clinical signs related to significant laryngeal collapse are at increased risk of complications and most likely have a poorer outcome overall.

A recent study developed a brachycephalic risk (BRisk) score for predicting complications following BOAS surgery (Tarricone et al. 2019). Dogs were given a 10-point score based on breed, previous surgery, concurrent procedures, body condition score, airway status and rectal temperature on admission. Dogs with a score of greater than 3 were significantly more likely to have a negative outcome. Although an interesting concept, the authors are wary of such prognostic tools as they may have an undue influence on decision-making in individual dogs (Jeffery and Tivers 2018).

CONCLUSIONS

There are a variety of techniques described for BOAS surgery, but for most dogs, staphylectomy and alarplasty are the mainstay of treatment. Recently, there has been considerable progress in refining treatment, particularly for dogs with more extreme BOAS. It is likely that more effective surgery in terms of removing more obstructing tissue will have a greater improvement in airway function and recent evidence does support this concept (Liu et al. 2017). However, there remains limited information on the comparative efficacy of each individual technique. Further research is needed to allow us to determine the optimal combination of surgical techniques. It is possible that this may vary between breeds and individuals based on the underlying conformation.

Overall, clinically affected dogs benefit from surgery. However, selecting these dogs is sometimes challenging, particularly due to poor understanding and recognition of clinical signs by owners. A 3-minute exercise test has shown potential as a way of selecting dogs that would benefit from surgery. Early treatment is recommended for clinically affected dogs to prevent further deterioration. However, surgery does not cure dogs, and many will have ongoing clinical signs, although most will be significantly improved in terms of function and quality of life. As many dogs remain affected following surgery, identifying those that would benefit from revision or additional surgery can be challenging. This is exacerbated by assessment of outcome typically being subjective and prone to the limitations mentioned above. WBBP offers an objective outcome measure, but its availability is very limited at present. The use of BOAS functional grading based on a 3-minute exercise test is very useful for assessing response to surgery, but further research is needed to assess its performance in different settings.

Although surgery can be used to improve dogs with BOAS, most remain affected to some degree and this is a welfare concern. Urgent progress needs to be made in breeding towards less extreme phenotypes to reduce the frequency and severity of BOAS. Education of the public is also needed to increase understanding of and therefore reduce demand for extreme brachycephalic animals and the need for the surgeries outlined in this chapter.

REFERENCES

Anderson, G. M. 2018. "Soft tissues of the oral cavity." In *Veterinary Surgery Small Animal*, edited by S. A. Johnston and K. M. Tobias, 1637–1652. Missouri: Elsevier.

Bini, G., E. Vettorato, C. De Gennaro, and F. Corletto. 2018. "A retrospective comparison of two analgesic strategies after uncomplicated tibial plateau levelling osteotomy in dogs." *Vet Anaesth Analg* 45 (4):557–65. doi:10.1016/j.vaa.2017.11.005.

Bird, F. G., R. Vallefuoco, G. Dupre, and H. Brissot. 2018. "A modified temporary tracheostomy in dogs: Outcome and complications in 21 dogs (2012 to 2017)." *J Small Anim Pract* 59 (12):769–76. doi:10.1111/jsap.12928.

Brdecka, D. J., C. A. Rawlings, A. C. Perry, and J. R. Anderson. 2008. "Use of an electrothermal, feedback-controlled, bipolar sealing device for resection of the elongated portion of the soft palate in dogs with obstructive upper airway disease." *J Am Vet Med Assoc* 233 (8):1265–9. doi:10.2460/javma.233.8.1265.

Bright, R.M., and L.G. Wheaton. 1983. "A modified surgical technique for elongated soft palate in dogs." *J Am Anim Hosp Assoc* 19:288.

Caccamo, R., P. Buracco, G. La Rosa, M. Cantatore, and S. Romussi. 2014. "Glottic and skull indices in canine brachycephalic airway obstructive syndrome." *BMC Vet Res* 10:12. doi:10.1186/1746-6148-10-12.

Cantatore, M., M. Gobbetti, S. Romussi, G. Brambilla, C. Giudice, V. Grieco, and D. Stefanello. 2012. "Medium term endoscopic assessment of the surgical outcome following laryngeal saccule resection in brachycephalic dogs." *Vet Rec* 170 (20):518. doi:10.1136/vr.100289.

Cook, D. A., P. A. Moses, and J. T. Mackie. 2015. "Clinical effects of the use of a bipolar vessel sealing device for soft palate resection and tonsillectomy in dogs, with histological assessment of resected tonsillar tissue." *Aust Vet J* 93 (12):445–51. doi:10.1111/avj.12384.

Darcy, H. P., K. Humm, and G. Ter Haar. 2018. "Retrospective analysis of incidence, clinical features, potential risk factors, and prognostic indicators for aspiration pneumonia in three brachycephalic dog breeds." *J Am Vet Med Assoc* 253 (7):869–76. doi:10.2460/javma.253.7.869.

Davidson, E. B., M. S. Davis, G. A. Campbell, K. K. Williamson, M. E. Payton, T. S. Healey, and K. E. Bartels. 2001. "Evaluation of carbon dioxide laser and conventional incisional techniques for resection of soft palates in brachycephalic dogs." *J Am Vet Med Assoc* 219 (6):776–81. doi:10.2460/javma.2001.219.776.

De Lorenzi, D., D. Bertoncello, and M. Drigo. 2009. "Bronchial abnormalities found in a consecutive series of 40 brachycephalic dogs." *J Am Vet Med Assoc* 235 (7):835–40. doi:10.2460/javma.235.7.835.

Dickerson, V. M., C. M. B. Dillard, J. A. Grimes, M. L. Wallace, J. F. McAnulty, and C. W. Schmiedt. 2020. "Dorsal offset rhinoplasty for treatment of stenotic nares in 34 brachycephalic dogs." *Vet Surg*. doi:10.1111/vsu.13504.

Downing, F., and S. Gibson. 2018. "Anaesthesia of brachycephalic dogs." *J Small Anim Pract* 59 (12):725–33. doi:10.1111/jsap.12948.

Dunie-Merigot, A., B. Bouvy, and C. Poncet. 2010. "Comparative use of CO(2) laser, diode laser and monopolar electrocautery for resection of the soft palate in dogs with brachycephalic airway obstructive syndrome." *Vet Rec* 167 (18):700–4. doi:10.1136/vr.c5107.

Dupre, G., and D. Heidenreich. 2016. "Brachycephalic Syndrome." *Vet Clin North Am Small Anim Pract* 46 (4):691–707. doi:10.1016/j.cvsm.2016.02.002.

Ellis, J., and E. A. Leece. 2017. "Nebulized adrenaline in the postoperative management of brachycephalic obstructive airway syndrome in a pug." *J Am Anim Hosp Assoc* 53 (2):107–10. doi:10.5326/JAAHA-MS-6466.

Ellison, G. W. 2004. "Alapexy: An alternative technique for repair of stenotic nares in dogs." *J Am Anim Hosp Assoc* 40 (6):484–9. doi:10.5326/0400484.

Farquharson, J., and K. W. Smith. 1942. "Resection of the soft palate in the dog." *J Am Vet Med Assoc* 100:427–430.

Fasanella, F. J., J. M. Shivley, J. L. Wardlaw, and S. Givaruangsawat. 2010. "Brachycephalic airway obstructive syndrome in dogs: 90 cases (1991–2008)." *J Am Vet Med Assoc* 237 (9):1048–51. doi:10.2460/javma.237.9.1048.

Fenner, J. V. H., R. J. Quinn, and J. L. Demetriou. 2020. "Postoperative regurgitation in dogs after upper airway surgery to treat brachycephalic obstructive airway syndrome: 258 cases (2013–2017)." *Vet Surg* 49 (1):53–60. doi:10.1111/vsu.13297.

Findji, L., and G. Dupre. 2008. "Folded flap palatoplasty for treatment of elongated soft palates in 55 dogs." *Vet. Med. Austria / Wien Tierarztl Mschr* 95:56–63.

Ginn, J. A., M. S. Kumar, B. C. McKiernan, and B. E. Powers. 2008. "Nasopharyngeal turbinates in brachycephalic dogs and cats." *J Am Anim Hosp Assoc* 44 (5):243–9. doi:10.5326/0440243.

Gobbetti, M., S. Romussi, P. Buracco, V. Bronzo, S. Gatti, and M. Cantatore. 2018. "Long-term outcome of permanent tracheostomy in 15 dogs with severe laryngeal collapse secondary to brachycephalic airway obstructive syndrome." *Vet Surg* 47 (5):648–653. doi:10.1111/vsu.12903.

Grimes, J. A., A. M. Davis, M. L. Wallace, A. A. Sterman, K. M. Thieman-Mankin, S. Lin, V. F. Scharf, K. C. Hlusko, B. M. Matz, K. K. Cornell, C. A. M. Vetter, and C. W. Schmiedt. 2019. "Long-term outcome and risk factors associated with death or the need for revision surgery in dogs with permanent tracheostomies." *J Am Vet Med Assoc* 254 (9):1086–1093. doi:10.2460/javma.254.9.1086.

Haimel, G., and G. Dupre. 2015. "Brachycephalic airway syndrome: A comparative study between pugs and French bulldogs." *J Small Anim Pract* 56 (12):714–9. doi:10.1111/jsap.12408.

Harvey, C. E. 1982. "Upper airway obstruction part IV: Partial laryngectomy in brachycephalic dogs." *J Am Anim Hosp Assoc* 18:548–550.

Harvey, C. E. 1983. "Review of results of airway obstruction surgery in the dog." *J Small Anim Practice* 24:555–559.

Huck, J. L., B. J. Stanley, and J. G. Hauptman. 2008. "Technique and outcome of nares amputation (Trader's technique) in immature shih tzus." *J Am Anim Hosp Assoc* 44 (2):82–5. doi:10.5326/0440082.

Hughes, J. R., B. M. Kaye, A. R. Beswick, and G. Ter Haar. 2018. "Complications following laryngeal sacculectomy in brachycephalic dogs." *J Small Anim Pract* 59 (1):16–21. doi:10.1111/jsap.127C63.

Jeffery, N., and M. S. Tivers. 2018. "Prognostic markers: What are they good for?" *J Small Anim Pract* 59 (6):321–323. doi:10.1111/jsap.12864.

Kirsch, M. S., D. Spector, S. R. Kalafut, G. E. Moore, and R. McDougall. 2019. "Comparison of carbon dioxide laser versus bipolar vessel device for staphylectomy for the treatment of brachycephalic obstructive airway syndrome." *Can Vet J* 60 (2):160–166.

Leonard, H. C. 1960. "Collapse of the larynx and adjacent structures in the dog." *J Am Vet Med Assoc* 137:360–363.

Lindsay, B., D. Cook, J. M. Wetzel, S. Siess, and P. Moses. 2020. "Brachycephalic airway syndrome: Management of post-operative respiratory complications in 248 dogs." *Aust Vet J*. doi:10.1111/avj.12926.

Liu, N. C., V. J. Adams, L. Kalmar, J. F. Ladlow, and D. R. Sargan. 2016. "Whole-body barometric plethysmography characterizes upper airway obstruction in 3 brachycephalic breeds of dogs." *J Vet Intern Med* 30 (3):853–65. doi:10.1111/jvim.13933.

Liu, N. C., M. A. Genain, L. Kalmar, D. R. Sargan, and J. F. Ladlow. 2019. "Objective effectiveness of and indications for laser-assisted turbinectomy in brachycephalic obstructive airway syndrome." *Vet Surg* 48 (1):79–87. doi:10.1111/vsu.13107.

Liu, N. C., G. U. Oechtering, V. J. Adams, L. Kalmar, D. R. Sargan, and J. F. Ladlow. 2017. "Outcomes and prognostic factors of surgical treatments for brachycephalic obstructive airway syndrome in 3 breeds." *Vet Surg* 46 (2):271–280. doi:10.1111/vsu.12608.

Liu, N. C., D. R. Sargan, V. J. Adams, and J. F. Ladlow. 2015. "Characterisation of brachycephalic obstructive airway syndrome in french bulldogs using whole-body barometric plethysmography." *PLoS One* 10 (6):e0130741. doi:10.1371/journal.pone.0130741.

Mann, F. A., C. Wagner-Mann, J. A. Allert, and J. Smith. 1992. "Comparison of intranasal and intratracheal oxygen administration in healthy awake dogs." *Am J Vet Res* 53 (5):856–60.

Michelsen, J. 2011. "Use of the harmonic scalpel for soft palate resection in dogs: A series of three cases." *Aust Vet J* 89 (12):511–4. doi:10.1111/j.1751–0813.2011.00846.x.

Monnet, E., and K. M. Tobias. 2018. "Larynx." In *Veterinary Surgery Small Animal*, edited by S. A. Johnston and K. M. Tobias, 1946–1963. Missouri: Elsevier.

Nicholson, I., and S. Baines. 2012. "Complications associated with temporary tracheostomy tubes in 42 dogs (1998 to 2007)." *J Small Anim Practice* 53 (2):108–114. doi:10.1111/j.1748-5827.2011.01167.x.

Occhipinti, L. L., and J. G. Hauptman. 2014. "Long-term outcome of permanent tracheostomies in dogs: 21 cases (2000–2012)." *Can Vet J* 55 (4):357–60.

Oechtering, G. 2010. "Brachycephalic syndrome – new information on an old congenital disease. " *Veterinary Focus* 20 (2):1–9.

Oechtering, G. U., S. Pohl, C. Schlueter, J. P. Lippert, M. Alef, I. Kiefer, E. Ludewig, and R. Schuenemann. 2016. "A novel approach to brachycephalic syndrome. 1. Evaluation of anatomical intranasal airway obstruction." *Vet Surg* 45 (2):165–72. doi:10.1111/vsu.12446.

Oechtering, G. U., S. Pohl, C. Schlueter, and R. Schuenemann. 2016. "A novel approach to brachycephalic syndrome. 2. Laser-assisted turbinectomy (LATE)." *Vet Surg* 45 (2):173–81. doi:10.1111/vsu.12447.

Packer, R., A. Hendricks, and C. Burn. 2012. "Do dog owners perceive the clinical signs related to confor-mational inherited disorders as "normal" for the breed? A potential constraint to improving canine welfare." *Anim Welf* 21:81–93.

Packer, R. M., A. Hendricks, M. S. Tivers, and C. C. Burn. 2015. "Impact of facial conformation on canine health: Brachycephalic obstructive airway syndrome." *PLoS One* 10 (10):e0137496. doi:10.1371/journal.pone.0137496.

Pink, J. J., R. S. Doyle, J. M. Hughes, E. Tobin, and C. R. Bellenger. 2006. "Laryngeal collapse in seven brachycephalic puppies." *J Small Anim Pract* 47 (3):131–5. doi:10.1111/j.1748-5827.2006.00056.x.

Pohl, S., F. S. Roedler, and G. U. Oechtering. 2016. "How does multilevel upper airway surgery influence the lives of dogs with severe brachycephaly? Results of a structured pre- and postoperative owner question-naire." *Vet J* 210:39–45. doi:10.1016/j.tvjl.2016.01.017.

Poncet, C. M., G. P. Dupre, V. G. Freiche, and B. M. Bouvy. 2006. "Long-term results of upper respiratory syndrome surgery and gastrointestinal tract medical treatment in 51 brachycephalic dogs." *J Small Anim Pract* 47 (3):137–42. doi:10.1111/j.1748-5827.2006.00057.x.

Poncet, C. M., G. P. Dupre, V. G. Freiche, M. M. Estrada, Y. A. Poubanne, and B. M. Bouvy. 2005. "Prevalence of gastrointestinal tract lesions in 73 brachycephalic dogs with upper respiratory syndrome." *J Small Anim Pract* 46 (6):273–9. doi:10.1111/j.1748-5827.2005.tb00320.x.

Ree, J. J., M. Milovancev, L. A. MacIntyre, and K. L. Townsend. 2016. "Factors associated with major com-plications in the short-term postoperative period in dogs undergoing surgery for brachycephalic airway syndrome." *Can Vet J* 57 (9):976–80.

Reeve, E. J., D. Sutton, E. J. Friend, and C. M. R. Warren-Smith. 2017. "Documenting the prevalence of hiatal hernia and oesophageal abnormalities in brachycephalic dogs using fluoroscopy." *J Small Anim Pract* 58 (12):703–708. doi:10.1111/jsap.12734.

Riecks, T. W., S. J. Birchard, and J. A. Stephens. 2007. "Surgical correction of brachycephalic syndrome in dogs: 62 cases (1991–2004)." *J Am Vet Med Assoc* 230 (9):1324–8. doi:10.2460/javma.230.9.1324.

Riggs, J., N. C. Liu, D. R. Sutton, D. Sargan, and J. F. Ladlow. 2019. "Validation of exercise testing and laryn-geal auscultation for grading brachycephalic obstructive airway syndrome in pugs, French bulldogs, and English bulldogs by using whole-body barometric plethysmography." *Vet Surg* 48 (4):488–496. doi:10.1111/vsu.13159.

Rondelli, V., A. Briganti, P. Centonze, F. Perini, F. Romano, A. Bufalari, V. De Monte, and F. Staffieri. 2016. "Respiratory effects of continuous positive airway pressure administered during recovery from general anaesthesia in brachycephalic dogs." *Vet Anaesth Analg* 43:A15.

Rutherford, L., L. Beever, M. Bruce, and G. Ter Haar. 2017. "Assessment of computed tomography derived cricoid cartilage and tracheal dimensions to evaluate degree of cricoid narrowing in brachycephalic dogs." *Vet Radiol Ultrasound* 58 (6):634–646. doi:10.1111/vru.12526.

Sarran, D., A. Caron, I. Testault, S. Segond, and J. P. Billet. 2018. "Position of maximal nasopharyngeal maxi-mal occlusion in relation to hamuli pterygoidei: Use of hamuli pterygoidei as landmarks for palatoplasty in brachycephalic airway obstruction syndrome surgical treatment." *J Small Anim Pract* doi:10.1111/jsap.12909.

Schmiedt, C. W. 2018. "Nasal planum, nasal cavity, and sinuses." In *Veterinary Surgery Small Animal*, edited by S. A. Johnston and K. M. Tobias, 1919–1934. Missouri: Elsevier.

Schuenemann, R., and G. Oechtering. 2014a. "Inside the brachycephalic nose: Conchal regrowth and mucosal contact points after laser-assisted turbinectomy." *J Am Anim Hosp Assoc* 50 (4):237–46. doi:10.5326/JAAHA-MS-6086.

Schuenemann, R., and G. U. Oechtering. 2014b. "Inside the brachycephalic nose: Intranasal mucosal contact points." *J Am Anim Hosp Assoc* 50 (3):149–58. doi:10.5326/JAAHA-MS-5991.

Schuenemann, R., S. Pohl, and G. U. Oechtering. 2017. "A novel approach to brachycephalic syndrome. 3. Isolated laser-assisted turbinectomy of caudal aberrant turbinates (CAT LATE)." *Vet Surg* 46 (1):32–38. doi:10.1111/vsu.12587.

Senn, D., N. Sigrist, F. Forterre, J. Howard, and D. Spreng. 2011. "Retrospective evaluation of post-operative nasotracheal tubes for oxygen supplementation in dogs following surgery for brachy-cephalic syndrome: 36 cases (2003–2007)." *J Vet Emerg Crit Care (San Antonio)* 21 (3):261–7. doi:10.1111/j.1476-4431.2011.00612.x.

Staffieri, F., A. Crovace, V. De Monte, P. Centonze, G. Gigante, and S. Grasso. 2014. "Noninvasive continuous positive airway pressure delivered using a pediatric helmet in dogs recovering from general anesthesia." *J Vet Emerg Crit Care (San Antonio)* 24 (5):578–85. doi:10.1111/vec.12210.

Stordalen, M. B., F. Silveira, J. V. H. Fenner, and J. L. Demetriou. 2020. "Outcome of temporary tracheostomy tube-placement following surgery for brachycephalic obstructive airway syndrome in 42 dogs." *J Small Anim Pract*. doi: 10.1111/jsap.13127.

Tamburro, R., B. Brunetti, L. V. Muscatello, C. Mantovani, and D. De Lorenzi. 2019. "Short-term surgical outcomes and histomorphological evaluation of thermal injury following palatoplasty performed with diode laser or air plasma device in dogs with brachycephalic airway obstructive syndrome." *Vet J* 253:105391. doi:10.1016/j.tvjl.2019.105391.

Tarricone, J., G. M. Hayes, A. Singh, and G. Davis. 2019. "Development and validation of a brachycephalic risk (BRisk) score to predict the risk of complications in dogs presenting for surgical treatment of brachycephalic obstructive airway syndrome." *Vet Surg* 48 (7):1253–1261. doi:10.1111/vsu.13291.

Ter Haar, G., and G. U. Oechtering. 2018. "Brachycephalic airway disease." In *BSAVA Manual of Canine and Feline Head, Neck and Thoracic Surgery*, 2nd edition, edited by D. J. Brockman, D. E. Holt and G. Ter Haar, 82–91. Gloucester: BSAVA.

Torrez, C. V., and G. B. Hunt. 2006. "Results of surgical correction of abnormalities associated with brachycephalic airway obstruction syndrome in dogs in Australia." *J Small Anim Pract* 47 (3):150–4. doi:10.1111/j.1748-5827.2006.00059.x.

Trader, R. 1949. "Nose operation." *J Am Vet Med Assoc* 114:210–211.

Trinterud, T., P. Nelissen, and R. A. White. 2014. "Use of silicone tracheal stoma stents for temporary tracheostomy in dogs with upper airway obstruction." *J Small Anim Pract* 55 (11):551–9. doi:10.1111/jsap.12267.

Trostel, C. T., and D. J. Frankel. 2010. "Punch resection alaplasty technique in dogs and cats with stenotic nares: 14 cases." *J Am Anim Hosp Assoc* 46 (1):5–11. doi:10.5326/0460005.

Villedieu, E., L. Rutherford, and G. Ter Haar. 2019. "Brachycephalic obstructive airway surgery outcome assessment using the 6-minute walk test: A pilot study." *J Small Anim Pract* 60 (2):132–5. doi:10.1111/jsap.12942.

White, R. N. 2012. "Surgical management of laryngeal collapse associated with brachycephalic airway obstruction syndrome in dogs." *J Small Anim Pract* 53 (1):44–50. doi:10.1111/j.1748-5827.2011.01156.x.

Worth, D. B., J. A. Grimes, D. A. Jimenez, A. Koenig, and C. W. Schmiedt. 2018. "Risk factors for temporary tracheostomy tube placement following surgery to alleviate signs of brachycephalic obstructive airway syndrome in dogs." *J Am Vet Med Assoc* 253 (9):1158–1163. doi:10.2460/javma.253.9.1158.

12 Ophthalmology in Practice for Brachycephalic Breeds

Màrian Matas Riera
Memvet – Centre de Referència

CONTENTS

Orbital, Periocular and Adnexal Conditions 201
  Orbital Conditions 202
    Lack of Enophthalmia 202
    Proptosis 202
    Euryblepharon 204
    Orbital Pneumatosis 205
    Nasal Folds 205
    Prolapse of the Nictitans Membrane Gland 206
  Eyelid Abnormalities 207
    Diamond Eye/Pagoda Eye 207
    Cilia Abnormalities 207
  Nasolacrimal System Abnormalities 209
    Dry Eye (Keratoconjunctivitis Sicca – KCS) 210
Corneal Conditions 211
  Dermoid 211
  Corneal Ulceration 211
  Melting Ulcer (keratomalacia) 212
  Corneal Abscess 213
  Corneal Pigmentation 213
  Corneal Neoplasia 213
Lens Abnormalities 214
  Cataracts 214
Glaucoma 215
Retinal Diseases 215
Conclusions 215
References 216

The extreme conformational changes selected for in brachycephalic breeds have resulted in a shortening of the orbital cavity, large palpebral apertures and prominent globes. This conformation in combination with reduction of corneal sensitivity in these breeds leads to severe ocular conditions in these patients. Most of the conditions seen in brachycephalic dogs can be related to exposure of the globe and its reduced protective measures. This chapter discusses common conditions seen in brachycephalic breeds to assist veterinarians managing these patients.

ORBITAL, PERIOCULAR AND ADNEXAL CONDITIONS

In canines with normal ocular conformation, some sclera (white of the eye) can be visualised adjacent to the lateral canthus. In some cases, this area of exposed sclera and overlying conjunctiva is

FIGURE 12.1   Note the marked difference of scleral show between a classical brachycephalic breed, a Pug, and a Golden Retriever. Note how the conjunctiva overlying the sclera in the Pug has become pigmented and is not seen as white, but as dark to brown in colour.

pigmented. However, in brachycephalic breeds scleral exposure may occur throughout nearly 360° of the cornea. This exposure is due to several conformational factors, the shortening of the orbit and the large palpebral fissure (Figure 12.1).

In this section, we will discuss some common periocular pathologies seen in brachycephalic breeds.

ORBITAL CONDITIONS

Lack of Enophthalmia

Enophthalmia is defined as a globe being positioned caudally into the orbital cavity leading to the nictitans membrane protrusion which partially covers the globe. Although not a condition *per se*, the author believes this leads to ocular problems in brachycephalic breeds. Enophthalmia is generally a sign of ocular pain and a way to protect the globe. The brachycephalic conformation and associated shortening of the orbital space restricts the mechanism of enophthalmia and protection of the globe. As a result, when an ocular surface condition or insult is present, the protection normally provided by the eyelids and nictitans membrane is absent.

Proptosis

Proptosis is defined as the rostral displacement of the globe where the eyelids are entrapped between the bony orbital rim and the globe (Figure 12.2). The key to diagnosing proptosis (as opposed to exophthalmia) is that the eyelid margin cannot be visualised as it is entrapped within the orbital bony rim and the rostrally displaced globe.

Brachycephalic breeds seem prone to traumatic proptosis; however, they seem to have better prognosis compared to dolichocephalic breeds when it occurs (Gilger et al. 1995). This is probably due to the greater intensity of the traumatic force needed for proptosis to occur in dolichocephalic versus brachycephalic breeds.

Proptosis may lead to extraocular muscle and even optic nerve avulsion. Rupture of two or more extraocular muscles is believed to be associated with a poorer visual prognosis (Ali and Mostafa 2019). Non-visual eyes can be retained for cosmetic reasons and will be surgically repositioned into the orbital cavity. When optic nerve avulsion accompanies, rupture of multiple extraocular muscles, enucleation is indicated.

Globe deviation is a common sequela after a proptosis. The medial rectus muscle which has the most rostral insertion is the most vulnerable to rupture. Hence, a common presentation of a patient with a proptosis is the presence of lateral globe deviation due to medial rectus muscle avulsion (Figure 12.3).

Pe'er et al. performed a retrospective evaluation of visual outcome after proptosis. The population in this study was 70% brachycephalic with Pekingese and Shih-Tzu being the most commonly

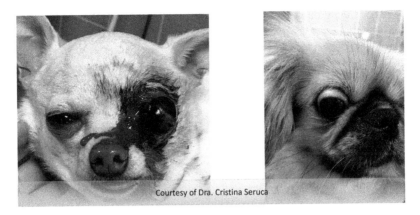

Courtesy of Dra. Cristina Seruca

FIGURE 12.2   Pictures of two patients with a left and right eye, respectively, proptosed. The picture on the left shows a severe proptosis, where the cornea is deviated dorsolaterally. This patient might even have the optic nerve avulsed. This eye will need to be enucleated. The picture on the right shows a brachycephalic dog, and the proptosis can be identified when the reader observes the absence of upper eyelid margin, in this case is traped behind the equator of the rostrally displaced globe. This globe has good visual prognosis if treated promptly.

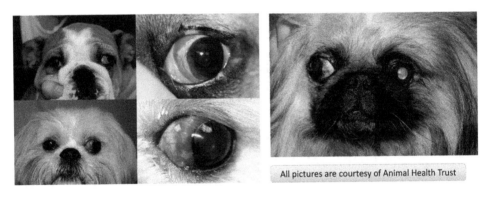

All pictures are courtesy of Animal Health Trust

FIGURE 12.3   These patients suffered from a proptosis of the left globe approximately one week before taking the pictures. These patients were referred to evaluate further damage of intraocular structures as they could not see well from the affected eye. Both patients had sequelae such as blindness and lateral globe deviation. In both globes, we can see conjunctival hyperaemia, and the bulbar conjunctiva can be seen torn in the English Bulldog. The patient on the right picture had a proptosis some weeks prior taking the picture, marked globe deviation can be seen. The English Bulldog and the Pekingese show some eyelid changes due to the temporal tarsorrhaphy placed by the referring veterinarian to control exposure of the globe while the orbital swelling reduced.

affected breeds. In this study, the presence of direct and consensual pupillary light reflexes (PLR) was the best visual prognostic indicator of the ophthalmic examination (Pe'er, Oron, and Ofri 2020). Interestingly in this study, breed was not a prognostic indicator of vision in proptotic globes in contrast to Gilger and colleagues' findings (Gilger et al. 1995).

When dealing with a proptosed globe, it is essential to assess the patient for other concomitant traumatic lesions, especially ruling out life-threatening conditions. Assessment of the globe is aimed to decide if replacement needs to be performed or enucleation is the best option for the patient. During this assessment and preparation of the instrumentation etc., it is recommended to constantly lubricate the globe. In case of doubt of the viability of the globe, it is always better to replace the globe into the orbital cavity and continue to monitor the globe function after the orbital and periocular swelling has reduced.

In order to replace the globe into the correct position, the patient should receive pain relief and be anaesthetised. Prepping the periocular area is advised. A temporal canthotomy is performed in order to enlarge the palpebral fissure; this will help eversion of the trapped eyelids, so the globe can be placed back into position. A generous flush with saline of the conjunctival sac is recommended and closure of the temporal canthotomy with a 4-0 or 6-0 suture performed. Fluorescein staining should be performed to assess for corneal ulceration and evaluate its size if present. Following this, a nictitans membrane flap and/or a large temporal tarsorrhaphy is performed. The author generally prefers the nictitans membrane flap as provides a more even pressure over the cornea. The patient will be recovered from the anaesthesia, pain relief will be continued (such as systemic non-steroidals) and systemic antibiosis considered depending on the presence of concurrent periocular lesions. If corneal ulceration is present, a topical broad spectrum antibiotic eyedrop – TID or ointment BID is generally recommended.

Euryblepharon

Euryblepharon is also known as macropalpebral fissure. Brachycephalic breeds tend to have large palpebral fissures leading to corneal exposure due to incomplete or inadequate closure of the eyes and therefore inadequate protection of the globe. In dolichocephalic dog breeds, we tend to see a small triangle of sclera laterally but no more than this. In brachycephalic breeds, the tendency is to see sclera nearly 360° around the cornea (Figure 12.4). This can lead to chronic irritation and secondary pigmentation of the sclera and cornea that becomes tinted with light to dark brown pigmentation (Figure 12.1).

Large palpebral fissures where exposed sclera is seen is associated with a nearly three times increased risk of corneal ulceration (Packer, Hendricks, and Burn 2015). From the results of this study, one could think that performing canthoplasties (narrowing of the palpebral fissure) would revert the risk of corneal ulceration; however, the association of large palpebral fissures with ulceration does not imply causality, in other words, does not imply that the only reason for this increased risk is the enlarged palpebral fissure. Therefore, it still needs to be demonstrated that shortening the palpebral fissure reduces the incidence of corneal ulceration. In the author's opinion, it is likely that this risk will not revert completely since the exposure of the globe is corrected but other anatomical and genetic traits of the breed are not.

Shortening of the palpebral fissure needs to be done under general anaesthesia. Brachycephalic breeds have a high risk to develop anaesthetic complications (Downing and Gibson 2018), so the author's personal opinion is not to proceed to surgery to shorten the palpebral fissure unless the patient is showing clear clinical signs. Once it has been scientifically demonstrated that this technique reduces the risk of corneal ulceration, offering this procedure to clients would be justified. However, when exposure is believed to be a cause of irritation, and other periocular conditions need

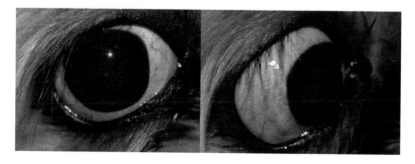

FIGURE 12.4   Pictures of the same patient's right eye. These pictures show large palpebral fissures and ocular surface exposure. In this case, the exposed conjunctiva is not pigmented as opposed to the patient in Figure 12.1.

to be surgically addressed, shortening of the palpebral fissure is a surgical technique that can be offered to clients at the same time as other procedures to reduce exposure risks.

There are many surgical techniques to reduce the palpebral fissure. The surgeon can either reduce the length by closing medial or lateral canthus or both in most severe cases. Resection of both eyelid margins at the canthi and closure in two layers is recommended. Although these techniques might seem relatively simple, the use of small suture diameter material (6-0) and the small surgical area requires magnification and referral is advised. Also, attention needs to be paid to the nasolacrimal system when performing medial canthoplasties. The author uses Allgoewer's medial canthoplasty technique (Allgoewer 2014), which solves medial canthal entropion and medial large palpebral fissure, and slits the lacrimal punta, moving it deeper and caudally within the conjunctival fornix, allowing for large closure of the medial palpebral aperture.

Orbital Pneumatosis

Orbital pneumatosis or emphysema is a rare orbital complication seen some weeks to months after enucleation. The few cases reported in the literature happen to be in brachycephalic breeds (Gornik, Pirie, and Alario 2015, Barros et al. 1984, Bedford 1979). The presentation is of an orbit filled with air following an otherwise unremarkable enucleation. Its pathophysiology is believed to be linked to patient's nasolacrimal system connecting the nasal cavity with the orbit and due to the increased inspiratory effort in brachycephalics, forcing air into the orbit (Figure 12.5). These cases can be diagnosed with palpation and/or aspiration of the content of the orbit which will be air. These patients need to undergo exploratory orbitotomy (ideally with magnification using surgical loupes) to identify conjunctival lining of the nasolacrimal puncta, removal of this epithelium and closure of these two puncta. This procedure should be curative (Gornik, Pirie, and Alario 2015).

Nasal Folds

The presence of large nasal folds tends to cause multiple ophthalmic problems. The skin hidden under the fold can be a cause of periocular skin infection and owners need to be advised regarding daily hygiene routines to clean this region to reduce/control infection and yeast/bacterial

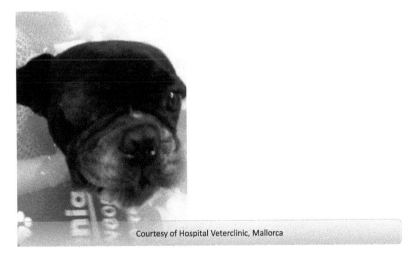

Courtesy of Hospital Veterclinic, Mallorca

FIGURE 12.5   This patient had the right eye enucleated due to a corneal perforation 3 months prior this picture. Swelling of the area was noticed by the veterinarian and mild emphysema could be palpated on the adjacent areas. On palpation, it was not warm nor the patient showed major discomfort. Air could be palpated (tympanic sound), and on aspiration of the cavity with a fine needle, air was retrieved. Surgery was performed, and tissue compatible with conjunctival lining was present around the upper nasal punctum; both puncta were closed with reabsorbable suture, and the skin was closed routinely.

overgrowth. In some cases, the nasal skin folds cover the external nares of the patient which can lead to further breathing difficulties.

Another problem that nasal folds can cause in long haired patients is trichiasis. The hairs of the fold can contact the ocular surface leading to corneal irritation or ulceration. In these patients, resecting the nasal fold may be an option to control the trichiasis. However, before considering this type of surgery is important to identify if there is concomitant medial canthal entropion or eury-blepharon so both can be treated surgically at the same time.

Even if none of these problems are detected during the physical examination, it is worth noticing that a study showed that dogs with nasal folds are nearly five times more likely to be affected by corneal ulcers than those without (Packer, Hendricks, and Burn 2015). The same comment as discussed in regard to euryblepharon (large palpebral fissure) would be applicable. This link between the presence of nasal folds with corneal ulceration does not imply direct causality, so a study would be needed to identify if resection of the fold reduces the risk of corneal ulceration.

Treatment to correct nasal folds is surgical. The author recommends resecting the fold if it is large, interfering with vision, causing severe dermatitis under the fold or causing trichiasis. Surgical technique is paramount for this resection to be cosmetically acceptable, reducing the visible scar in the skin of the face of these patients. Owners commonly hesitate to accept this type of surgery as it will change the appearance of the face of their pet. So, aiming to have a minimal scar and good cosmetic outcome is important in these cases to improve owner satisfaction and promote these types of surgeries when needed within the population (Figure 12.6).

Prolapse of the Nictitans Membrane Gland

Prolapse of the nictitans membrane gland (PNMG) is informally called 'cherry eye'. The nictitans membrane or third eyelid has an inverted triangular shape. At the ventral aspect of this membrane, there is a lacrimal gland which in some patients becomes dislodged from its normal anatomical position and protrudes ventro-medially through the palpebral fissure.

PNMG is not exclusive of brachycephalic breeds. Mazzurcheli et al. reported the French Bulldog, Shar Pei, Great Dane, English Bulldog and Cane Corso as the breeds that are more commonly

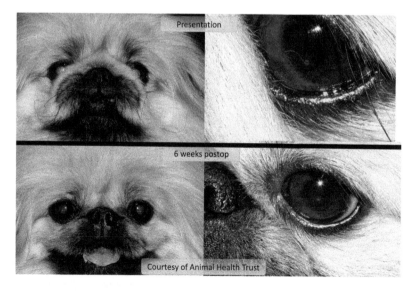

FIGURE 12.6   Three-year-old Pekingese. Pictures on top show presence of a large nasal fold and trichiasis (hairs from nasal fold touching the left cornea) leading to visual impairment, epiphora, chronic irritation. This patient underwent nasal fold removal to correct the trichiasis, and pictures at the bottom are 5 weeks after the surgery showing no trichiasis and minimal interference with the patient's visual axis.

affected by PNMG bilaterally (although not always at the same time) (Mazzucchelli et al. 2012). Another paper reported English Bulldogs, Cane Corsos and crossbreeds as the first three most common breeds to develop PNMG (Multari et al. 2016).

The classical treatment technique for these patients is the Morgan and Moore technique (the commonly called 'Pocket technique') (Morgan, Duddy, and McClurg 1993, Moore, Frappier, and Linton 1996) either alone or combined with an anchoring technique to the orbital rim. Several different surgical approaches are reported in the literature, and different veterinary ophthalmologists will have their preferences. The author usually prefers the Morgan and Moore technique as the initial procedure in most cases, finding it to be the least traumatic and generating good outcomes. This is true except in very enlarged glands in large breeds or relapses of the condition, where combining the Morgan and Moore technique with an orbital rim anchoring technique should be considered. Other techniques can be found in the literature such as suture anchor placement technique around the insertion of the ventral rectus muscle by Sapienza et al. or the perilimbal pocket technique by Prémont et al. (Sapienza, Mayordomo, and Beyer 2014, Prémont et al. 2012). The author uses topical treatment with an eye ointment that combines a broad-spectrum antibiotic and steroid anti-inflammatory twice or three times daily 2–3 days prior to surgery, and the same regime is continued for 5–7 days after the surgery. All patients will have their corneas fluorescein stained prior discharge, a test that should be used after any periocular surgical procedure.

Eyelid Abnormalities

The most common eyelid abnormalities seen in brachycephalic breeds tend to be medial canthal entropion. Extra eyelashes will be discussed later in the chapter (section 'Cilia Abnormalities'), and larger palpebral fissures have been discussed before (section 'Euryblepharon').

Diamond Eye/Pagoda Eye

This condition is most commonly seen in large mesocephalic and brachycephalic breeds including the St. Bernard, Clumber Spaniel and Rottweiler. This condition is due to a large palpebral fissure and an excess of loose skin on the head which results in a combination of entropion and ectropion in both upper and lower eyelids. The excess skin periocularly and from the forehead can cause difficulties opening the eyes and the eyelids rolling onto the surface of the eye leading to entropion (Figure 12.7). Ectropion of the central lower eyelid, due to excess skin, and its weight is also commonly seen. This type of conformation leads to eyelids unable to glide over the surface of the cornea, with areas of entropion and ectropion. Surgical treatment is recommended in all patients where entropion or ocular surface irritation is identified. These patients generally require a combination of several eyelid shortening and entropion techniques in order to restore good eyelid function. These patients would ideally benefit from referral to an ophthalmologist as the eyelid reconstruction can be an extremely difficult surgery with variable outcomes depending on the dog's conformation and the owner expectations need to be managed.

Cilia Abnormalities

This section will discuss the ophthalmic issues associated with aberrant periocular cilia and clarify the often-confusing terminology used to describe the various conditions commonly seen in practice (Figure 12.8). Although these conditions do not exclusively occur in brachycephalic breeds, they are commonly affected.

Distichiasis

Distichiasis occurs when hairs emerge from the meibomian gland opening and can be visualised when examining the eyelid margin. These cilia generally appear as multiple cilia along the meibomian gland openings in both upper and lower eyelids of the patients. The author evaluates how much discomfort these appear to cause, by examining the surface of the cornea for pigmentation or

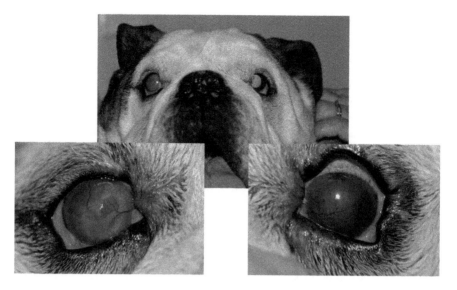

FIGURE 12.7 English Bulldog with *mild* pagoda eye or diamond eye conformation where a combination of entropion and ectropion occurs. This patient also had chronic dry eye that was non-responsive to treatment. The left eye shows scarring from chronic corneal surface disease (combination of dry eye and eyelid abnormalities) in addition to a spontaneous chronic corneal epithelial defect (formerly known as Indolent ulcer) that caused this marked fibrosis and neovascularisation. The left eye shows clearly how the upper medial eyelid cilium are touching the cornea. There is also corneal opacity, superficial neovascularisation secondary to the ocular surface problem this patient has.

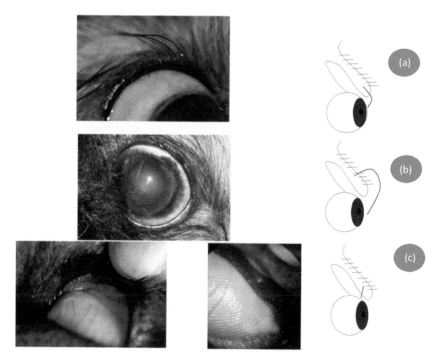

FIGURE 12.8 In this figure, we try to depict the three common cilia abnormalities seen in brachycephalic dogs (a). Distichiasis, hair coming out of the meibomian gland opening on the eyelid margin (b). Trichiasis: hairs from a normal position but their length or direction leads to corneal contact and irritation (c).

erosive changes detected with fluorescein stain. Also, local anaesthetic is applied to evaluate if discomfort is reduced. Another approach is to epilate the distichia under local anaesthetic and evaluate if the clinical signs improve during the first weeks. The owners need to be informed that this is a temporary treatment and cilia will regrow after 4–6 weeks. If clinical signs re-appear some weeks after the epilation, then treatment is indicated. Several treatments have been described to treat distichiasis: cryosurgery, LASER treatment, electrolysis and diathermia. The selection of treatment depends on the ophthalmologist preference and experience. To date, there is not a sole treatment recommended for this condition, and many of them are associated with regrowth to some degree. This regrowth may or may not be associated with clinical signs necessitating a second treatment.

A recent treatment has been published, describing partial tarsal plate excision via a transconjunctival approach (Palella Gómez et al. 2020). This treatment involves removing a strip of conjunctiva where the hair follicles of the distichia originate. This treatment seems to have a good long-term outcome, although the study showed effects on tear film quality shortly after the procedure. This is believed to be due to damage of the meibomian glands at the time of the excision (Palella Gómez et al. 2020).

Ectopic cilia

Ectopic cilia occurs when cilia emerge through the conjunctiva. Generally, these are seen on the upper eyelid approximately 4–6 mm from the eyelid margin. They are generally identified in young dogs, and the reason for consultation tends to be corneal ulceration (generally affecting the upper half cornea) that does not heal with usual corneal ulceration treatment (broad spectrum topical antibiosis). Everting the upper eyelid and examining with magnification the surface of the conjunctiva will help identifying an ectopic cilium. Close examination will be needed as sometimes more than one foci of cilia can be seen. These have to be removed en bloc under general anaesthesia. A chalazion clamp is placed to facilitate haemostasis and a stable surface to work on. A circular or square shape resection of the cilia and its follicle is performed with a punch biopsy or surgical knife, respectively (D'Anna et al. 2007).

Trichiasis

Trichiasis is where cilia grow in a normal position but in a direction that leads to irritation of the ocular surface. These cilia arising from a normal location but causing discomfort or corneal lesions need to be addressed. Treatment varies according to the origin of the cilia. Sometimes, shortening the palpebral fissure can help reducing the exposure. Training owners to try and groom periocular hairs away from the ocular surface, with the help of an oil base ophthalmic lubricant (as if it was hair gel), can sometimes help reducing the effects of trichiasis.

Nasolacrimal System Abnormalities

The nasolacrimal system is a combination of approximately 1 mm diameter ducts that drain tears away from the ocular surface to the nose. There is an upper and lower lacrimal puncta, which are the openings of the system, located on the medial canthus 4–6 mm from the medial canthi. The puncti tend to have a pigmented border which can sometimes help identifying them. Both ducts meet in the lacrimal sac which lies in the lacrimal fossa (indentation of the lacrimal bone, just medially over the maxillary bone). From the lacrimal sac, the lacrimal duct extends to the nose opening at the nasal ostium. The nasal ostium can be seen in large breed dogs, although its identification can be challenging. Puncta abnormalities include

1. Micropunctum: Small dorsal or ventral puncti can be seen, and these are difficult to cannulate as only fine cannula might be inserted. (Figure 12.9)
2. Imperforated punctum: The duct has been formed, but an epithelial lining of conjunctiva is covering the opening.

FIGURE 12.9   Picture on the left, green circle indicating the lower micropunctum. This patient had all four puncta smaller in size. This lead to chronic epiphora. Cannulation under topical anaesthesia was difficult due to the small size of the openings. The picture on the right is depicting in green how large the author would have expected the opening to be in this patient, so the reader can compare.

In both situations, treatment will be recommended if the patient is showing clinical signs such as epiphora or periocular wetting. Given that these clinical signs are highly non-specific, evaluation of the nasolacrimal duct drainage with a Jones test is recommended. This test is performed applying a drop of fluorescein stain on the ocular surface. It is considered positive if fluorescein is seen coming out of the nose or in very few cases in the oropharynx. The Jones test is negative if fluorescein is seen on the ocular surface and periocular area but not from the ipsilateral nostril or oropharynx. Generally, the author waits 5–10 minutes for concluding the test.

A further consideration for brachycephalic breeds is the compression of the nasolacrimal system (together with the muzzle) in the longitudinal axis which may impede normal nasolacrimal drainage even in cases with normally formed (patent) nasolacrimal systems. A study in feline brachycephalic cats demonstrating the anatomy of the nasolacrimal system by Schlueter et al. showed how the nasolacrimal system in these breeds leads to a V-shaped nasolacrimal duct and therefore an inefficient function (Schlueter et al. 2009). To the author's knowledge, anatomical and advance imaging evaluation of the nasolacrimal system in brachycephalic dogs has not been performed. Many brachycephalic breeds have negative Jones test results, but when the upper punctum is cannulated and flushed with saline, patency of the duct can be demonstrated. Given this, the abnormal anatomy of the duct may be the cause of abnormal nasolacrimal drainage in some cases with the lack of nasolacrimal drainage being functional rather than secondary to obstruction.

Dry Eye (Keratoconjunctivitis Sicca – KCS)

Dry eye is defined as a reduction of normal tear production leading to ocular surface disease such as increased mucoid discharge, conjunctival hyperaemia and corneal neovascularisation. Normal tear readings with the Schirmer tear test (STT) is from 15 to 25 mm/1 min per eye. Readings below this range in patients with clinical signs are considered diagnostic for KCS and immediate treatment is recommended. Patients with clinical signs and readings on the low side of the normal range need to be monitored very closely.

Causes of KCS can vary, including congenital, drug-induced (e.g. topical atropine), drug toxicity (e.g. paracetamol or trimethoprim sulfamides), immune-mediated, irradiation, neurogenic, surgically induced/iatrogenic, systemic disease (e.g. distemper, diabetes mellitus). The most common cause of KCS in dogs is immune-mediated, especially in breeds believed to be predisposed to develop this problem. Many brachycephalic breeds are included in the list of breeds reported to be predisposed to immune-mediated dry eye: Cavalier King Charles Spaniel, English Bulldog, Lhasa Apso, Shih-Tzu, Pug, Pekingese and Boston Terrier (Giuliano and Moore 2007).

Brachycephalic breeds have lower tear production compared with non-brachycephalic breeds (Bolzanni et al. 2020). The predisposition to develop immune-mediated dry eye together with the increased evaporation of the tear film due to the large palpebral fissure is likely to be responsible for the chronic ocular surface disease in many of these patients (Figure 12.7).

CORNEAL CONDITIONS

DERMOID

This lesion is defined as an area where ectopic normal tissue is present in an abnormal location. Dermoids, due to their ectodermal origin, can have skin, hair follicles and glands. They can appear affecting the cornea exclusively or affecting the lateral canthus and lateral conjunctiva. Although any breed can be affected, it seems that French Bulldogs might be predisposed to develop ocular/periocular dermoids.

The diagnosis is made generally within the first weeks or months of age, as they appear obvious to owners, especially those dermoids with long hairs emerging from them. Surgery to resect the dermoid is recommended, referral should be advised, especially for corneal dermoids, as some of these patients might need a grafting procedure in addition to resection of the dermoid.

CORNEAL ULCERATION

Due to their conformation, shallow orbits and exposed globes, brachycephalic dogs are often seen in practice with corneal disease. In an epidemiological study of prevalence and predispositions of corneal ulcers, breeds with the highest prevalence of corneal ulcers included the Pug (5.42% of the breed affected), Boxer (4.98%), Shih-Tzu (3.45%), Cavalier King Charles Spaniel (2.49%) and English Bulldog (2.41%). Overall, brachycephalic breeds had 11.18 times the odds for corneal ulcers compared with crossbreds (O'Neill et al. 2017).

Corneal ulceration can occur and progress rapidly in any breed, however due to the multifactorial predispositions to ulceration in brachycephalic dogs (including increased exposure of the cornea, incomplete palpebral closure, poor tear film stability and medial canthal entropion), rapid deterioration of corneal ulceration is often observed, with melting ulcers potentially leading to perforation within 24 hours. For this reason, owners should be advised to seek veterinary advice if ocular discomfort occurs in brachycephalic breeds. Brachycephalic breeds should be examined by a veterinary surgeon every 24–48 hours for the first 2–3 days after diagnosing an ulcer, to make sure there is no deterioration (Figure 12.10).

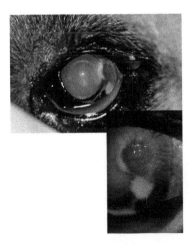
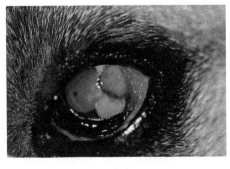

FIGURE 12.10   This patient has developed a SCCEDs: spontaneous chronic corneal epithelial defect. This type of ulcer is due to a hyaline membrane deposited over the stroma interfering the epithelium normal healing. These ulcers are the only ones requiring corneal debridement once ectopic cilia or foreign bodies have been ruled out. Picture on the left shows the initial fluorescein stain and how there was fluorescein underrunning the edge of the epithelium. Picture on the right, after the corneal debridement.

Ulcers can be complicated by infection including bacteria (rods or cocci) or far less commonly fungi (hyphae or yeasts). Corneal cytology might help to identify organisms within the ulcer bed and treat accordingly. Cytology results performed as an in-house technique are immediate and can provide veterinary surgeons with initial information to decide and create a treatment plan for corneal ulceration depending on severity, clinical signs and cytology results.

Any ulceration where there is fluorescein uptake needs to be covered with an antibiotic, such as fusidic acid ointment BID, chloramphenicol ointment BID or drops TID, and tetracycline eye ointment BID. When infiltration of the ulcer occurs with loss of stromal transparency and cream to yellow infiltrates, then the corneal cytology will help us identify if there is an infectious agent complicating the ulceration. Treatment will be tailored depending on this in-house test. Corneal microbiology is indicated to confirm the growth and sensitivity of the microbe. In presence of an infiltrated complicated ulcer: tobramycin, ofloxacin or gentamycin will be used topically at intensive rates in order to control bacterial growth. In those cases, the use of antimetalloproteases will certainly be beneficial in an attempt to control the secondary damage of the corneal structures (see section below).

Melting Ulcer (keratomalacia)

Keratomalacia is an acute corneal condition where enzymatic activity breaks down the collagen of the corneal stroma; this destruction can be exogenous (enzymatic activity from bacteria) or endogenous (enzymes released by the patient). Brachycephalic breeds, due to globe exposure, appear to be predisposed to corneal melting, with the Shih-Tzu, Pekingese and Pugs appearing to be predisposed to severe corneal melting (Dubielzig et al. 2010). Some of these patients can present with rapidly progressing corneal melting disease that will need intensive or even surgical management to try and preserve vision (Figure 12.11).

These patients generally benefit from corneal cytology to try and differentiate between a septic or an aseptic melting ulcer. Corneal melting can occur due to bacterial infection with *Pseudomonas* spp. typically reported in the literature, but development of the corneal melting is not exclusively associated with this bacterium. Corneal melting can also be aseptic in origin where the enzymatic activity is endogenous from epithelial cells and keratocytes. This differentiation helps to focus the treatment plan on targeting the infection and controlling the enzymatic activity, or treatment efforts can be focused, mainly controlling the enzymatic activity.

Many antiproteases have been described in the literature; the most commonly used are serum from the same patient or from a healthy donor, EDTA 1%, *N*-acetylcysteine 5%–10%. Doxycycline does have antimetalloprotease activities, and some ophthalmologists use this, as it has been shown to be secreted through the meibomian glands; therefore, it probably reaches the tear film. Doxycycline is used at regular doses in these cases.

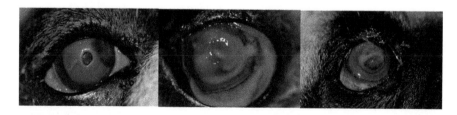

FIGURE 12.11　These three different patients presented with (a) Descemetocele affecting the axial cornea. (b) Melting ulcer per-acute. (c) Melting ulcer with perforation: note that the iris is in contact with the cornea indicating a loss of aqueous humour.

Corneal Abscess

This is a relatively uncommon condition that is seen in dogs, where there is an accumulation of inflammatory cells within the corneal stroma. Secondary to this lesion, the patient develops reflex uveitis and secondary corneal oedema of the same eye. The clinical presentation includes blepharospasm, conjunctival hyperaemia, corneal oedema and miosis. As the lesion is located within the stroma of the cornea, many times the lesion is fluorescein negative. The veterinary surgeon will need to carefully examine the cornea to identify an area or denser opacity, which will be the abscess. Examination with magnification in these cases is highly advised. A differential diagnosis for a lesion within the stroma is an epithelial inclusion cyst; however, this condition does not tend to cause discomfort or reflex uveitis as a corneal abscess does. Patients diagnosed with a corneal abscess will need to be prescribed an antibiotic that penetrates the epithelium of the cornea, such as chloramphenicol, tobramycin or ofloxacin and applied from four to six times daily. Atropine to relax the ciliary muscle spasm once daily and oral anti-inflammatories systemically can be recommended. If medical treatment is not successful within a few days, then surgical intervention might be needed, and referral is advised.

These patients should improve within 24–48 hours where corneal oedema will reduce in intensity and the patient will show signs to be more comfortable.

Corneal Pigmentation

Pigment migration over the cornea is generally believed to occur in chronic inflammatory conditions such as chronic superficial keratitis (Pannus or Uberreiter syndrome), a lymphoplasmacytic infiltration of the ventrolateral quadrant of the cornea and its limbus. This condition is typically described in German Shepherd Dogs and Greyhounds, although it can affect many other breeds.

However, pigment can be seen in brachycephalic breeds without a clinically obvious associated inflammatory response. In these patients, the pigment tends to be an under-diagnosed change in which the superficial cornea progressively develops a pigment layer, which can be of variable density. This pigment deposition can be an incidental finding on a routine examination or can cause serious visual deficits. This will depend on how dense the pigment is and if there are other concomitant ophthalmic issues interfering with sight (which might not be identifiable due to the corneal changes). Visual compromise may vary from negligible in those with faint corneal pigmentation, to complete blindness in patients where the pigment is dense, particularly if the visual axis is involved. Examination of the intraocular structures may be completely obscured in these severe cases (Figure 12.12).

Corneal Neoplasia

Squamous cell carcinoma (SCC), although a highly uncommon condition, is seen in chronic dry eye cases that have had long-term immunosuppressive topical treatment prior to the diagnosis. (Dreyfus, Schobert, and Dubielzig 2011). Of these 26 cases published in 2011, 20 were brachycephalic breeds (Pugs, Cavalier King Charles Spaniels, English Bulldogs). Many of these brachycephalic breeds are predisposed to chronic ocular surface inflammation due to exposure, which can be aggravated by dry eye which often require medication with immunosuppressants topically, either cyclosporine or tacrolimus. Although the trigger for corneal SCC is not clear, it appears that patients with chronic corneal inflammatory conditions and/or topical immunosuppressive treatment may be at risk of developing corneal SCC.

The appearance of SCC is of a proliferative lobular irregular mass on the corneal surface. A corneal cytology can help with the diagnosis. Treatment requires referral to an ophthalmologist for superficial keratectomy and adjunctive treatment such as cryosurgery, plesiotherapy with Strontium 90, or photodynamic therapy has been described. Metastatic disease appears to be rare (Dreyfus, Schobert, and Dubielzig 2011).

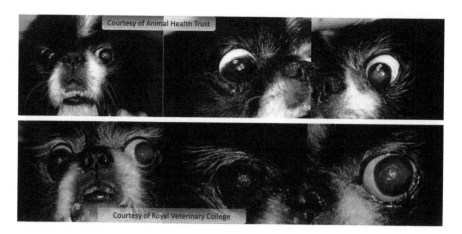

FIGURE 12.12 Series of pictures on top from a 9-year-old Pekingese, diagnosed with dry eye, trichiasis and euryblepharon. The patient was not followed up by the ophthalmologist but continued with treatment of cyclosporine 0.2% BID, until seen in another referral centre for another non-ocular condition 4 years later. Notice how there is marked corneal pigmentation of the right eye with keratosis of the cornea; this eye was blind. The left eye shows the same periocular changes as before but no pigment on the cornea. Dry discharge was adherent on the axial surface of the cornea typically seen in patients with dry eye.

Hemangiosarcoma appears infrequently in the literature as case reports. Shank et al. (2019) retrospectively reviewed the histopathology of cases with vascular corneal tumours in dogs. Fourteen dogs were identified, and although in the article discussion, the authors mention that dogs with chronic surface disease can be predisposed to vascular tumours, such as brachycephalic breeds, no brachycephalic breeds were within the 14 cases described in the article (one dog was of unknown breed) (Shank, Teixeria, and Dubielzig 2019).

The next part of the chapter includes abnormalities associated with brachycephalic breeds; however, these abnormalities are unlikely to be linked to the extreme conformation of these breeds.

LENS ABNORMALITIES

Lens abnormalities are generally described in two categories. First, and most commonly seen, is loss in transparency of part of, or a complete area of the lens. Any loss of transparency is called cataract. Second are those abnormalities due to a malpositioning of the lens, generally due to lesions on the filaments that support the lens in its 360°, called zonula filaments.

Cataracts

Two forms of cataracts, early and late onset, are confirmed heritable in Boston Terriers (Mellersh et al. 2006, 2007), and cataracts are suspected to be heritable in the French Bulldog (Chaudieu et al. 2011). Other brachycephalic breeds where cataracts are suspected hereditary are the Shih-Tzu, Pug, Boxer and English Bulldog (Davidson and Nelms 2013). This list probably will increase as more is known about the natural history of this condition in other breeds.

At the moment, cataract development is an irreversible condition and the only treatment available to regain vision in an eye that has developed a cataract is by performing cataract surgery, also called phacoemulsification of the lens. Those patients that might not be candidates to undergo cataract surgery are also advised to seek veterinary ophthalmology attention since untreated cataracts can lead to multiple ocular conditions including lens induced uveitis, chronic glaucoma, retinal detachment and lens luxation all of which can be sources of pain. Patients that cannot undergo cataract removal are recommended to have an annual visit to the ophthalmologist in order to identify any secondary ophthalmic problems. The author elects to prescribe a long-term topical NSAID (ketorolac) once or twice daily as a palliative measure to reduce the intraocular inflammation cataracts can cause.

Although genetic testing is a highly important and useful tool in breeding decisions, it is important to understand its limitations when used in our daily patients. The genetic test will tell us if the patient is carrier, affected or free of the mutation of the specific test; however, the patient can have another type of cataracts that might not be genetic, i.e. traumatic or senile. This is a useful tool for breeders, although it is important to remind ourselves that genetic testing cannot overcome the ophthalmic examination by a board-certified ophthalmologist to make sure no other diseases are present prior considering generating offspring from a dog.

GLAUCOMA

Glaucoma is a complex disease where vascular perfusion of the globe, intraocular pressure (IOP) and optic nerve alterations are seen. The main tool to estimate the intraocular pressure is the tonometer; the commonly used in clinical settings are TonoVet and Tonopen. There is a certain IOP in each patient that can lead to damage of the nerve fibre layer which forms the axons of the optic nerve and reaches the central nervous system. Primary glaucoma is believed to be developed when there is an abnormality of the drainage of the aqueous humour from the eye, leading to high intraocular pressure and reduced perfusion of the ocular tissues. Secondary glaucoma is generally caused by chronic inflammation of the eye or by intraocular tumours.

Chihuahua, Dandie Dinmont Terrier and Shih-Tzu seem to be predisposed to develop primary glaucoma. Boston Terrier, Lhasa Apso and Pekinges seem to have high prevalence of primary glaucoma.

Glaucoma is a complex disease, and the role of the ophthalmologist is to diagnose and treat accordingly in order to maintain the intraocular pressures below 20 mmHg. This is generally achieved either by hypotensive medication, or ablation surgeries (surgical destruction of part of the ciliary processes that produce aqueous humour), or by increasing the outflow of aqueous humour out of the eye (surgically placing shunts that allow aqueous humour to exit the anterior chamber to an area under the conjunctiva, called gonioimplant). These surgeries need improvement to be successful, but these together with the topical medication are currently the only tools to manage this condition. Without treatment, glaucoma will lead to a blind and potentially painful eye. Coaching the owners and controlling expectations are important when managing this frustrating disease.

RETINAL DISEASES

There is an increasing number of genetic abnormalities leading to retinal disease (commonly retinal degeneration) in many breeds of dogs. A recent genetic abnormality has been identified in the Lhasa Apso which leads to progressive retinal atrophy (Hitti-Malin et al. 2020). However, little clinical information is available for this disease at the time of writing this chapter.

CONCLUSIONS

The orbit and palpebral fissures protect the globe, and the consequences of selecting for aberrant variants of these essential structures are far reaching for ophthalmological health in the dog. Brachycephalic breeds face a lifetime of increased risk to a range of uncomfortable, painful and/ or blinding disorders due to their desired conformational features. Many of the disorders described in this chapter likely require conformational changes to reduce their prevalence of, e.g. for corneal ulcers, shorter eyelid openings and reduced or absent nasal folds. In the future, these more moderate conformations should be achieved via ethical breeding decisions rather than necessitating surgical intervention. Prompt diagnosis and treatment of many of the disorders described in this chapter is essential, and veterinary professionals should strive to ensure owners of brachycephalic breeds are aware of common ophthalmic disorders in their dog at a young age, to promote patient welfare.

REFERENCES

Ali, K. M., and A. A. Mostafa. 2019. "Clinical findings of traumatic proptosis in small-breed dogs and complications associated with globe replacement surgery." *Open Veterinary Journal* 9 (3):222–229.

Allgoewer, I. 2014. "A simplified medial canthoplasty technique." Annual Scientific Meeting of the European College of Veterinary Opthalmologists, Barcelona, Spain, May 16–19, 2013.

Barros, M., J.M. Matera, J. Alvarenga, and M. Iwasaki. 1984. "Orbital pneumatosis in a dog." *Modern Veterinary Practice* 65 (1):38.

Bedford, P.G.C. 1979. "Orbital pneumatosis as an unusual complication to enucleation." *Journal of Small Animal Practice* 20 (9):551–555.

Bolzanni, H., A.P. Oriá, A.C.S. Raposo, and L. Sebbag. 2020. "Aqueous tear assessment in dogs: Impact of cephalic conformation, inter-test correlations, and test-retest repeatability." *Veterinary Ophthalmology* 23 (3):534–543.

Chaudieu, G., P.H. Pilorge, S. Chahory, J.P. Jegou, C. Mellersh, and A. Thomas. 2011. "Primary cataract in the french bulldog: A preliminary report." Annual Meeting of the European College of Veterinary Ophthalmologists, Berlin, Germany, May 28–29.

D'Anna, N., J.S. Sapienza, A. Guandalini, and A. Guerriero. 2007. "Use of a dermal biopsy punch for removal of ectopic cilia in dogs: 19 cases." *Veterinary Ophthalmology* 10 (1):65–67.

Davidson, M.G., and S.R Nelms, eds. 2013. *Diseases of the Lens and Cataract Formation*. Edited by Kirk N Gelatt, Brian C Gilger and Thomas J Kern, *Essentials of Veterinary Ophthalmology*. John Wiley & Sons, Hoboken, NJ.

Downing, F., and S. Gibson. 2018. "Anaesthesia of brachycephalic dogs." *Journal of Small Animal Practice* 59 (12):725–733.

Dreyfus, J., C.S. Schobert, and R.R. Dubielzig. 2011. "Superficial corneal squamous cell carcinoma occurring in dogs with chronic keratitis." *Veterinary Ophthalmology* 14 (3):161–168.

Dubielzig, R.R., K.L. Ketring, G.J. McLellan, and D.M. Albert. 2010. Veterinary Ocular *Pathology: A Comparative Review*. Elsevier Health Sciences, Edinburgh.

Gilger, B.C., H.L. Hamilton, D.A. Wilkie, A. Van der Woerdt, S.A. McLaughlin, and R.D. Whitley. 1995. "Traumatic ocular proptoses in dogs and cats: 84 cases (1980–1993)." *Journal of the American Veterinary Medical Association* 206 (8):1186–1190.

Giuliano, E.A., and C.P. Moore. 2007. "Diseases and surgery of the lacrimal secretory system." *Veterinary Ophthalmology* 4:633–61.

Gornik, K.R., C.G. Pirie, and A.F. Alario. 2015. "Orbital and subcutaneous emphysema following enucleation and respiratory distress in a Japanese Chin." *Journal of the American Animal Hospital Association* 51 (6):413–418.

Hitti-Malin, R.J., L.M. Burmeister, S.L. Ricketts, T.W. Lewis, L. Pettitt, M. Boursnell, E.C. Schofield, D. Sargan, and C.S. Mellersh. 2020. "A LINE-1 insertion situated in the promoter of IMPG2 is associated with autosomal recessive progressive retinal atrophy in Lhasa Apso dogs." *BMC Genetics* 21 (1):100. doi:10.1186/s12863-020-00911-w.

Mazzucchelli, S., M.D. Vaillant, F. Wéverberg, H. Arnold-Tavernier, N. Honegger, G. Payen, M. Vanore, L. Liscoet, O. Thomas, and B. Clerc. 2012. "Retrospective study of 155 cases of prolapse of the nictitating membrane gland in dogs." *Veterinary Record* 170 (17):443–443.

Mellersh, C.S., K.T. Graves, B. McLaughlin, R.B. Ennis, L. Pettitt, M. Vaudin, and K.C. Barnett. 2007. "Mutation in HSF4 associated with early but not late-onset hereditary cataract in the Boston terrier." *Journal of Heredity* 98 (5):531–533.

Mellersh, C.S., L. Pettitt, O.P. Forman, M. Vaudin, and K.C. Barnett. 2006. "Identification of mutations in HSF4 in dogs of three different breeds with hereditary cataracts." *Veterinary Ophthalmology* 9 (5):369–378.

Moore, C.P., B.L. Frappier, and L.L. Linton. 1996. "Distribution and course of ducts of the canine third eyelid gland: Effects of two surgical replacement techniques." *Veterinary & Comparative Ophthalmology* 6:258–264

Morgan, R.V., J.M. Duddy, and K. McClurg. 1993. "Prolapse of the gland of the third eyelid in dogs: A retrospective study of 89 cases (1980 to 1990)." *The Journal of the American Animal Hospital Association* 29:56–60.

Multari, D., A. Perazzi, B. Contiero, G. De Mattia, and I. Iacopetti. 2016. "Pocket technique or pocket technique combined with modified orbital rim anchorage for the replacement of a prolapsed gland of the third eyelid in dogs: 353 dogs." *Veterinary Ophthalmology* 19 (3):214–219.

O'Neill, D.G., M.M. Lee, D.C. Brodbelt, D.B. Church, and R.F. Sanchez. 2017. "Corneal ulcerative disease in dogs under primary veterinary care in England: Epidemiology and clinical management." *Canine Genetics and Epidemiology* 4 (1):5. doi:10.1186/s40575-017-0045-5.

Packer, R.M.A., A. Hendricks, and C.C. Burn. 2015. "Impact of facial conformation on canine health: Corneal ulceration." *PLoS One* 10 (5):e0123827. doi:10.1371/journal.pone.0123827.

Palella Gómez, A., S. Mazzucchelli, E. Scurrell, K. Smith, and R.P. de Lacerda. 2020. "Evaluation of partial tarsal plate excision using a transconjunctival approach for the treatment of distichiasis in dogs." *Veterinary Ophthalmology* 23 (3):506–514.

Pe'er, O., L. Oron, and R. Ofri. 2020. "Prognostic indicators and outcome in dogs undergoing temporary tarsorrhaphy following traumatic proptosis." *Veterinary Ophthalmology* 23 (2):245–251.

Prémont, J.E., S. Monclin, F. Farnir, and M. Grauwels. 2012. "Perilimbal pocket technique for surgical repositioning of prolapsed nictitans gland in dogs." *Veterinary Record*. doi:10.1136/vr.100582.

Sapienza, J.S., A. Mayordomo, and A.M. Beyer. 2014. "Suture anchor placement technique around the insertion of the ventral rectus muscle for the replacement of the prolapsed gland of the third eyelid in dogs: 100 dogs." *Veterinary Ophthalmology* 17 (2):81–86.

Schlueter, C., K.D. Budras, E. Ludewig, E. Mayrhofer, H.E. Koenig, A. Walter, and G. Oechtering. 2009. "Brachycephalic feline noses: CT and anatomical study of the relationship between head conformation and the nasolacrimal drainage system." *Journal of Feline Medicine and Surgery* 11 (11):891–900.

Shank, A.M.M., L.B.C. Teixeria, and R.R. Dubielzig. 2019. "Canine, feline, and equine corneal vascular neoplasia: A retrospective study (2007–2015)." *Veterinary Ophthalmology* 22 (1):76–87.

13 Dermatological Problems in the Brachycephalic Patient

Hilary Jackson
Dermatology Referral Service

Deborah Gow
The University of Edinburgh

CONTENTS

Introduction...219
Hereditary Skin Disease...220
   Ichthyosis..220
Demodicosis...221
Clinical Signs...222
Diagnosis..223
Management..223
Atopic Dermatitis/Adverse Food Reactions 223
Intertriginous Dermatitis..224
Interdigital Draining Tracts..224
Otitis Externa 226
   Clinical Signs 228
   Treatment 229
Primary Secretory Otitis Media (PSOM)..230
   Clinical Signs 230
   Treatment 230
   Non-pruritic Alopecia..231
   Dermatophytosis in Persian Cats 232
Conclusion 232
References..232

INTRODUCTION

Although disorders of the skin and ears are commonly diagnosed in general practice (O'Neill, Baral, Church, Brodbelt, et al. 2018), the unique conformations of brachycephalic dogs predisposes these animals to increased risks and severity of certain skin conditions.

Dermatological disorders were reported in 17.9% of French Bulldogs under primary veterinary care in the UK in 2013. The most common disorders reported were skin fold dermatitis, pyoderma, pododermatitis and atopic dermatitis (O'Neill, Baral, Church, Brodbent, et al. 2018). Also using the VetCompass data source, diseases of the skin and ears were common in Pugs and Bulldogs (O'Neill et al. 2016) (O'Neill, Skipper, et al. 2019). It should be noted that VetCompass reports on diagnosis levels by first opinion veterinarians for the wider dog population (Programme, VetCompass). Small cohorts of Bulldogs have also been examined by board-certified dermatologists in Finland and the USA. Of 27 English Bulldogs assessed, the Finnish study identified that all had some form of dermatological disease although this went unrecognised by 37% of their owners (Seppanen, Kaimio,

Schildt, Lilja-Maula, Hyytiaunen, et al. 2019). Similarly in the USA, all 34 English Bulldogs examined while participating at a show were assessed to have dermatological disease, although only 16 of these dogs had been previously diagnosed with cutaneous abnormalities (Webb Milum, Griffin, and Blessing 2018). This highlights that dermatological disease in these breeds is commonly unrecognised or may be normalised by owners and breeders.

The narrow gene pool associated with popular brachycephalic breeds can result in increased incidence of common disorders known to have an element of hereditary basis such as atopic dermatitis (Bizikova et al. 2015) and possibly demodicosis (O'Neill, Turgoose, et al. 2019). Selection for brachycephalic traits has resulted in 'wrinkly skin', screw tail formation and a narrow conformation of the external ear canal, all of which promote intertriginous dermatitis that can be further exacerbated by inflammatory disorders of the integument.

Less obviously associated with brachycephaly in respect to cutaneous disease, the criteria described with the breed standards for many brachycephalic breeds have promoted orthopaedic and gait abnormalities which affect weight bearing on the paws. This results in the so-called 'false pad' formation (discussed later in this chapter) and painful interdigital draining sinus tracts (IDDSTs) which significantly affects the individual's quality of life.

In the authors' experience, many brachycephalic dogs exhibit reduced expression of behaviours commonly associated with pruritus such as licking and scratching. However, it is unclear whether these dogs genuinely have a blunted sensation of pruritus or this observation is related to conformational changes which limit their ability to reach certain areas of their body. It is also possible that the presence of respiratory compromise may further limit their exertional ability to express these behaviours.

In contrast, some of these dogs may manifest less conventional pruritic behaviours such as 'rubbing and scooting' instead, and the clinician should be alert to these during history-taking. Likewise, pedal discomfort due to skin inflammation may manifest as lameness or reluctance to walk that may be misinterpreted as an orthopaedic abnormality.

HEREDITARY SKIN DISEASE

ICHTHYOSIS

Ichthyotic diseases are rare and result in the inability to form the outermost layer of the skin, the stratum corneum, which consists of overlapping layers of corneocytes surrounded by lipid. Congenital or acquired defects in this layer lead to impaired skin barrier function and predispose affected animals to secondary skin infections (Vahlquist and Torma 2020). A generalised ichthyosis has been characterised in the American Bulldog which clinically presents with diffuse scaling around 1–2 weeks of age and persists into adulthood. Affected dogs are predisposed to secondary Malassezia dermatitis. The disease is autosomal recessive and arises from a mutation in the gene coding NIPAL4 which plays a role in epidermal lipid metabolism (Mauldin et al. 2015).

Another ichthyotic disease is seen in the Cavalier King Charles Spaniel in association with keratoconjunctivitis sicca. Affected dogs are born with an abnormally rough coat and develop hyperkeratosis of the footpads and onychodystrophy. The disorder is autosomal recessive and associated with a base pair deletion in FAM83H (Forman et al. 2012).

Management of ichthyotic disorders is directed at improving epidermal barrier. Topical application of products containing ceramides and sphingolipids may be beneficial used alongside sebolytic shampoos.

A localised histologically parakeratotic hyperkeratosis affecting the face, pinnae and pressure points has been described in young Boston Terriers (Lee et al. 2016) and is also seen in French Bulldogs. Clinically, this has similarities to the zinc-responsive dermatosis seen in Nordic breeds, and some cases may respond to oral zinc supplementation.

Footpad hyperkeratosis in the Dogue de Bordeaux (French Mastiff) is an autosomal recessive mutation in the gene coding for keratin 16 (Plassais et al. 2015). Affected dogs develop frond-like thickening of the footpads. Painful fissures can develop which can cause lameness. Topical therapy with hydrating agents such as propylene glycol or plant oils can alleviate this condition.

Shar Pei dogs are characterised by thick and wrinkly skin associated with high dermal mucin content. In some dogs, mucinotic vesicles can develop, often called mucinosis. It has been demonstrated that there is an increased transcription of hyaluronan synthetase two possibly leading to increased activity of hyaluronic acid synthetase in this breed (Zanna et al. 2009). Severe cases may response to short-term glucocorticoid therapy.

Idiopathic facial dermatitis is seen in young Persian and Himalayan cats. They develop a highly inflammatory dermatitis primarily affecting the face which becomes secondarily infected with bacteria and/or Malassezia. The condition can be pruritic and causes considerable discomfort (Fontaine and Heimann 2004). Affected cats should be assessed for secondary infections and treated accordingly. The primary condition may be managed successfully with ciclosporin, but long-term treatment is usually required.

DEMODICOSIS

Canine demodicosis is reported to affect 0.5% of dogs overall aged under 2 years, with VetCompass evidence suggesting that many brachycephalic breeds may be predisposed (O'Neill, Turgoose, et al. 2019). The disease is caused by a variety of demodex species; *Demodex canis* is the most commonly isolated species in the dog, *D. injai* and another unnamed mite, possibly a morphological variant of *D. canis* is also found (Sastre et al. 2013) (Figure 13.1).

Demodex mites are commensal organisms, which are transferred from the mother to the young shortly after birth. Most dogs carry small numbers of these mites as commensal organisms. Transient patchy alopecia around 4–5 months of age can be seen in some dogs, and demodex mites demonstrated on skin scrapes. This localised disease will usually resolve without specific treatment. Generalised demodicosis can potentially occur in any breed with an increased incidence of

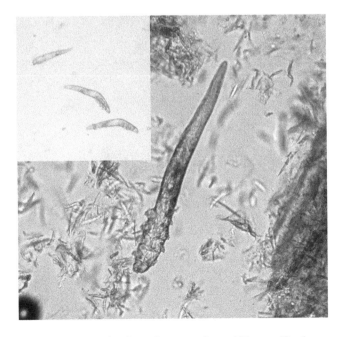

FIGURE 13.1   *D. injai* with *D. canis* mites inset for comparison ×100 magnification.

both juvenile and adult onset demodicosis in short-haired breeds (Day 1997), many of which are brachycephalic. There is some evidence of a genetic predisposition to developing juvenile onset generalised disease (Mueller et al. 2012). In the UK, predisposed breeds for developing demodicosis include popular brachycephalic breeds such as the British Bulldog, Staffordshire Bull Terrier, French Bulldog, Pug and Boxer, and the less popular breeds such as the Chinese Shar Pei and Dogue de Bordeaux (O'Neill, Baral, Church, Brodbelt, et al. 2018, O'Neill, Skipper, et al. 2019, O'Neill, Turgoose, et al. 2019).

Demodicosis describes the disease caused by mite proliferation. The precise immune dysregulation from which this arises is unknown. Once the mite proliferation is advanced, reduced interleukin-2 and high interleukin-10 (IL-10) and transforming growth factor-β production (TGF-β) by lymphocytes can be demonstrated in affected dogs characteristic of *T*-cell exhaustion (Felix et al. 2013, Kumari et al. 2017, Ferrer, Ravera, and Silbermayr 2014). This is normalised after effective acaricidal treatment. In this regard, it is interesting to note that cytokines such as IL-10 and TNF-α are upregulated in brachycephalic dogs with concurrent brachycephalic obstructive airway syndrome (BOAS) (Rancan et al. 2013). Whether this represents a common pathway of immune dysregulation in these breeds or is a consequence of their structural abnormalities is not clear at this time.

In adult dogs, demodicosis is more commonly seen in dogs with an underlying systemic illness such as hyperadrenocorticism or hypothyroidism, or in many cases, long-term immunosuppressive treatment. In up to 30% of cases, however, a specific condition may not be identified (Duclos, Jeffers, and Shanley 1994).

Demodicosis, the disease associated with proliferation of these mites, is classically divided into juvenile onset (Chang et al.) and adult onset disease. *Demodex injai* is associated with a different clinical presentation (see the 'Clinical Signs' section below).

CLINICAL SIGNS

Clinical signs associated with *D. canis* proliferation can be varied but cases will often present with multifocal areas of alopecia, erythema, papules, comedones, scale and follicular casts (Figure 13.2).

This may be localised or generalised. Pruritus can be variable but is usually present if secondary infection is established. Alopecic areas often take on a grey hue. Secondary bacterial infection can

FIGURE 13.2   Eight-month-old Pug with generalised demodicosis. Note the alopecia and scaling.

result in papules and pustules. A deep pyoderma can be associated with folliculitis and furunculosis clinically presenting as ulcerations and draining tracts. Secondary *Malassezia* dermatitis is associated with enhanced scaling. In dogs with pododemodicosis, the paws can become very painful and swollen leading to reluctance to ambulate. Most dogs affected with generalised demodicosis will also show regional lymphadenopathy.

Demodex injai is generally associated with two different clinical presentations which are both pruritic: a seborrhoea along the dorsum (more common in Terriers) (Ordeix et al. 2009) and in brachycephalic dogs such as Shih Tzu and the Lhasa Apso, pruritus of the face, which can often be intense (Forsythe, Auxilia, and Jackson 2009). *Demodex injai* is usually seen in adult dogs and is not associated with immunosuppression although comorbidities such as atopic dermatitis are not uncommon.

DIAGNOSIS

Demodectic mites are typically demonstrated with deep skin scrapings or trichograms. In affected animals, the mites are usually easily found with a good scraping technique. However, these tests may be inadequate in cases of pododemodicosis and in certain breeds such as the Shar Pei where skin biopsies are often required to demonstrate the mites.

Demodex injai mites conversely are often challenging to find. Dogs with facial pruritus should be sedated to collect samples, and where the skin is thickened or scarred, skin biopsies should be taken. These mites have a longer tail than those of *D. canis*.

MANAGEMENT

The treatment for generalised juvenile, adult onset demodicosis and *Demodex injai* is similar. Management includes treatment of secondary infections, acaricides and, in cases of adult onset demodicosis, identification and treatment of the underlying condition or cessation of immunosuppressive treatment.

The most commonly used licensed therapies are the systemic isoxazolines such as sarolaner, fluralaner and afoxolaner. Topical therapy with an imidacloprid and moxidectin spot-on may be efficacious in mild cases. Care should be taken when using isoxazolines in animals with a prior history of seizure activity (see product-specific data sheet), in which their use is contraindicated. Brachycephalic breeds may be at increased risk of developing seizure activity, either as a separate condition or as part of their BOAS (Wiles et al. 2017). A full history of the case will be required before instigating therapy. However, seizures may still occur in animals without a prior history of fits (Gaens et al. 2019).

Treatment should be continued until clinical signs have resolved and skin scrapings and trichograms are negative for mites on two separate occasions, each a month apart and then continued treatment for at least 4 weeks beyond this point. In cases where mites were initially difficult to find (*D. injai* and pododemodicosis), the clinician must rely on resolution of clinical signs. Many dogs with pododemodicosis and some with adult onset demodicosis may require lifelong therapy. Long-term systemic therapy with isoxazolines that additionally offer routine flea and ectoparasite prophylaxis is often practised.

ATOPIC DERMATITIS/ADVERSE FOOD REACTIONS

Canine atopic dermatitis (CAD) is a genetically predisposed pruritic and inflammatory disorder characterised by the development of IgE antibodies to environmental allergens (Olivry et al. 2015). Food allergens may also trigger flares of CAD. Some affected dogs have been shown to have specific defects in the skin barrier and impaired local immune defence that render them more susceptible to secondary skin infections with bacteria and, or, *Malassezia*. Breed predispositions are generally inferred by comparing affected dogs to the general hospital population. A Swiss study in 2008 assessed the breed incidence of CAD against the general dog population and identified that French Bulldogs and boxers along with other mesocephalic breeds were predisposed (Picco et al. 2008). Boxers and Pugs were over-represented in the food-induced CAD group in this study. The boxer breed was also over-represented for CAD in a Swedish study (Nodtvedt, Engenvall, and Bergvall 2006).

Dogs with atopic dermatitis typically present with pruritus affecting the paws, limbs, axillae, groin, face and ears. Onset is usually between 6 months and 5 years of age. Pruritus may be seasonal or non-seasonal. Management should be focussed on a balance of controlling long-term pruritus, acute flares, prevention and treatment of secondary infections and minimising side effects of therapy. A detailed discussion is beyond the scope of this chapter, and the reader is referred to excellent review articles addressing this disorder (Hensel et al. 2015, Olivry et al. 2015). Specific problems that can be exacerbated by concurrent atopic dermatitis are cheilitis, otitis, skin fold dermatitis (affecting the face and limbs) and pododermatitis.

INTERTRIGINOUS DERMATITIS

Skin fold or intertriginous dermatitis occurs at any site where contiguous sections of skin are in contact. Typically, in brachycephalic breeds, this occurs around the facial folds, lips, vulval folds and often the tail where it is particularly short or 'screw tail' conformation exists. Commensal bacteria or *Malassezia* spp. can multiply in these areas and, in chronic cases, the bacterial overgrowth can include opportunists such as *Enterococci*, *E. coli* and *Pseudomonas* spp. Clinically, this is associated with discomfort, pruritus and malodour. *Pseudomonas* infections often become ulcerated. Affected dogs may rub their faces and, where the tail or genital areas are involved, they are often seen to 'scoot', rubbing their perineal area along the ground.

Skin fold dermatitis is also increasingly recognised as a problem for brachycephalic rabbit breeds such as the Lion Head and Netherland Dwarf lop bred to have shorter facial features. Additional health issues related to the skin that may also arise in brachycephalic rabbits include dental health/overgrown teeth that may result in subcutaneous abscesses and blockage of the tear duct resulting in epiphora and skin irritation (Figures 13.3 and 13.4).

The organisms involved in intertriginous dermatitis can be identified using cytology. Effective management consists of daily topical therapy with wipes, shampoo or foams containing antibacterial and antifungal agents. Owners with newly acquired puppies should be counselled about regular prophylactic cleaning of skin folds because this is a lifelong problem to manage. Intertriginous dermatitis is often exacerbated by obesity, so weight control becomes an important feature of management as dogs' age. The inflammation associated with hypersensitivity disorders such as atopic dermatitis and flea-bite hypersensitivity is also an important contributing factor and was reported in 95% of dogs of dogs with cheilitis (inflammation of the lips) (Doelle et al. 2016). For animals responding poorly to topical therapy, surgical ablation or reduction of skin folds may be required.

INTERDIGITAL DRAINING TRACTS

The development of IDDSTs in dogs has multifactorial causes. Once the lesions become established, affected dogs show pain and discomfort, can become intermittently lame and avoid walking on rough ground. This not only reduces their quality of life but can also significantly impact on the client/pet relationship. A quality of life survey for dogs and their owners ranked pododermatitis as second only to sarcoptic ascariasis as causing poor quality of life (Banovic, Olivry, and Linder 2014).

The evolution from comedone formation to rupture of the interdigital folliculosebaceous units before finally developing into interdigital draining tracts on the dorsal interdigital skin has been elegantly described (Duclos, Hargis, and Hanley 2008). Briefly, swelling and inflammation on the ventral interdigital haired skin results in follicular plugging and comedone formation. The

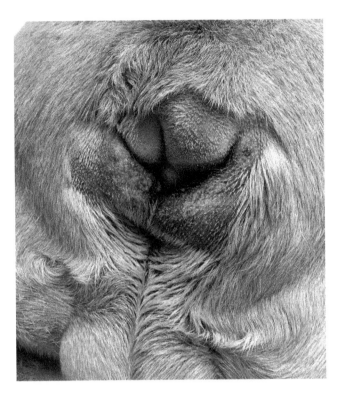

FIGURE 13.3 Peri-anal excoriations in a young Bulldog associated with intertriginous dermatitis of the tail fold.

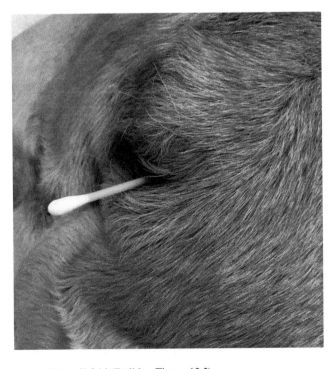

FIGURE 13.4 The depth of the tail fold (Bulldog Figure 13.3).

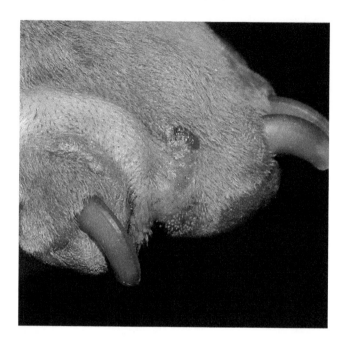

FIGURE 13.5　Interdigital swelling and draining tract.

blocked follicles dilate and rupture, causing an inflammatory response and the development of draining tracts that exit on the dorsal interdigital surface with fibrosis and scarring of the affected tissue. Chronic inflammation and discomfort lead to altered distribution of weight bearing, and the development of so-called 'false paw pads' formed from haired interdigital skin (Figures 13.5 and 13.6).

Interdigital draining tracts are not pathognomonic for any one disease but can follow common many primary diseases causing inflammation and pruritus of the pedal skin such as demodicosis, atopic dermatitis and cutaneous adverse food reactions. Orthopaedic disease, obesity and poor conformation are also contributory factors. Localised lesions may be associated with foreign bodies, neoplasia or deep infections. Brachycephalic dogs are predisposed to chronic pododermatitis, especially English Bulldogs (Seppanen, Kaimio, Schildt, Lilja-Maula, Hyytiaunen, et al. 2019, Webb Milum, Griffin, and Blessing 2018) and French Bulldogs (O'Neill, Baral, Church, Brodbent, et al. 2018).

Affected interdigital skin may become secondarily infected, but response to antibiotics alone is only partial. Successful management requires identification and treatment of all the primary and contributing diseases. Judicial anti-inflammatory therapy may be required, and cyclosporine is especially effective in this situation. Fluorescence biomodulation may be a useful adjunctive therapy (Marchegiani et al. 2019). A technique of tissue ablation using a CO_2 laser is used by the authors, and an alternative approach also using laser surgery is described by Duclos and others (Duclos, Hargis, and Hanley 2008).

OTITIS EXTERNA

Otitis externa (Doelle et al.) is one of the most common conditions presented to clinicians in general practice (Angus 2004, O'Neill et al. 2014), and brachycephalic breeds are over-represented (O'Neill, Baral, Church, Brodbent, et al. 2018).

A useful clinical approach when faced with cases of recurrent otitis externa is to consider the classification system which has been developed describing the primary, predisposing, secondary and perpetuating factors. Primary factors include atopic dermatitis, cutaneous adverse reaction to

FIGURE 13.6   Marked hyperplasia of the ventral interdigital skin displacing the weight-bearing pads. Note the broken hairs and comedone formation.

food, ectoparasites and systemic disease. Atopic dermatitis is the most common cause of development of OE in dogs (Saridomichelakis et al. 2007), and as previously discussed, brachycephalic breeds are predisposed to this condition.

Significant predisposing factors in the brachycephalic breeds are narrow ear canals (Seppanen, Kaimio, Schildt, Lilja-Maula, Hyytiainen, et al. 2019) and in the Shar Pei, in particular, the 'letter box' configuration of the ear (Muller 1990) (Figure 13.7).

Conformational predisposing factors do not directly cause otitis externa but can result in reduced air circulation and alter the local environment of the ear canal. Reduced external ear canal diameter additionally disrupts the normal ear cleaning mechanism, which reduces expulsion the normal skin cells leading to keratin and ceruminous material accumulation. In dogs that then develop inflammatory otitis externa due to atopic dermatitis, the local environment is readily predisposed to the development of secondary bacterial and, or *Malassezia* spp. infections.

Perpetuation of ear disease generally results from failure to manage the predisposing and primary problems because the secondary infections only are treated in many cases. Repeated episodes of inflammation and infection will lead to chronic changes in the ear canals which further perpetuate the disease.

From a clinical standpoint, the narrow ear canals in many brachycephalic dog and rabbit breeds (especially lop-eared rabbits) present various problems (Johnson and Burn 2019). Thorough otoscopic examination of the vertical and horizontal canals is challenging, and in many cases, the tympanic membrane cannot be visualised at all. For lop-eared rabbits, the aural examination is also associated with more discomfort compared to erect-eared rabbits. Consequently, collection of appropriate cytological samples is difficult and may not be representative of deeper infections,

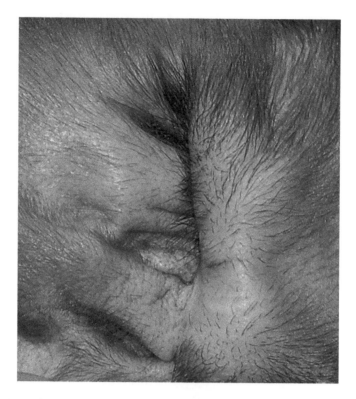

FIGURE 13.7   Normal Shar Pei. Note narrow auditory orifice.

often resulting in inappropriate topical treatment selection. Additionally, even if appropriate topical treatment is selected, response to therapy may be poor because owners may struggle to administer topical medication and/or cleaners into the ear canal.

Without adequate delivery of topical therapy, the perpetuating factors will lead to chronic changes that progressively reduce the ear canal diameter even further. This commonly results in dogs with long-term painful, infected and inflamed ears which has a massive impact on their longer-term welfare and quality of life. An additional consequence of chronic OE is the development of otitis media (OM), where the infection extends from the external ear canal, past the tympanic membrane and into the middle ear cavity (bulla) (Figure 13.8).

Collectively, brachycephalic breeds carry higher potential for OE to be mismanaged/misdiagnosed and thus progress to chronic OE with perpetuating factors (hyperplasia, stenosis, mineralisation) that may become irreversible. Although the breeds most commonly presented for total ear canal ablation surgery have not been reported, anecdotally, the authors feel the brachycephalic breeds are over-represented.

CLINICAL SIGNS

As discussed above, due to their typical brachycephalic breed features (presence of respiratory compromise, overweight, limb conformation, spinal malformation), these dogs may be unable to achieve ear scratching which is often a feature of OE in many dogs. The more subtle signs such as pinnal erythema, discharge and smell may then either go unnoticed by the owner or be considered of minor significance or even 'normal' for the breed (Seppanen, Kaimio, Schildt, Lilja-Maula, Hyytiainen, et al. 2019). Where possible, every brachycephalic dog presented for veterinary care should undergo otoscopic examination, even when ear pruritus is not reported. Sedation may be required for thorough otoscopic examination because breed-associated narrowed ear canals can

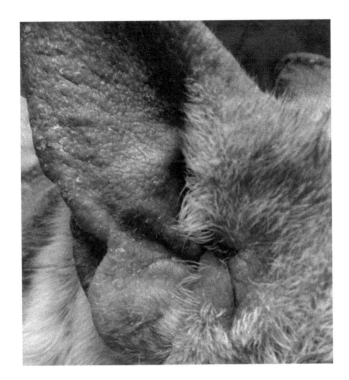

FIGURE 13.8   Chronic otitis externa causes further inflammation and swelling of the pinna and auditory orifice.

make this an uncomfortable procedure, even in healthy ears. However, the addition of sedation in these patients adds safety concerns and potential complications (unprotected airway, increased risk of oesophageal gastric reflux) that need to be considered. Cytology samples should be routinely collected from as deep as possible within the canal to identify the infectious agents present for appropriate therapy. In chronic cases, particularly where rod-shaped bacteria are present on cytology, bacterial culture and sensitivity is recommended. Ideally, the tympanic membrane should be appraised to determine whether it is intact before topical therapy is started. However, it is the authors' experience that, even under anaesthesia with the use of a video-otoscope, complete assessment of the tympanic membrane in brachycephalic breeds is often impossible.

TREATMENT

General principles of OE management should be implemented for these cases. The primary triggers should be investigated, any predisposing factors managed where possible, and secondary infections and perpetuating changes prevented or reversed. Owner compliance for diagnosis and therapy can be poor as there may be poor recognition of the significance of ear disease on their dog's quality of life, partly due to the stoic nature of the breeds involved and reduced manifestation of overt, ear-directed clinical signs. Therefore, time spent educating owners on the importance of diligent treatment, and due to the narrowed canals, effective application of any topical medication will be well spent.

For both chronic OE, with significant narrowing of the ear canals, and the development of subsequent OM, total ear canal ablation with a bulla osteotomy may be necessary. The narrow ear anatomies in brachycephalic dogs may predispose such breeds to develop cholesteatoma after middle ear surgery because complete removal of all inflammatory and epithelial tissue can be more difficult than in other breeds (Schuenemann and Oechtering 2012).

PRIMARY SECRETORY OTITIS MEDIA (PSOM)

Primary secretory otitis media (PSOM) has been well described in the Cavalier King Charles Spaniel (CKCS), with up to 40% of CKCS reportedly affected (Cole 2012). More recently, the boxer (Paterson 2017) and several other brachycephalic breeds (Cole 2012) have also been reported with this condition. A possible explanation for a brachycephalic predisposition may be the poor drainage of middle ear secretions due to their abnormal anatomy of the auditory tube (Hayes, Friend, and Jeffery 2010). Skull shape alterations of brachycephalic breeds, combined with bulla malformations and soft tissue changes to the soft palate and nasopharynx, may impair drainage of mucus from the middle ear cavity which has a respiratory epithelial lining (Hayes, Friend, and Jeffery 2010, Cole 2012, Mielke, Lam, and Ter Haar 2017). Mucus accumulation in the tympanic bulla puts pressure on the tympanic membrane and oval window of the inner ear cavity resulting in pain and neurological signs such as vestibular symptoms and hearing loss.

CLINICAL SIGNS

As with OE, early signs of PSOM can be subtle and are easily missed until the mucoid effusion builds up to a critical level that causes pain, vocalisation, hearing loss and neurological deficits.

Clinical signs can be variable: head shaking or pressing/rubbing the affected ear along surfaces, vestibular disturbances, ataxia, pain opening mouth or when yawning and inappetence. Affected dogs often have disturbed sleep patterns due to positional redistribution of the fluid in the bullae. Phantom scratching is also possible that may mimic that seen in syringomyelia and can pose a diagnostic dilemma (Cole et al. 2015). Magnetic resonance image (MRI) screening of clinically normal CKCS for Chiari malformation and syringomyelia has detected fluid in the tympanic bullae in approximately 50% of cases (Hayes, Friend, and Jeffery 2010), so the co-occurrence of syringomyelia with PSOM can make attribution of clinical signs to underlying pathology difficult.

If fluid in the bullae is considered an 'incidental' finding on imaging and no clinical signs are present, owners should be counselled that PSOM may become a problem later. However, no long-term evaluation on the evolution of this disease has been performed.

In animals with compatible clinical signs, the presence of a bulging tympanic membrane (pars flaccida) on otoscopic examination is pathognomonic for this disorder (Cole et al. 2015). However, the absence of this sign does not rule PSOM out. Furthermore, as previously discussed, visualisation of the tympanic membrane in its entirety with a handheld otoscope in these breeds is often impossible. Confirmation of the diagnosis generally requires video-otoscopic examination of the ear canal and tympanic membrane under full anaesthetic. In cases where there are compatible clinical signs, the presence of fluid in the bulla seen on imaging but no bulging pars flaccida, a myringotomy is advisable to relieve the pressure. In some cases, the pressure of the fluid in the middle ear cavity will have already caused rupture of the tympanic membrane (Figure 13.9).

TREATMENT

The most appropriate therapy requires performing a myringotomy, puncturing the ventro-caudal part of the tympanic membrane (pars tensa) to allow the mucus to be removed by suction and flushing with sterile saline. In many cases, the mucus is very viscous and can be difficult to extract.

Cytological assessment and culture of the extracted mucus should be performed. The majority of middle ear bacterial cultures are sterile (Cole et al. 2019). Treatment of these cases is therefore targeted at reducing the local inflammation with a short course of anti-inflammatory oral glucocorticoids as determined by the clinician on a case by case basis. However, affected dogs may need repeated myringotomies to alleviate recurring discomfort. Theoretically, prophylactic use of mucolytic agents such as N-acetylcysteine or bromhexine could reduce mucous viscosity and aid auditory drainage, but no prospective studies have been reported to demonstrate such efficacy in this

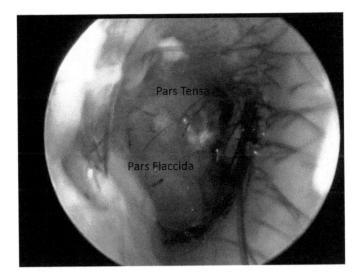

FIGURE 13.9   Bulging pars flaccida due to mucous accumulation in the middle ear cavity.

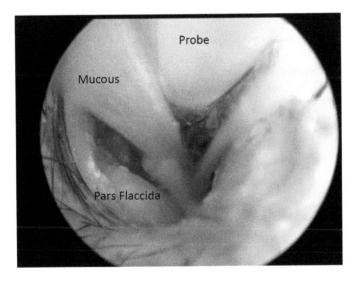

FIGURE 13.10   Mucous evacuation after performing a video-otoscopic guided myringotomy.

disease. The authors routinely prescribe oral mucolytics using *N*-acetylcysteine (600 mg every 12 or 24 hours) for brachycephalic dogs with PSOM to try and extend the interval between myringotomies, although there is no published evidence base to support this approach (Figure 13.10).

NON-PRURITIC ALOPECIA

Endocrine disorders such as hypothyroidism and hyperadrenocorticism are common in dogs in general but, apart from the boxer with a predisposition to hypothyroidism (Nesbitt et al. 1980), brachycephalic breeds have not been reported at increased risk for endocrine diseases over the general canine population. However, seasonal flank alopecia is a poorly understood condition, which presents with non-pruritic symmetrical hair loss on the flanks, that is commonly seen in brachycephalic dogs and indeed was initially reported in boxers and Airedale Terriers (Miller and Dunstan 1993). Affected dogs lose hair in a symmetrical pattern on the flanks usually in the autumn (in the Northern Hemisphere) with regrowth occurring in the spring. The condition is not associated with any demonstrable endocrine abnormalities and is cosmetic in nature. It is more common at higher latitudes. An association with the photoperiod is postulated, and oral melatonin 3–6 mg three times daily may arrest or shorten the period of hair loss. Response to therapy however may be unpredictable.

DERMATOPHYTOSIS IN PERSIAN CATS

Dermatophytosis in cats is most commonly caused by *Microsporum canis*. The Persian breed is particularly predisposed to this infection, and a carrier status may often develop (Hnilica and Medleau 2002). Persian cats also appear to be at increased risk of developing dermatophytic pseudomycetomas (Chang et al. 2011, Miller 2010). These are subcutaneous infections with dermatophytes which usually present clinically with nodules and draining tracts.

Diagnosis of dermatophytosis relies on a combination of screening with an ultraviolet light source (Woods lamp), trichoscopy, fungal culture and PCR. Optimal treatment is a combination of systemic and topical triazole antifungals. Dermatophyte pseudomycetomas often require additional surgical excision. The reader is directed to an excellent recent review paper on this subject. (Moriello et al. 2017). While dermatophytosis is a zoonosis, serious complications in humans following infection acquired from animals are not common.

CONCLUSION

Due to breeding practices, brachycephalic breeds are genetically predisposed to many common dermatological conditions. However, it should be noted that the conformational changes associated with these pets result in an early onset of skin disease and often an escalation in the severity of common problems such as otitis and pododermatitis. Early recognition and intervention from the veterinary community is key alongside the counselling of pet owners in the appropriate awareness and management of such conditions.

REFERENCES

Angus, J. C. 2004. "Otic cytology in health and disease." *Vet Clin North Am Small Anim Pract* 34 (2):411–24. doi:10.1016/j.cvsm.2003.10.005.

Banovic, F., T. Olivry, and K. Linder. 2014. "Ciclosporin therapy for canine generalized discoid lupus erythematosus refractory to doxycycline and niacinamide." *Veterinary Dermatology* 25 (5):483-e79.

Bizikova, P., C. M. Pucheu-Haston, M. Eisenschenk, R. Marsella, T. Nuttall, and D. Santoro. 2015. "Review: Role of genetics and the environment in the pathogenesis of canine atopic dermatitis." *Veterinary Dermatology* 26 (2):95-e26. doi:10.1111/vde.12198.

Chang, S. C., J. W. Liao, C. L. Shyu, W. L. Hsu, and M. L. Wong. 2011. "Dermatophytic pseudomycetomas in four cats." *Vet Dermatol* 22 (2):181–7. doi:10.1111/j.1365-3164.2010.00937.x.

Cole, L. K. 2012. "Primary secretory otitis media in Cavalier King Charles spaniels." *Vet Clin North Am Small Anim Pract* 42 (6):1137–42. doi:10.1016/j.cvsm.2012.08.002.

Cole, L. K., P. J. Rajala-Schultz, G. Lorch, and J. B. Daniels. 2019. "Bacteriology and cytology of otic exudates in 41 cavalier King Charles spaniels with primary secretory otitis media." *Vet Dermatol* 30 (2):151-e44. doi:10.1111/vde.12724.

Cole, L. K., V. F. Samii, S. O. Wagner, and P. J. Rajala-Schultz. 2015. "Diagnosis of primary secretory otitis media in the cavalier King Charles spaniel." *Vet Dermatol* 26 (6):459–66, e106-7. doi:10.1111/vde.12248.

Day, M. J. 1997. "An immunohistochemical study of the lesions of demodicosis in the dog." *J Comp Pathol* 116 (2):203–16. doi:10.1016/s0021-9975(97)80077-1.

Doelle, M., A. Loeffler, K. Wolfe, V. Kostka, and M. Linek. 2016. "Clinical features, cytology and bacterial culture results in dogs with and without cheilitis and comparison of 3 sampling techniques." *Vet Dermatol* 27:140–e37.

Duclos, D., A. M. Hargis, and P. W. Hanley. 2008. "Pathogenesis of canine interdigital palmar and plantar comedones and follicular cysts and their response to laser surgery." *Vet Dermatol* 19:134–141.

Duclos, D. D., J. G. Jeffers, and K. J. Shanley. 1994. "Prognosis for treatment of adult-onset demodicosis in dogs: 34 cases (1979–1990)." *J Am Vet Med Assoc* 204 (4):616–9.

Felix, A. O., E. G. Guiot, M. Stein, S. R. Felix, E. F. Silva, and M. O. Nobre. 2013. "Comparison of systemic interleukin 10 concentrations in healthy dogs and those suffering from recurring and first time Demodex canis infestations." *Vet Parasitol* 193 (1–3):312–5. doi:10.1016/j.vetpar.2012.11.012.

Ferrer, L., I. Ravera, and K. Silbermayr. 2014. "Immunology and pathogenesis of canine demodicosis." *Vet Dermatol* 25 (5):427-e65. doi:10.1111/vde.12136.

Fontaine, J., and M. Heimann. 2004. "Idiopathic facial dermatitis of the Persian cat: Three cases controlled with cyclosporine." *Vet Dermatol* 15:64. doi:10.1111/j.1365-3164.2004.00414_70.x.

Forman, O.P., J. Penderis, C. Hartley, L.J. Hayward, S.L. Ricketts, and C.S. Mellersh. 2012. "Parallel mapping and simultaneous sequencing reveals deletions in BCAN and FAM83H associated with discrete inherited disorders in a domestic dog breed." *PloS Genet* 8. doi:10.1371/journal.pgen.1002462.

Forsythe, P.J., S.T. Auxilia, and H.A. Jackson. 2009. "Intense facial pruritus associated with Demodex injai infestation: A report of 10 cases." In: *Proceedings from the 24th North American Veterinary Dermatology Forum; Savannah, GA.*

Gaens, D., C. Rummel, M. Schmidt, M. Hamann, and J. Geyer. 2019. "Suspected neurological toxicity after oral application of fluralaner (Bravecto(R)) in a Kooikerhondje dog." *BMC Vet Res* 15 (1):283. doi:10.1186/s12917-019-2016-4.

Hayes, G. M., E. J. Friend, and N. D. Jeffery. 2010. "Relationship between pharyngeal conformation and otitis media with effusion in Cavalier King Charles spaniels." *Vet Rec* 167 (2):55–8. doi:10.1136/vr.b4886.

Hensel, P., D. Santoro, C. Favrot, P. Hill, and C. Griffin. 2015. "Canine Atopic dermatitis: Detailed guidelines for diagnosis and allergen identification." *BMC Vet Res* 11:196. doi:10.1186/s12917-015-0515-5.

Hnilica, K. A., and L. Medleau. 2002. "Evaluation of topically applied enilconazole for the treatment of dermatophytosis in a Persian cattery." *Vet Dermatol* 13 (1):23–8. doi:10.1046/j.0959-4493.2001.00282.x.

Johnson, J. C., and C. C. Burn. 2019. "Lop-eared rabbits have more aural and dental problems than erect-eared rabbits: A rescue population study." *Vet Rec* 185 (24):758. doi:10.1136/vr.105163.

Kumari, P., R. Nigam, A. Singh, U. P. Nakade, A. Sharma, S. K. Garg, and S. K. Singh. 2017. "Demodex canis regulates cholinergic system mediated immunosuppressive pathways in canine demodicosis." *Parasitology* 144 (10):1412–16. doi:10.1017/S0031182017000774.

Lee, F.F., C.W. Bradley, C.L. Cain, S.D. White, C.A. Outerbridge, L.A. Murphy, and E.A. Mauldin. 2016. "Localized parakeratotic hyperkeratosis in sixteen Boston terrier dogs." *Vet Dermatol* 27:384–e96.

Marchegiani, A., A. Spaterna, M. Cerquetella, A. M. Tambella, A. Fruganti, and S. Paterson. 2019. "Fluorescence biomodulation in the management of canine interdigital pyoderma cases: a prospective, single-blinded, randomized and controlled clinical study." *Vet Dermatol* 30:371-e109.

Mauldin, E. A., P. Wang, E. Evans, C. A. Cantner, J. D. Ferracone, K. M. Credille, and M. L. Casale. 2015. "Autosomal recessive congenital ichthyosis in American bulldogs is associated with NIPAL4 (Ichthyin) deficiency." *Vet Pathol* 52 (4):654–662.

Mielke, B., R. Lam, and G. Ter Haar. 2017. "Computed tomographic morphometry of tympanic bulla shape and position in brachycephalic and mesaticephalic dog breeds." *Vet Radiol Ultrasound* 58 (5):552–558. doi:10.1111/vru.12529.

Miller, M. A., and R. W. Dunstan. 1993. "Seasonal flank alopecia in boxers and Airedale terriers: 24 cases (1985–1992)." *J Am Vet Med Assoc* 203 (11):1567–72.

Miller, R. I. 2010. "Nodular granulomatous fungal skin diseases of cats in the United Kingdom: a retrospective review." *Vet Dermatol* 21 (2):130–5. doi:10.1111/j.1365-3164.2009.00801.x.

Moriello, K. A., K. Coyner, S. Paterson, and B. Mignon. 2017. "Diagnosis and treatment of dermatophytosis in dogs and cats: Clinical consensus guidelines of the world association for veterinary dermatology." *Vet Dermatol* 28 (3):266-e68. doi:10.1111/vde.12440.

Mueller, R. S., E. Bensignor, L. Ferrer, B. Holm, S. Lemarie, M. Paradis, and M. A. Shipstone. 2012. "Treatment of demodicosis in dogs: 2011 clinical practice guidelines." *Vet Dermatol* 23 (2):86–96, e20-1. doi:10.1111/j.1365-3164.2011.01026.x.

Muller, G.H. 1990. "Skin diseases of the Chinese Shar-Pei." *Vet Clin North Am Small Anim Pract* 20 (6):1655–1670. doi:10.1016/s0195-5616(90)50166-7

Nesbitt, G. H., J. Izzo, L. Peterson, and R. J. Wilkins. 1980. "Canine hypothyroidism: a retrospective study of 108 cases." *J Am Vet Med Assoc* 177 (11):1117–22.

Nodtvedt, A., A. Engenvall, and K. Bergvall. 2006. "Incidence of and risk factors for atopic dermatitis in a Swedish population of insured dogs." *Vet Rec* 159:241–6.

O'Neill, D. G., L. Baral, D. B. Church, D.C. Brodbent, and R. M. A. Packer. 2018. "Demography and disorders of the French Bulldog population under primary veterinary care in the UK in 2013." *Canine Genet Epidemiol* 5 (3):2–12. doi:10.1186/s40575-018-0057-9.

O'Neill, D. G., E. C. Darwent, D. B. Church, and D. C. Brodbent. 2016. "Demography and health of pugs under primary veterinary care in England." *Canine Genet Epidemiol* 3 (1):1–12.

O'Neill, D. G., A. M. Skipper, J. Kadhim, D. B. Church, D. C. Brodbelt, and R. M. A. Packer. 2019. "Disorders of Bulldogs under primary veterinary care in the UK in 2013." *PLoS One* 14 (6):e0217928. doi:10.1371/journal.pone.0217928.

O'Neill, D. G., E. Turgoose, D. B. Church, D. C. Brodbelt, and A. Hendricks. 2019. "Juvenile-onset and adult-onset demodicosis in dogs in the UK: prevalence and breed associations." *J Small Anim Pract.* doi:10.1111/jsap.13067.

O'Neill, D.G., D.B. Church, P.D. McGreevy, P.C. Thomson, and D.C. Brodbelt. 2014. "Prevalence of disorders recorded in dogs attending primary-care veterinary practices in England." *PLOS One* 9:1–16 doi:10.1371/journal.pone.0090501.

Olivry, T., D. DeBoer, C. Favrot, H. Jackson, R. Mueller, T. Nuttal, and P. Prelaud. 2015. "Treatment of canine atopic dermatitis: 2015 updated giudeline from the International Committee on Allergic Diseases of Animals." *BMC Vet Res* 11:210. doi:10.1186/s12917-015-0514-6.

Ordeix, L., M. Bardagi, F. Scarampella, L. Ferrer, and A. Fondati. 2009. "Demodex injai infestation and dorsal greasy skin and hair in eight wirehaired fox terrier dogs." *Vet Dermatol* 20 (4):267–72. doi:10.1111/j.1365-3164.2009.00755.x.

Paterson, S. 2017. "Otitis media with effusion in the boxer: A report of seven cases." *J Small Anim Pract.* doi:10.1111/jsap.12801.

Picco, F., E. Zini, C. Nett, C. Naegeli, B. Bigler, S. Rufenacht, et al. 2008. "A prospective study on canine atopic dermatitis and food induced dermatitis in Switzerland." *Vet Dermatol* 19 (3):150–155.

Plassais, J., E. Guaguere, L. Lagoutte, A. S. Guillory, C. D. de Citres, F. Degorce-Rubiales, M. Delverdier, A. Vaysse, P. Quignon, C. Bleuart, A. Hitte, A. Fautrel, C. Kaerle, P. Bellaud, E. Bensignor, G. Queney, E. Bourrat, A. Thomas, and C. Andre. 2015. "A spontaneous KRT16 mutation in a dog breed: A model for human focal non-epidermolytic palmoplantar keratoderma (FNEPPK)." *J Invest Dermatol* 135 (4):1187–1190. doi:10.1038/jid.2014.526.

Programme, VetCompass: VetCompass™.

Rancan, L., S. Romussi, P. Garcia, M. Albertini, E. Vara, and M. Sanchez de la Muela. 2013. "Assessment of circulating concentrations of proinflammatory and anti-inflammatory cytokines and nitric oxide in dogs with brachycephalic airway obstruction syndrome." *Am J Vet Res* 74 (1):155–60. doi:10.2460/ajvr.74.1.155.

Saridomichelakis, M. N., R. Farmaki, L. S. Leontides, and A. F. Koutinas. 2007. "Aetiology of canine otitis externa: a retrospective study of 100 cases." *Vet Dermatol* 18 (5):341–7. doi:10.1111/j.1365-3164.2007.00619.x.

Sastre, N., I. Ravera, D. Ferreira, L. Altet, A. Sanchez, M. Bardagi, O. Francino, and L. Ferrer. 2013. "Development of a PCR technique specific for Demodex injai in biological specimens." *Parasitol Res* 112 (9):3369–72. doi:10.1007/s00436-013-3531-z.

Schuenemann, R. M., and G. Oechtering. 2012. "Cholesteatoma after lateral bulla osteotomy in two brachycephalic dogs." *J Am Anim Hosp Assoc* 48 (4):261–8. doi:10.5326/JAAHA-MS-5760.

Seppanen, R. T. K., M. Kaimio, K. J. M. Schildt, L. Lilja-Maula, H. K. Hyytiainen, S. Molsa, M. Morelius, M. M. Rajamaki, A. K. Lappalainen, and M. Rantala. 2019. "Skin and ear health in a group of English bulldogs in Finland - a descriptive study with special reference to owner perceptions." *Vet Dermatol* 30 (4):307-e85. doi:10.1111/vde.12752.

Seppanen, R., M. Kaimio, K. Schildt, L. Lilja-Maula, H. Hyytiaunen, S. Molsa, M. Morelius, M. Rajamaki, A. Lappalainen, and M. Rantala. 2019. "Skin and ear health in a group of English Bulldogs in Finland- a descriptive study with special reference to owner perceptions." *Veterinary Dermatology* 30:307–313.

Vahlquist, A., and H. Torma. 2020. "Ichthyosis: A road model for skin research." *Acta Derm Venereol.* doi:10.2340/00015555-3433.

Webb Milum, A. N., C. E. Griffin, and K. S. Blessing. 2018. "A cross-sectional study of show English bulldogs in the United States: Evaluating paw lesions, cytological findings, pruritic behaviours and gastrointestinal signs." *Vet Dermatol* 29 (5):395–401. doi:10.1111/vde.12676.

Wiles, B. M., A. M. Llewellyn-Zaidi, K. M. Evans, D. G. O'Neill, and T. W. Lewis. 2017. "Large-scale survey to estimate the prevalence of disorders for 192 Kennel Club registered breeds." *Canine Genet Epidemiol* 4:8. doi:10.1186/s40575-017-0047-3.

Zanna, G., M. J. Docampano, D. Fondevila, M. Bardagí, A. Bassols, and L. Ferrer. 2009. "Hereditary cutaneous mucinosis in shar pei dogs is associated with increased hyaluronan synthase-2 mRNA transcription by cultured dermal fibroblasts." *Vet Dermatol* 20: 377–382.

14 Dental and Oral Health for the Brachycephalic Companion Animal

Fraser A. Hale
Hale Veterinary Clinic

CONTENTS

Brachycephalic Dental/Oral Issues 237
Malocclusion with Abnormal Contacts..238
Caudal Buccal Traumatic Granulomas 240
Dental Crowding and Rotation 241
Under-Eruption and Pericoronitis 241
Unerupted Teeth and Dentigerous Cysts..243
Furry Palate 245
Wide, Loose Symphysis..245
Increased Risk for Mandibular Fracture 246
Gingival Hyperplasia 247
Conclusion 247
References..247

Note to reader: Many of the resources cited in this chapter are freely available on the Old CUSP Articles page of the author's website, www.toothvet.ca. These resources offer more in-depth discussions of the concepts explored in this chapter and are well illustrated with drawings, clinical photographs and radiograph.

The inclusion of a chapter on the management of oral health impacts from brachycephalism in companion animals is critically important for any text that aims to provide a holistic welfare perspective. The increasing popularity of, and demand for, companion animals with brachycephalism means that there has never been a greater need for proactive and reactive care provided by veterinary dental teams, primary care veterinarians and owners. Oral health care in companion animals has a history of being a neglected discipline in veterinary medicine, although this situation may be improving over recent decades. However, many veterinarians sadly still graduate having received only basic instruction in dental and oral anatomy, physiology, pathology and treatment (Kerr 2018, Anderson et al. 2017). This relative paucity of undergraduate dental training may lead to deep-seated but erroneous beliefs that dentistry is less important or interesting than other veterinary disciplines (Perry 2014). For interested veterinarians, there are many continuing education resources available (BVDA 2020, EVDS 2020, AVDC 2020a), but it is still largely up to individual practitioners to prioritise their own limited continuing education time and budget following graduation. Consequently, there exists a wide spectrum of interest, understanding and competence among practitioners regarding the management of oral health issues. This chapter aims to narrow this range, specifically in relation to the aspects of oral care commonly associated with brachycephalism. The breeder also has a key role to play in relation to selectively breeding away from poorer dental conformation, and the owner plays a pivotal role when choosing which breed to promote when getting a pet and then again when electing to institute good dental care measures.

This chapter cannot cover every aspect of veterinary dentistry and therefore focuses on aspects that are especially relevant for brachycephalic companion animals. The individual veterinary practitioner must judge their own personal status across the spectrum of oral health knowledge and skill from resources such as the American Animal Hospital Association's (AAHA) '2019 AAHA Dental Care Guidelines for Dogs and Cats' (Bellows et al. 2019) and the longer set of 'Global Dental Guidelines' published by the World Small Animal Veterinary Association (WSAVA) (Niemiec et al. 2020). These excellent resources provide information on common oral and dental pathologies, diagnostic procedures, treatments, anaesthesia and pain management, home dental care, nutritional information and recommendations on the role of the universities in improving veterinary dentistry. Importantly, negative effects from undiagnosed or untreated dental disease on the welfare of our patients are also highlighted. These resources are also useful for the dedicated breeder and owner. I encourage all veterinary practitioners to read and become familiar with these guidelines. Pet owners can also use the information in these guidelines to help them to understand the level of dental care and expertise offered by their veterinary practice.

Specifically in relation to brachycephalic pets, the normalisation phenomenon has become a huge barrier for optimum oral care for these animals (Ravetz 2017). Acceptance of the notion that the craniofacial exaggerations and deformities associated with brachycephaly are 'normal for the breed' and therefore are normal in general suggests that these features either require no assessment or intervention, or that the problems are unfixable (Hale 2013b, 2014). It is not just pet owners who may succumb to the normalisation fallacy. All veterinarians practicing at the time of writing have grown up in a world where animals with brachycephalism pre-existed the development of our own critical thinking skills or concerns about companion animal welfare. Consequently, issues and pathologies associated with brachycephalism in companion animals have become like white noise; always there but barely noticed. Fortunately, more and more evidence is being published regarding the health concerns associated with brachycephalism such as BOAS (Packer et al. 2015), corneal ulceration (O'Neill, Lee, et al. 2017) and dystocia (O'Neill, O'Sullivan, et al. 2017), for example. However, relatively little professional literature has been published regarding oral health impacts of brachycephalism. But the absence of evidence is not evidence of absence. Some papers report prevalence data based on pet insurance claims (Feng et al. 2017, Fawcett et al. 2018), but these are often limited in two main respects. First, the problem in question must be covered by the insurance policy before a claim can be registered; many pet insurance policies do not cover dental disease or treatments. Second, the primary care veterinarian must recognise the pathology and need for treatment, and the owner must consent to that treatment before the claim can even be submitted to the insurance company in the first place. Therefore, epidemiological studies based on insurance data may significantly underestimate the degree of dental disease in animals with brachycephalism.

Another barrier to appropriate oral care is that evolutionary survival mechanisms have rendered our pet dogs and cats as masters at masking pain and distress (Hellyer et al. 2007, Price and Dussor 2014). This effect is compounded by our limited ability as owners and veterinarians to judge of how animals are feeling (Hale 2020a).

The following is a quote from a 1991 paper on the veterinarian's responsibility in assessing and managing pain (Johnson 1991):

> The veterinary profession's understanding and management of pain is yet in its infancy, but it has left the dark ages of the early 1900s as evidence in this 1906 surgery text: "Anesthesia in veterinary surgery today is a means of restraint and not an expedient to relieve pain. So long as an operation can be performed by forcible restraint... the thought of anesthesia does not enter into the proposition..."

Veterinary science has come a long way since 1906, and even further since 1991, but it remains true that pet dogs and cats often suffer while their owners refute that their animal has oral pain, citing that the animal is still eating as evidence of oral comfort (Bellows et al. 2019). The willingness to eat does not discount oral pain because the alternative to eating is starvation; so animals are driven to eat, even in the face of considerable oral pain. In my referral practice, I have seen dogs with jaw fractures that healed naturally while their owners were completely unaware that there had ever been a problem. Again, the absence of evidence is not evidence of absence. Therefore, veterinary practitioners and pet owners must learn to recognise the key signs of dental disease and then actively search for these. An effective examination requires much more than a quick flip of the lip in the consultation room; dental pathology is often hidden from view below the gumline (Hale 2020d, Verstraete, Kass, and Terpak 1998, Verstraete 1998, Döring, Arzi, Hatcher, et al. 2017, Döring, Arzi, Barich, et al. 2017). A complete dental examination requires general anaesthesia for a detailed examination of the entire oral cavity, a tooth-by-tooth evaluation with probing and charting, and whole-mouth intra-oral dental radiographs and/or cone beam CT scan.

My hope is that this chapter will raise awareness of the possible pathologies lurking behind the flat-faced exteriors of brachycephalic heads, and offer some suggestions about what we all can, and should, do to redress the effects of these poor conformational designs such that animals with brachycephalism might enjoy lives free from oral pain and infection (Mestrinho et al. 2018, Gyles 2017).

Although few of the issues discussed in this chapter are exclusive to animals with brachycephalism, these problems are generally much more common in brachycephalics. And not every animal with brachycephalism will have all of these problems (O'Neill et al. 2018, 2019, 2016). Finally, this chapter does not provide an exhaustive list of every oral problem of pet dogs and cats. Instead, I aim to include those problems that I regularly see at my dental referral practice in the mouths of my brachycephalic dog and cat patients. These problems may be either a direct outcome from their distorted facial anatomy or may result from other genetic faults associated with the brachycephalic genotype. The focus will be on dogs, but cats may also be affected by some of these, though generally to a lesser degree (Mestrinho et al. 2018).

BRACHYCEPHALIC DENTAL/ORAL ISSUES

By definition of the conformational changes that have been selected in the brachycephalic skull, every brachycephalic mouth is likely to show several dental abnormalities that will vary in their relative severity and progression. It is challenging to discuss these individually in isolation because there is a lot of interaction between the oral and dental structures. Consequently, there will be some redundancy and crossover in this chapter as each issue is explored. Specific issues that I see very commonly in dogs (and in some cats) with brachycephalism in my dental referral practice include

- Malocclusion with abnormal tooth-to-tooth and tooth-to-soft tissue contacts causing trauma to hard and soft tissues and sometimes pulp necrosis in teeth [More on brachycephalism] (Hale 2017b)
- Caudal buccal traumatic granulomas from abnormal tooth-to-soft tissue contacts [Gum chewer syndrome] (Hale 2013a, 2008c)
- Crowding and rotation of teeth predisposing to periodontal disease [Focus on microdogs] (Hale 2012b, 2021)
- Under-erupted teeth predisposing to pericoronitis [Pericoronitis] (Hale 2011a)
- Unerupted teeth predisposing to dentigerous cyst formation [Dentigerous cysts] (Hale 2007b)
- Fur entrapment in deep palatal fold and grooves [Furry Palate video] (Hale 2016a, Regalado Ibarra and Legendre 2019)
- Wide, loose mandibular symphysis [Focus on the mandibular symphysis] (Hale 2005b)
- Increased risk for mandibular fracture (Hale 2017a)
- Gingival hyperplasia especially in Boxer dogs [Gingival hyperplasia] (Hale 2020b)

MALOCCLUSION WITH ABNORMAL CONTACTS

By definition, virtually every animal with brachycephaly has been selectively bred to have a class III malocclusion (upper jaw too short relative to the lower jaws); this is often even defined in their breed standard (Hale 2018, AVDC 2020b, American Kennel Club 2020, The Kennel Club 2020). As well as the upper jaw being relatively too short, the lower jaw is also often too short relative to the size of the animal and its teeth, i.e., both upper and lower jaws are too short but the upper jaw more dramatically so.

Figure 14.1 is of a young dog with normal occlusion (AVDC 2020b). Both jaws are appropriate lengths in relation to the size of the dog and in relationship to each other. The teeth are properly proportioned and located within those proper jaws. Everything is as it should be in this photo.

The only place in a dog's mouth where tooth-to-tooth contact is acceptable is where the two upper molars occlude against the distal 1/3 of the lower first molar and the second and third lower molars. Any tooth-to-tooth contact elsewhere in the mouth is abnormal and undesirable. There should be no tooth-to-soft tissue contact anywhere in the mouth.

Almost all dogs with a class III malocclusion show abnormal tooth-to-tooth and/or tooth-to-soft tissue contacts because the upper incisor teeth come down on the lingual aspect of the lower incisor and canine teeth. These mouths may show other abnormal contacts elsewhere as well. These animals are effectively biting themselves each time they close their mouth. It is usually necessary to extract some teeth to alleviate traumatic contacts and dental crowding issues, but this raises the challenging question of which tooth (teeth) to remove? The answer is based on several criteria. I generally aim to sacrifice less important teeth in order to improve the prognosis for more important teeth and to preserve as much form and function as possible. Paradoxically, as we remove some teeth, the risk is that the mouth may close further so that new abnormal contacts are now generated. Each mouth should be managed on its own merits, with a careful assessment of the live patient in three dimensions [Treat or Extract] (Hale 2011b).

Figure 14.2 of a mature Boxer dog shows irritated contact points in the oral mucosa on the floor of the mouth as well as gingival recession and attrition (wear from tooth-to-tooth contact) on the lingual aspect of the lower canine teeth. The upper incisors have been digging into the floor of the mouth and abrading the lower canine teeth since the adult teeth erupted. Additionally, there is gingival hyperplasia around the lower incisor teeth in this image and that will be discussed later.

Disorders in general are not binary (healthy/sick) but tend to show degrees of abnormality. On the mild end of the malocclusion spectrum, the tooth-to-tooth contact is minor enough and at an angle such that no intervention is indicated. At the more severe end, selective extraction of either the upper or lower tooth (removing either the hammer or the anvil) to alleviate traumatic contact is indicated. Sometimes, crown reduction and vital pulp therapy are indicated (Hale 2007a).

Figure 14.2 shows an example where chronic trauma from the upper incisors has already moved the soft tissues out of the way to accommodate the situation in an acceptable manner so removing

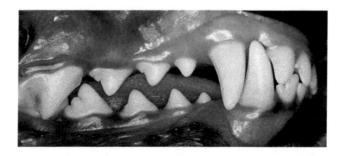

FIGURE 14.1   Photograph of the right side of a mesaticephalic dog's mouth showing proper proportioning and spacing of teeth and occlusal relationships between upper and lower teeth.

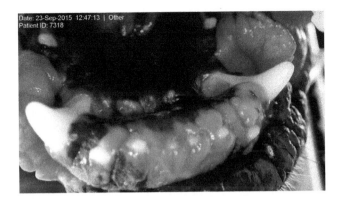

Date: 23-Sep-2015 12:47:13 | Other
Patient ID: 7318

FIGURE 14.2 Photograph of the rostral mandible of a Boxer dog showing traumatic occlusal pits in the mucosa of the floor of the mouth and attrition (tooth-on-tooth wear) and gingival recession, all caused by traumatic contacts from the maxillary incisor teeth secondary to the class III malocclusion.

the upper incisors offered no further benefit. However, in other cases, the upper incisors will repeatedly pack foreign debris into the traumatic pits and subgingival pockets around the lower canines and incisors leading to active inflammation and extraction of the upper incisors and debridement of the mandibular defects will be indicated. Careful clinical and intra-oral dental radiographic evaluation is required to make that assessment for each individual case (Hale 2016b).

Depending on the severity of the class III malocclusion, the lower adult canine teeth may become trapped palatally to the upper second and third incisors, resulting in repeated traumatic contact with the palate. This not only causes trauma to the palate (and possibly through to the roots of the incisors) but may also mechanically prevent the complete eruption of the lower canine teeth, predisposing these to [pericoronitis] (Hale 2011a), which will be discussed in more detail later. The abnormal contacts and impediments to eruption are managed by selective extraction that aims to allow the animal to close its mouth without trauma and to allow important teeth to erupt without impediment. NOTE – this is a time-sensitive problem. If we want the lower canine teeth to erupt fully, we must perform the appropriate extraction at an early age while the teeth still have the potential to erupt (by 7 months of age if at all possible and certainly before a year of age).

Some veterinarians and owners are reluctant to opt for extraction of what they perceive as healthy teeth. However, we need to remember that the upper incisors have become functionally useless and, even worse, are actively causing problems in a class III malocclusion. The overriding objective is to optimise oral health and comfort [Things I tell clients] (Hale 2010). In malformed mouths, this will almost always require proactive extraction of healthy teeth to redress the poor design. While abnormal occlusal contacts may seem a relatively insignificant problem, imagine spending your life walking around with a pebble in your shoe. You could do it, but you would rather not have to. Removing this constant irritant would improve your quality of life.

So far, we have focused on how the upper incisors cause problems with the mandibular structure. It is also not uncommon to find evidence of pulp necrosis in the maxillary incisor teeth of dogs with class III malocclusion (Hale 2017a). This is typically evidenced by coronal discolouration and/or larger than normal pulp chambers in the affected teeth. For more on the diagnosis of endodontic disease, I would refer you to this [Endodontic diagnosis] (Hale 2008b). Repeated concussive trauma to the upper incisors from contacting the mandibular structures may cause a traumatic pulpitis and pulp necrosis, offering another reason to remove the upper incisors at an early age to spare the pet having to endure that painful process.

As well as abnormal contacts between the upper incisors and various mandibular hard and soft tissues, abnormal contacts can also occur further back in the mouth, and therefore a thorough, whole-mouth examination at an early age is indicated for all dogs and cats with brachycephaly. When substantial abnormal contact is found, selective dental extraction is typically the best approach. When two teeth are in contact with each other, it is logical to remove the less important tooth in favour of preserving the more important tooth. In other cases, crown reduction and vital pulp therapy may be an option, depending on access to a practitioner trained in this advanced procedure [Crown reduction] (Hale 2007a).

In this section, I have been talking about extracting permanent (adult) teeth to alleviate traumatic contacts. Dogs with brachycephalism will usually also have abnormal contacts associated with their primary (baby) teeth. In a perfect world, the same approach of selective extraction would be appropriate. However, a cost versus benefit analysis needs to be considered. Extracting the primary teeth in a dog that is genetically programmed to be brachycephalic is very unlikely to result in improved occlusion as the animal grows. Furthermore, extraction of the primary teeth provides only temporary relief until the permanent teeth erupt, at which time reassessment and more intervention will be required.

CAUDAL BUCCAL TRAUMATIC GRANULOMAS

The cheeks of dogs (and some cats) with brachycephalism often are very tight to the teeth, preventing full closure of their mouths without biting their cheek lining (the inside of the cheek is known as the buccal surface). This results in trauma to the buccal mucosa, which can become inflamed and swollen, thereby creating a vicious circle leading to further biting and trauma. As humans, we know how much it hurts when we accidentally bite the inside of our own cheeks. Figure 14.3 shows an older Shih Tzu with significant caudal buccal traumatic granulomas (cheek-chewer lesions) from chronic biting on the right side.

This case was managed by surgical resection of the excess tissue and suturing, although reoccurrence is always possible because the essential anatomy of the mouth has not changed. However, many of these lesions are too broad-based for resection and wound closure without restricting the ability of the dog to open its mouth. In such cases, selective dental extraction is required to alleviate the constant soft tissue trauma as in the case of a one-and-half-year-old Brussels Griffon who had already developed some significant lesions. She was missing her upper fourth premolar and had an under-erupted second molar. I removed both upper molars (the hammers) so that she could close her mouth without traumatising the soft tissue further. As typical, the problem was bilateral and required the same approach on both sides of her mouth. By removing the teeth that are causing the trauma, we can expect a permanent resolution of the problem.

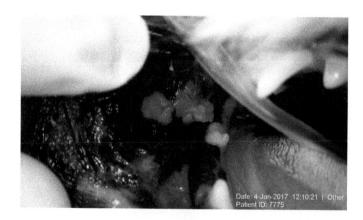

FIGURE 14.3   Photograph of the right caudal buccal pouch mucosa showing the proliferative 'cheek-chewer' lesions resulting from this tissue being traumatized by getting pinched between upper and lower molars.

DENTAL CROWDING AND ROTATION

From an evolutionary perspective, dogs have 20 upper teeth: 10 per side (three incisors, one canine, four premolars and two molars). These teeth should be proportioned and spaced as in the normal example of Figure 14.1. Animals with brachycephalism have malformed heads with upper jaws that are too short to accommodate all 20 teeth with proper spacing and alignment. To understand the impact this has on oral health, you can review [Periodontal anatomy and physiology] (Hale 2003, 2020, 2020c). In short, each tooth should be surrounded completely by an intact collar of gingiva attached to a thin band of cementum-covered root that is above the top of the bony socket. It is the attachment of the gingiva to the suprabony cementum that is the physical barrier between the bacteria in the mouth and the bone that surrounds the roots, i.e. the physical barrier against periodontal disease (Hale 2020c).

Figure 14.4 shows the right maxilla (upper jaw) of a typical Pug where the second and third premolars have rotated 90° to fit in. There is no space (no gum tissue) between the first and second premolars or between the third and fourth premolars. These changes led to periodontal disease affecting all four of these teeth because of infection effectively rotting away the bone around the tooth roots while being hidden from view below the gumline [Periodontal disease is hidden] (Hale 2020d, 2021).

While the upper jaw in dogs with brachycephalism is almost always too short compared to the lower jaw, the lower jaws are also often too short compared to the size of the dog and its teeth. Consequently, dental crowding also often increases the risk for periodontal disease of the mandibular teeth.

Selective extraction offers the only solution to these crowding issues, removing less important teeth to alleviate the over-crowding and improve the prognosis for more important teeth. Again this should be done at an early age as a proactive measure [Proactive dental care] (Hale 2015).

UNDER-ERUPTION AND PERICORONITIS

Figure 14.5 illustrates the ideal relationship between the tooth and its surrounding periodontal tissues (Hale 2020c). If the tooth has erupted properly, the cementoenamel junction is above the top of the bony socket. Periodontal ligament fibres run from bone to cementum within the socket. The gingiva is firmly attached to the bone up to the top of the socket and then attached to the cementum between the top of the socket and the cementoenamel junction. The free gingiva extends further above the cementoenamel junction to lie passively against the enamel. The space between the free gingiva and the enamel is called the gingival sulcus. This is crucial to understand – periodontal

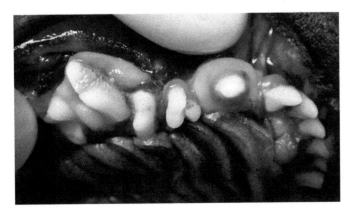

FIGURE 14.4   Photograph of the right maxilla of a Pug showing the dental crowding and rotation of the premolar teeth which led to end-stage periodontal disease.

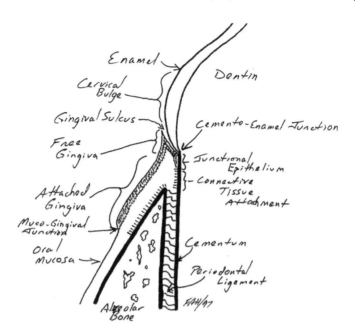

FIGURE 14.5 A line drawing close-up cross section through a tooth and its periodontal tissues demonstrating their desired anatomy and relationships.

ligament and gingiva will not attach well to enamel! How deep should the gingival sulcus be? Well that depends on the size of the animal and the tooth. The sulcus should be no more than 1 mm deep for the small teeth of small dogs, whereas up to 3 mm might be acceptable for the larger teeth of larger dogs [Probing depth] (Hale 2009a, 2020c).

When a tooth fails to erupt fully, more of the enamel-covered crown is trapped below the gumline and maybe even within the bony socket. Since the gingiva and periodontal ligament will not attach to the buried enamel, the effective 'gingival sulcus' is far deeper than it should be, creating a haven for bacteria. Pericoronitis describes the inflammation incited by subgingival bacteria around the buried portion of the crown of an under-erupted tooth. Pericoronitis can allow periodontal disease to progress along the root of the tooth, deeper into the socket.

As discussed earlier, under-eruption may result from mechanical impediment to eruption when the lower canines are in traumatic contact with the palatal aspect of some of the upper incisors. In those cases, timely (6–7 months of age hopefully) alleviation of the mechanical impediment through selective extraction can allow (but will not cause) the canines to erupt acceptably.

However, under-eruption can sometimes happen for no clear reason. The canine teeth of small dog breeds with brachycephaly (e.g. Boston Terrier, Pug, French Bulldog, Shih Tzu) are often dramatically under-erupted despite a lack of any obvious mechanical obstruction to eruption. They simply fail to erupt properly and so there are no proactive actions we can take to encourage eruption. In some cases, it may be possible to surgically expose the crown properly (crown lengthening surgery), but those are advanced periodontal surgeries [Not for the casual operator] (Hale 2008d, 2001, 2012a).

Figure 14.6 of a 9-month-old Shih Tzu shows the lower left fourth premolar (a) and first (b) and second (c) molar teeth. The less-important fourth premolar has erupted well, but the mesial (forward) aspect of the first molar is under-erupted. The green line represents the cementoenamel junction. The red line is the gingival margin. The blue line is highlighting a 'V-shaped' cleft in the bone at the mesial aspect of the first molar.

Also notice how deep into the mandible the mesial root of that molar extends – there is very little solid bone ventral to that root tip. Let's take a short detour here.

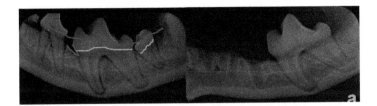

FIGURE 14.6 (a) A preoperative radiograph of the left mandibular fourth premolar tooth and first and second molar teeth of a 9-month-old Shih Tzu demonstrating the under-eruption of the mesial aspect of the first molar and all of the second molar. The (red) line represents the free gingival margin, the (green) line is the cementoenamel junction and the (blue) line is the cleft in the bone mesial to the first molar, predisposing to the early onset and rapid progression of periodontal disease. (b) A post-op radiograph of the same patient with the periodontal anatomy (and future) of the first molar greatly improved.

Each mandible (left and right) is a three-dimensional beam of bone with substantial length and height but not much width (from lip side to tongue side). This narrow beam is perforated with holes (sockets) to accommodate the tooth roots; these creates weak spots that are subject to fracture. In a large dog with well-proportioned teeth, the roots are relatively short compared to the height of the bone (Gioso et al. 2001, Scherer et al. 2018), and the considerable uninterrupted bone below the tooth roots gives the mandible its strength. Now look at the radiograph of that Shih Tzu again and see how the mesial root of the first molar tooth extends almost fully through the ventral cortex (the dense layer of bone along the bottom of the jaw).

If the under-eruption of that molar were left unmanaged, there is a real risk that periodontal disease would develop at the bottom of that cleft (blue V in Figure 14.6) and work down the root, destroying bone along the way, progressively weakening the jaw. Even minor stress might then fracture this mandible, and surgical repair would be very challenging because of the deficiency of bone available for fixation and healing (Hale 2002, 2020c). To reduce the risk of such a catastrophic event, the fourth premolar and second molar (it was also under-erupted with respect to the gingiva) were removed, the bone mesial to the first molar recontoured so that the cementoenamel junction was above the level of the bone (where it should be) and the mesial aspect of that crown was out in the open. As a Shih Tzu, this dog was still at risk for periodontal disease and jaw fracture, but the risk has been reduced (more on jaw fractures below).

In other cases, the best action is simply to extract the under-erupted tooth. Remember that a tooth that has failed to erupt properly is functionally useless and may just be a liability. To optimise longer term oral health, extraction is often indicated.

UNERUPTED TEETH AND DENTIGEROUS CYSTS

Dentigerous cysts are fluid-filled structures that can develop around the crown of an unerupted tooth [Dentigerous cysts] (Hale 2007b). During tooth development, the crown of the tooth is surrounded by the enamel organ, composed of the outer enamel epithelium, the inner enamel epithelium and the stellate reticulum between the two. The inner enamel epithelium is lined by the ameloblasts that form the enamel covering the crown of the tooth. When the enamel organ has finished the task of forming the enamel that covers the crown of the tooth, the stellate reticulum atrophies and the outer enamel epithelium collapsed onto the inner enamel epithelium to become the reduced enamel epithelium. As the tooth erupts, the reduced enamel epithelium is torn away except for a small portion at the cementoenamel junction that remains as the junctional epithelium of the gingival attachment (MacGee, Pinson, and Shaiken 2012).

When a tooth fails to erupt, the reduced enamel epithelium can remain intact, surrounding the crown of the tooth. Cells shed from the inner lining of this sac into the space between the sac epithelium and the crown of the tooth to set up an osmotic gradient that pulls fluid into the space. Over time, as the space fills with more and more fluid, it expands at the expense of the surrounding bone. The cysts start out sterile and asymptomatic, but eventually they can cause sufficient bone loss to result in mobility of teeth, ingress of oral bacteria (so now sepsis) and pathological fractures with very little remaining bone to facilitate surgical repair.

While dentigerous cysts can occur in any breed, they are far more common in brachycephalic breeds and their hybrids (MacGee, Pinson, and Shaiken 2012, Baxter 2004, D'Astous 2011, Thatcher 2017, Okuda et al. 2007). In the author's referral practice, 76 out of most recent 103 dentigerous cysts treated were in dogs with brachycephaly as follows:

Boxer	27	Shit Tzu cross	3
Shih Tzu	20	Lhasa Apso	1
Pug	14	Boston Terrier cross	1
Boston Terrier	5	Boxer cross	1
Lhasa Apso cross	3	Pug cross	1

Figure 14.7 shows the right rostral mandible of a 6.5-year-old Shih Tzu with a dramatic dentigerous cyst centred around an unerupted lower right first premolar tooth. There was no obvious sign of trouble visually in the live animal other than the apparent absence of the lower first premolar.

The teeth look very clean because the referring veterinarian had recently carried out a cleaning and had (quite properly) performed whole-mouth intra-oral dental radiography which identified the cyst. The unerupted first premolar was present, and the cyst had destroyed the bone supporting the incisors, the canine and back to the third premolar. All of those teeth required extraction combined with curettage of the cyst lining prior to wound closure.

Detection and extraction of the unerupted lower first premolar when, for example, this dog was anesthetised for neutering at 6 months of age, would have prevented this dentigerous cysts. It should be the habit of every practitioner to do a thorough oral examination and dental inventory at about 6 months of age to identify abnormalities. Apparent absence of any adult tooth is a definite indication to anesthetise the patient for intra-oral dental radiography. Teeth confirmed as absent through diagnostic intra-oral dental radiographs should be recorded as such in the animal's permanent dental record, and no further clinical action is required. However, teeth that are present but unerupted should be removed immediately, along with the remains of the reduced enamel epithelium, to prevent the formation of a cyst. Had that been done in the case above, the dog would have only had to lose its unerupted (and therefore non-functional) first premolar tooth, rather than losing three incisors, a canine and three premolars with the added risk of a mandibular fracture due to bone loss.

As an aside, Boxers seem especially prone to having supernumerary upper first premolar teeth, often having three copies of this tooth on each side. Sometimes, one or more of these supernumerary first premolars remain unerupted, giving rise to a cyst. Therefore, in Boxers especially, even if a visual examination indicates that all 42 adult teeth have erupted, whole-mouth intra-oral dental radiography is still indicated to check for unerupted supernumerary teeth.

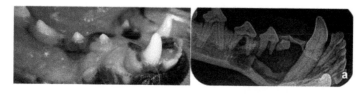

FIGURE 14.7   Clinical photograph (a) and pre-operative intra-oral dental radiograph (b) showing a very large dentigerous cyst in the right mandible of a 6.5-year-old Shih Tzu. The cyst formed around the crown of the unerupted lower right first premolar tooth.

FURRY PALATE

I use the term Furry Palate colloquially because fur entrapment is a very common feature of the condition [Furry Palate] (Hale 2016a) (Regalado Ibarra and Legendre 2019). This condition should more accurately be called 'foreign-body-induced ulcerative palatitis'. The roof of the canine mouth (the hard palate) is covered by tough, dense soft tissue known as palatal mucosa that has ridges known as palatal rugae. In a mesaticephalic dog, these look like waves on the water with crests and troughs. When the upper jaw is an appropriate length, the waves are spaced well apart and the troughs are open to prevent debris becoming trapped. In dogs with brachycephaly, the palatal mucosa is bunched up (like a closed concertina) so that the waves are close together and the troughs become deep slits with opposing side walls. These closed palatal rugae accumulate fur from when dogs groom themselves, food and other debris as well as bacteria. This accumulation often results in an ulcerative palatitis along the trough floors. Figure 14.8a and b shows the palates (a) of an Australian Shepherd dog with normal, healthy palatal mucosa and rugae compared with (b) a Pug with rugae bunched together (among other issues), predisposing to entrapment of debris.

Furry palate tends to happen most commonly in the troughs between the upper incisors and the incisive papilla (the prominent bump just behind the upper first incisor teeth) into the incisive ducts around the papilla and in the rugae between the premolars. Further back in the mouth, between the molars, the troughs tend to be shallower, with reduced predisposition to accumulations of debris.

I cannot offer a cure for furry palate. We cannot remove the palatal mucosa nor can we stretch the maxilla to a desirable length. All I can suggest to owners of dogs with brachycephaly is that they also sweep the roof of the mouth to remove debris from these narrow troughs when they are brushing their dog's teeth (daily). A recent paper in the *Journal of Veterinary Dentistry* offers a treatment involving cautery of the trough floors in an attempt to cause tissue shrinkage and scarification. At time of writing, the author of this text has too little clinical experience with this procedure to comment on its efficacy, but it shows some promise and the paper is an excellent review of palatal mucosal anatomy (Regalado Ibarra and Legendre 2019).

WIDE, LOOSE SYMPHYSIS

Every mammal has two mandibles, one right and one left. In pigs and primates, the mandibles fuse at the midline during puberty. In all other domestic mammals, the mandibles remain as two distinct and separate bones that are associated with each other at the symphysis. The plate of

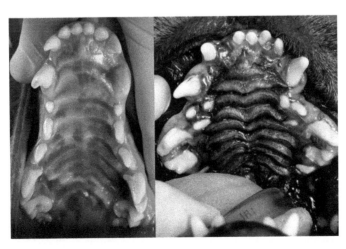

FIGURE 14.8 Photographs of the palates of (a) an Australian Shepherd and (b) a Pug, showing the good palatal mucosal anatomy (wide open troughs between the crests of the palatal rugae) on the one hand and poor anatomy (rugae bunched together to trap food/fur/debris) on the other.

fibrocartilaginous tissue between the rostral ends of the two mandibles is radiolucent, so radiographs of the rostral mandible in dogs and cats will always show a lucent line running down the mid-line.

In many dogs, particularly large-sized, mesaticephalic and dolichocephalic dogs, the fibrocartilaginous plate is thin and firm, and the bone on either side is folded so that the sides interdigitate to create a relatively firm union with only micro-movement between the mandibles. In smaller dogs, especially small-sized brachycephalic breeds, the symphysis may be quite wide and show varying degrees of mobility.

Over time (years), the fibrocartilaginous plate may stretch and become flaccid. Once some mobility develops, continued chewing activities tend to stretch the symphysis further, thus increasing mobility in a self-perpetuating cycle.

When chewing hard food, treats or toys on one side of the mouth, both mandibles are initially pulled closed by equal muscle tension, but then one mandible meets the resistance provided by the food/treat/toy, while the other mandible has its movement restricted at the symphysis. Thus, chewing activities place a strain on a loose symphysis that will further stretch and loosen it over time.

Many dogs with a loose symphysis will show no bony fractures or evidence of osteolytic tumour or infection on intra-oral dental radiography. In many of these cases, the best clinical approach is to DO NOTHING. Do not wrap cerclage wire around the base of the canine teeth or circumferentially around the mandibles just behind the canine teeth. Such a procedure will just cause pain and tissue damage but bring no clinical benefits. Remember, a loose symphysis had no trauma to stimulate healing but just represents a flabby fibrocartilaginous plate.

Wiring a loose symphysis does no good and can do harm. Whether the wire is wrapped around the base of the canines or around the mandibles, it will usually pull the mandibles too close together and alter their alignment when tightened. This can cause the mandibular canine teeth to malocclude with the maxilla and put strain on the temporomandibular joints. The physical presence of the wire in the mouth or running through the soft tissues behind the canines also causes tissue trauma and can lead to infection of the tissues.

If a loose symphysis is deemed 'loose enough' to actually be a clinical problem for the patient (I have never seen such a case), the optimum way to tighten it would be through arthrodesis. A preferable and conservative approach for animals with a clinically relevant loose symphysis would be to have them on a soft diet and withhold hard toys and treats.

INCREASED RISK FOR MANDIBULAR FRACTURE

While not necessarily associated with brachycephaly itself, dogs with brachycephaly do seem over-represented in the group of dogs with mandibular fractures, particularly fractures occurring following relatively minor duress (Gioso et al. 2001, Scherer et al. 2018). Any dog, regardless of size or anatomy, may suffer a mandibular fracture following a motor vehicle accident or similarly violent incident. However, small dogs (and many of the brachycephalic breeds are also breeds of small stature) may additionally suffer mandibular fractures resulting from seemingly minor trauma as mentioned in [Radiographic interpretation] (Hale 2009b, 2008a) and [Focus on mandibular fractures] (Hale 2005a). In a review of the author's records, 29 of 46 (63%) mandibular fractures were in dogs with brachycephaly (21 in Shih Tzus alone) and a further 12 were in dogs under 5 kg [Microdogs] (Hale 2012b).

Small dogs have proportionally much larger teeth than large dogs; hence, the roots of the mandibular teeth, especially the canines and first molars, get much larger relative to the mandible in an inverse relationship with bodyweight (Gioso et al. 2001, Scherer et al. 2018). Consequently, the mandibles tend to be structurally weaker around the mandibular canine teeth and first molars. The dental crowding associated with brachycephalism combined with small stature of many of these breeds brings increased risk for development of periodontal disease which results in bone loss, further weakening an already compromised mandible. The author has seen mandibular fractures in many small dogs or dogs with brachycephaly (or both) that occurred iatrogenically during extraction of diseased teeth, accidentally after falling off of a couch or grooming table, or pathologically after chronic periodontal disease had destroyed so much bone that the jaw simply collapsed (Hale 2005a).

Reducing the risk of mandibular fractures in dogs with brachycephaly starts with the awareness of the risk factors. We cannot make the teeth smaller or the mandibles larger, but we can alleviate crowding and under-eruption. We can develop good, proactive programs to prevent periodontal disease [Whole mouth extraction for everyone] (Hale 2019). And we can inform pet owners of the risks of fracture and advise avoiding activities that might result in fracture such as hard chew toys and tug-of-war games.

It is beyond the scope of this chapter to discuss management of mandibular fractures, but many fine references exist in the professional literature (Verstraete, Lommer, and Bezuidenhout 2012).

GINGIVAL HYPERPLASIA

Gingival hyperplasia is not specifically associated with the distorted brachycephalic craniofacial anatomy, but it is included in this chapter as it is a very common problem in the Boxer dog and the English Bulldog [Gingival hyperplasia] (Hale 2020b). Although the author is unaware that the specific genetics have been elucidated, it seems logical that problems that are so prevalent (while not exclusive) in a single breed, or a few breeds, do have a genetic basis.

The clinical relevance of gingival hyperplasia is similar to that of under-erupted teeth. The excess and redundant gingival tissue lying against the crowns of the teeth creates deep false-pockets that entrap food, fur, other debris and bacteria that combine to promote deeper periodontal infection. The mass effect of the excess tissue can also cause teeth to shift and result in abnormal and painful tooth-to-soft tissue contacts. The dogs literally chew on the excess tissue as they close their mouths.

Treatment for gingival hyperplasia has been well described in the literature (Verstraete, Lommer, and Bezuidenhout 2012, Hale 2020b). In short, the gingiva is surgically recontoured back to a normal (desirable) anatomy. The condition is likely to recur in time, but may theoretically be slowed by maintaining good, daily home plaque control and oral hygiene.

CONCLUSION

Dogs, and to some degree cats, with brachycephaly are at high risk for several oral pathologies that have a negative impact on health and quality of life. It is incumbent upon pet owners and the veterinary profession to recognise and actively seek out these abnormalities and to exert all reasonable efforts to alleviate them. The challenges involved include the 'normalisation phenomenon', the fact that much of the pathology is hidden from view on conscious examination and that animals are masters of masking their distress. Additionally, current under-graduate training for veterinarians with respect to dental and oral health generally does not prepare new graduates well to identify and manage these conditions. We must all do our best to clear those hurdles so that these animals may enjoy optimum oral health.

REFERENCES

American Kennel Club. 2020. "Dog Breeds: This is the official list of all American Kennel Club dog breeds." AKC Global Services, accessed August 10. http://www.akc.org/breeds/index.cfm.

Anderson, J. G., G. Goldstein, K. Boudreaux, and J. E. Ilkiw. 2017. "The state of veterinary dental education in North America, Canada, and the Caribbean: A descriptive study." *Journal of Veterinary Medical Education* 44 (2):358–363. doi:10.3138/jvme.1215-204R.

AVDC. 2020a. "American Veterinary Dental College." American Veterinary Dental College (AVDC), accessed August 8. https://avdc.org/.

AVDC. 2020b. "AVDC Nomenclature." American Veterinary Dental College, accessed August 13. https://avdc.org/avdc-nomenclature/.

Baxter, C. J. K. 2004. "Bilateral mandibular dentigerous cysts in a dog." *Journal of Small Animal Practice* 45 (4):210–212. doi:10.1111/j.1748-5827.2004.tb00227.x.

Bellows, J., M. L. Berg, S. Dennis, R. Harvey, H. B. Lobprise, C. J. Snyder, A. E. S. Stone, and A. G. Van de Wetering. 2019. "2019 AAHA dental care guidelines for dogs and cats." *Journal of the American Animal Hospital Association* 55 (2):49–69.

BVDA. 2020. "BVDA Vet Dentistry." British Veterinary Dental Association (BVDA), accessed August 8. https://www.bvda.co.uk/.

D'Astous, J. 2011. "An overview of dentigerous cysts in dogs and cats." *The Canadian Veterinary Journal* 52 (8):905–907.

Döring, S., B. Arzi, C. R. Barich, D. C. Hatcher, P. H. Kass, and F. J. M. Verstraete. 2017. "Evaluation of the diagnostic yield of dental radiography and cone-beam computed tomography for the identification of anatomic landmarks in small to medium-sized brachycephalic dogs." *American Journal of Veterinary Research* 79 (1):54–61. doi:10.2460/ajvr.79.1.54.

Döring, S., B. Arzi, D. C. Hatcher, P. H. Kass, and F. J. M. Verstraete. 2017. "Evaluation of the diagnostic yield of dental radiography and cone-beam computed tomography for the identification of dental disorders in small to medium-sized brachycephalic dogs." *American Journal of Veterinary Research* 79 (1):62–72. doi:10.2460/ajvr.79.1.62.

EVDS. 2020. "European Veterinary Dental Society." European Veterinary Dental Society (EVDS), accessed August 8. https://www.evds.org/.

Fawcett, A., V. Barrs, M. Awad, G. Child, L. Brunel, E. Mooney, F. Martinez-Taboada, B. McDonald, and P. McGreevy. 2018. "Consequences and management of canine brachycephaly in veterinary practice: Perspectives from australian veterinarians and veterinary specialists." *Animals* 9 (1):3.

Feng, T., C. McConnell, K. O'Hara, J. Chai, and G. Spadafori. 2017. Nationwide Brachycephalic Breed Disease Prevalence Study. edited by NationwideDVM.com. http://nationwidedvm.com/wp-content/uploads/2017/03/NWBrachycelphalicStudy0317.pdf.

Gioso, M. A., F. Shofer, P. S. M. Barros, and C. E. Harvey. 2001. "Mandible and mandibular first molar tooth measurements in dogs: Relationship of radiographic height to body weight." *Journal of Veterinary Dentistry* 18 (2):65–68. doi:10.1177/089875640101800202.

Gyles, C. 2017. "Brachycephalic dogs - time for action." *The Canadian Veterinary Journal (La revue vétérinaire canadienne)* 58 (8):777–780.

Hale, F.A. 2001. "Crown lengthening for mandibular and maxillary canine teeth in the dog." *Journal of Veterinary Dentistry* 18 (4):219–221.

Hale, F. A. 2012a. "Crown lengthing." In *Oral and Maxillofacial Surgery in Dogs and Cats*, edited by F. J. M. Verstraete, M. J. Lommer and A. J. Bezuidenhout, 193–199. Edinburgh; New York: Saunders/Elsevier.

Hale, F. A. 2002. "Management of bilateral, pathologic, mandibular fractures in a dog." *Journal of Veterinary Dentistry* 19 (1):22–24.

Hale, F. 2003. "The anatomy and physiology of the periodontium." Fraser Hale, accessed August 9. http://www.toothvet.ca/PDFfiles/PerioAnat&Physio.pdf.

Hale, F. 2005a. "Focus On: Mandibular Fractures." Fraser Hale, accessed August 10. http://www.toothvet.ca/PDFfiles/MandFx.pdf.

Hale, F. 2005b. "Focus On: The Mandibular Symphysis." Fraser Hale, accessed August 9. http://www.toothvet.ca/PDFfiles/TheSymphysis.pdf.

Hale, F. 2007a. "Crown Reduction for Malocclusions." Fraser Hale, accessed August 9. http://www.toothvet.ca/PDFfiles/CrownReduction.pdf.

Hale, F. 2007b. "Dentigerous Cysts: An Avoidable Catastrophe." Fraser Hale, accessed August 8. http://www.toothvet.ca/PDFfiles/dentigerouscysts.pdf.

Hale, F. 2008a. "The Chevron Sign." Fraser Hale, accessed August 10. http://www.toothvet.ca/PDFfiles/Chevron.pdf.

Hale, F. 2008b. "Diagnosis of Endodontic Disease." Fraser Hale, accessed August 9. http://www.toothvet.ca/PDFfiles/endo_dx.pdf.

Hale, F. 2008c. "Gum chewer syndrome: Introduction to Traumatic Granulomas." Fraser Hale, accessed August 8. http://www.toothvet.ca/PDFfiles/Gum_Chewer.pdf.

Hale, F. 2008d. "NFCO (Not For Casual Operators) - A Cautionary Tale." Fraser Hale, accessed August 10. http://www.toothvet.ca/PDFfiles/nfco.pdf.

Hale, F. 2009a. "Probing depth." Fraser Hale, accessed August 10. http://www.toothvet.ca/PDFfiles/probing.pdf.

Hale, F. 2009b. "Radiographic Interpretation: A Case Study." Fraser Hale, accessed August 10. http://www.toothvet.ca/PDFfiles/Rad_Interp_1.pdf.

Hale, F. 2010. "Things I Tell Our Clients." Fraser Hale, accessed August 8.

Hale, F. 2011a. "Pericoronitis." Fraser Hale, accessed August 8. http://www.toothvet.ca/PDFfiles/pericoronitis.pdf.

Hale, F. 2011b. "Treat or Extract?" Fraser Hale, accessed August 8. http://www.toothvet.ca/PDFfiles/treat_or_extract.pdf.

Hale, F. 2012b. "Focus On:"Micro-Dogs"." Fraser Hale, accessed August 8. http://www.toothvet.ca/PDFfiles/microdogs.pdf.

Hale, F. 2013a. "Feline Gum Chewer Syndrome; Feline Caudal Buccal Traumatic Granulomas." Fraser Hale, accessed August 8. http://www.toothvet.ca/PDFfiles/Feline_Gum_Chewer.pdf.

Hale, F. 2013b. "Stop brachycephalism, now!" *The Canadian Veterinary Journal (La Revue Veterinaire Canadienne)* 54:185–186.

Hale, F. 2014. "Stop Brachycephalism, Now! pdf." Hale Veterinary Clinic, accessed August 8. http://www.toothvet.ca/PDFfiles/Stop_Brachy.pdf.

Hale, F. 2015. "Proactive dental care - A tale of Two Brothers." Fraser Hale, accessed August 10. http://www.toothvet.ca/PDFfiles/proactive.pdf.

Hale, F. 2016a. Furry Palate - Video.

Hale, F. 2016b. "Minor brachycephalism in a bull mastiff." Fraser Hale, accessed August 13. http://www.toothvet.ca/facebook_pdfs/mastiff%20MAL3.pdf.

Hale, F. 2017a. "Brachycephalic Dental/Oral Issues." Fraser Hale, accessed August 13. http://www.toothvet.ca/PDFfiles/Stop_Brachy_2.pdf.

Hale, F. 2017b. "More on brachycephalism: Why do I say "Stop Brachycephalism, Now!"?" Fraser Hale, accessed August 8. http://www.toothvet.ca/PDFfiles/Stop_Brachy_2.pdf.

Hale, F. 2018. "Malocclusions - What to do about them, when and why." Fraser Hale, accessed August 13. http://www.toothvet.ca/PDFfiles/malocclusions.pdf.

Hale, F. 2019. "Whole Mouth Extraction for Everyone." accessed August 10. http://www.toothvet.ca/PDFfiles/WME_for_all.pdf.

Hale, F. 2020. "Periodontal Disease in Dogs & Cats; Part 1. Normal anatomy & progression of disease." accessed April 29, 2021. https://youtu.be/uRExyTGheAM.

Hale, F. 2020a. "Dental Mythology." Fraser Hale, accessed August 8. http://www.toothvet.ca/PDFfiles/mythology.pdf.

Hale, F. 2020b. "Focus On: Gingival Hyperplasia." Fraser Hale, accessed August 9. http://www.toothvet.ca/PDFfiles/gingival_hyperplasia.pdf.

Hale, F. 2020c. Periodontal Disease in Dogs & Cats; Part 1. Normal anatomy & progression of disease - Video.

Hale, F. 2020d. "Periodontal Disease: Outta Sight!". Fraser Hale, accessed August 8. http://www.toothvet.ca/PDFfiles/perio_hidden.pdf.

Hale, F. 2021. "Dental Crowding and its Effect on Periodontal Health." acccessed April 29, 2021. http://www.toothvet.ca/PDFfiles/Crowding.pdf.

Hellyer, P., I. Rodan, J. Brunt, R. Downing, J. E. Hagedorn, and S. A. Robertson. 2007. "AAHA/AAFP pain management guidelines for dogs and cats." *Journal of Feline Medicine and Surgery* 9 (6):466–480. doi:10.1016/j.jfms.2007.09.001.

Johnson, J.M. 1991. "The veterinarian's responsibility: Assessing and managing acute pain in dogs and cats. Part 1." *The Compendium on Continuing Education for the Practicing Veterinarian (USA): Small Animals* 13 (5):804–809.

Kerr, D. J. 2018. "Change is needed to correct the lack of veterinary dental education at the undergraduate level." *The Canadian Veterinary Journal (La revue veterinaire canadienne)* 59 (3):314–314.

MacGee, S., D.M. Pinson, and L. Shaiken. 2012. "Bilateral dentigerous cysts in a dog." *Journal of Veterinary Dentistry* 29 (4):242–249. doi:10.1177/089875641202900405.

Mestrinho, L.A., J.M. Louro, I.S. Gordo, M.M.R.E. Niza, J.F. Requicha, J.G. Force, and J.P. Gawor. 2018. "Oral and dental anomalies in purebred, brachycephalic Persian and Exotic cats." *Journal of the American Veterinary Medical Association* 253 (1):66–72. doi:10.2460/javma.253.1.66.

Niemiec, B., J. Gawor, A. Nemec, D. Clarke, K. McLeod, C. Tutt, M. Gioso, P.V. Steagall, M. Chandler, G. Morgenegg, and R. Jouppi. 2020. "World small animal veterinary association global dental guidelines." *Journal of Small Animal Practice* 61 (7):E36–E161. doi:10.1111/jsap.13132.

O'Neill, D. G., A. M. O'Sullivan, E. A. Manson, D. B. Church, A. K. Boag, P. D. McGreevy, and D. C. Brodbelt. 2017. "Canine dystocia in 50 UK first-opinion emergency-care veterinary practices: Prevalence and risk factors." *Veterinary Record* 181 (4). doi:10.1136/vr.104108.

O'Neill, D. G., L. Baral, D.B. Church, D. C. Brodbelt, and R.M. A. Packer. 2018. "Demography and disorders of the French Bulldog population under primary veterinary care in the UK in 2013." *Canine Genetics and Epidemiology* 5 (1):3. doi:10.1186/s40575-018-0057-9.

O'Neill, D. G., E.C. Darwent, D. B. Church, and D. C. Brodbelt. 2016. "Demography and health of Pugs under primary veterinary care in England." *Canine Genetics and Epidemiology* 3 (1):1–12. doi:10.1186/s40575-016-0035-z.

O'Neill, D. G., A. M. Skipper, J. Kadhim, D. B. Church, D.C. Brodbelt, and R.M. A. Packer. 2019. "Disorders of Bulldogs under primary veterinary care in the UK in 2013." *PLoS One* 14 (6):e0217928. doi:10.1371/journal.pone.0217928.

O'Neill, D.G., M. M. Lee, D. C. Brodbelt, D. B. Church, and R. F. Sanchez. 2017. "Corneal ulcerative disease in dogs under primary veterinary care in England: Epidemiology and clinical management." *Canine Genetics and Epidemiology* 4 (1):5. doi:10.1186/s40575-017-0045-5.

Okuda, A., T. Eguchi, Y. Kishikawa, N. Ichihara, and M. Asari. 2007. "Review of 117 impacted (98 permanent and 19 deciduous) teeth in 46 dogs." *Proceedings, 21st Ann Vet Dental Forum.*

Packer, R. M. A., A. Hendricks, M. S. Tivers, and C. C. Burn. 2015. "Impact of facial conformation on canine health: Brachycephalic obstructive airway syndrome." *PLoS One* 10 (10):e0137496. doi:10.1371/journal.pone.0137496.

Perry, R. 2014. "Final year veterinary students' attitudes towards small animal dentistry: A questionnaire-based survey." *Journal of Small Animal Practice* 55 (9):457–464. doi:10.1111/jsap.12258.

Price, T. J., and G. Dussor. 2014. "Evolution: The advantage of 'maladaptive' pain plasticity." *Current Biology* 24 (10):R384–R386. doi:10.1016/j.cub.2014.04.011.

Ravetz, G. 2017. "Stop normalising suffering: Vets speaking out about brachys." British Veterinary Association, accessed April 8. https://www.bva.co.uk/news-and-blog/blog-article/stop-normalising-suffering-vets-speaking-out-about-brachys/.

Regalado Ibarra, A.M., and L. Legendre. 2019. "Anatomy of the brachycephalic canine hard palate and treatment of acquired palatitis using CO_2 laser." *Journal of Veterinary Dentistry* 36 (3):186–197. doi:10.1177/0898756419893127.

Scherer, E., C.J. Snyder, J. Malberg, B. Rigby, S. Hetzel, and K. Waller. 2018. "A volumetric assessment using computed tomography of canine and first molar roots in dogs of varying weight." *Journal of Veterinary Dentistry* 35 (2):131–137. doi:10.1177/0898756418777861.

Thatcher, G. 2017. "Oral surgery: Treatment of a dentigerous cyst in a dog." *The Canadian Veterinary Journal (La revue veterinaire canadienne)* 58 (2):195–199.

The Kennel Club. 2020. "Breed Information Centre." The Kennel Club Limited, accessed August 10. http://www.thekennelclub.org.uk/services/public/breed/.

Verstraete, F. J. 1998. "Diagnostic value of full-mouth radiography in cats." *American Journal of Veterinary Research* 59 (6):692.

Verstraete, F. J., P. H. Kass, and C. H. Terpak. 1998. "Diagnostic value of full-mouth radiography in dogs." *American Journal of Veterinary Research* 59 (6):686–691.

Verstraete, Frank J. M., Milinda J. Lommer, and A. J. Bezuidenhout. 2012. *Oral and Maxillofacial Surgery in Dogs and Cats.* Edinburgh; New York: Saunders/Elsevier.

15 Brain Disorders Associated with Brachycephaly

Clare Rusbridge
Fitzpatrick Referrals
University of Surrey

Susan Penelope Knowler
University of Surrey

CONTENTS

Introduction 251
   Morphogenesis 253
   Cerebrospinal Fluid Physiology 256
      CSF Production and Reabsorption 256
      The Glymphatic System and Aquaporins 256
Clinical Syndromes 257
   Hydrocephalus 257
      Clinical Signs 258
      Pathophysiology 259
      Diagnosis 260
      Treatment and Prognosis 264
   Supracollicular Fluid Collections and Quadrigeminal Cistern Expansion 267
      Clinical Signs 268
      Pathogenesis 268
      Diagnosis 269
      Treatment and Prognosis 270
   Chiari-Like Malformation and syringomyelia 270
      Pathogenesis 270
      Clinical Signs 273
      Diagnosis 273
      Treatment and Prognosis 279
      Genetic Factors and Breeding Advice 281
Conclusions 282
Acknowledgement 282
Appendix: Differential Diagnosis of Brain Disorders in Brachycephalic Small Dog Breeds 282
References 284

INTRODUCTION

Brain disorders of the brachycephalic patient can be broadly divided into conditions that are directly related to a brachycephalic skull and conditions with suspected inherited tendency. The focus of the chapter is diseases associated with a brachycephalic skull. Breed dispositions to neurological disorders that are important differential diagnoses are covered in the Appendix.

Brachycephalic dog breeds have a shortened rostrum, wide zygomatic arches, a rounded neuro-cranium, dorsally rotated rostrum (airorhynchy) and a wide, flattened or convex palate (Geiger and Haussman 2016). Like hand in glove, brain conformation mirrors the cranial cavity, reflecting not only rostrocaudal shortening but rotation on the medial lateral axis with ventral displacement and reduction of the olfactory bulbs (Roberts, McGreevy, and Valenzuela 2010, Knowler, Cross, et al. 2017). (Figure 15.1)

Brachycephaly has the potential to impact the nervous system in three ways:

1. Alter cerebrospinal fluid (CSF) absorption and movement predisposing to CSF disorders such as ventriculomegaly and hydrocephalus, quadrigeminal cistern expansion, Chiari malformation (CM) and syringomyelia (SM).
2. Altered brain and blood vessel conformation which may reduce compliance and increase intracranial pressure contributing to the tendency for CSF disorders.

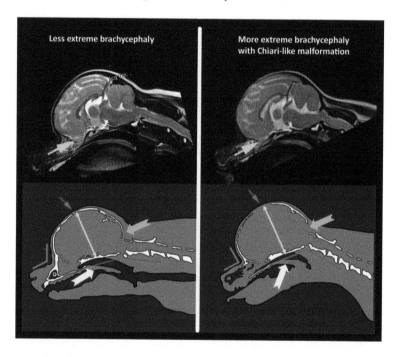

FIGURE 15.1　Hand in glove – neuroparenchymal and skull changes with more extreme brachyceph-aly. Midsagittal T2W brain MRI (top row) and line drawing (bottom row) made from the CT of two female sibling Chihuahuas. The 1.6 kg dog on the right has more extreme brachycephaly than her 2.5 kg sibling on the left. In addition to skull base shortening, the angle of stop (blue arrow) becomes more acute. With more extreme brachycephaly, brain rotation on the medial lateral axis with ventral displacement and reduction of the olfactory bulbs is more pronounced (orange arrow). The cranium increases in height (green arrow) and the molera (persistent bregmatic fontanelle; red arrow) is wider. Altered occipital bone conformation changes the angle between the skull base and the cervical vertebrae resulting in a cervical flexure (red line) and dorsal tipping of the dens into the spinal cord. The oropharynx is displaced caudally (yellow arrow). There is also a Chari-like malformation, and the supraoccipital bone is shorter and straighter (pink arrow). It has failed to ossify ventrally. The atlas is closer to the skull resulting in craniocervical junction overcrowding.

Video 1 **Hand in glove – neuroparenchymal and skull changes with more extreme brachycephaly**. A short video demonstrates Figure 15.1 in action (available from https://www.routledge.com/ 9780367207243 - Support Material tab). It shows the neuroparenchymal and skull changes with more extreme brachycephaly. The morph from the line drawings of Figure 15.1 illustrates the morphology of extreme brachycephaly in a dynamic manner. To enable the morph, the images have been adjusted to the same size – the 1.6kg sibling is smaller that her 2.5kg sister.

3. The third, unproven, under investigated impact is that brachycephaly may alter brain development and white matter integrity. Rabbits with familial coronal suture craniosynostosis have altered white matter tissue microstructure (Bonfield et al. 2015). Humans with brachycephalic craniofacial disorders have an increased prevalence of corpus callosum agenesis, hippocampal hypoplasia, septal defects and ventricular dilatation. Hippocampal abnormalities are proposed to be a consequence of distortion of the brain by the calvarial deformity (Tokumaru et al. 1996). Humans with psychosis are more likely to be brachycephalic with wider skull bases, shorter lower facial heights and more protruding ears; it is hypothesised that differences in temporal lobe development accompany the facial dysmorphia (McGrath et al. 2002).

Morphogenesis

Embryologically the skull is a complex structure consisting of three parts – the phylogenetically oldest facial skeleton (viscerocranium) which joins the neurocranium of the cranial base (skull base or chondrocranium) and cranial vault (calvarium; dermatocranium). The calvarium, facial bones and squamous occipital bone develop by membranous ossification, whereas the cranial base, the trough in which the brain sits, differentiates from cartilage by endochondral ossification (Table 15.1 and Figure 15.2). The skull and all the contained tissues within must have a coordinated development; selection for certain facial features jeopardises that coordination. Brachycephaly in dogs and cats is a consequence of selecting for juvenile characteristics of flattened face and a 'domed' or 'apple' head (Stockyard 1941).

The ultimate shape of the skull is determined by the growth of the cranial sutures (fibrous joints between calvarial bones). Sutures develop at the site of dural reflections, where the inner dural layer descends into the brain forming septa which divide into the brain into compartments (for example, the falx cerebri and the tentorium cerebelli). The dura has a mechanosensory capacity, and it responds to mechanical stress from the expanding brain and in so doing controls skull growth. The dura also secretes morphogenetic soluble factors, for example peptide growth factors and chemokines (Di Ieva et al. 2013, Tubbs, Bosmia, and Cohen-Gadol 2012). If one or more sutures close

---

TABLE 15.1
Embryological Origin and Ossification of the Skull Bones

Skull	Bones	Embryological Origin	Ossification
Cranial base	**Rostral to sella turcica** ethmoid, presphenoid and basisphenoid bones	Neural crest	Endochondral ossification of cartilage precursors – mesenchymal condensation, chondrification and ossification
	Caudal to sella turcica basisphenoid, basioccipital, exoccipitals, ventral supraoccipital, petrous and mastoid temporal	Paraxial mesoderm (First five somites)	
Cranial vault	Frontal, parietal, squamous temporal, supraoccipital (dorsal, interparietal and preinterparietal parts)	Mesenchyme from cranial neural crest	Intramembranous – no cartilaginous stage, mesenchyme forms a connective tissue capsule in which bony islands develop that ultimately form bones
Facial skeleton	Nasal, maxillae, premaxillae, zygomatic, mandible	First three branchial arches pharyngeal arches (mesenchyme core enclosed by neural crest)	Intramembranous ossification

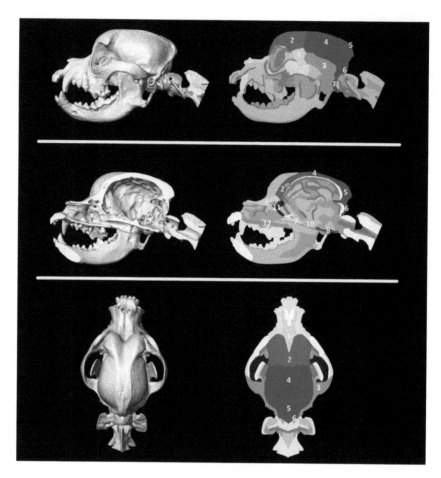

FIGURE 15.2 Canine skull bones. Reconstructed CT and line diagrams illustrating the skull bones from a clinically normal male Cavalier King Charles Spaniel. (Top) Lateral skull; (Middle) midsagittal skull; (Bottom) dorsal aspect of skull. Individual skull bones are numbered as follows: 1. nasal, 2 frontal, 3. temporal, 4. parietal, 5. interparietal, 6. supraoccipital, 7. exoccipital, 8 basioccipital, 9. presphenoid, 10. basisphenoid, 11. ethmoid (cribriform plate) and 12. hard palate (palatine and maxillae).

prematurely (craniosynostosis), then compensatory growth occurs in a plane parallel to the fused suture with minimal growth in a perpendicular plane (Virchow 1851). Hence, premature closure of skull-base sutures (basispheno-presphenoid synchondrosis and spheno-occipital synchondrosis) will result in a calvarial doming, the neonatal characteristic that humans find so appealing. However, brachycephalic dogs and cats also have premature closure of other cranial and facial sutures when compared to other skull types (Geiger and Haussman 2016, Schmidt et al. 2017). Skull suture closure also differs between domestic dogs and the wolf; indeed, it is postulated that the pattern of suture and synchondrosis closure determines the myriad of skull shapes in domestic dogs (Geiger and Haussman 2016). With more extreme brachycephaly and compound craniosynostosis, the ability of the skull to compensate may be overcome resulting in neuroparenchymal overcrowding. In addition, premature closure of skull base sutures has far reaching consequences. The skull base forms a platform on which the rest of the skull grows; the structures of the upper respiratory tract and oropharynx hang off it; the foramina for the cranial nerves and vessels develop within it; the foramen magnum accommodates the craniospinal junction; and the occipital condyles connect the skull to the cervical spine. Anatomical consequences of skull base shortening are described in Table 15.2; also see Figure 15.1.

TABLE 15.2

Anatomical Consequences of Skull Base Suture Craniosynostosis (Brachycephaly)

Anatomical Feature	Potential Consequences	References and Further Reading
Calvaria	Expansion and increased height	(Stockyard 1941, Geiger and Haussman 2016)
Skull base angulation	Sphenoid flexure – the skull base is flexed dorso-caudally which is mirrored by a corresponding 'concertina' brain flexure.	(Knowler, Galea, and Rusbridge 2018)
Cribriform plate	Ventral rotation.	(Dubrul and Laskin 1961)
Foramen magnum, occipital condyles, nuchal crest	Rostral displacement and ventral rotation.	(Dubrul and Laskin 1961, Hoyte 1997)
Cervical vertebral	Altered occipital condyle conformation affects the conformation of craniocervical junction which contributes to cervical flexure (ventral deviation of the neck), odontoid peg angulation and craniospinal junction conformation changes in dogs with SM secondary to CM and comparable to basilar invagination in humans.	(Knowler, Galea, and Rusbridge 2018, Donnally, Munakomi, and Varacallo 2020, Shoja et al. 2018)
Upper respiratory tract and oropharynx	The upper respiratory tract (for example, larynx and hyoid) is suspended from the cranial base. Skull base shortening may displace these tissues caudally and may affect swallowing and respiration especially if there are other conformation problems reducing airway space. In humans these conformational changes are associated with obstructive sleep apnoea and sleep disordered breathing.	(Reidenberg and Laitman 1991, Neelapu et al. 2017)
Ear	Skull base shortening may change the angle and rostrally displace the tympanic bullae within the temporal bone and distort the ear canals. Dogs with CM are predisposed to otitis media with effusion (OME; primary secretary otitis media), as are humans with syndromic and complex craniosynostosis. This is presumed to be due altered palatal anatomy, reduced eustachian tube opening and poor middle ear ventilation.	(Dubrul and Laskin 1961, Wijnrocx et al. 2017, Hayes, Friend, and Jeffery 2010, Bluestone and Swarts 2010)
Cranial foramina	Potentially small or malformed. The jugular foramina (between the temporal and occipital bone; providing passage to cranial nerves IX, X, XI, sigmoid sinus and lymphatics) is reduced in Cavalier King Charles Spaniels (CKCS) with SM associated with CM. The sigmoid sinus drains the transverse sinus, and theoretically, stenosis could disrupt CSF absorption and elevate intracranial pressure. Resorption of CSF occurs through the lymphatics.	(Schmidt et al. 2012)
Vasculature	CKCS with SM associated with CM have reduced volume of the caudal cranial fossa dorsal sinuses. Humans with and mouse models of craniosynostosis may have stenosis of the dural sinuses (cerebral veins) independent to the skull malformation. These persist after skull expansion surgery and may lead to elevated intracranial pressure. It has been shown that the bone morphogenic protein (BMP) signalling pathways from skull preosteoblasts and periosteal dura influence vascular development, and it is hypothesised that genetic mutations that affect BMP pathways may influence cerebral vein development and physiology.	(McGonnell and Akbareian 2019, Tischfield et al. 2017, Fenn et al. 2013)
Lymphatics	Reabsorption of CSF occurs through the lymphatics especially those passing through the skull base foramen (perineural and meningeal). Foremen stenosis, hypothetically, could impede lymphatic drainage. Olfactory perineural drainage through the cribriform plate is also paramount for CSF resorption. Reduction in surface area of the ethmoid bone and turbinals associated with airorhynchy could theoretically impede drainage.	(Ahn et al. 2019, Da Mesquita, Fu, and Kipnis 2018, Ma et al. 2017, Wagner and Ruf).

CEREBROSPINAL FLUID PHYSIOLOGY

Brachycephaly in dogs and cats predisposes disorders of CSF. To understand the pathogenesis of this requires an understanding of the function, production, movement and absorption of this body fluid. CSF was first documented by the Egyptians (The Edwin Smith Papyrus) between 3000 and 2500BC, and the first description of its production and flow from the ventricles is attributed to Galen of Pergamon (130–200) (Compston 2015, Deisenhammer 2015). Disorders of CSF and hydrocephalus are among the earliest recognised neurological conditions in dogs (Lindsay 1871), and the predisposition of toy breed brachycephalic dogs to ventriculomegaly and hydrocephalus is evident from the time that they first became fashionable, as evidenced by skulls in museum collections (Taylor 1830). However, despite significant gain of collective insight into CSF physiology and pathology, we still lack understanding on how CSF is produced and circulated (Miyajima and Arai 2015, Brinker et al. 2014, Bulat and Klarica 2011). Traditional thinking is that for dogs and cats, approximately 40% of CSF is secreted by the choroid plexus into the ventricular spaces, whereupon it circulates through the ventricular system, exits into the subarachnoid space through the lateral apertures and (variably) the ventriculus terminalis, being absorbed through the arachnoid villi and venous sinuses with a minor quantity through the lymphatics associated with the perineurial sheaths of cranial and spinal nerves (DeLahunta, Glass, and Kent 2014, Marin-Garcia et al. 1995, Orešković and Klarica 2010, Sato et al. 1975). However, this old-style description leaves poignant unanswered questions:

- What is the mechanism of ventriculomegaly in brachycephalic animals? For decades, this was attributed to mesenchymal aqueduct stenosis, but with the benefit of MRI, it is apparent that most brachycephalic animals with ventriculomegaly have normal or even a dilated mesencephalic aqueduct.
- Although the choroid plexus has a large surface area, how does a gland measuring a few millimetres actively secrete 40% of 35–50 µL (dog)/14–21 µL (cat) of fluid per minute (Orešković and Klarica 2010)?
- Why does experimental removal of the lateral ventricle choroid plexi not cure hydrocephalus or even affect the volume and composition of CSF produced (Milhorat 1972)?
- Why does experimental obstruction of the mesencephalic aqueduct in cats and dogs not change lateral ventricular pressure – suggesting that the formation and the absorption of CSF are balanced in normal animals (Bulat and Klarica 2011, Rekate 2011)?

CSF Production and Reabsorption

Recently, there has been a shift in understanding from a choroid-plexus-arachnoid-granulation-centric model of CSF production and absorption to an interstitial-fluid-lymphatic-vessel-centric model. This shift in understanding, summarised in Table 15.3, is particularly relevant to the brachycephalic dog. The irony is that the twenty-first-century hypothesis is supported by nineteenth-century experiments in dogs. Gustav Schwalbe injected Berlin blue dye into the canine subarachnoid space and showed that the CSF drained through the lymphatics (Schwalbe 1869). There is now overwhelming evidence that the bulk of CSF is absorbed through perineurial lymphatics (olfactory bulbs, cranial and spinal nerves) and meningeal lymphatic vessels of the skull and sacral spine (Ma et al. 2019, 2017). By contrast, CSF absorption through the arachnoid granulations is insignificant at normal CSF pressure and more likely serves as an 'overflow valve mechanism' when there is elevated CSF pressure (Bulat and Klarica 2011).

The Glymphatic System and Aquaporins

The fluid of the CSF and the brain interstitial space is in continuous flux. In the ventricles, CSF diffuses through the ependyma, and in the subarachnoid space, CSF drains into perivascular (Virchow–Robin) spaces, from where it may exchange with the interstitial fluid before emptying into lymphatic system (Casaca-Carreira et al. 2018, Ma et al. 2017). The exchange with interstitial fluid is via and

TABLE 15.3

Traditional versus Modern Understanding of CSF Physiology

CSF	Traditional Interpretation	Current Interpretation
Main site production	Choroid plexus.	Still under debate but contribution of brain interstitial fluid, leptomeningeal and parenchymal capillaries and ependyma considered greater than previously.
Main site reabsorption	Venous sinuses via arachnoid granulations.	Lymphatics (perineural and meningeal).
Movement through ventricular system	Choroid plexus CSF pump.	Respiration.
Movement through brain	Insignificant/irrelevant.	Glymphatic system aided by arterial pulsation.
Role[a]	Hydraulic cushion.	Hydraulic cushion.
	Buffers changes in intracranial pressure and reduces risk of volume shifts or herniation.	Buffers changes in intracranial pressure and reduces risk of volume shifts or herniation.
	Sink for nervous system waste.	CSF pressure affects brain development.
		Choroid plexus secretes proteins into CSF that regulate neural stem cells.
		Choroid plexus is the port of entry of immune cells into CNS.
		Glymphatic system delivers nutrients and helps clear toxins and metabolic waste products (such as amyloid-β).

[a] Reviewed by Lun, Monuki, and Lehtinen (2015).

controlled by aquaporins – water transporting pores in the foot processes of astrocytes wrapped around capillaries (Bulat and Klarica 2011, Jessen et al. 2015, Albertini and Bianchi 2010). This glymphatic movement of fluids facilitates elimination of waste protein, including β-amyloid and hydroxylated cholesterol, and metabolites and distribution of neurotransmitters, amino acids, glucose and lipid (Thomas 2019).

How brachycephaly could affect CSF movement and absorption and predisposed CSF disorders is described in Table 15.4.

CLINICAL SYNDROMES

HYDROCEPHALUS

Hydrocephalus is defined as an active distension of the ventricular system of the brain resulting from inadequate passage of CSF from its point of production within the cerebral ventricles to its point of absorption into the systemic circulation (Rekate 2008). It is classified as **obstructive** due to impediment of CSF channels and **communicating** due to imbalance in CSF production and elimination. However, distribution of individual cases into this classification is controversial and inconsistent. For example, for many, a true communicating hydrocephalus is when there is no resistance to flow including drainage into the lymphatics and arachnoid granulations. Therefore, most cases are obstructive (Rekate 2011). This lack of consistency is not least in part because these terms were first coined in the early twentieth century by Dandy using limited investigative techniques (injection of phenolsulphonephthalein into the lateral ventricle then immediately testing for it in spinal CSF; absence of the dye confirmed diagnosis of obstructive hydrocephalus) (Dandy and Blackfan 1913). The overall expert consensus opinion is that most forms of hydrocephalus can be classified according to an identifiable anatomical site of CSF flow obstruction or impairment (Del Bigio and Di Curzio 2016, Rekate 2011).

TABLE 15.4

Effect of Brachycephaly and Airorhynchy on CSF Absorption and Dynamics

Reduction of the cribriform plate and nasal mucosa

There is a direct continuity between the subarachnoid space and the perineurial spaces of the olfactory nerve fibres penetrating the cribriform plate. These nerve fibres are encircled by lymphatics and provide one of the main drainage routes of CSF (Zakharov, Papaiconomou, and Johnston 2004). If the surface area of lymphatic vessels is reduced, for example, by airorhynchy, then CSF absorption may be compromised (Sokolowski et al. 2018). When this route was blocked in sheep (external ethmoidectomy, removal of the olfactory nerves and mucosa, cribriform bone surface sealed with tissue glue) in addition to restricting the craniocervical CSF channels (ligature around the thecal sac at C1/C2), there was elevation in intracranial pressure; elevation in pulse pressure amplitude (and thereby reduction in parenchymal compliance); and impairment of intracranial pressure accommodation. In comparison, the sham surgery group (C1/C2 ligature and exposure only) did not have a significant change in intracranial pressure or pulse pressure amplitude (Mollanji et al. 2002, Mollanji et al. 2001). In this experimental sheep model, blockage of cranial cervical CSF channels in addition to the cribriform plate was necessary as 25% of global CSF transport occurs in the spinal subarachnoid compartment. This can be compared to SM in the dog where a combination of craniofacial hypoplasia and craniospinal junction abnormalities predisposes spinal cord cavitation (Knowler et al. 2020).

Reduction of skull base foramen

In addition to olfactory perineural drainage, basal meningeal lymphatics and cranial nerve perineural lymphatics are important for CSF drainage (Ahn et al. 2019, Da Mesquita, Fu, and Kipnis 2018, Ma et al. 2017). The lymphatic vessels pass through skull base foreman, and, hypothetically, reduction in the skull base could result in stenosis of the foramens and impede drainage.

Effect on venous drainage

Craniosynostosis in human and rodent models have stenosis of the dural sinus stenosis which is independent to the skull malformation and which persists after skull expansion surgery. CKCS with SM associated with CM have reduced volume of the caudal cranial fossa dorsal sinuses. It is hypothesised that this may be due to an effect on bone morphogenic protein (BMP) signalling pathways which influence cerebral vein development and physiology (McGonnell and Akbareian 2019, Tischfield et al. 2017, Fenn et al. 2013).

Brachycephalic obstructive airway disease

Respiration drives CSF movement through the ventricular system. The systolic pulse contributes to a lesser degree; the pulse of choroid plexus secretion is not relevant (Yamada et al. 2013). There is a bidirectional flow with rostral movement of CSF during deep inhalation and caudal movement during deep exhalation (Bothwell, Janigro, and Patabendige 2019). Apnoea impedes CSF movement and apnoea induced negative intrathoracic pressure will reduce CSF drainage through the venous sinuses (Sugita et al. 1985, Román et al. 2019). This has obvious implications for brachycephalic dogs with BOAS especially if they have sleep disordered breathing. Intracranial pressure increases during sleep and disordered sleep will also affect the glymphatic drainage (Jessen et al. 2015).

Clinical Signs

Animals with hydrocephalus are most commonly presented with forebrain signs relating to lateral ventricle enlargement including change in behaviour often with irritability; reduced vision or central blindness; depressed auditory acuity; seizures; circling and pleurotonus (head or neck turning to one side); and postural deficits ipsilateral to visual deficits and contralateral to circling. There may be head or cervical pain which worsens or is elicited on rapid changes of head position. Head carriage may be low. Neurological examination may also find evidence of more widespread brain disease such as ataxia which has a cerebellar or vestibular quality. There may be more obvious cerebellovestibular disease if there is enlargement of the fourth ventricle or a supracollicular fluid collection. If there is an associated SM, then there may be signs localising to the central spinal cord particularly thoracic limb weakness. Clinical hydrocephalus in brachycephalic dogs is often developmental in which case there is marked doming of the calvarium and enlarged fontanelle especially the bregmatic fontanelle (molera) at the junction of the sagittal, coronal and frontal sutures. There is often a ventral strabismus (often referred to as sunset or setting sun eyes) (Figure 15.3). Most clinically affected brachycephalic animals are presented as immature animals reflecting the developmental

FIGURE 15.3 **Dysmorphia of developmental hydrocephalus**. There is cavarial doming and downward rotation of the eyeballs (sunset eyes). This may be accompanied by a retraction of the upper eyelids and 'raising of the brow'. The aetiology of ventral eye position is debated with the most accepted explanation being that rostral expansion of the calvaria distorts the orbits and displaces the eyeball ventrally (DeLahunta, Glass, and Kent 2014). See also Figure 15.7. Some cases re-establish normal eye position after surgery; typically, these are juvenile animals, and this improvement may reflect skull remodelling. The alternative explanation for the strabismus is an upward gaze palsy from compression of the tectal and pretectal region from the distended third ventricle (Chattha and Delong 1975).

aetiology. Adult-onset hydrocephalus in brachycephalic animals is most likely decompensation of a 'compensated' developmental hydrocephalus or because of an acquired blockage, for example, following a bleed or arachnoid web (Figure 15.4).

Pathophysiology

Hydrocephalus is a consequence of abnormalities in the flow (obstruction) or resorption of CSF. The most common sites of obstruction are areas of narrowing in the CSF channels, for example, at the interventricular foramen, mesencephalic aqueduct, lateral apertures or craniocervical junction. The authors hypothesise that brachycephalic animals are predisposed to hydrocephalus due to reduced absorption of CSF through lymphatics (see under 'Cerebrospinal Fluid Physiology') which may be the primary or contributory cause of the ventricular dilatation. The other important cause of tri-ventricular hydrocephalus in brachycephalic dogs is aqueductal stenosis associated with mesencephalic developmental anomalies including fusion of the rostral colliculi (DeLahunta, Glass, and Kent 2014). Other, yet undefined, genetic contributions could be important. For example, a primary ciliary dyskinesia, a hereditary defect of motile cilia, associated with hydrocephalus and chronic respiratory infections has been characterised in Alaskan Malamutes (Anderegg et al. 2019) and in a single Golden Retriever (Reichler et al. 2001). The hydrocephalus is a consequence of dysfunction of the ependymal cilia and impaired CSF flow (Lee 2013). Lack of ependymal flow causes closure of the aqueduct and subsequent formation of tri-ventricular hydrocephalus during early postnatal brain development (Zhang, Williams, and Rigamonti 2006). There are many mouse models of inherited hydrocephalus, mostly resulting in defects of proteins involved in cellular signal pathways during early brain development, for example, affecting TGF-β (Zhang, Williams, and Rigamonti 2006). Theoretically hydrocephalus could occur due to failure of aquaporin-4 clearance of excess brain water. However, most studies report increased aquaporin-4 expression in hydrocephalus including in dogs, where CSF concentration of aquaporin decreased following ventriculoperitoneal (VP) shunt surgery to levels comparable to the control population (Verkman et al. 2017, Schmidt et al. 2016). This supports a theory that in some brachycephalic dogs, there is reduced drainage of CSF via the lymphatic route resulting in a compensatory upregulation in aquaporin activity.

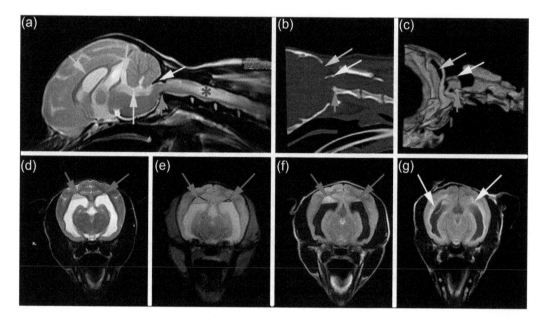

FIGURE 15.4 **Decompensated acute hydrocephalus and SM in a 5-year-old female neutered Yorkshire Terrier following intraventricular haemorrhage.** The dog was presented due to acute onset seizures with no other relevant medical history except that she had been vigorously shaking her head, while playing with a toy 10 minutes before the first seizure. She was noted to have a hypermetric thoracic limb gait which the owner stated was usual for her. She had a scleral haemorrhage. The dog was positioned in dorsal recumbency for imaging. (a) T2W midsagittal MRI brain and cranial cervical spinal cord. There is dilatation of all ventricles and the quadrigeminal cistern (**green arrow**) with a syringomyelia (SM: red asterixis), with elevation and thinning of the corpus callosum (**orange arrow**) and fluid signal-void sign suggesting rapid or turbulent CSF flow (**blue arrow**). There is Chiari-like malformation (CM) with a marked craniocervical junction abnormality with craniospinal junction compression by a dorsally angled odontoid peg (**pink arrow**). This is the likely cause of the hypermetric gait. The atlas is rostral with respect to the occipital crest (**yellow arrow**). (b) midsagittal reformatted and (c) midsagittal three-dimensional (3D) reconstructed CT of the craniocervical junction. Marked craniocervical junction abnormality with craniospinal junction compression by a dorsally angled odontoid peg (**pink arrow**) with atlantoaxial subluxation. The atlas is rostral with respect to the occipital crest (**yellow arrow**). Much of the supraoccipital bone and some temporal bone has failed to ossify (**grey arrow**) likely because of pressure from the developing brain. (d–f) Transverse brain MRI at the level of the midbrain demonstrating recent haemorrhage in the lateral ventricle (**red arrow**). There is fluid which is hypointense on T2W (d), hyperintense on gradient echo (e) and hyperintense on FLAIR (f). The blood is denser than the CSF and has sunk with respect to that fluid. There was also evidence of recent haemorrhage in the fourth ventricle at the level of the lateral apertures (g) Transverse FLAIR brain at the level of the midbrain obtained 3 days after (f). The final diagnosis was a pre-existing CM with atlantoaxial instability, ventriculomegaly and SM which had acutely decompensated associated with interventricular haemorrhage and obstruction of the lateral apertures (later confirmed by post-mortem). The dog was managed with corticosteroids but was represented 3 days later following further seizures and with depression, postural pain, negative menace response, tetraparesis and ataxia. Repeat MRI suggested worsening hydrocephalus with transependymal periventricular oedema (**white arrows**) and syringobulbia (extension of the syrinx into the brain stem).

Diagnosis

Diagnosis and therefore treatment of hydrocephalus in brachycephalic animals can be challenging because ventriculomegaly is a common clinical finding and not always clinically relevant. Clinical hydrocephalus is associated with neurological signs, usually localising to the forebrain, and is not necessarily based on ventricle size. It is also important to rule out other brain diseases especially

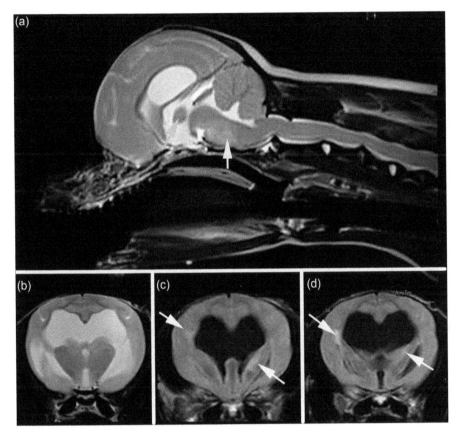

FIGURE 15.5 **Important differentials for hydrocephalus. Meningoencephalitis of unknown origin (MUO)** A 4-year-old female neutered Chihuahua presented with multifocal signs including turning to left, visual deficits localising to the right optic tract and hemiparesis on the right. (a) Midsagittal T2W MRI suggests dilatation of all ventricles most likely related to obstruction of the lateral apertures and foramen magnum due to CM. There are changes consistent with chronic hydrocephalus including elevation and thinning of the corpus callosum (**pink arrow**) and flattening of the intra-thalamic adhesion (**orange arrow**), but it is difficult to ascertain if this is clinically relevant. There is an area of hyperintensity on the pontine region suggesting increased oedema (**yellow arrow**). (b) Transverse T2W MRI at the level of the intra-thalamic adhesion. Dilatation of the lateral ventricles makes it difficult to appreciate subtle pathology. (c and d) Transverse FLAIR at the level of the caudate lobe. FLAIR imaging nulls the signal from CSF allowing appreciation of patchy periventricular oedema (**white arrows**). The multifocal pattern is most consistent with inflammatory disease or lymphoma. The Chihuahua was treated successfully with a prolonged course of immunosuppressive chemotherapy that included cytosine arabinoside and prednisolone.

meningoencephalomyelitis of unknown origin (MUO; Figure 15.5) and atrophic ventriculomegaly associated with ageing and cognitive decline (Figure 15.6; the Appendix).

Magnetic Resonance Imaging

Although ventricular dilatation can be appreciated with ultrasound through an open fontanelle or with computer tomography (CT), MRI provides a better assessment of the CSF channels and any potential obstruction. A minimum diagnostic protocol to evaluate ventricular enlargement is illustration in Table 15.5.

Features that indicate clinically relevant ventriculomegaly are listed in Table 15.6. With the exception of transependymal oedema and white matter tears, these MRI findings may be seen in

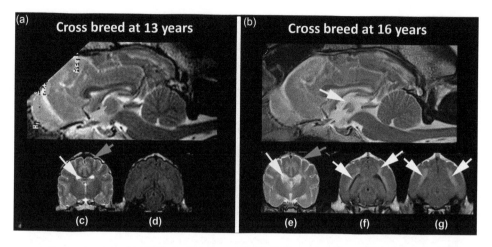

FIGURE 15.6 **Important differentials for hydrocephalus; atrophic ventriculomegaly** – brain MRI from the same male neutered crossbreed dog aged 13 years old (image on left) and 16 years old (images on right). (a) TW2 midsagittal brain aged 13 years. (b) TW2 midsagittal brain aged 16 years. (c) T2W transverse brain at the level of the thalamus aged 13 years. (d) FLAIR transverse brain at the level of the midbrain aged 13 years. (e) T2W transverse brain at the level of the thalamus aged 16 years. (f) FLAIR transverse brain at the level of the midbrain aged 16 years. (g) FLAIR transverse brain at the level of the occipital lobes aged 16 years. In the 16-year-old dog, there is dilatation of the lateral and third ventricles (yellow arrows) and atrophy of the intra-thalamic adhesion (orange arrow) due to loss of cortical tissue. Unlike hydrocephalus, the corpus callosum is not elevated (pink arrow), and the subarachnoid space is dilated (blue arrow). Leukoaraiosis (white arrows) is appreciable as hyperintensity in the white matter dorsal to the lateral ventricles.

TABLE 15.5

Minimum MRI Diagnostic Protocol for Ventriculomegaly

Sequence	Section	Notes
T2-weighted (T2W)	Three planes (i.e., dorsal sagittal and transverse) including cranial cervical spinal cord on sagittal.	The aim is to determine the extent and cause of the ventricular enlargement, i.e. which CSF spaces are enlarged and where is the CSF pathway obstruction? The presence of SM or pre-SM (spinal cord oedema) should be established. If a syrinx is present, then spinal cord imaging may be necessary (see under *Chiari malformation and syringomyelia*).
Fluid-attenuated inversion recovery (FLAIR)	At least two planes typically transverse and dorsal.	This sequence suppresses CSF, making it possible to see subtle lesions such as periventricular hyperintensity.
T1-weighted (T1W) pre- and post- paramagnetic contrast	At least two planes typically transverse and dorsal, before and after contrast.	The aim is to rule out other causes of CSF channel obstruction including tumours and inflammatory disease.
Balanced steady-state free precession (for example FIESTA- or 3D-CISS)	Sagittal.	In instance of supracollicular fluid collection to identify the membranous separation between the third ventricle and quadrigeminal cistern if not discernible on other imaging.
	Craniocervical junction.	May be indicted if the cause of CSF channel disruption is not apparent and to improve detection of arachnoid webs and diverticula.

TABLE 15.6

MRI Signs of Clinically Relevant Hydrocephalus

MRI Feature	Best MRI Sequence to Visualise	Comment
Transependymal (periventricular) oedema	Transverse and dorsal FLAIR	Indicates an acute obstructive hydrocephalus and is consequence of CSF forced through compromised and torn ependyma (Figure 15.4).
		The differential diagnosis for this appearance is leukoariosis associated with ageing associated cognitive decline – this likely also represents failure of brain's 'glymphatic' interstitial fluid drainage system. Leukoariosis (Figure 15.6) is characterised by hyperintensity of T2W within the centrum semiovale (white matter dorsal to the lateral ventricles and corpus callosum).
CSF-filled clefts	Dorsal T2W	The transependymal migrating CSF dissects between white matter tracts forming diverticula, clefts and tears particularly between the caudal pole of the caudate nucleus and internal capsule (Higgins, Vandevelde, and Braund 1977, Laubner et al. 2015).
Intraventricular septations of white matter	Any T2W	The tearing and dissection of the white matter eventually results in separation of bands of white matter – especially the internal capsule (Figure 15.7). Dissection into the nervous tissue can continue to form intraparenchymal cavities communicating with the ventricles.
Intraventricular fluid signal-void sign (FSVS) from CSF movement	Any T2W	FSVS is appreciated on T2W images as 'dark' hypointense regions within the 'white' hyperintense syringe and within the ventricular spaces. It suggests rapid or turbulent CSF flow with reduced intracranial compliance (Figure 15.4 and 15.8).
Reduction of subarachnoid space	Any T2W	Effacement of the sulci helps distinguish the transependymal (periventricular) oedema from leukoariosis. In the latter, the subarachnoid space is wide (Figure 15.6).
Ventricle/brain (v/b) index greater than 0.6	Dorsal T2W	Laubner et al. (2015) found v/b index useful to discriminate clinically relevant hydrocephalus. The v/b index is defined as the maximum continuous distance between the internal borders of the ventricles divided by the maximum width of the brain parenchyma in the same imaging. This could be more simply described as – if maximum width of ventricle in a dorsal plane is greater than 60% of the maximum brain width in same image, then clinical hydrocephalus is more likely. However, v/b index is elevated in many clinically normal brachycephalic dogs.
Elevated corpus callosum	Sagittal T2W	Stretching and dorsal displacement of the corpus callosum (Figures 15.7 and 15.8). Thinning (atrophy) of the corpus callosum is also feature of canine cognitive dysfunction but without elevation (Figure 15.6).
Dilatation olfactory recesses	Any (dorsal easiest)	The olfactory ventricles or recesses are a rostral extension of the lateral ventricle that extends into the olfactory bulb (Figure 15.7).
Dorsoventral flattening of the intra-thalamic adhesion	Transverse or sagittal T2W	Thinning of the intra-thalamic adhesion must be distinguished from atrophy seen in canine cognitive dysfunction. However, in the former the corpus callosum is elevated and the third ventricle expanded with a concave floor (Figures 15.6 and 15.7).
Optic nerve sheath diameter?	T2W	Enlargement of the optic nerve sheath diameter is 90% accurate for detection of raised intracranial pressure in humans, but this is not yet a practical test for dogs and cats. Mean optic nerve sheath diameter for all dogs is 3 mm (1–4 mm) (Scrivani et al. 2013).
SM/presyringomyelia	Cervical sagittal and transverse T2W	May be seen in addition to hydrocephalus especially if there is CSF pathway obstruction at the lateral apertures or foramen magnum (Figures 15.7 and 15.8).

clinically normal dogs with compensated hydrocephalus. Diagnosis is based on appropriate clinical signs in conjunction with several consistent MRI features.

Ultrasonography

When presented with a puppy with forebrain signs, a domed calvaria with open fontanelle (molera), ultrasound through the skull opening is a low-cost method of confirming ventricular dilatation. If the height of the ventricle is greater than 25% of the height of the brain, then a significant ventriculomegaly is present; if it is greater than 65%, then clinical hydrocephalus is likely (Saito et al. 2003). It is also possible to assess the basilar artery through the fontanelle or foramen magnum (positioning as for a cisternal CSF sampling) which has the potential to be a non-invasive technique to detect intracranial hypertension (Saito et al. 2003, Sasaoka et al. 2018). In humans, ultrasonographic retrobulbar optic nerve sheath diameter is a non-invasive method estimate intracranial pressure and may prove useful for dogs in the future (Smith et al. 2018, Liu et al. 2017).

Treatment and Prognosis

Although there are both medical and surgical treatment options, prognosis is guarded as failure rates for both are high.

Medical Management

In the short term, animals with clinical hydrocephalus will often make a good response to medical management, but a sustained response is unlikely due to lack of efficacy and adverse effect of drugs. Options are listed in Table 15.7.

Surgical Management

The mainstay of surgical management of hydrocephalus in small animals is ventricular to peritoneal (VP) shunts. In humans, endoscopic third ventriculostomy is the preferred surgical procedure by which a hole is made through the floor of the third ventricle into the subarachnoid space. Access to the third ventricle is via an endoscope placed in the lateral ventricle and passed through the interventricular foramen (Yadav et al. 2012). Small brain size and risk of haemorrhage in veterinary patients makes this a difficult procedure. The authors' attempt at this in a single patient, with the aid of an experienced human neurosurgeon, was unsuccessful because of postoperative haematoma (Marsh 2018). The principle in VP shunts is to place a catheter (proximal shunt) into the lateral ventricle which is connected to a distal catheter in the peritoneal space of the abdomen (or in any tissue with epithelial cells capable of absorbing the incoming CSF). A one-way valve connects the two catheters. These valves may be fixed parameter or adjustable (Hoshide et al. 2017). Fixed-parameter valves open at a set pressure difference between the intracranial and the drainage space, for example, medium pressure valves range between 7 and 10 cm H_2O water opening pressure. Once the intracranial pressure decreases below the opening pressure, the valve will close (note there are many valves; some are described by closing pressure). In contrast, adjustable pressure valves can be adjusted based on clinical response or complications and therefore replacement valve surgery is less likely. However, adjustable values are more expensive and more likely to have mechanical failures (Hoshide et al. 2017). In animals, the proximal shunt is placed at the level of the ectosylvian or suprasylvian gyrus via a rostrotentorial approach and a 5 mm burr hole. The depth and the angle of insertion is determined pre-operatively from the diagnostic imaging. The distal catheter is placed in the peritoneal cavity caudal to the last rib and secured with a Chinese finger-trap suture and then passed subcutaneously to connect to the valve in the cervical region (Kolecka et al. 2015). Complication rates for VP shunt surgery are high; this and the expense of the catheters are the main contraindications for the procedure. In humans, the 5-year complication rate was approximately 48% in children and 27% in adults (Wu et al. 2007, Del Bigio and Di Curzio 2016). Another study showed a 33% shunt failure in 3 years in children (Riva-Cambrin et al. 2016). The complication rate in veterinary medicine is likely underreported due to paucity of studies. Complications are

TABLE 15.7

Medical Management of Hydrocephalus

Drug	Action	Notes
Acetazolamide 4–10 mg/kg every 8–12 hours.	Choroid plexus (carbonic anhydrase inhibitor) inhibits aquaporin-mediated water conductance and has anti-oxidant actions and affects vascular smooth muscle causing vasodilation (Swenson 2016).	Kolecka et al. (2015) found no effect on lateral ventricle size in a group of dogs which were subsequently and successfully managed with surgical therapy. In adult humans with hydrocephalus, the long-term therapeutic effect is considered negligible, but positive response to acetazolamide bolus is useful for predicting shunt response (Del Bigio and Di Curzio 2016).
Furosemide 3.75–5.5 mg/kg IV IM	Diuretic agent that inhibits the Na-K-2Cl symporter in kidney distal tubules. Based on early experimental studies unlikely to affect CSF formation in healthy cats and dogs (Domer 1969, McCarthy and Reed 1974).	Emergency use only. In humans used in combination with acetazolamide for acute hydrocephalus management.
Corticosteroids For acute management **dexamethasone** 0·1–0.2 mg/kg IV or **methylprednisolone sodium succinate** 15–30 mg/kg IV **Prednisone/Prednisolone/ methylprednisolone** 0.5 mg/kg PO SID then decrease to the lowest possible ideally alternate day dose that controls signs	Reduces CSF production by effect on mineralocorticoid receptor on choroid plexus epithelial cells. Also has neuroprotective effect including protection against increases in vascular endothelial and blood–brain barrier permeability and suppression of inflammatory cytokines and reactive oxygen species production. Corticosteroids may also reduce fibrosis in the subarachnoid compartment, for example, associated with haemorrhage (Del Bigio and Di Curzio 2016).	There is a strong argument for short-term use – in normal dogs and in dogs with kaolin-induced hydrocephalus, intravenous dexamethasone or methylprednisolone resulted in a 40% reduction in CSF flow and the effect lasts for approximately 6 hours (Del Bigio and Di Curzio 2016, Weiss and Nulsen 1970, Sato 1967). However, iatrogenic hyperadrenocorticism and other complications are the inevitable consequence of long-term corticosteroids.
Intravenous osmotic diuretic **Mannitol** 0.25–0.5 g/kg over 15–20 minutes	Osmotic diuretics create a concentration gradient between the intravascular and interstitial and intracellular fluids at the blood–brain barrier encouraging movement of fluid into the intravascular space. Also has an osmotic potential in the kidney tubule. Contraindicated with dehydration and can enhance hypotension with hypovolemia.	Emergency use only.
Oral *osmotic* diuretic **Isosorbide (*hydronol*)** 3 g/kg (experimental) every 8–12 hours	Isosorbide (Ismotic, Hydronol) is a dihydric alcohol formed by the removal of two molecules of water from sorbitol. It is absorbed rapidly from the gastrointestinal tract and is excreted unchanged in the urine. Adverse effects include diarrhoea, vomiting, nausea, dehydration and hypernatremia. Should not be confused with isosorbide mononitrate or dinitrate used to treat angina.	Oral isosorbide transiently reduces CSF pressure in healthy adult dogs (Wise, Mathis, and Wright 1966), and a similar effect was demonstrated in hydrocephalic children; however, the overall consensus is that the any benefit is temporary and this and adverse effects preclude long-term use (Del Bigio and Di Curzio 2016).

most likely during the first 6 months after shunt placement (Gradner, Kaefinger, and Dupré 2019). Gradner, Kaefinger, and Dupré (2019) reviewed multiple studies (73 cats and dogs) and found that 13.7% died because of shunt-related complications including obstruction, infection, over-shunting, disconnection, pain and neurological deterioration especially seizures. Animals that had seizures prior to having shunt placement are likely to remain epileptic and require ongoing anti-epileptic therapy (de Stefani et al. 2011). Decrease in ventricular volume and increase in brain parenchyma after VP shunting are associated with improvement in clinical signs (Figure 15.8) (Schmidt et al. 2019). In the authors' experience, the most common complication is over drainage which can result in subdural haematoma and signs of postural pain and nausea (Figure 15.9).

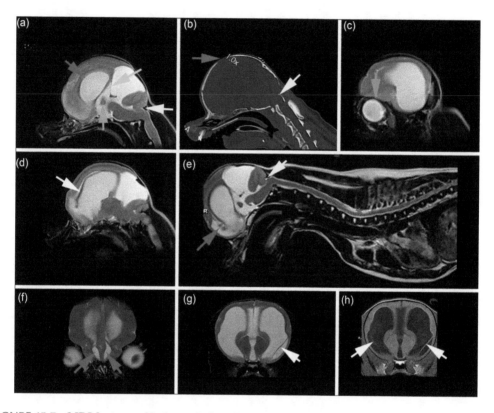

FIGURE 15.7   **MRI features of hydrocephalus with supracollicular fluid collection and CM.** A 1-year-old 1.8 kg male Chihuahua presented with acute onset pain. Clinical examination found a dome-shaped cranium with an open fontanelle with a bilateral ventrolateral strabismus. There were behavioural signs suggesting head and cranial neck pain and a hypermetric gait suggesting spinocerebellar tract or cerebellar dysfunction: (a) T2W midsagittal brain and craniocervical junction. (b) Midsagittal reformatted CT of the skull and cranial cervical spine. (c) T2W parasagittal brain at the level of the eye ball. (d) T2W parasagittal brain at the level of the caudate lobe. (e) T2W midsagittal brain and vertebral column to the level of L1. (f) T2W transverse brain at the level of the eye ball. (g) T2W transverse brain at the level of the rostral tectal plate. (h) FLAIR transverse brain at the level of the rostral tectal plate. **Large pink arrow** – elevated and stretched corpus callosum; **orange arrow** – dorsoventral flattening of the intra-thalamic adhesion; **green arrow** – quadrigeminal cistern expansion and separation from the third ventricle by a thin membrane of pia mater and ependyma. **Red arrow** – large persistent fontanelle; **yellow arrow** – craniocervical junction abnormality consistent with CM; **purple arrow** – possible mechanical displacement of the globe by pressure from the orbital roof plate. This has been postulated as a possible pathogenesis of sunset eye strabismus; **white arrows** – tearing and dissection of the white matter eventually results in separation of bands of white matter (internal capsule); **blue arrow** – dilatation of the olfactory recesses; **red asterix** – developing cervicothoracic syrinx.

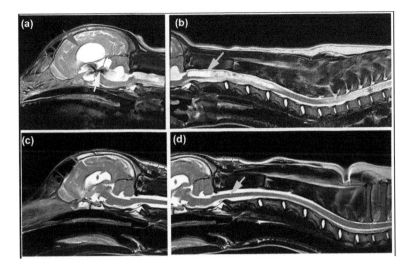

FIGURE 15.8 **VP shunt to manage hydrocephalus and SM.** TW2 MRI midsagittal brain (left) and cervical (right) from a male Border Collie presenting aged 5 months with abnormal behaviour, visual deficits and weakness. (a) Hydrocephalus with dilatation of all four ventricles and the mesencephalic aqueduct. The corpus callosum is elevated and stretched (pink arrow), and the intrathalamic adhesion is flattened (orange arrow). Fluid void signal (yellow arrow) indicates high velocity fluid in the third ventricle and through the mesencephalic aqueduct; (b) there is a large SM of the cervical spinal cord (green arrow) and pre-SM in the thoracic spinal cord (red asterixis). (c) 5 weeks after placement of a VP shunt (red arrow). There is a considerable reduction in the size of the ventricles and lowering of the corpus callosum (pink arrow) and a more normal intra thalamic adhesion (orange arrow). (d) When then factors predisposing SM have resolved, there is rapid and sustained collapse of the syringe (green arrow) and pre-SM (red asterixis). If the syrinx remains expanded within the spinal cord postoperatively, then active filling is still occurring.

Intraventricular Pressure and Valve Selection

Pressure measurements of the CSF spaces are not routine in veterinary medicine as they are in human medicine. The variation of CSF pressure in skull size, position, age and the effect of anaesthesia complicates accurate assessment, making it more difficult to select appropriate intraventricular shunts. The physiological intraventricular pressure (IVP) in normal dogs varies between 5 and 12 mmHg. One study found that the mean IVP in hydrocephalic dogs was 8.8 mm Hg ranging from 3 to 18 mm Hg (Kolecka et al. 2019). Chihuahuas had a mean IVP of 7.5 mm Hg, brachycephalic dogs a mean of 8.9 mm Hg and mesocephalic dogs a mean of 10.33 mm Hg with no significant difference between the groups. However, animals with a longer duration of signs had a lower IVP suggesting compensation of the hydrocephalus (Kolecka et al. 2019). This would suggest a high or medium opening pressure valve as the most appropriate choice in selecting a fixed-parameter valve.

SUPRACOLLICULAR FLUID COLLECTIONS AND QUADRIGEMINAL CISTERN EXPANSION

A supracollicular fluid accumulation describes an expansion of CSF filled spaces dorsal to the colliculi and tectal plate (quadrigeminal plate; tectum) and rostral to the cerebellum (Figure 15.10). This is either a dorsocaudal expansion of the third ventricle or quadrigeminal cistern expansion or both. Supracollicular fluid collections have also been referred to in the literature as quadrigeminal (arachnoid or subarachnoid) cysts or quadrigeminal diverticulum. The quadrigeminal cistern is a midline dilatation of the subarachnoid space connected to the pericerebellar subarachnoid space caudally and medial cerebral hemisphere subarachnoid space laterally. It is dorsal to the third ventricle and mesencephalic aqueduct but separated from the ventricular system by a thin membrane of pia mater and ependyma (Bertolini, Ricciardi, and Caldin 2016). Expansion of the quadrigeminal

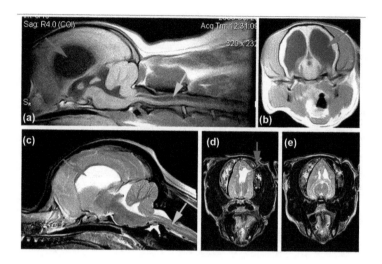

FIGURE 15.9 **Bilateral subdural haematoma following VP shunting for management of SM.** (a) TW1 midsagittal brain MRI prior to placement of a VP shunt in a 3-year-old male neutered CKCS. Sequential MRI obtained since he was 1 year of age had revealed a progressive expanding syrinx (green arrow) associated with ventriculomegaly (pink arrow). (b) TW1 transverse brain MRI at the level of the occipital cortex demonstrating dilated lateral ventricles (pink arrow). One month after placement of the VP shunt, the dog suffered a rapid and severe neurological decline to stupor and non-ambulatory tetraparesis. MRI suggested bilateral subdural haematomas. He was managed with change to a higher-opening pressure valve and supportive care and eventually made a good recovery. He had yearly MRI until developing clinically significant mitral valve disease; the last imaging being 7 years after placement of the shunt. (c) 7 years postoperative TW2 midsagittal brain MRI. The dilated lateral ventricle is indicated by the purple arrow. Interestingly and despite the severe postoperative complications the syrinx cavity (green arrow) remained collapsed. (d and e) T2W MRI transverse brain showing marked distortion by longstanding subdural haematomas (yellow stars). The ventricular shunt (red arrow) is likely non-functional as the tip is within the cortical tissue.

cistern is a common incidental finding in brachycephalic animals and may cause clinical signs when sufficiently large to compress the adjacent brain structures (Matiasek et al. 2007).

Clinical Signs

Clinical signs are related to forebrain compression (focal and generalised seizures) or cerebellar compression (vestibular signs). If there is associated hydrocephalus, there may be other signs for example aggression, reduced mentation and postural pain. Shih Tzus, Chihuahuas, Maltese and the Persian Cat are predisposed (Bertolini, Ricciardi, and Caldin 2016, Schmidt et al. 2017, Lowrie et al. 2009)

Pathogenesis

Pathogenesis for third ventricle expansion is described under *Hydrocephalus*. Pathogenesis of quadrigeminal cistern expansion is unproven. There is evidence that there is a connection between the quadrigeminal cistern and the third ventricle via the velum interpositum and with the fourth ventricle via the dorsal medullary velum (Fenstermacher et al. 1997). These valae are subarachnoid extensions and are hypothesised to function as an alternative CSF route in circumstances of abnormal CSF circulation. In a chronic obstructive hydrocephalus rodent model (kaolin injected into the cisterna magna causing obstruction of the lateral apertures), CSF flows from the third ventricle into the quadrigeminal cistern and from the lateral ventricle into the ambient cistern (Yoon et al. 2015). The authors hypothesise that for certain brain and CSF pathways conformation, a pressure differential is created to allow one-way valve 'balloon' expansion of the quadrigeminal cistern.

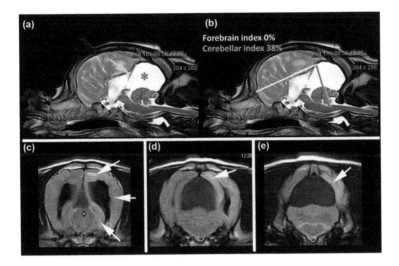

FIGURE 15.10   **Quadrigeminal cistern enlargement in 4-month-old male English Bulldog presenting with seizures.** (a) T2W midsagittal brain. There is an expansion of the quadrigeminal cistern (**red asterixis**) which is separated from the third ventricle by a thin membrane (green arrow). (b) To assess the space-occupying effect of supracollicular fluid. Matiasek et al. 2007 recommend taking two measurements from the midsagittal MRI. In this study, the most reproducible was the forebrain index (**yellow line**). If the supracollicular fluid encompasses more that 14% of a line drawn between the rostral tip of the osseous tentorium cerebelli and the point where the cerebral cortex and the fila olfactoria join, then seizure activity is more likely. The seizures are thought to be a consequence of compression of the occipital lobe. The cerebellum index was calculated from a line drawn between the tentorium cerebelli and obex (**red line**) and is the actual diameter of the cerebellum (**blue line**) versus the expected diameter of the cerebellum (**blue line**); however, this measurement was poorly correlated with the presence of clinical signs (Matiasek et al. 2007). However, these measurements have proved less useful in practice because they rely on the presence of the osseous tentorium cerebelli as a landmark, and this membranous bone may fail to develop, fully ossify or become more horizontal with the expansion of fluid filled spaces during development (Bertolini, Ricciardi, and Caldin 2016). For example, in this dog, presenting with seizures the forebrain compression index was 0%, and the cerebellar compression index was 38.2%. (c–e) Transverse FLAIR in at the level of the occipital lobes. There is oedema through the white matter (**white arrows**). This oedema was thought to be post-ictal supporting the hypothesis that seizures occur secondary to compression of the occipital lobes by the supracollicular fluid. The oedema could also be transependymal in origin. Cisternal CSF analysis and pressure were normal.

Diagnosis

Magnetic Resonance Imaging

Diagnosis is based on advantaged diagnostic imaging and clinical signs. As for hydrocephalus, MRI is the preferred imaging modality; especially, this allows identification of the thin membranous separation between the quadrigeminal cistern and third ventricle (Table 15.5). A classification system based on which cavity is expanded is suggested (Bertolini, Ricciardi, and Caldin 2016, Matiasek et al. 2007):

1. Quadrigeminal cistern expansion (Figures 15.7 and 15.10)
2. Dorsocaudal expansion of the third ventricle in conjunction with quadrigeminal cistern enlargement (Figure 15.4).
3. Dorsocaudal expansion of the third ventricle; quadrigeminal cistern not enlarged. May be in conjunction with lateral ventricle expansion and mesencephalic aqueduct obstruction, i.e. hydrocephalus.

Ultrasonography

It is possible to identify supracollicular fluid collections using ultrasound through the foramen magnum or persistent fontanelle (bregmatic or mastoid), and this may be a useful low-cost screening test of puppies or for monitoring purposes (Saito, Olby, and Spaulding 2001).

Treatment and Prognosis

Prognosis and medical management options are for hydrocephalus. Surgical management options for clinically relevant supracollicular fluid collections include

- Placement of a shunt device directly into the fluid cavity draining to the peritoneum or other epithelial lined cavity (Lowrie et al. 2009)
- VP shunting as for hydrocephalus
- Endoscopic fenestration of the thin membrane separating the quadrigeminal cistern and third ventricle although the authors have found this method challenging due to small size of the patient.

For all treatment options, seizures may continue necessitating ongoing anti-epilepsy therapy.

CHIARI-LIKE MALFORMATION AND SYRINGOMYELIA

CM and SM is a disorder of CSF, characterised by pain and myelopathy due to fluid cavitation of the spinal cord. Often compared to Chiari type I and 0 malformation in humans (CM-I, CM-0), this development malformation is closer to the hindbrain herniation seen with syndromic and complex craniosynostosis where facial and cranial shortening combined with occipital bone insufficiency and craniocervical vertebrae malformation results in neuroparenchymal disproportion and obstruction of the CSF channels. Due to the association with brachycephaly with airorhynchy, an alternative term 'BOCCS' (brachycephalic obstructive cerebrospinal fluid channel syndrome), was proposed to reflect the similarity to brachycephalic airway obstruction (Knowler, Galea, and Rusbridge 2018). Predisposed toy brachycephalic breeds have a high prevalence of malformation, but it is not always clinically relevant; the spectrum covers four variations (Table 15.8 and Figure 15.11).

Pathogenesis

CM

Brachycephalic toy breeds and crosses, where the fashionable and modern head shape is for a shortened muzzle and pronounced stop, are predisposed to pain associated with CM (CM-P) and SM. CM-SM

TABLE 15.8

Spectrum of Chiari-Like Malformation and SM in Predisposed Toy Breed Dogs

CMSM Spectrum	Features
CM-N	Clinically **normal** dogs with an MRI appearance of CM but with no behavioural or clinical signs of pain
CM-P	Dog has behavioural and clinical signs of **pain** with an MRI appearance of CM. SM may or may not be present
SM-M	Dog has **mild** SM but no SM-specific signs. Typically, the syrinx is symmetrical, central and has a maximum transverse diameter of < 4 mm (for CKCS – other breeds not quantified). Dog can be CM-N or have CM-P
SM-S	Dog has SM-**specific signs** relating to spinal cord damage (phantom scratching, scoliosis, weakness or ataxia). Localisation of the syrinx is consistent with the neuro-localisation; for example, thoracic limb weakness is associated with a large syrinx in the C6-T2 spinal segments. Dog can be CM-N or have CM-P. In the CKCS, the syrinx typically has a maximum transverse diameter ≥4 mm

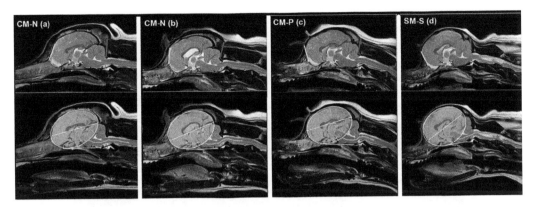

FIGURE 15.11 **Spectrum of CM and SM ranging from clinically unaffected to affected.** TW2 midsagittal MRI of head and cranial cervical region. (a) **Normal CKCS (CM-N)**, a 3.6-year-old male neutered 17-kg CKCS presented with lumbar pain due to L3/L4 intervertebral disc (IVD) extrusion. MRI had included brain because of owner concern for breed predisposition to CMSM. (b) **Normal CKCS (CM-N)**, this 7-year-old male entire 16 kg CKCS illustrates the grey area between normal and abnormal. This dog was presented with acute onset cervical pain related to a C3/C4 IVD extrusion. There was no previous history consistent with CM-P and the cervical pain resolved after surgical decompression. The appearance of the skull, craniocervical junction and brain is only subtly different from the clinically affected dogs, and there is a cerebellar herniation. (c) **CKCS with CM-P**, this 7-year-old 11 kg female neutered CKCS was presented with repeated vocalisation as in in pain when falling asleep and then unwillingness to rise. In addition, she was averse to her ears being touched, walked slowly up the stairs, and often would yelp when being picked up from under the sternum. (d) **CKCS with SM-specific signs (SM-S)**, this 7.5-year 7.3 kg female neutered CKCS presented with a scratching action towards the shoulder area without making skin contact and when walking on a leash, mostly with the right pelvic limb but sometimes with the left. **Red line** – outline of muzzle, stop and 'forehead'. The angle of stop and size of frontal sinus progressively reduces with clinical affectedness. **Green line** – progressive rostral flattening of the forebrain with clinical affectedness. **Aqua shading** – progressive reduction and ventral displacement of the olfactory bulbs with clinical affectedness. **Purple shading** – progressive shortening and reduction of the basicranium especially the presphenoid bone with clinical affectedness. **Blue arrows** – In CM-P and SM-S, the forebrain is displaced caudally. The space for the hindbrain is compromised rostrally by the forebrain and caudally by the small caudal skull. The cerebellum loses its rounded shape and is pushed out of the foramen magnum. **Purple line** – with increasing clinical affectedness, the supraoccipital bone becomes flatter and shorter, and the opisthion (dorsal foramen magnum) becomes rostral with respect to the occipital crest. **Orange star** – progressive reduction in bony tissue of the occipital crest as the dog becomes more clinically affected; however, sexual dimorphism may also influence. **Green and pink shading** – with increasing clinical affectedness, the rough outline of the forebrain changes from a rugby (CM-N) to a football shape (SM-S) because of increasing brachycephaly. **White line** – progressive ventral rotation of the axis of the brain as the dog becomes more clinically affected. **Yellow line** – progressive shortening and tilting of cranial base as the dog becomes more clinically affected. **Pink star** – syringomyelia. **Turquoise line** – change in the conformation of the craniospinal junction because of cranioverte-bral junction (occiput, atlas and axis) malformation. The atlas is closer to the skull with cervical flexure and acute angulation of the odontoid peg resulting in kinking or elevation of the neuroparenchyma.

is most commonly reported in CKCS, King Charles Spaniels, Griffon Bruxellois, Affenpinschers, Chihuahuas, Yorkshire Terriers, Maltese and Pomeranians (Knowler et al. 2019, 2020). Less consistently affected breeds are French Bulldogs, Boston Terriers and Pug dogs; cerebellar vermal herniation is not a usual feature in these breeds. Perhaps surprising, many toy breeds with extreme facial foreshortening such as the Pekinese, Japanese chin and Shi Tzu are not as predisposed reflecting different skull shape, brain size and presumably genetic heritage. Brachycephalic cats, especially the Persian, can have CM (Huizing, Sparkes, and Dennis 2017). Associated SM has not been reported possibility because they do not have the craniovertebral junction abnormalities common in dogs (Figure 15.12).

The feature common to all is brachycephaly due to skull base shortening combined with craniofacial and occipital bony tissue reduction (Knowler, Kiviranta, et al. 2017, Knowler et al. 2020). The reduction in the rostral cranial fossa results in rostrotentorial crowding giving the rostral forebrain

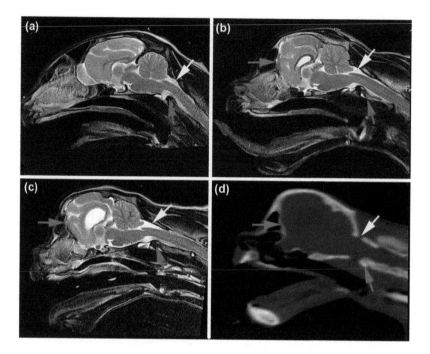

FIGURE 15.12   **Chiari-like malformation associated with coronal suture craniosynostosis in the Persian cat.** (a) Midsagittal T2W brain MRI from a cat described as a Persian but with a long muzzle. (b) Midsagittal T2W brain MRI from neurologically normal Persian. (c) Midsagittal T2W brain MRI from neurologically normal Persian. (d) Reformatted midsagittal CT from a neurologically normal Persian. **Blue arrow** – there is considerable foreshortening of the cranium and distortion of the forebrain. **Yellow arrow** – there is a cerebellar vermal herniation, but unlike many SM-affected dogs, the cisterna magnum is wide. **Pink arrow** brachycephalic cats not appear to have the craniocervical junction malformations observed in dogs. This may protect them from SM. (This figure is available in full colour at https://www.routledge.com/9780367207243.)

a flattened appearance with reduced and ventrally displaced olfactory bulbs. The caudal fossa is reduced, rostrally by the displaced forebrain and caudally by a short vertically flat supraoccipital bone. Some predisposed breeds such as the CKCS have comparatively big brains (Cross, Cappello, and Rusbridge 2009, Shaw et al. 2012). The cerebellum is flattened against the supraoccipital bone resulting in vermal indentation and herniation into or through the foramen magnum. Craniospinal junction overcrowding can be compounded by craniocervical junction deformation including change in angulation of the dens (cervical flexure; see Table 15.1) and increased proximity of the atlas to the skull and loss of the cisterna magna (often referred to as atlanto-occipital overlapping). Overcrowding of the craniospinal junction results in kinking or elevation of the medulla (Knowler, Kiviranta, et al. 2017, Cerda-Gonzalez, Olby, and Griffith 2015). This feature is particularly relevant and correlated to clinical signs in miniature toy breeds such as the Chihuahua and Pomeranian (Kiviranta et al. 2017, Marino et al. 2012). In these miniature breeds, cerebellar deformation and herniation may not be present especially if there is occipital dysplasia. Occipital dysplasia is a failure of ossification of the membranous supraoccipital bone is thought to be secondary to pressure of the cerebellum against the developing bone.

CM-P

Dogs with CM-P have more extreme brachycephaly, i.e. shorter cranial base, more craniofacial hypoplasia with greater neuroparenchymal disproportion and overcrowding when compared to CM-N (Knowler, Cross, et al. 2017) (Figure 15.11). However, based on MRI appearance, there is a considerable 'grey area' between CM-N and CM-P; therefore, this diagnosis is made based on appropriate clinical signs, diagnostic imaging and elimination of other explanations.

SM

An entirely satisfactory explanation of how fluid cavities develop in the spinal cord following CSF pathway obstruction has yet to be described. It is also not proved whether syrinx fluid is CSF or is closer to extracellular fluid. The most accepted theory is that subarachnoid space obstruction results in a mismatch in timing between the arterial pulse peak pressure and CSF pulse peak pressure. The perivascular space changes in size during the cardiac cycle and is widest when spinal arteriole pressure is low. Earlier arrival of peak CSF pressure compared to peak spinal arterial pressure encourages flow of CSF into the perivascular space which acts as a 'leaky' one-way valve. From the perivascular space, fluid flows into the central canal ultimately resulting in a syrinx (Stoodley 2014). SM can occur because of any obstruction to CSF pathways and has been reported in a variety of disorders ranging from acquired cerebellar herniation secondary to intracranial masses to spinal arachnoid diverticulum. However, in veterinary medicine by far the most common cause is CM.

SM associated with CM is predisposed by two morphological phenotypes (or combination of): the first by more extreme brachycephaly, as for CM-P, and the second by more extreme craniocervical junction deformation (Knowler, Cross, et al. 2017, Knowler, Kiviranta, et al. 2017). Other factors likely important in the development of CM-associated SM are reduced CSF absorption through nasal lymphatics (Knowler, Galea, and Rusbridge 2018), reduced venous drainage (Tables 15.2 and 15.4), altered neuroparenchymal compliance and conformation of the spinal canal and cord (Cirovic et al. 2018, Sparks, Robertson, and Olby 2019).

Clinical Signs

Dogs can be presented at any age although most dogs with clinically relevant disease will show signs of CM-P or SM-S before 4 years of age.

CM-P

Pain in CM-P is thought to relate to obstruction of CSF pathways. The most common sign is vocalisation when being lifted or during movement especially when recumbent, hypothesised to be a failure to equilibrate intracranial pressure due to obstruction of CSF pathways and micro-gravitational effects when rapidly elevating the head. Classically, human patients with the analogous conditions CM-0 or CM-I have an occipital or suboccipital headache which is exacerbated by Valsalva manoeuvres, i.e. when there is a brief increase in intrathoracic pressure, for example, when coughing or with abdominal straining. Affected dogs may also avoid exercise perhaps because systolic pressure increases intracranial pressure and therefore pain (Haykowsky et al. 2003). Signs of canine CM-P are listed in Table 15.9.

SM

Dogs with SM may show signs of CM-P (Rusbridge, McFadyen, and Knower 2019); specific signs of SM (SM-S) are listed in Table 15.10. Dogs with a narrow syrinx may not have associated signs, and the clinician should be cautious not to overinterpret especially in breeds where CM is ubiquitous. SM results in a central spinal cord syndrome reflecting grey matter damage. The signs correspond to the localisation of the syrinx; for example, a syrinx involving the caudal cervical and cranial thoracic segments will result in thoracic limb and spinal weakness. A wide mid-cervical syrinx with superficial dorsal horn involvement is associated with phantom scratching and cervicothoracic torticollis (Rusbridge, McFadyen, and Knower 2019, Nalborczyk et al. 2017) (Figure 15.15). Pelvic limb proprioceptive deficits and paresis can be surprisingly mild even with an extensive syrinx. If there is a non-ambulatory tetraparesis or severe paraparesis, then other differentials should be considered, for example intervertebral disc disease, degenerative myelopathy, vertebral malformations or constrictive myelopathy (see the chapter on spinal cord diseases).

Diagnosis

CM is commonly reported in predisposed breeds, and SM can be an incidental finding; therefore, care must be taken not to over diagnose (Figure 15.11). Diagnostic results should be related to historical and clinical findings.

TABLE 15.9

Behavioural and Clinical Signs of CM-P

Clinical Sign	Behavioural Signs
Postural (Valsalva) pain	Vocalisation when moving, while lying resting or asleep, when lifted under sternum, on rising or described by owner as 'out of nowhere'.
	Vocalisation/hesitating/refusing/difficulty on jumping or doing stairs.
	Change in greeting behaviour – vocalising during or refusal to get up and greet owner.
	Elevated or extended head posture when resting (rule out brachycephalic obstructive airway syndrome as alternative explanation).
	Lowered head/neck posture when awake.
	Vocalisation while straining to defecate.
Head pain	Rubbing or scratching at head or ears (when there is no evidence of skin disease and failure to respond to trial therapy for skin/ear disease).
	Vocalisation while scratching head or ears.
	Avoids touch or is averse to grooming of the head and ears while being happy to be petted or grooming elsewhere.
	Owner describes a grimacing facial expression suggesting pain (Figure 15.13; pain face).
	Photophobia – describes as eyes closing in bright light and other explanations such as keratoconjunctivitis sicca ruled out.
Spinal pain	Reactive to palpation in the cervical, thoracolumbar or caudal lumbar/lumbosacral region.
Activity change	Less willing to exercise or described as exercise intolerant, unwilling to exercise or lethargic or more likely to sleep during day.
Sleep disruption	Restless or frequent waking during the night. Other explanations such as skin disease and brachycephalic obstructive airway syndrome ruled out.
Change in behaviour or demeanour	Increased nervousness or anxiety, having uncharacteristic hostility to other dogs/people or becoming more withdrawn (Figure 15.14).
Forelimb hypermetria	A subtle tendency to overshoot because of increased proximal joint movement.
Flank, sternum or limb sensitivity	Avoids touch or is averse to grooming of this body part, while being happy to be petted or groomed elsewhere.
	Licking paws without evidence of skin or orthopaedic disease and failure to respond to appropriate management of skin disease.

FIGURE 15.13 **Pain face in a 9-year-old male neutered CKCS with MRI characteristics of CM-P.** First presented when 1 year old, other clinical signs included a tendency to rub at his nose and face, being averse to his head being touched or groomed, neck sensitivity and vocalising during the night in apparent pain. Clinical signs were mostly controlled by medication, but occasionally, there was a breakdown in pain control. On the day the photo was taken, the atmospheric pressure had dropped, and thunderstorms were forecasted – weather conditions have been hypothesised to influence pain in dogs and humans. Picture acknowledgement – K. Barnard.

FIGURE 15.14 **Uncharacteristic aggression in a 4-year-old female neutered CKCS with MRI characteristics of CM-P.** Presented because of a sudden-onset behavioural change, described as growling when owner went to stroke her, specifically when stroking was directed at the back of the head and more likely in the morning or during the night. The owner also reported a tendency for the dog to scratch at and rub the back of her head especially when excited. Clinical signs resolved with analgesia.

TABLE 15.10

Behavioural and Clinical Signs of Clinically Relevant MS (Syringomyelia-Specific Signs; SM-S)

Clinical Signs	Neuro-localisation and Details
Phantom Scratching	Large mid-cervical syringe extending to the superficial dorsal horn ipsilateral to the scratching (Figure 15.15).
	A characteristic scratching action that can be induced by rubbing the dermatome associated with the superficial dorsal horn involvement. The rhythmic scratching action is towards the neck or sternum, but not contacting the skin, together with a curvature of the body and neck towards the foot. Can also be triggered by excitement, anxiety and exercise. The aetiology of the scratching is thought to be disruption of sensory input to the scratching central pattern generator and a hyperactive scratch reflex (Nalborczyk et al. 2017).
Cervicothoracic torticollis (scoliosis)	Large syringe extending to the superficial dorsal horn.
	A 'corkscrew' twisting of the neck results in a contralateral ventral head tilt, and the shoulder is 'pushed out' ipsilateral to the superficial dorsal horn involvement. The aetiology of the abnormal posture is thought to be loss or disruption of neck proprioceptive input (Figure 15.15).
Weakness	Thoracic limb and paraspinal muscle weakness associated with a large cervicothoracic syringe (Figure 15.16).
Thoracic limb muscle atrophy	Wide C6-T2 spinal segment syringe. Typically, atrophy is more apparent in the distal limb and triceps muscle.
Proprioceptive deficits	Typically associated with a large C6-T2 spinal segment syringe. Thoracic limbs more affected than pelvic limbs. As demonstrated by 'hopping', 'hemi-walking' and 'correction of knuckled-over paw' testing (Figure 15.16).

Advanced Diagnostic Imaging (CT and MRI)

MRI is the most sensitive test for assessing the CSF spaces and detecting SM. It is possible to detect cerebellar herniation and a large syrinx using CT, but it is not possible to appreciate small syringes and spinal cord oedema (pre-SM or presyrinx). However as the primary abnormality in CM is bony insufficiency, it is likely that in the future, CT will play a role in screening at risk dogs using morphological measurements or a machine learning algorithm (Spiteri et al. 2019). A minimum diagnostic protocol for MRI assessment of CM and SM is indicated in Table 15.11. Additional imaging may be required; for more details, see the review by Rusbridge and others (2018).

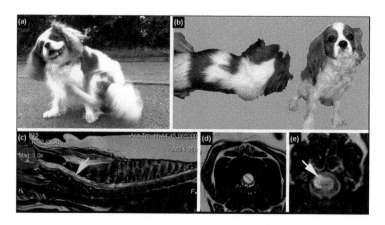

FIGURE 15.15 **Syringomyelia-specific signs (SM-S) associated with wide mid-cervical syringes and superficial dorsal horn involvement.** (a) **Phantom scratching in a CKCS with MRI characteristics of SM-S.** SM-associated 'phantom' scratching is a specific but not universal sign of SM. It is characterised by a rhythmic scratching action, without contacting the skin, together with a curvature of the body and neck towards the foot. The scratching is to the shoulder/neck region and is one-sided although can progress to both sides in severe cases. It can be induced by lightly rubbing a specific area of skin, typically on the neck. This light touch induced scratch reflex makes walking on a collar difficult. Phantom scratching is also triggered by excitement or anxiety. It is associated with a large syrinx in the mid-cervical spinal cord that extending to the region of the superficial dorsal horn on the same side as the scratching. Phantom scratching is not necessarily associated with pain or paraesthesia (abnormal sensations). Dogs with CMSM may also rub or scratch their heads and/or ears – this is **not** phantom scratching but may be a sign of CM-P. SM **does not** cause generalised pruritus – if the dog is scratching at the abdomen or nibbling the back feet, then allergic or infectious skin disease is a much more likely cause. Picture acknowledgement – passionate productions. (b) **Cervicothoracic torticollis (scoliosis) in a CKCS with syringomyelia.** Cervicothoracic torticollis is associated with wide syringes of the dorsal spinal cord. Affected dogs have a corkscrew deviation so that the head is twisted ventrally contralateral to, and the shoulder twisted and pushed out ipsilateral to the side of spinal cord superficial dorsal horn involvement. Cervicothoracic torticollis is thought to be due to asymmetrical damage of the dorsal grey column, over several spinal cord segments, resulting in an imbalance of afferent proprioceptive information from the cervical neuromuscular spindles. When viewed from the front, the head tilt is more obvious and could be confused with vestibular disease. (c) **TW2 MRI midsagittal cervical spine demonstrating a clinically relevant syringomyelia.** The spinal cord outline is expanded by the syrinx indicating active filling. Hypointensity within the syrinx cavity (green arrow; fluid signal void sign) indicates pulsatile or turbulent flow. (d) **TW2 MRI transverse cervical spine at the level of C2.** For assessment of syrinx severity, transverse images of the widest part of the syrinx are obtained (red line). Myelopathic signs in CKCS are associated with a syrinx transverse width of 4 mm or more. (e) **TW2 MRI transverse cervical spine at the level of C3.** Extension of the syrinx to the edges of spinal cord in the region of the superficial dorsal horn (2 and 10 o'clock position). This finding is associated with phantom scratching (same side) and scoliosis (opposite side).

For predisposed breeds, the microchip number (confirmed by the veterinary surgeon) and Kennel Club registration number should be included on the DICOM registration even if CM-P or SM is not the main a differential. This is to permit submission to an official CMSM health scheme should the owner request this following the imaging (BVA 2013). This canine health scheme accepts a three-sequence imaging protocol for healthy breeding dogs; this limited protocol of the cauda fossa and cranial cervical spinal cord is not enough for the diagnostic investigation of a dog with head or cervical pain as many differentials could be missed.

Diagnosis of CM-P   CM-P is a diagnosis of exclusion made based on

1. Appropriate clinical signs (Table 15.9)
2. Appropriate MRI features (Table 15.12 and Figure 15.11)
3. Ruling out other causes of discomfort.

FIGURE 15.16 **Weakness and proprioceptive deficits in CKCSs with MRI characteristics of SM-S.** Weakness can develop with wide syringes in the cervicothoracic region. (a) Trauma to the dorsum of the paw because of failure to place correctly when walking. (b) Lordosis because of thoracic paraspinal muscle weakness. (c–e) Weakness of the right thoracic limb (ipsilateral to syrinx asymmetry) with a tendency for the limb to buckle and collapse. Broken skin on the carpus because of repeated trauma.

TABLE 15.11
Minimum MRI Diagnostic Protocol for CM and SM

Sequence	Region	Notes
Sagittal T2-weighted (T2W)[a]	Brain to caudal extent of syrinx/presyrinx	The aim is to determine the cause of the SM (i.e. CSF pathway obstruction) and the extent of the spinal cord involvement. Correlated with the neuro-localisation
Sagittal T1-weighted (T1W)[b]	At least one region of the spinal cord (typically cervical)	Determine that the signal characteristics of the fluid filled cavity are identical to CSF to help rule out other causes of T2W hyperintensity, for example neoplasia or oedema associated with meningoencephalomyelitis of unknown origin (MUO)
Transverse spinal cord (T2W for high field, T1W for low field)	Widest parts of the syringes in C1-C5, C6-T2, T3-L3 and L4-S3 spinal segments	Determine the maximum transverse width and position within the spinal cord including dorsal horn involvement. Correlated with the neuro-localisation
Transverse head (T2W for high field, T1W for low field)	Brain, nasal passages and ears	Assessment of other brain lesions, for example ventriculomegaly and rule out other differentials, e.g. MUO. Also, to assess other comorbidities, e.g. middle ear effusion (see the Appendix)
Paramagnetic contrast enhancement	Where indicated	Indicated if evidence of a mass as intramedullary tumours can be cystic and confused with SM. Cause of the CSF channel obstruction is not apparent There is a presyrinx only to rule out other causes of spinal cord oedema
Steady-state sequences (for example FIESTA- or 3D-CISS)	Where indicated	Indicted if the cause of CSF channel disruption is not apparent and to improve detection of arachnoid webs and diverticula

[a] To reduce scanning time, it may be possible to use an ultra-fast spin echo sequence (for example HASTE – HAlf Fourier Single-shot Turbo spin-Echo) which provides a single slice MR myelogram sequence allowing rapid assessment of the CSF spaces and the longitudinal extent of the syrinx and whether more detailed imaging is required.

[b] For a low-field MRI machine, it may be preferable to use TW1 sagittal sequences to determine the extent of the syrinx with however at least one TW2 sequence should be obtained to improve interpretation of the findings (for example detection of oedema).

TABLE 15.12

MRI Features of CM-P

Morphological Change	Craniofacial and MRI Features
Craniofacial hypoplasia	Muzzle which is short and length with a well-defined or indented stop. The overall appearance from the front is a dog with centrally placed features (eyes and nose) and a large round 'forehead' On MRI (or CT), this 'forehead' can be appreciated as the frontal bone overlying the brain with absent or minuscule frontal sinuses. The junction between nasal and maxilla bones forms an angle rather than a slope (see also Figure 15.1)
Rostrotentorial overcrowding	The neuroparenchyma mirrors the bony changes resulting in a rostral flattening of the forebrain with a ventral rotation of the olfactory bulbs which are reduced in size. There is an increased cranial height, especially in the occipital region Compared to normal dogs, the brain is rounder and more like a soccer football than a rugby ball shape
Reduction/obstruction of CSF channels	Dilatation of ventricles and cisterns, except the cisterna magna which is reduced. Reduction in cranial and spinal subarachnoid space
Skull base shortening	Shortening of the basicranium especially the presphenoid bone (see Table 15.2 for associated features)
Caudotentorial overcrowding	The functional caudotentorial space is reduced rostrally by the forebrain which is displaced dorsocaudally and caudally by the short and vertical supraoccipital bone. Opisthion (dorsal foramen magnum) rostral with respect to the occipital crest
Cerebellar vermal indentation and foramen magnum herniation	Cerebellum loses its rounded shape and may be displaced into the foramen magnum

Diagnosis of Clinically Relevant SM (SM-S)　　　Fluid-filled cavities in the spinal cord develop following disturbance of CSF flow, spinal cord tethering, with an intramedullary tumour or cyst and following severe spinal cord injury (malacia); therefore, if a spinal cord cavity is identified, then the cause should be ascertained (Rusbridge, Stringer, and Knowler 2018). SM is a disorder of CSF, and it is not an appropriate description for myelomalacia or cystic lesions. Presyrinx describes a condition of non-inflammatory spinal cord oedema which is a prequel to SM. Presyrinx most commonly affects the dorsal and ventral columns of the spinal cord. Infectious and non-infectious CNS inflammatory diseases are the most important differentials for presyrinx and therefore identifying this change on MRI prompts other MRI sequences (Table 15.11) and, variably, CSF analysis. As SM can be an incidental finding an assessment should be made as to whether the location and severity of the syrinx would account for the signs. The MRI features of SM associated with CM are described in Table 15.13 and Figures 15.1, 15.11 and 15.15.

Other Diagnostic Tests

If there is a suspicion of CNS inflammatory disease, then CSF analysis should be performed preferably from a lumbar puncture especially if the cisternal space is small. Index of suspicious for CNS inflammatory disease is increased if (1) there is spinal cord oedema (presyrinx) rather than distant cavities, (2) there is marked pain especially if multifocal or induced on cervical ventroflexion with a normal atlantoaxial junction and (3) there is other evidence of inflammation, for example elevated C-reactive protein. CSF analysis results may be difficult to interpret because CM and SM is often associated with a marginal elevation in cell counts including the presence of neutrophils (Whittaker et al. 2011) and CSF protein is often elevated due to stagnation of flow (Froin's syndrome).

TABLE 15.13
MRI Features of SM Associated with CM

MRI Feature	Notes
More extreme brachycephaly	See Table 15.12
Craniovertebral junction malformation	Cervical flexure resulting in acute angulation of the odontoid peg relative to the skull base. Atlas is closer to the skull (Table 15.2 and Figure 15.1)
Craniospinal junction kinking or elevation	Concertina like flexure of nervous tissue at the junction between skull and spine due to rostrocaudal shortening and reduction in volume of the craniovertebral junction (Figure 15.11)
Fluid signal-void sign (FSVS)	FSVS is appreciated on T2W images as 'dark' hypointense regions within the 'white' hyperintense syringe (Figures 15.8 and 15.15)
	FSVS within syrinx cavity indicates pulsatile or turbulent flow and a sign of an 'active' and filling syrinx more likely to expand.
	FSVS may also be seen within the ventricular spaces and suggests rapid or turbulent CSF flow with reduced intracranial compliance
Spinal cord outline	A quiescent syrinx is centrally located, elliptical on sagittal images and symmetrical, usually circular on transverse images and results in little or no change to the outline of the spinal cord.
	If the spinal cord outline is expanded, then the syrinx is filling actively and more likely clinically relevant (Figure 15.15)
Syrinx maximum transverse width	Myelopathic signs in CKCS are associated with a syrinx transverse width of four millimetres or more (SM-S)
	Measurement is made on a transverse slice from the widest syringes identified on the sagittal images (Figure 15.15)
Syrinx position in spinal cord and neurological signs	**Phantom scratching** – mid-cervical syrinx and extension of the syrinx into the superficial dorsal horn (2 or 10 o'clock position on the outer rim of the spinal cord) ipsilateral to the scratching and associated with the dermatome that can be rubbed to induce the scratching action in the conscious dog (Figure 15.15)
	Cervicotorticollis – mid-cervical syrinx and extension of the syrinx into the superficial dorsal horn contralateral to the head tilt/twist.
	Proprioceptive deficits – wide syrinx that involves the dorsal funiculus
	Thoracic limb paresis – wide cervicothoracic junction syrinx that extends into ventral horn
	Thoracic lordosis – wide thoracic syrinx (Figure 15.16)

Treatment and Prognosis

Prognosis

Although syrinx width increases over time, the rate of increase is not constant. Many syringes appear to rapidly expand within months (or even days) and then can remain remarkably unchanged over years having achieved a hydrodynamic equilibrium. Dogs that are presented with SM-S before 3 years of age have a poorer prognosis and are more likely to develop severe weakness which is more difficult to manage. Cervicotorticollis can improve slowly despite persistence of the syrinx.

Clinical signs will progress in approximately 75% of dogs and approximately 15% will be euthanised because of CM-P and SM-S (Plessas et al. 2012, Thofner et al. 2015). However, despite progressive signs, many dogs with signs of pain and phantom scratching respond to medical management and are considered by their owners to have an acceptable quality of life.

Management of CM-P and SM-S

There is a paucity of studies on the medical and surgical management of CM-SM with small cohorts and short follow-up periods. Treatment guidelines based on the available evidence and the authors' experience are detailed in Figure 15.17.

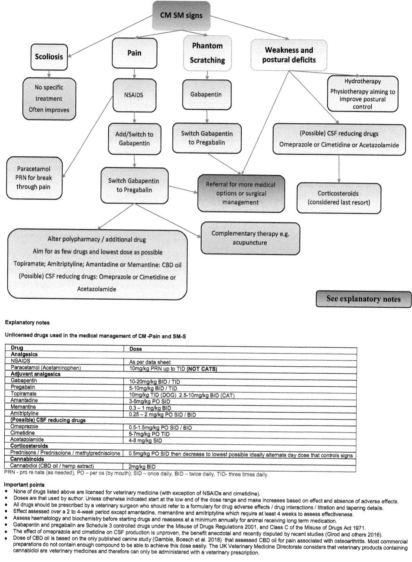

FIGURE 15.17   Treatment algorithm for CM-P and SM-S.

Surgical Management   Surgical options for management of CM-P and SM-S are summarised in Table 15.14. Although the few surgical reports have all reported a good clinical improvement, no surgical series has provided MRI evidence of sustained postoperative syringe collapse. Active syrinx filling is occurring if the postoperative MRI reveals a syrinx expanding the spinal cord outline. Documenting that syrinx is static is not proof of surgical efficacy as many syringes achieve a hydrodynamic equilibrium. Likewise, surgical efficacy is not proved by stabilisation or improvement of clinical signs as this can occur with medical management which many postoperative patients receive.

Medical Management   From the limited studies on medical management, one can conclude that it is unlikely CM-P will improve with prescription of carprofen (or other non-steroidal anti-inflammatory drugs) alone; however, a combination of carprofen and gabapentin is more

TABLE 15.14

Indication for Surgical Management of CM and SM

Surgery	Indication	Notes
Craniocervical decompression surgery (foramen magnum decompression)	CM-P (with or without SM) with a poor response to medical management	Removal of the supraoccipital bone with a C1 rostral dorsal laminectomy. Surgery is successful in reducing pain in approximately 80% of cases; however, SM is likely to persist along with associated clinical signs (for example phantom scratching or weakness) Recurrence of signs of pain can occur, often attributed to fibrous tissue adhesions over the foramen magnum. There is no convincing evidence that this is less likely when the bony defect is reconstructed with an implant as fibrous adhesions develop following blood contamination which is common to all surgery
VP shunt	SM associated with ventricular dilation	Appears to be more successful in facilitating syrinx collapse. However, shunting procedures have a higher complication rate especially subdural haematoma, infection and shunt blockage (see under *Hydrocephalus*)
Syringo to pleural or subarachnoid shunting	When other surgeries are inappropriate, e.g. SM secondary to arachnoid webs and persistence of SM following other surgical procedures (Tauro and Rusbridge 2020)	Involves placement of a tube directly into the syrinx allowing fluid to drain into the pleural or other epithelial lined cavity or subarachnoid space. Shunting procedures have a higher complication rate especially shunt blockage

likely to be successful, and there is reasonable evidence for improvement with pregabalin alone (Plessas et al. 2015, Sanchis-Mora et al. 2019, Plessas et al. 2012). There is also evidence that pregabalin will reduce SM-associated phantom scratching (Thoefner et al. 2020). If an animal remains in pain, then other adjuvant analgesics, acupuncture, physical and physiotherapy may be useful. Anecdotally, a positive response to antacids such as cimetidine or omeprazole is reported. The principle is that these drugs may reduce CSF production, thus reducing the driving force contributing to CM-P and SM; however, there is poor evidence that these drugs achieve a therapeutic choroid plexus concentration or affect CSF production (Girod et al. 2016, 2019). Excessive weight gain should be discouraged and exercise, to within the dog's limitations, should be encouraged. Hydrotherapy can be useful for some patients especially those with weakness or proprioceptive deficits. A possible treatment algorithm is indicated in Figure 15.17.

Genetic Factors and Breeding Advice

CM is a complex trait, and the tendency for SM involves additional genetic factors (Kibar et al. 2007, Lemay et al. 2014, Ancot et al. 2018). MRI or CT screening for CM-P is likely to require a machine leaning solution to determine an objective measure using a simple artificial intelligence tool (Spiteri et al. 2019). Screening for SM using MRI is available, and many European Kennel Clubs enforce or recommend SM screening of predisposed breeds and breeding according to guidelines (BVA 2012). The late onset nature means that MRI screening is performed from 12 months at age and before breeding. Ideally breeding dogs would have additional MRI screening after 5 years old. This is because dogs free of SM at 12 months of age do not necessarily stay free of SM – data in the CKCS breed suggests the prevalence increases from 25% at 12 months to 70% at 72 months (Parker et al. 2011). If there was breed-wide screening and contribution to an estimated breeding value scheme, then this could be used to predict the risk of SM in descendants. In countries where there is a law-enforced obligatory screening program, then there is reduced prevalence in offspring generation by generation (Laterveer 2020).

CONCLUSIONS

The skull protects the brain, enables ingestion and respiration, and encloses the visual, auditory and olfactory systems. The consequences of shortening the box for these essential organs is far reaching. The main impact of brachycephaly on the nervous system is disruption of CSF movement and absorption. The bulk of CSF is absorbed through perineurial lymphatics (olfactory bulbs, cranial and spinal nerves) and meningeal lymphatic vessels of the skull and sacral spine. Skull shortening reduces the area of the cribriform plate, nasal mucosa surface and skull base foramen volume. Thus, CSF absorption is compromised, predisposing ventriculomegaly or SM particularly if there is caudal obstruction of CSF channels, for example, at the craniospinal junction. Furthermore, respiration drives CSF movement through the ventricular system and CSF movement and drainage is impeded by apnoea. This may explain the tendency of brachycephalic animals to ventriculomegaly, hydrocephalous, quadrigeminal cistern expansion, Chiari-like malformation and SM. All these disorders are associated with pain with variable other neurological signs including forebrain signs and seizures (hydrocephalous and quadrigeminal cistern expansion) and spinal cord disease including phantom scratching and scoliosis (SM). Confirmation of diagnosis is by MRI, but this is complicated by an indistinct boundary between normal and clinically relevant disease. Surgical options have limited success, and many animals require lifelong medication including pain management.

ACKNOWLEDGEMENT

The authors are grateful to Thomas Rusbridge for his artistic skills in creating Figures 15.1 and 15.2.

APPENDIX

DIFFERENTIAL DIAGNOSIS OF BRAIN DISORDERS IN BRACHYCEPHALIC SMALL DOG BREEDS

Condition	Breed	Notes
	Seizures, Paroxysmal Events and Other Forebrain Disease	
Idiopathic epilepsy	All	Seizures, normal interictal neurological examination.
Paroxysmal dyskinesia (epileptoid cramping syndrome)	Any but particularly Boston Terrier Maltese CKCS	Episodes of abnormal movement that are self-limiting with long periods of normality between episodes. Abnormal movements may include tremor, alternate limb lifting/flexion and weaving head movements with variable gastrointestinal signs. Pet remains conscious and able to respond to owner, for example, may try to walk to owner. Possible connection to gastrointestinal dysfunction and gluten sensitivity. Episodes may reduce with a hypoallergenic or gluten free diet. Treatment – anti-epilepsy drugs, acetazolamide

(Continued)

Condition	Breed	Notes
Fly catching disorder	CKCS	Behaviour where the dog acts as if watching/catching a fly (although ignore actual flies). Some may also behave as if their ears or feet are irritated, and some can also tail chase. May be more likely with certain light intensities and emotional states.
		Possible connection to gastrointestinal dysfunction. Episodes may reduce with a hypoallergenic or low protein diet.
		Neurochemical imbalance has been suggested as some dogs respond to anti-epilepsy drugs or selective serotonin re-uptake inhibitors.
Idiopathic head tremors	Bulldog breeds	Sudden onset vertical or horizontal rhythmic head movements lasting from a few seconds to several hours
Myoclonus	CKCS	Brief jerking of the head often with buckling of a thoracic limb when the dog is standing or sitting.
		Older CKCS (greater than 5 years old). Initially relatively benign but can be progressive over years and the myoclonic jerks can cause the dog to fall or stumble.
Primary or secondary brain tumour	French Bulldog Boston Terrier Bulldog	Glioma is most common tumour (Mayousse et al. 2017, Song et al. 2013, Kishimoto et al. 2020)
Neuronal ceroid lipofuscinosis	Chihuahua	Visual loss, behavioural changes including changes in personality (e.g. development of aggressive behaviour) and loss of learned behaviours, tremors, cerebellar ataxia, cognitive and motor decline, sleep disturbance and seizures (Faller et al. 2016)
Hepatic encephalopathy	Any	Mild lethargy to ataxia, disorientation and coma
Vestibular disease		
Idiopathic vestibular disease	Any	Acute onset vestibular signs (head tilt, nystagmus, positional strabismus, asymmetrical ataxia) with no proprioceptive deficits or weakness.
		In the CKCS, idiopathic vestibular disease has two variations: one in geriatric dogs and the other in middle aged dogs often in conjunction with idiopathic facial paresis.
Otitis media and interna	Bulldog breeds	Narrowed ear canals and poor middle ear ventilation likely contribute to predisposition (Seppänen et al. 2019, Mayousse et al. 2017)
Rostral cerebellar artery infarction	Any but particularly CKCS	Acute onset paradoxical vestibular signs with other signs suggesting a cerebellar location (menace response deficit, intention tremor, decerebellate posture). Often improve with supportive care.
Primary or secondary brain tumour	French Bulldog Boston Terrier Bulldog	Glioma is most common tumour (Mayousse et al. 2017, Song et al. 2013, Kishimoto et al. 2020)
(Ruptured) Quadrigeminal dermoid or epidermoid cyst	Any (rare)	Differential for supracollicular fluid collection associated with a mass. Rupture can result in a chemical meningitis (Beard, Munro, and Gow 2011)

(Continued)

Condition	Breed	Notes
Multifocal (Combination Signs That May Include Seizures, Vestibular, Pain, and Spinal Cord Disease)		
Meningoencephalomyelitis of unknown origin (MUO)	All but particularly Yorkshire Terrier, French Bulldog, Pug, Maltese and Chihuahua.	Spinal pain with variable neurological signs relating to immune-mediated inflammation of the brain or spinal cord (Figure 15.5). For French Bulldogs more likely in females less than 5.5 years (Mayousse et al. 2017). For review on breed-specific MRI features, see (Flegel 2017)
Extension of otitis media/interna to skull base osteomyelitis, subdural empyema, bacterial meningitis	Bulldog breeds	Narrowed ear canals and poor middle ear ventilation may increase likelihood of intracranial extension of infection.
Canine neuroaxonal dystrophy	Papillon Chihuahua	Visual loss and cerebellar signs. missense mutation in *PLA2G6* gene
Other Cranial Nerve Disorders		
Idiopathic facial paresis	CKCS	Inability to close the eye or move lips, ears or other facial muscles. Facial sensation (trigeminal nerve) normal. In CKCS may be bilateral (not necessarily simultaneously) or associated with vestibular signs.

REFERENCES

Ahn, J. H., H. Cho, J.-H. Kim, S. H. Kim, J.-S. Ham, I. Park, S. H. Suh, S. P. Hong, J.-H. Song, Y.-K. Hong, Y. Jeong, S.-H. Park, and G. Y. Koh. 2019. "Meningeal lymphatic vessels at the skull base drain cerebrospinal fluid." *Nature* 572 (7767):62–66. doi:10.1038/s41586-019-1419-5.

Albertini, R., and R. Bianchi. 2010. "Aquaporins and glia." *Curr Neuropharmacol* 8 (2):84–91. doi:10.2174/157015910791233178.

Ancot, F., P. Lemay, S. P. Knowler, K. Kennedy, S. Griffiths, G. B. Cherubini, J. Sykes, P. J. J. Mandigers, G. A. Rouleau, C. Rusbridge, and Z. Kibar. 2018. "A genome-wide association study identifies candidate loci associated to syringomyelia secondary to Chiari-like malformation in Cavalier King Charles Spaniels." *BMC Genet* 19 (1):16. doi:10.1186/s12863-018-0605-z.

Anderegg, L., M. Im Hof Gut, U. Hetzel, E. W. Howerth, F. Leuthard, K. Kyöstilä, H. Lohi, L. Pettitt, C. Mellersh, K. M. Minor, J. R. Mickelson, K. Batcher, D. Bannasch, V. Jagannathan, and T. Leeb. 2019. "NME5 frameshift variant in Alaskan Malamutes with primary ciliary dyskinesia." *PLoS Genet* 15 (9):e1008378. doi:10.1371/journal.pgen.1008378.

Beard, P. M., E. Munro, and A. G. Gow. 2011. "A quadrigeminal dermoid cyst with concurrent necrotizing granulomatous leukoencephalomyelitis in a Yorkshire Terrier dog." *J Vet Diagn Invest* 23 (5):1075–8. doi:10.1177/1040638711416630.

Bertolini, G., M. Ricciardi, and M. Caldin. 2016. "Multidetector computed tomographic and low-field magnetic resonance imaging anatomy of the quadrigeminal cistern and characterization of supracollicular fluid accumulations in dogs." *Vet Radiol Ultrasound* 57 (3):259–68. doi:10.1111/vru.12347.

Bluestone, C. D., and J. Douglas Swarts. 2010. "Human evolutionary history: consequences for the pathogenesis of otitis media." *Otolaryngol Head Neck Surg* 143 (6):739–744. doi:10.1016/j.otohns.2010.08.015.

Bonfield, C. M., L. M. Foley, S. Kundu, W. Fellows-Mayle, T. Kevin Hitchens, G. K. Rohde, R. Grandhi, and M. P. Mooney. 2015. "The influence of surgical correction on white matter microstructural integrity in rabbits with familial coronal suture craniosynostosis." *Neurosurg Focus* 38 (5):E3. doi:10.3171/2015.2.FOCUS14849.

Bothwell, S. W., D. Janigro, and A. Patabendige. 2019. "Cerebrospinal fluid dynamics and intracranial pressure elevation in neurological diseases." *Fluids Barriers CNS* 16 (1):9. doi:10.1186/s12987-019-0129-6.

Brinker, T., E. Stopa, J. Morrison, and P. Klinge. 2014. "A new look at cerebrospinal fluid circulation." *Fluids Barriers CNS* 11 (1):10. doi:10.1186/2045-8118-11-10.

Bulat, M., and M. Klarica. 2011. "Recent insights into a new hydrodynamics of the cerebrospinal fluid." *Brain Res Rev* 65 (2):99–112. doi:10.1016/j.brainresrev.2010.08.002.

BVA. 2013. "Chiari Malformation/Syringomyelia Scheme (CM/SM Scheme)." British Veterinary Association accessed 8th July.

BVA. 2012. "Appendix 1 Breeding recommendations until relevant EBVs are available." [PDF]. BVA, The, accessed 8th July.

Casaca-Carreira, J., Y. Temel, S. A. Hescham, and A. Jahanshahi. 2018. "Transependymal Cerebrospinal fluid flow: Opportunity for drug delivery?" *Mol Neurobiol* 55 (4):2780–2788. doi:10.1007/s12035-017-0501-y.

Cerda-Gonzalez, S., N. J. Olby, and E. H. Griffith. 2015. "Medullary position at the craniocervical junction in mature cavalier king charles spaniels: Relationship with neurologic signs and syringomyelia." *J Vet Intern Med* 29 (3):882–6. doi:10.1111/jvim.12605.

Chattha, A. S., and G. R. Delong. 1975. "Sylvian aqueduct syndrome as a sign of acute obstructive hydrocephalus in children." *J Neurol Neurosurg Psychiatry* 38 (3):288–296. doi:10.1136/jnnp.38.3.288.

Cirovic, S., R. Lloyd, J. Jovanovik, H. A. Volk, and C. Rusbridge. 2018. "Computer simulation of syringomyelia in dogs." *BMC Vet Res* 14 (1):82. doi:10.1186/s12917-018-1410-7.

Compston, A. 2015. "Cerebrospinal fluid had come a long way since Galen of Pergamon (130–200) declared the ventricles to be reservoirs for the animal spirits." *Brain* 138 (Pt 6):1759–63.

Cross, H. R., R. Cappello, and C. Rusbridge. 2009. "Comparison of cerebral cranium volumes between cavalier King Charles spaniels with Chiari-like malformation, small breed dogs and Labradors." *J Small Anim Pract* 50 (8):399–405. doi:10.1111/j.1748-5827.2009.00799.x.

Da Mesquita, S., Z. Fu, and J. Kipnis. 2018. "The meningeal lymphatic system: A new player in neurophysiology." *Neuron* 100 (2):375–388. doi:10.1016/j.neuron.2018.09.022.

Dandy, W. E., and K. D. Blackfan. 1913. "An experiemental and clinical study of internal hydrocephalous." *J Am Med Assoc* 61 (25):2216–2217. doi:10.1001/jama.1913.04350260014006.

de Stefani, A., L. de Risio, S. R. Platt, L. Matiasek, A. Lujan-Feliu-Pascual, and L. S. Garosi. 2011. "Surgical technique, postoperative complications and outcome in 14 dogs treated for hydrocephalous by ventriculoperitoneal shunting." *Vet Surg* 40 (2):183–191. doi:10.1111/j.1532-950X.2010.00764.x.

Deisenhammer, F. 2015. "The history of cerebrospinal fluid." In *Cerebrospinal Fluid in Clinical Neurology*, edited by F. Deisenhammer, F. Sellebjerg, C. E. Teunissen and H. Tumani, 3–16. Cham: Springer International Publishing.

Del Bigio, M. R., and D. L. Di Curzio. 2016. "Nonsurgical therapy for hydrocephalous: A comprehensive and critical review." *Fluids Barriers CNS* 13:3. doi:10.1186/s12987-016-0025-2.

DeLahunta, A., E. Glass, and M. Kent. 2014. "Cerebrospinal fluid and hydrocephalous." In *Veterinary Neuroanatomy and Clinical Neurology*, 78–101. St Louis, MO: Elsevier.

Di Ieva, A., E. Bruner, J. Davidson, P. Pisano, T. Haider, S. S. Stone, M. D. Cusimano, M. Tschabitscher, and F. Grizzi. 2013. "Cranial sutures: A multidisciplinary review." *Childs Nerv Syst* 29 (6):893–905. doi:10.1007/s00381-013-2061-4.

Domer, F. R. 1969. "Effects of diuretics on cerebrospinal fluid formation and potassium movement." *Exp Neurol* 24 (1):54–64. doi:10.1016/0014-4886(69)90005-3.

Donnally III, C. J., S. Munakomi, and M. Varacallo. 2020. "Basilar Invagination." In StatPearls. Treasure Island (FL): StatPearls Publishing Copyright © 2020, StatPearls Publishing LLC.

Dubrul, E. L., and D. M. Laskin. 1961. "Preadaptive potentialities of the mammalian skull: An experiment in growth and form." *Am J Anat* 109:117–32. doi:10.1002/aja.1001090203.

Faller, K. M., J. Bras, S. J. Sharpe, G. W. Anderson, L. Darwent, C. Kun-Rodrigues, J. Alroy, J. Penderis, S. E. Mole, R. Gutierrez-Quintana, and R. J. Guerreiro. 2016. "The Chihuahua dog: A new animal model for neuronal ceroid lipofuscinosis CLN7 disease?" *J Neurosci Res* 94 (4):339–47. doi:10.1002/jnr.23710.

Fenn, J., M. J. Schmidt, H. Simpson, C. J. Driver, and H. A. Volk. 2013. "Venous sinus volume in the caudal cranial fossa in Cavalier King Charles spaniels with syringomyelia." *Vet J* 197 (3):896–7. doi:10.1016/j.tvjl.2013.05.007.

Fenstermacher, J. D., J. F. Ghersi-Egea, W. Finnegan, and J. L. Chen. 1997. "The rapid flow of cerebrospinal fluid from ventricles to cisterns via subarachnoid velae in the normal rat." *Acta Neurochir Suppl* 70:285–7 doi:10.1007/978-3-7091-6837-0_88.

Flegel, T. 2017. "Breed-specific magnetic resonance imaging characteristics of necrotizing encephalitis in dogs." *Front Vet Sci* 4 (203). doi:10.3389/fvets.2017.00203.

Geiger, M., and S. Haussman. 2016. "Cranial suture closure in domestic dog breeds and its relationships to skull morphology." *Anat Rec* 299 (4):412–420. doi:10.1002/ar.23313.

Girod, M., F. Allerton, K. Gommeren, A. C. Tutunaru, J. de Marchin, I. Van Soens, E. Ramery, and D. Peeters. 2016. "Evaluation of the effect of oral omeprazole on canine cerebrospinal fluid production: A pilot study." *Vet J* 209:119–24. doi:10.1016/j.tvjl.2015.10.045.

Girod, M., F. Allerton, E. Vangrinsven, A. C. Tutunaru, J. de Marchin, C. Gómez-Fernández-Blanco, A. Ruiz-Nuño, A. Wojnicz, F. Farnir, K. Gommeren, and D. Peeters. 2019. "CSF omeprazole concentration and albumin quotient following high dose intravenous omeprazole in dogs." *Res Vet Sci* 125:266–271 doi:10.1016/j.rvsc.2019.07.009.

Gradner, G., R. Kaefinger, and G. Dupré. 2019. "Complications associated with ventriculoperitoneal shunts in dogs and cats with idiopathic hydrocephalous: A systematic review." *J Vet Intern Med* 33 (2):403–412. doi:10.1111/jvim.15422.

Hayes, G. M., E. J. Friend, and N. D. Jeffery. 2010. "Relationship between pharyngeal conformation and otitis media with effusion in Cavalier King Charles spaniels." *Vet Rec* 167 (2):55–8. doi:10.1136/vr.b4886.

Haykowsky, M. J., N. D. Eves, D. E. R. Warburton, and M. J. Findlay. 2003. "Resistance exercise, the valsalva maneuver, and cerebrovascular transmural pressure." *Med Sci Sports Exerc* 35 (1):65.

Higgins, R. J., M. Vandevelde, and K. B. Braund. 1977. "Internal hydrocephalous and associated periventricular encephalitis in young dogs." *Vet Pathol* 14 (3):236–46. doi:10.1177/030098587701400306.

Hoshide, R., H. Meltzer, C. Dalle-Ore, D. Gonda, D. Guillaume, and C. C. Chen. 2017. "Impact of ventricular-peritoneal shunt valve design on clinical outcome of pediatric patients with hydrocephalous: Lessons learned from randomized controlled trials." *Surg Neurol Int* 8:49. doi:10.4103/sni.sni_11_17.

Hoyte, D. A. N. 1997. "Growth of the cranial base." In *Fundamentals of Craniofacial Growth* edited by A. D. Dixon, D. A. N. Hoyte, and O. Ronning, 258–333. Boca Raton, FL: CRC Press LLC.

Huizing, X., A. Sparkes, and R. Dennis. 2017. "Shape of the feline cerebellum and occipital bone related to breed on MRI of 200 cats." *J Feline Med Surg* 19 (10):1065–1072. doi:10.1177/1098612x16676022.

Jessen, N. A., A. S. Munk, I. Lundgaard, and M. Nedergaard. 2015. "The glymphatic system: A beginner's guide." *Neurochem Res* 40 (12):2583–99. doi:10.1007/s11064-015-1581-6.

Kibar, Z., M. P. Dubé, S. P. Knowler, C. Rusbridge, and G. Rouleau. 2007. "Preliminary results from syringomyelia (SM) genome wide scans in cavalier King Charles spaniel kindred and directions for future research." *Chiari-like malformation and Syringomyelia in the Cavalier King Charles Spaniel*:179.

Kishimoto, T. E., K. Uchida, J. K. Chambers, M. K. Kok, N. V. Son, T. Shiga, M. Hirabayashi, N. Ushio, and H. Nakayama. 2020. "A retrospective survey on canine intracranial tumors between 2007 and 2017." *J Vet Med Sci* 82 (1):77–83. doi:10.1292/jvms.19-0486.

Kiviranta, A. M., C. Rusbridge, O. Laitinen-Vapaavuori, A. Hielm-Bjorkman, A. K. Lappalainen, S. P. Knowler, and T. S. Jokinen. 2017. "Syringomyelia and craniocervical junction abnormalities in chihuahuas." *J Vet Intern Med* 31 (6):1771–1781. doi:10.1111/jvim.14826.

Knowler, S. P., L. Gillstedt, T. J. Mitchell, J. Jovanovik, H. A. Volk, and C. Rusbridge. 2019. "Pilot study of head conformation changes over time in the Cavalier King Charles spaniel breed." *Vet Rec* 184 (4):122. doi:10.1136/vr.105135.

Knowler, S. P., C. Cross, S. Griffiths, A. K. McFadyen, J. Jovanovik, A. Tauro, Z. Kibar, C. J. Driver, R. M. L. Ragione, and C. Rusbridge. 2017. "Use of morphometric mapping to characterise symptomatic chiari-like malformation, secondary syringomyelia and associated brachycephaly in the cavalier king charles spaniel." *PLoS One* 12 (1). doi:10.1371/journal.pone.0170315.

Knowler, S. P., A.-M. Kiviranta, A. K. McFadyen, T. S. Jokinen, R. M. La Ragione, and C. Rusbridge. 2017. "Craniometric analysis of the hindbrain and craniocervical junction of chihuahua, affenpinscher and cavalier king charles spaniel dogs with and without syringomyelia secondary to chiari-like malformation." *PLoS One* 12 (1). doi:10.1371/journal.pone.0169898.

Knowler, S. P., E. Dumas, M. Spiteri, A. K. McFadyen, F. Stringer, K. Wells, and C. Rusbridge. 2020. "Facial changes related to brachycephaly in Cavalier King Charles Spaniels with Chiari-like malformation associated pain and secondary syringomyelia." *J Vet Intern Med* 34 (1):237–246. doi:10.1111/jvim.15632.

Knowler, S. P., G. L. Galea, and C. Rusbridge. 2018. "Morphogenesis of canine chiari malformation and secondary syringomyelia: Disorders of cerebrospinal fluid circulation." *Front Vet Sci* 5 (171). doi:10.3389/fvets.2018.00171.

Kolecka, M., D. Farke, K. Failling, M. Kramer, and M. J. Schmidt. 2019. "Intraoperative measurement of intraventricular pressure in dogs with communicating internal hydrocephalous." *PLoS One* 14 (9):e0222725. doi:10.1371/journal.pone.0222725.

Kolecka, M., N. Ondreka, A. Moritz, M. Kramer, and M. J. Schmidt. 2015. "Effect of acetazolamide and subsequent ventriculo-peritoneal shunting on clinical signs and ventricular volumes in dogs with internal hydrocephalous." *Acta Vet Scand* 57 (1):49. doi:10.1186/s13028-015-0137-8.

Laterveer, M. 2020. "The effect of phenotypic selection on 14 years of chiari-like malformation and syringomyelia MRI scanning in Cavalier King Charles Spaniels in the Netherlands." Master Master Thesis, Faculty of Veterinary Medicine, Utrecht University.

Laubner, S., N. Ondreka, K. Failing, M. Kramer, and M. J. Schmidt. 2015. "Magnetic resonance imaging signs of high intraventricular pressure–comparison of findings in dogs with clinically relevant internal hydrocephalous and asymptomatic dogs with ventriculomegaly." *BMC Vet Res* 11:181. doi:10.1186/s12917-015-0479-5.

Lee, L. 2013. "Riding the wave of ependymal cilia: Genetic susceptibility to hydrocephalous in primary ciliary dyskinesia." *J Neurosci Res* 91 (9):1117–32. doi:10.1002/jnr.23238.

Lemay, P., S. P. Knowler, S. Bouasker, Y. Nedelec, S. Platt, C. Freeman, G. Child, L. B. Barreiro, G. A. Rouleau, C. Rusbridge, and Z. Kibar. 2014. "Quantitative Trait Loci (QTL) study identifies novel genomic regions associated to chiari-like malformation in griffon bruxellois dogs." *PLoS One* 9 (4):e89816. doi:10.1371/journal.pone.0089816.

Lindsay, W. L. 1871. "Insanity in the lower animals." *Br Foreign Med Chir Rev* 48 (95):178–213.

Liu, D., Z. Li, X. Zhang, L. Zhao, J. Jia, F. Sun, Y. Wang, D. Ma, and W. Wei. 2017. "Assessment of intracranial pressure with ultrasonographic retrobulbar optic nerve sheath diameter measurement." *BMC Neurol* 17 (1):188. doi:10.1186/s12883-017-0964-5.

Lowrie, M., A. Wessmann, D. Gunn-Moore, and J. Penderis. 2009. "Quadrigeminal cyst management by cystoperitoneal shunt in a 4-year-old Persian cat." *J Feline Med Surg* 11 (8):711–3. doi:10.1016/j.jfms.2009.01.007.

Lun, M. P., E. S. Monuki, and M. K. Lehtinen. 2015. "Development and functions of the choroid plexus-cerebrospinal fluid system." *Nat Rev Neurosci* 16 (8):445–57. doi:10.1038/nrn3921.

Ma, Q., Y. Decker, A. Müller, B. V. Ineichen, and S. T. Proulx. 2019. "Clearance of cerebrospinal fluid from the sacral spine through lymphatic vessels." *J Exp Med* 216 (11):2492–2502. doi:10.1084/jem.20190351.

Ma, Q., B. V. Ineichen, M. Detmar, and S. T. Proulx. 2017. "Outflow of cerebrospinal fluid is predominantly through lymphatic vessels and is reduced in aged mice." *Nat Commun* 8 (1):1434. doi:10.1038/s41467-017-01484-6.

Marin-Garcia, P., J. Gonzalez-Soriano, P. Martinez-Sainz, J. Contreras-Rodriguez, C. Del Corral-Gros, and E. Rodriguez-Veiga. 1995. "Spinal cord central canal of the German shepherd dog: morphological, histological, and ultrastructural considerations." *J Morphol* 224 (2):205–12. doi:10.1002/jmor.1052240209.

Marino, D. J., C. A. Loughin, C. W. Dewey, L. J. Marino, J. J. Sackman, M. L. Lesser, and M. B. Akerman. 2012. "Morphometric features of the craniocervical junction region in dogs with suspected Chiari-like malformation determined by combined use of magnetic resonance imaging and computed tomography." *Am J Vet Res* 73 (1):105–11. doi:10.2460/ajvr.73.1.105.

Marsh, H. 2018. *Admissions: A Life in Brain Surgery.* London: Weidenfeld & Nicolson.

Matiasek, L. A., S. R. Platt, S. Shaw, and R. Dennis. 2007. "Clinical and magnetic resonance imaging characteristics of quadrigeminal cysts in dogs." *J Vet Intern Med* 21 (5):1021–6.

Mayousse, V., L. Desquilbet, A. Jeandel, and S. Blot. 2017. "Prevalence of neurological disorders in French bulldog: A retrospective study of 343 cases (2002–2016)." *BMC Vet Res* 13 (1):212. doi:10.1186/s12917-017-1132-2.

McCarthy, K. D., and D. J. Reed. 1974. "The effect of acetazolamide and furosemide on cerebrospinal fluid production and choroid plexus carbonic anhydrase activity." *J Pharmacol Exp Ther* 189 (1):194–201.

McGonnell, I. M., and S. E. Akbareian. 2019. "Like a hole in the head: Development, evolutionary implications and diseases of the cranial foramina." *Semin Cell Dev Biol* 91:23–30 doi:10.1016/j.semcdb.2018.08.011.

McGrath, J., O. El-Saadi, V. Grim, S. Cardy, B. Chapple, D. Chant, D. Lieberman, and B. Mowry. 2002. "Minor physical anomalies and quantitative measures of the head and face in patients with psychosis." *Arch Gen Psychiatry* 59 (5):458–464. doi:10.1001/archpsyc.59.5.458.

Milhorat, T. H. 1972. "Cerebrospinal fluid physiology". In *Hydrocephalus and the Cerebrospinal Fluid*, 1–41. Baltimore, MD: Williams & Wilkins.

Miyajima, M., and H. Arai. 2015. "Evaluation of the production and absorption of cerebrospinal fluid." *Neurol Med Chir (Tokyo)* 55 (8):647–56. doi:10.2176/nmc.ra.2015-0003.

Mollanji, R., R. Bozanovic-Sosic, I. Silver, B. Li, C. Kim, R. Midha, and M. Johnston. 2001. "Intracranial pressure accommodation is impaired by blocking pathways leading to extracranial lymphatics." *Am J Physiol Regul Integr Comp Physiol* 280 (5):R1573–R1581. doi:10.1152/ajpregu.2001.280.5.R1573.

Mollanji, R., R. Bozanovic-Sosic, A. Zakharov, L. Makarian, and M. G. Johnston. 2002. "Blocking cerebrospinal fluid absorption through the cribriform plate increases resting intracranial pressure." *Am J Physiol Regul Integr Comp Physiol* 282 (6):R1593–9. doi:10.1152/ajpregu.00695.2001.

Nalborczyk, Z. R., A. K. McFadyen, J. Jovanovik, A. Tauro, C. J. Driver, N. Fitzpatrick, S. P. Knower, and C. Rusbridge. 2017. "MRI characteristics for "phantom" scratching in canine syringomyelia." *BMC Vet Res* 13 (1):340. doi:10.1186/s12917-017-1258-2.

Neelapu, B. C., O. P. Kharbanda, H. K. Sardana, R. Balachandran, V. Sardana, P. Kapoor, A. Gupta, and S. Vasamsetti. 2017. "Craniofacial and upper airway morphology in adult obstructive sleep apnea patients: A systematic review and meta-analysis of cephalometric studies." *Sleep Med Rev* 31:79–90 doi:10.1016/j.smrv.2016.01.007.

Orešković, D., and M. Klarica. 2010. "The formation of cerebrospinal fluid: Nearly a hundred years of interpretations and misinterpretations." *Brain Res Rev* 64 (2):241–262. doi:10.1016/j.brainresrev.2010.04.006.

Parker, J. E., S. P. Knowler, C. Rusbridge, E. Noorman, and N. D. Jeffery. 2011. "Prevalence of asymptomatic syringomyelia in Cavalier King Charles spaniels." *Vet Rec* 168 (25):667. doi:10.1136/vr.d1726.

Plessas, I. N., C. Rusbridge, C. J. Driver, K. E. Chandler, A. Craig, I. M. McGonnell, D. C. Brodbelt, and H. A. Volk. 2012. "Long-term outcome of Cavalier King Charles spaniel dogs with clinical signs associated with Chiari-like malformation and syringomyelia." *Vet Rec* doi:10.1136/vr.100449.

Plessas, I. N., H. A. Volk, C. Rusbridge, A. E. Vanhaesebrouck, and N. D. Jeffery. 2015. "Comparison of gabapentin versus topiramate on clinically affected dogs with Chiari-like malformation and syringomyelia." *Vet Rec* 177 (11). doi:10.1136/vr.103234.

Reichler, I. M., A. Hoerauf, F. Guscetti, O. Gardelle, M. H. Stoffel, B. Jentsch, H. Walt, and S. Arnold. 2001. "Primary ciliary dyskinesia with situs inversus totalis, hydrocephalous internus and cardiac malformations in a dog." *J Small Anim Pract* 42 (7):345–8. doi:10.1111/j.1748-5827.2001.tb02471.x.

Reidenberg, J. S., and J. T. Laitman. 1991. "Effect of basicranial flexion on larynx and hyoid position in rats: an experimental study of skull and soft tissue interactions." *Anat Rec* 230 (4):557–69. doi:10.1002/ar.1092300416.

Rekate, H. L. 2008. "The definition and classification of hydrocephalous: A personal recommendation to stimulate debate." *Cerebrospinal Fluid Res* 5:2. doi:10.1186/1743-8454-5-2.

Rekate, H. L. 2011. "A consensus on the classification of hydrocephalous: Its utility in the assessment of abnormalities of cerebrospinal fluid dynamics." *Childs Nerv Syst* 27 (10):1535–41. doi:10.1007/s00381-011-1558-y.

Riva-Cambrin, J., J. R. W. Kestle, R. Holubkov, J. Butler, A. V. Kulkarni, J. Drake, W. E. Whitehead, J. C. Wellons, C. N. Shannon, M. S. Tamber, D. D. Limbrick, C. Rozzelle, S. R. Browd, and T. D. Simon. 2016. "Risk factors for shunt malfunction in pediatric hydrocephalous: A multicenter prospective cohort study." 17 (4):382. doi:10.3171/2015.6.Peds14670.

Roberts, T., P. McGreevy, and M. Valenzuela. 2010. "Human induced rotation and reorganization of the brain of domestic dogs." *PLoS One* 5 (7):e11946. doi:10.1371/journal.pone.0011946.

Román, G. C., R. E. Jackson, S. H. Fung, Y. J. Zhang, and A. K. Verma. 2019. "Sleep-disordered breathing and idiopathic normal-pressure: Recent pathophysiological advances." *Curr Neurol Neurosci Rep* 19 (7):39. doi:10.1007/s11910-019-0952-9.

Rusbridge, C., A. K. McFadyen, and S. P. Knower. 2019. "Behavioral and clinical signs of Chiari-like malformation-associated pain and syringomyelia in Cavalier King Charles spaniels." *J Vet Internal Med* 33 (5):2138–2150. doi:10.1111/jvim.15552.

Rusbridge, C., F. Stringer, and S. P. Knowler. 2018. "Clinical application of diagnostic imaging of chiari-like malformation and syringomyelia." *Front Vet Sci* 5 (280). doi:10.3389/fvets.2018.00280.

Saito, M., N. J. Olby, K. Spaulding, K. Muñana, and N. J. Sharp. 2003. "Relationship among basilar artery resistance index, degree of ventriculomegaly, and clinical signs in hydrocephalic dogs." *Vet Radiol Ultrasound* 44 (6):687–94. doi:10.1111/j.1740-8261.2003.tb00532.x.

Saito, Miyoko, Natasha J. Olby, and Kathy Spaulding. 2001. "Identification of arachnoid cysts in the quadrigeminal cistern using ultrasonography." *Vet Radiol Ultrasound* 42 (5):435–9. doi:10.1111/j.1740-8261.2001.tb00966.x.

Sanchis-Mora, S., Y. M. Chang, S. M. Abeyesinghe, A. Fisher, N. Upton, H. A. Volk, and L. Pelligand. 2019. "Pregabalin for the treatment of syringomyelia-associated neuropathic pain in dogs: A randomised, placebo-controlled, double-masked clinical trial." *Vet J* 250:55–62 doi:10.1016/j.tvjl.2019.06.006.

Sasaoka, K., K. Nakamura, T. Osuga, T. Morita, N. Yokoyama, K. Morishita, N. Sasaki, H. Ohta, and M. Takiguchi. 2018. "Transcranial doppler ultrasound examination in dogs with suspected intracranial hypertension caused by neurologic diseases." *J Vet Internal Med* 32 (1):314–323. doi:10.1111/jvim.14900.

Sato, O. 1967. "The effect of dexamethasone on cerebrospinal fluid production rate in the dog." *No To Shinkei* 19 (5):485–92.

Sato, O., E. A. Bering Jr., M. Yagi, R. Tsugane, M. Hara, Y. Amano, and T. Asai. 1975. "Bulk flow in the cerebrospinal fluid system of the dog." *Acta Neurol Scand* 51 (1):1–11. doi:10.1111/j.1600-0404.1975.tb01354.x.

Schmidt, M. J., A. Hartmann, D. Farke, K. Failling, and M. Kolecka. 2019. "Association between improvement of clinical signs and decrease of ventricular volume after ventriculoperitoneal shunting in dogs with internal hydrocephalous." *J Vet Intern Med* 33 (3):1368–75. doi:10.1111/jvim.15468.

Schmidt, M. J., M. Kampschulte, S. Enderlein, D. Gorgas, J. Lang, E. Ludewig, A. Fischer, A. Meyer-Lindenberg, A. R. Schaubmar, K. Failing, and N. Ondreka. 2017. "The relationship between brachycephalic head features in modern persian cats and dysmorphologies of the skull and internal hydrocephalous." *J Vet Intern Med* 31 (5):1487–1501. doi:10.1111/jvim.14805.

Schmidt, M. J., N. Ondreka, C. Rummel, H. Volk, M. Sauerbrey, and M. Kramer. 2012. "Volume reduction of the jugular foramina in Cavalier King Charles Spaniels with syringomyelia." *BMC Vet Res* 8 (1):158. doi:10.1186/1746-6148-8-158.

Schmidt, M. J., C. Rummel, J. Hauer, M. Kolecka, N. Ondreka, V. McClure, and J. Roth. 2016. "Increased CSF aquaporin-4, and interleukin-6 levels in dogs with idiopathic communicating internal hydrocephalous and a decrease after ventriculo-peritoneal shunting." *Fluids Barriers CNS* 13 (1):12. doi:10.1186/s12987-016-0034-1.

Schwalbe, G. 1869. "Die Arachnoidalraum, ein Lymphraum und sein Zusammenhang mit dem Perichoriordalraum." *Zbl med Wiss Zentralblatt fur die medizinischen Wissenschaften.* 7:465–467.

Scrivani, P. V., D. J. Fletcher, S. D. Cooley, A. J. Rosenblatt, and H. N. Erb. 2013. "T2-Weighted magnetic resonance imaging measurements of optic nerve sheath diameter in dogs with and without presumed intracranial hypertension." *Vet Radiol Ultrasound* 54 (3):263–70. doi:10.1111/vru.12023.

Seppänen, R. T. K., M. Kaimio, K. J. M. Schildt, L. Lilja-Maula, H. K. Hyytiäinen, S. Mölsä, M. Morelius, M. M. Rajamäki, A. K. Lappalainen, and M. Rantala. 2019. "Skin and ear health in a group of English bulldogs in Finland - a descriptive study with special reference to owner perceptions." *Vet Dermatol* 30 (4):307-e85. doi:10.1111/vde.12752.

Shaw, T. A., I. M. McGonnell, C. J. Driver, C. Rusbridge, and H. A. Volk. 2012. "Increase in cerebellar volume in Cavalier King Charles Spaniels with Chiari-like malformation and its role in the development of syringomyelia." *PLoS One* 7 (4):e33660. doi:10.1371/journal.pone.0033660.

Shoja, M. M., R. Ramdhan, C. J. Jensen, J. J. Chern, W. Jerry Oakes, and R. Shane Tubbs. 2018. "Embryology of the craniocervical junction and posterior cranial fossa, part I: Development of the upper vertebrae and skull." *Clin Anat* 31 (4):466–487. doi:10.1002/ca.23049.

Smith, J. J., D. J. Fletcher, S. D. Cooley, and M. S. Thompson. 2018. "Transpalpebral ultrasonographic measurement of the optic nerve sheath diameter in healthy dogs." *J Vet Emerg Crit Care (San Antonio)* 28 (1):31–38. doi:10.1111/vec.12677.

Sokolowski, W., N. Czubaj, M. Skibniewski, K. Barszcz, M. Kupczynska, W. Kinda, and Z. Kielbowicz. 2018. "Rostral cranial fossa as a site for cerebrospinal fluid drainage - volumetric studies in dog breeds of different size and morphotype." *BMC Vet Res* 14 (1):162. doi:10.1186/s12917-018-1483-3.

Song, R. B., C. H. Vite, C. W. Bradley, and J. R. Cross. 2013. "Postmortem evaluation of 435 cases of intracranial neoplasia in dogs and relationship of neoplasm with breed, age, and body weight." *J Vet Intern Med* 27 (5):1143–52. doi:10.1111/jvim.12136.

Sparks, C. R., I. Robertson, and N. J. Olby. 2019. "Morphometric analysis of spinal cord termination in Cavalier King Charles Spaniels." *J Vet Intern Med* 33 (2):717–725. doi:10.1111/jvim.15437.

Spiteri, M., S. P. Knowler, C. Rusbridge, and K. Wells. 2019. "Using machine learning to understand neuromorphological change and image-based biomarker identification in Cavalier King Charles Spaniels with Chiari-like malformation-associated pain and syringomyelia." *J Vet Intern Med.* doi:10.1111/jvim.15621.

Stockyard, C. R. 1941. "The genetic and endocrinic basis for differences in form and behaviour." In *Anatomical Memoirs*, 40–357. Philadelphia, PA: Wistar Institute of Anatomy and Biology.

Stoodley, M. 2014. "The filling mechanism." In *Syringomyelia: A Disorder of CSF Circulation*, edited by G. Flint and C. Rusbridge, 87–101. Berlin, Heidelberg: Springer.

Sugita, Y., S. Iijima, Y. Teshima, T. Shimizu, N. Nishimura, T. Tsutsumi, H. Hayashi, H. Kaneda, and Y. Hishikawa. 1985. "Marked episodic elevation of cerebrospinal fluid pressure during nocturnal sleep in patients with sleep apnea hypersomnia syndrome." *Electroencephalogr Clin Neurophysiol* 60 (3):214–9. doi:10.1016/0013-4694(85)90033-1.

Swenson, E. R. 2016. "Pharmacology of acute mountain sickness: Old drugs and newer thinking." *J Appl Physiol* 120 (2):204–15. doi:10.1152/japplphysiol.00443.2015.

Tauro, A., and C. Rusbridge. 2020. "Syringopleural shunt placement in a pug with a cervical spinal diverticulum and associated syringomyelia." *Clin Case Rep.* doi:10.1002/ccr3.2845.

Taylor, R. 1830. Catalogue of the Hunterian Collection in the Museum of the Royal College of Surgeons in London. In *Part II the Pathological Preparations in a Dried State.* London Royal College of Surgeons.

Thoefner, M. S., L. T. Skovgaard, F. J. McEvoy, M. Berendt, and O. J. Bjerrum. 2020. "Pregabalin alleviates clinical signs of syringomyelia-related central neuropathic pain in Cavalier King Charles Spaniel dogs: a randomized controlled trial." *Vet Anaesth Analg* 47 (2):238–248. doi:10.1016/j.vaa.2019.09.007.

Thofner, M. S., C. L. Stougaard, U. Westrup, A. A. Madry, C. S. Knudsen, H. Berg, C. S. Jensen, R. M. Handby, H. Gredal, M. Fredholm, and M. Berendt. 2015. "Prevalence and heritability of symptomatic syringomyelia in Cavalier King Charles Spaniels and long-term outcome in symptomatic and asymptomatic littermates." *J Vet Intern Med* 29 (1):243–50. doi:10.1111/jvim.12475.

Thomas, J. H. 2019. "Fluid dynamics of cerebrospinal fluid flow in perivascular spaces." *J R Soc Interface* 16 (159):20190572. doi:10.1098/rsif.2019.0572.

Tischfield, M. A., C. D. Robson, N. M. Gilette, S. M. Chim, F. A. Sofela, M. M. DeLisle, A. Gelber, B. J. Barry, S. MacKinnon, L. R. Dagi, J. Nathans, and E. C. Engle. 2017. "Cerebral vein malformations result from loss of twist1 expression and BMP signaling from skull progenitor cells and dura." *Dev cell* 42 (5):445–461.e5. doi:10.1016/j.devcel.2017.07.027.

Tokumaru, A. M., A. J. Barkovich, S. F. Ciricillo, and M. S. Edwards. 1996. "Skull base and calvarial deformities: Association with intracranial changes in craniofacial syndromes." *AJNR Am J Neuroradiol* 17 (4):619–30.

Tubbs, R. S., A. N. Bosmia, and A. A. Cohen-Gadol. 2012. "The human calvaria: A review of embryology, anatomy, pathology, and molecular development." *Child's Nervous Syst* 28 (1):23–31. doi:10.1007/s00381-011-1637-0.

Verkman, A. S., L. Tradtrantip, A. J. Smith, and X. Yao. 2017. "Aquaporin Water Channels and Hydrocephalous." *Pediatr Neurosurg* 52 (6):409–16. doi:10.1159/000452168.

Virchow, R. 1851. "Uber den cretinismus, namentilich in franken, und uber pathologische schadelformen." *Verh Phys Med Ges Wurzburg* 2:230–256.

Wagner, F., and I. Ruf. 2021. ""Forever young"—Postnatal growth inhibition of the turbinal skeleton in brachycephalic dog breeds (Canis lupus familiaris)." *Anat Rec* 304:154–189 doi:10.1002/ar.24422.

Weiss, M. H., and F. E. Nulsen. 1970. "The effect of glucocorticoids on CSF flow in dogs." *J Neurosurg* 32 (4):452–8. doi:10.3171/jns.1970.32.4.0452.

Whittaker, D. E., K. English, I. M. McGonnell, and H. A. Volk. 2011. "Evaluation of cerebrospinal fluid in Cavalier King Charles Spaniel dogs diagnosed with Chiari-like malformation with or without concurrent syringomyelia." *J Vet Diagn Invest* 23 (2):302–7.

Wijnrocx, K., L. W. L. Van Bruggen, W. Eggelmeijer, E. Noorman, A. Jacques, N. Buys, S. Janssens, and P. J. J. Mandigers. 2017. "Twelve years of chiari-like malformation and syringomyelia scanning in Cavalier King Charles Spaniels in the Netherlands: Towards a more precise phenotype." *PLoS One* 12 (9):e0184893. doi:10.1371/journal.pone.0184893.

Wise, B. L., J. L. Mathis, and J. H. Wright. 1966. "Experimental use of isosorbide: An oral osmotic agent to lower cerebrospinal pressure and reduce brain bulk." *J Neurosurg* 25 (2):183–8. doi:10.3171/jns.1966.25.2.0183.

Wu, Y., N. L. Green, M. R. Wrensch, S. Zhao, and N. Gupta. 2007. "Ventriculoperitoneal Shunt Complications in California: 1990 TO 2000." *Neurosurgery* 61 (3):557–563. doi:10.1227/01.Neu.0000290903.07943.Af.

Yadav, Y. R., V. Parihar, S. Pande, H. Namdev, and M. Agarwal. 2012. "Endoscopic third ventriculostomy." *J Neurosci Rural Pract* 3 (2):163–73. doi:10.4103/0976-3147.98222.

Yamada, S., M. Miyazaki, Y. Yamashita, C. Ouyang, M. Yui, M. Nakahashi, S. Shimizu, I. Aoki, Y. Morohoshi, and J. G. McComb. 2013. "Influence of respiration on cerebrospinal fluid movement using magnetic resonance spin labeling." *Fluids Barriers CNS* 10 (1):36. doi:10.1186/2045-8118-10-36.

Yoon, J. S., T. K. Nam, J. T. Kwon, S. W. Park, and Y. S. Park. 2015. "CSF flow pathways through the ventricle-cistern interfaces in kaolin-induced hydrocephalous rats-laboratory investigation." *Childs Nerv Syst* 31 (12):2277–81. doi:10.1007/s00381-015-2901-5.

Zakharov, A., C. Papaiconomou, and M. Johnston. 2004. "Lymphatic vessels gain access to cerebrospinal fluid through unique association with olfactory nerves." *Lymphat Res Biol* 2 (3):139–46. doi:10.1089/lrb.2004.2.139.

Zhang, J., M. A. Williams, and D. Rigamonti. 2006. "Genetics of human hydrocephalous." *J Neurol* 253 (10):1255–66. doi:10.1007/s00415-006-0245-5.

16 Vertebral Malformations and Spinal Disease in Brachycephalic Breeds

Steven De Decker
University of London

Rodrigo Gutierrez-Quintana
University of Glasgow

CONTENTS

Hemivertebra..292
    Prevalence of Hemivertebra in Brachycephalic Dogs..294
    Development of Clinical Signs in Dogs with Hemivertebra......................................295
    Diagnosis and Treatment of Hemivertebra with Kyphosis 297
    Should Hemivertebra Be Considered Clinically Irrelevant in the Majority of Cases? 298
    Screening Programs for Hemivertebrae 299
Transitional Vertebrae 300
Thoracic Vertebral Canal Stenosis 300
Caudal Articular Process Dysplasia and Meningeal Fibrosis 300
    Diagnosis and Treatment of Caudal Articular Process Dysplasia............................302
Spinal Arachnoid Diverticula...303
    Diagnosis and Treatment of Spinal Arachnoid Diverticula.....................................304
Spina Bifida, Meningocele and Meningomyelocele 305
Intervertebral Disc Disease 306
Conclusions..307
References...308

Small brachycephalic dogs, such as the French Bulldog and Pug, are commonly affected by spinal conditions and are regularly presented to veterinary practitioners for further evaluation of a gait abnormality. This is illustrated by a study that evaluated owner-based questionnaires and video footage of 550 Pugs registered by the Swedish Kennel Club (Rohdin et al., 2018a). This study revealed that gait abnormalities were commonly observed with 30.7% of owners perceiving their dog to have an abnormal gait. Evaluation of video footage suggested that the nature of observed gait abnormalities was most suggestive for an underlying neurologic cause. An abnormal gait was further listed as the most common cause of death in this cohort of Pugs (Rohdin et al., 2018a). Another study revealed that 18.7% of all French Bulldogs admitted at an academic referral institution in France were seen for evaluation of neurologic signs (Mayousse et al. 2017). The majority of these dogs suffered from spinal disease with intervertebral disk herniation being the most common spinal disorder, followed by spinal arachnoid diverticula (SAD) and congenital vertebral malformations. Congenital vertebral malformations are commonly seen on diagnostic imaging studies of small brachycephalic dogs, such as French Bulldogs, English Bulldogs, Pugs and Boston Terriers. It is however important to realise that vertebral malformations occur commonly in clinically normal dogs and are only rarely directly associated with clinical signs (Ryan et al., 2017; Gutierrez-Quintana and De Decker, 2021).

Although often grouped together, results of recent studies suggest that different small brachycephalic breeds are predisposed to different spinal conditions. More specifically, it appears that French Bulldogs are predisposed to acute intervertebral disc disease (IVDD), while Pugs seem more sensitive for development of complex spinal malformations, such as hemivertebra with kyphosis, articular process dysplasia and meningeal fibrosis (Fisher et al., 2013; Aikawa et al., 2014; Mayousse et al., 2017; De Decker et al., 2019; Rohdin et al., 2020). Clear genetic differences exist between Pugs on the one hand and French Bulldogs, English Bulldogs and Boston Terriers on the other hand. Whereas French Bulldogs, English Bulldogs and Boston Terriers represent genetically closely related breeds, Pugs have a different ancestry (Parker, 2012). Because of their similar external morphological characteristics, these breeds are occasionally referred to collectively as 'screw-tailed' brachycephalic breeds. Pugs actually don't have a truncated, shortened and kinked tail and the genetic mutation responsible for this 'screw-tail' morphology, a frameshift mutation in the *DISHEVELLED 2* gene, is not present in Pugs (Mansour et al., 2018). Although other factors cannot be excluded, these distinct genetic differences could potentially explain why Pugs are affected by different spinal conditions compared to 'screw-tailed' brachycephalic dogs.

In this chapter, we will discuss the most common spinal disorders in brachycephalic dogs with an emphasis on the prevalence and clinical relevance of vertebral and spinal malformations.

One group of conditions typically associated with small brachycephalic dogs are congenital vertebral malformations. Although the used terminology is somewhat controversial and potentially confusing, they have been classified as defects in segmentation (block vertebra) and defects in formation (hemivertebra) (Westworth and Sturges, 2010). Several studies have demonstrated a high prevalence of congenital vertebral malformations in small brachycephalic dogs (Moissonnier et al., 2011, Gutierrez-Quintana et al., 2014, Ryan et al., 2017; Rohdin et al., 2018b). A study evaluating the prevalence of thoracic vertebral malformations in French Bulldogs, Pugs and English Bulldogs demonstrated that 80.7% of dogs were affected by at least one malformation with hemivertebra being the most common (Ryan et al., 2017). Similarly, another study evaluating the prevalence of lumbosacral and caudal vertebral malformations found that 51% of small brachycephalic dogs had at least one vertebral malformation. Transitional vertebra was the most common malformation in this anatomic region (Bertram et al., 2019). Other vertebral and spinal malformations commonly seen in small brachycephalic dogs are thoracic vertebral canal stenosis, caudal articular process dysplasia, meningeal fibrosis, SAD and spina bifida, which will be discussed in this chapter.

HEMIVERTEBRA

Hemivertebrae or vertebral body formation defects result from failure of vertebral ossification centres, which can lead to uneven growth between the two halves of a vertebral body during embryologic development. This can result in incomplete fusion between the two halves of the vertebra, resulting in a wedge-shaped or even an absent vertebral body (Dewey et al., 2016). Based on human classification systems, multiple hemivertebra subtypes have been recognised. The two most important criteria to differentiate hemivertebra subtypes are severity of the vertebral body defect and the location of the defect within the vertebral body (Gutierrez-Quintana et al., 2014). A radiographic classification system proposed six subtypes of hemivertebra: (1) ventral aplasia or 'dorsal hemivertebra', (2) ventral hypoplasia or 'wedged vertebra', (3) lateral aplasia or 'lateral hemivertebra', (4) lateral hypoplasia or 'lateral wedged vertebra', (5) ventrolateral aplasia or 'dorsolateral hemivertebrae', and (6) ventral and median aplasia or 'butterfly vertebra' (Gutierrez-Quintana et al., 2014). More recently, detailed evaluation of thoracic hemivertebra by computed tomography (CT) allowed identification of three more hemivertebra subtypes: ventrolateral hypoplasia, ventral and median hypoplasia, and symmetrical hypoplasia or 'short vertebra' (Ryan et al., 2019).

Hemivertebra are most often found in the mid-thoracic vertebral column between T6 and T10. Although imaging studies of affected dogs can demonstrate a single abnormal vertebra, it is common to find multiple hemivertebrae along the vertebral column (Dewey et al., 2016). Hemivertebra can result in an abnormal curvature of the vertebral column, which is referred to as kyphosis (i.e. abnormal dorsoventral curvature) or scoliosis (i.e. abnormal lateral curvature). The presence of both kyphosis and scoliosis is referred to as kyphoscoliosis (Figure 16.1). The degree of spinal curvature can be determined by the Cobb angle, which measures the vertebral angulation caused by the abnormal vertebra. The higher the Cobb angle, the more severe the angulation of the vertebral column (Guevar et al., 2014).

Although the exact aetiology of hemivertebra is unclear, it is assumed to be hereditary (Schlensker and Distl, 2016; Mansour et al., 2018). In French Bulldogs, there is an association between thoracic hemivertebra and the desirable tail morphology; a higher number and more severe grade of thoracic hemivertebra is seen in French Bulldogs with shorter tails (Schlensker and Distl, 2016). This 'screw-tail' morphology is also sought after in English Bulldogs and Boston Terriers. Although it is considered a desirable phenotypic trait, this tail morphology should be considered a complex vertebral malformation in itself and is characterised by a variable number of malformed and fused vertebrae and absence of 8–15 caudal vertebrae, which normally form the canine tail (Figure 16.2) (Mansour et al., 2018). The 'screw-tailed' morphology is caused by a frameshift mutation in the *DISHEVELLED 2* gene and has been demonstrated to segregate with thoracic vertebral malformations, supporting further the role of hereditary factors in the aetiology of thoracic hemivertebra (Mansour et al., 2018). As mentioned above, Pugs should not be considered 'screw-tailed'

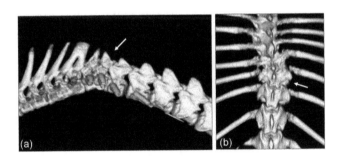

FIGURE 16.1   (a) Sagittal and (b) dorsal reconstructed CT study of 3-year-old neurologically normal French Bulldog. The vertebral column has both (a) an abnormal dorsal curvature and (b) an abnormal lateral curvature. This CT study therefore demonstrates kyphoscoliosis of the vertebral column.

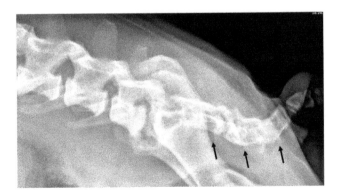

FIGURE 16.2   The 'screw-tail' phenotype should be considered a complex vertebral malformation consisting of several malformed, fused and missing caudal vertebrae (arrows).

brachycephalic dogs and do not carry the aforementioned genetic mutation (Mansour et al., 2018). It is therefore possible that the aetiology and pathophysiology of thoracic hemivertebra are different in Pugs compared to 'screw-tailed' brachycephalic breeds.

PREVALENCE OF HEMIVERTEBRA IN BRACHYCEPHALIC DOGS

The prevalence of thoracic hemivertebrae in neurologically normal brachycephalic dogs is surprisingly high. This is especially well documented in the French Bulldog. The prevalence of hemivertebra in neurologically normal French Bulldogs varies from 75% to 94%, depending on the study population and used diagnostic modality (Moissonnier et al., 2011; Schlensker and Distl, 2016; Ryan et al., 2017). Although radiography is routinely used and widely available for veterinary practitioners, it should be considered inferior to CT for the detection and characterisation of thoracic hemivertebra (Brocal et al., 2018a). Compared to radiography, evaluation of CT studies results in a significantly higher detection of thoracic hemivertebra and a more accurate classification into different hemivertebra subtypes (Brocal et al., 2018a). Although the prevalence of thoracic hemivertebrae is best characterised in French Bulldogs, this vertebral anomaly is also common in other brachycephalic breeds. One study that evaluated CT studies of neurologically normal Pugs, French and English Bulldogs demonstrated that one or more thoracic hemivertebrae were present in 94% of French Bulldogs, 73.2% of English Bulldogs and 17.6% of included Pugs (Ryan et al., 2017). As mentioned above, vertebral malformations are also commonly observed in the lumbosacral region of small brachycephalic dogs. Lumbosacral hemivertebra have been reported in 32% of neurologically normal French Bulldogs, 24.3% of English Bulldogs and 1.7% of neurologically normal Pugs (Bertram et al., 2019). Thoracic and lumbosacral hemivertebrae occur significantly less common in neurologically normal Pugs compared to French and English Bulldogs (Ryan et al., 2017; Bertram et al., 2019). The prevalence of thoracic and lumbosacral hemivertebra in neurologically normal Boston Terriers is currently unknown.

The prevalence of hemivertebra is not the only difference in spinal morphology between brachycephalic breeds, there are also breed-specific anatomical characteristics. Neurologically normal Pugs are almost exclusively affected by one hemivertebra subtype; ventral hypoplasia or 'wedge-shaped vertebra'. In contrast, imaging studies of neurologically normal French and English Bulldogs reveal a wider variation in hemivertebra subtypes with ventral and median aplasia ('butterfly vertebra') and ventral and median hypoplasia being the most common hemivertebra subtypes in both breeds (Ryan et al., 2019). The ventral hypoplasia hemivertebra subtype is significantly more likely to result in kyphosis, and neurologically normal Pugs with thoracic hemivertebra are therefore more likely to demonstrate kyphosis compared to neurologically normal French and English Bulldogs with a thoracic hemivertebra (Ryan et al., 2019). Although the clinical relevance of these findings is currently unclear, we will discuss below that both the occurrence of hemivertebra in Pugs and the presence of severe kyphosis are considered risk factors for the development of clinical signs associated with thoracic hemivertebra (Guevar et al., 2014; Ryan et al., 2017; De Decker et al., 2019).

From the above, it is clear that the occurrence of a thoracic or lumbosacral hemivertebra on diagnostic imaging studies is common and should be considered an incidental finding in the majority of cases (Figure 16.3). Even in dogs with spinal disease, considerations should be given to other, more common, disorders. This is especially true in French Bulldogs; despite the high prevalence of spinal disease and the extremely high prevalence of hemivertebra in this breed, neurological signs directly caused by hemivertebra in French Bulldogs is rare. This is illustrated in the aforementioned study reporting the prevalence of neurological disorders in French Bulldogs. In this study, compressive vertebral malformations represented only 8.6% of all French Bulldogs diagnosed with a spinal condition (Mayousse et al., 2017). Another study revealed that the overall prevalence of a clinical diagnosis of thoracic hemivertebra was 4.7% of all Pugs, 0.95% of all French Bulldogs and 0.0% of all English Bulldogs that presented to an academic institution in the United Kingdom for a variety of neurological and non-neurological conditions (Ryan et al., 2017). The clinical importance

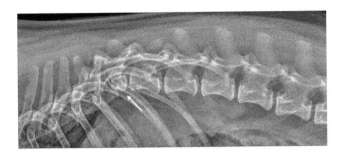

FIGURE 16.3   Lateral radiograph of a 3-year-old French Bulldog with spinal hyperaesthesia and paraparesis. Although a thoracic hemivertebra with moderate kyphosis can be readily observed (arrow), this was not the cause of the dog's clinical signs. The dog was eventually diagnosed with a thoracolumbar intervertebral disc extrusion. Although thoracic hemivertebrae are very common in French Bulldogs, they should only rarely be considered the direct cause of clinical signs.

of congenital vertebral malformations was also questioned in a study that compared the presence of thoracic and lumbar vertebral malformations in Pugs with and without neurological deficits (Rohdin et al., 2018b). Although 96% of Pugs presented with one or more congenital vertebral malformations, these anomalies were equally common in Pugs with and without neurological signs. It is unclear why only a minority of dogs with thoracic hemivertebra will develop gait abnormalities, while most affected dogs will never develop clinical signs. It therefore remains challenging to correctly quantify the clinical importance of vertebral malformations in small brachycephalic dogs. Several studies have attempted to identify clinical and diagnostic variables that could aid in differentiating between clinically relevant and irrelevant thoracic hemivertebra (Moissonnier et al., 2011; Guevar et al., 2014, Gutierrez-Quintana et al., 2014; Ryan et al., 2017; De Decker et al., 2019). There is also increasing evidence that thoracic hemivertebra with kyphosis can accelerate or contribute to degenerative processes and development of other spinal conditions along the vertebral column (Aikawa et al., 2014; Faller et al., 2014; Inglez de Souza et al., 2018).

DEVELOPMENT OF CLINICAL SIGNS IN DOGS WITH HEMIVERTEBRA

Although thoracic hemivertebrae are not directly associated with clinical signs in the majority of dogs, a proportion of affected dogs will develop signs of progressive pelvic limb ataxia and paraparesis. Although most affected dogs are younger than 1 year of age, clinical signs can also occur later in life. Obvious spinal hyperaesthesia is only rarely present (Wyatt et al., 2018). Development of clinical signs is considered to be multifactorial with static and dynamic factors, including vertebral canal stenosis, instability and subluxation involved (Moissonnier et al., 2011; Dewey et al., 2016). As discussed above, the high prevalence of thoracic hemivertebra complicates differentiating clinically relevant from irrelevant diagnostic findings. Several studies have identified clinical and diagnostic criteria that could potentially aid in making a clinical diagnosis of hemivertebra with kyphosis (Moissonnier et al., 2011; Guevar et al., 2014; Gutierrez-Quintana et al., 2014; Ryan et al., 2017; De Decker et al., 2019). The two most consistent variables associated with a higher likelihood of clinical signs are the 'Pug breed' and 'severe spinal kyphosis' (Guevar et al., 2014; Ryan et al., 2017; De Decker et al., 2019).

Pugs are strongly represented in published case series evaluating the clinical presentation and treatment of thoracic hemivertebra with kyphosis (Charalambous et al., 2014; Mathiesen et al., 2018; Wyatt et al., 2018). We have discussed above that hemivertebrae occur less commonly in neurologically normal Pugs compared to neurologically normal French and English Bulldogs (Ryan et al., 2017; Bertram et al., 2019). Paradoxically, two studies have demonstrated that Pugs are significantly over-represented with a clinical diagnosis of thoracic hemivertebrae compared to French and English Bulldogs (Ryan et al., 2017; De Decker et al., 2019). These findings illustrate that the presence of thoracic hemivertebra on diagnostic imaging studies should be considered a clinically more important problem in Pugs compared to other brachycephalic breeds.

As discussed above, thoracic hemivertebra can be associated with an abnormal dorsoventral curvature of the vertebral column or kyphosis. Several studies have suggested that the severity of kyphosis should be considered an important risk factor for development of clinical signs (Figure 16.4) (Moissonnier et al., 2011; Guevar et al., 2014; De Decker et al., 2019). As discussed above, the degree of spinal curvature can be measured by the Cobb angle. One study identified that dogs with a Cobb angle exceeding 35° had an increased likelihood of neurological signs (Guevar et al., 2014). Another study confirmed this finding and demonstrated that a Cobb angle of 34.5° was associated with the highest combined sensitivity (85%) and specificity (93%) to differentiate between dogs with and without neurological signs (De Decker et al., 2019). Kyphosis is associated with vertebral instability and alterations in the biomechanical properties of the vertebral column (Moissonnier et al., 2011; Aikawa et al., 2014). Repetitive microtrauma associated with severe kyphosis should therefore be considered a major factor in the development of clinical signs associated with hemivertebra (Moissonnier et al., 2011; De Decker et al., 2019).

Hemivertebrae can be classified into subtypes based on the severity and location of the vertebral body defect. Ventral aplasia or 'dorsal' hemivertebra is considered the 'most severe' form of hemivertebra and is characterised by a complete absence of the vertebral body (Gutierrez-Quintana et al., 2014). Dogs with ventral aplasia often have neurological signs, while this hemivertebra subtype is only rarely seen in dogs without clinical signs (Charalambous et al., 2014; Gutierrez-Quintana et al., 2014; De Decker et al., 2019). Ventral aplasia has been associated with the Pug breed, a higher Cobb angle and more severe vertebral canal stenosis (De Decker et al., 2019). Although it seems therefore logical to consider the ventral aplasia ('dorsal hemivertebra') hemivertebra subtype a strong risk factor for development of clinical signs, this could not be confirmed in one study (De Decker et al., 2019).

Other factors that have been associated with a higher likelihood of clinical signs are the presence of vertebral subluxation and the number of hemivertebra along the vertebral column (Moissonnier et al., 2011; De Decker et al., 2019). More specifically, it has been suggested that fewer instead of more hemivertebra along the vertebral column are more likely associated with presence of clinical signs (De Decker et al., 2019). This finding has potential important consequences for the development of radiographic screening programs, which will be discussed briefly below. Interestingly, several studies have suggested that vertebral canal stenosis should not be considered a primary factor in the development of clinical signs associated with hemivertebra (Moissonnier et al., 2011; De Decker et al., 2019).

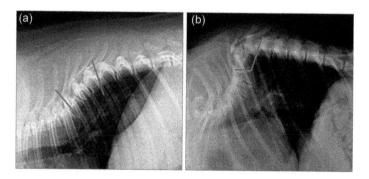

FIGURE 16.4   Survey radiographs of (a) a 2-year-old Pug with a thoracic hemivertebra without neurological signs and (b) an 8-month-old Pug with neurological signs caused by hemivertebra. The degree of kyphosis is considered an important risk factor for development of clinical signs. It is clear that the clinically affected dog has much more severe kyphosis compared to the unaffected dog.

Although the variables discussed above can aid in differentiating clinically relevant from irrelevant thoracic hemivertebra, identification of these factors also provides an opportunity to improve our understanding of the pathophysiology and selection of treatment strategies for this challenging spinal problem. The increasing evidence that the severity of spinal kyphosis is a key factor in development of clinical signs, supports the role of instability in the pathophysiology of hemivertebra and supports consequentially the importance of vertebral stabilisation in the surgical treatment of hemivertebra with kyphosis. In contrast, the lack of strong evidence that vertebral canal stenosis should be considered a primary factor in the development of clinical signs, questions the importance of surgical decompression in affected dogs. It should however be emphasised that the factors discussed above cannot completely explain development of clinical signs and there is a substantial overlap between radiographic and CT studies of brachycephalic dogs with and without neurological signs associated with hemivertebra. This is illustrated by the fact that not every Pug with hemivertebra will develop neurological signs and that neurologically normal brachycephalic dogs with hemivertebra can also display kyphosis exceeding 35° (Gutierrez-Quintana et al., 2014; Guevar et al., 2014; Ryan et al., 2017, 2019; Inglez de Souza et al., 2018). These finding are additional complications for the development of reliable radiographic screening programs.

DIAGNOSIS AND TREATMENT OF HEMIVERTEBRA WITH KYPHOSIS

Although hemivertebra with kyphosis can be readily observed on survey radiographs and CT, it is clear that the mere presence of this malformation on diagnostic imaging studies is not sufficient to make a definitive diagnosis of hemivertebra with kyphosis (Guttierez-Quintana and De Decker, 2021). The diagnostic modality of choice is magnetic resonance imaging (MRI) (Figure 16.5) (Dewey et al., 2016). The two most important reasons to perform MRI are to evaluate spinal cord compression at the level of a hemivertebra and to assess possible concurrent spinal conditions, such

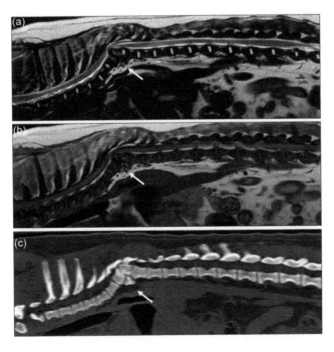

FIGURE 16.5 (a) Sagittal T2-weighted magnetic resonance image, (b) sagittal T1-weighted magnetic resonance image, and (c) sagittal reconstructed CT image of a 9-month-old Pug with a T8 hemivertebra and severe spinal kyphosis (arrow) causing spinal cord compression with progressive ataxia and paresis of the pelvic limbs.

as IVDD or spinal arachnoid diverticulum. It is not uncommon to observe multiple spinal abnormalities on one diagnostic imaging study.

Although hemivertebra with kyphosis is an uncommon cause of spinal disease, treatment of this vertebral malformation is challenging (Gutierrez-Quintana and De Decker, 2021). It has originally been suggested that young dogs with mild clinical signs could be treated medically. It was hypothesised that clinical signs could then possibly stabilise when affected dogs would become skeletally mature (Dewey et al., 2016). A more recent study has however suggested that medical management of hemivertebra is associated with a poor prognosis (Wyatt et al., 2018). This study assessed the outcome of non-surgical management for hemivertebra in 13 small brachycephalic dogs. All dogs experienced progression of neurological signs after medical management was started. More specifically, four of the 13 dogs were euthanised and two other dogs underwent surgery because of progressive neurological deterioration. The remaining seven dogs were still alive (at least 6 months after starting treatment) despite neurological deterioration. Six of these seven dogs required assistance to ambulate and three had completely or partially lost faecal and/or urinary continence (Wyatt et al., 2018).

It is therefore likely that surgery is the treatment modality of choice in dogs with severe neurological signs. Although initial surgical reports have described the combination of decompressive surgery with vertebral stabilisation, it has been debated if decompressive surgery is actually indicated and should perhaps even be avoided (Aikawa et al., 2007; Charalambous et al., 2014; Mathiesen et al., 2018). Several studies have suggested that vertebral canal stenosis is not a primary factor in the pathophysiology of hemivertebra with kyphosis and it can be hypothesised that creating a bony defect by a laminectomy or hemilaminectomy could contribute to already present vertebral instability. Several studies have reported positive outcomes after stabilisation without additional decompression (Charalambous et al., 2014; Mathiesen et al., 2018). The role of partial realignment of the vertebral column is currently unclear. Surgery for thoracic hemivertebra is however challenging for several reasons: (1) placing surgical implants in the thoracic vertebral column is technically demanding; (2) affected dogs are typically small and skeletally immature; and (3) the anatomy of the thoracic vertebral column in affected dogs is by definition abnormal. Two surgical approaches that have been associated with successful outcomes are a transthoracic approach to the ventral and lateral aspects of the vertebral column (Mathiesen et al., 2018) and a dorsal approach requiring more extensive dissection of the paravertebral muscles (Charalambous et al., 2014). The use of 3D-printing technology has been suggested to aid in surgical planning and accurate implant placement (Figure 16.6) (Elford et al., 2020).

SHOULD HEMIVERTEBRA BE CONSIDERED CLINICALLY IRRELEVANT IN THE MAJORITY OF CASES?

Although hemivertebra should only rarely be considered the primary cause of clinical signs in dogs with a gait abnormality, their occurrence should not necessarily be considered a benign finding. There is increasing evidence that thoracic hemivertebra with kyphosis alter spinal biomechanics and contribute to the development or progression of degenerative spinal processes. Neurologically normal French Bulldogs with kyphosis have slightly altered gait variables compared to French Bulldogs without kyphosis. More specifically, French Bulldogs with kyphosis have a different distribution of their bodyweight with a tendency to shift some of their bodyweight from the pelvic into their thoracic limbs (Wyatt et al., 2019). Thoracic hemivertebrae with kyphosis have been associated with accelerated intervertebral disc degeneration adjacent to the kyphotic vertebral segments (Faller et al., 2014) and an altered distribution of thoracolumbar intervertebral disc extrusions along the vertebral column. French Bulldogs with kyphosis or scoliosis have more caudally located intervertebral disc extrusions compared to French Bulldogs without kyphosis. (Aikawa et al., 2014; Inglez de Souza et al., 2018). French Bulldogs with kyphosis have also an increased risk of acute intervertebral disc extrusions. More specifically, kyphosis was associated with twice the odds of being diagnosed with thoracolumbar intervertebral disc extrusion (Inglez de Souza et al., 2018).

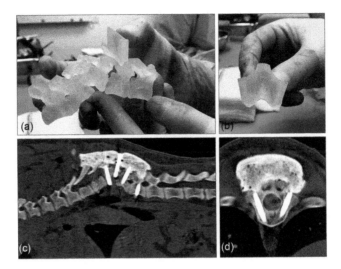

FIGURE 16.6   (a) 3D-printed models and (b) 3D-printed drill guides can facilitate correct implant placement in dogs with thoracic hemivertebra. (c) Sagittal reconstructed and (d) transverse postoperative CT demonstrating near perfect implant placement.

It seems therefore fair to assume that the high prevalence of thoracic vertebral malformations in 'screw-tailed' brachycephalic breeds poses a potential health and welfare concern.

SCREENING PROGRAMS FOR HEMIVERTEBRAE

Radiographic screening programs for hemivertebrae have been considered, proposed and utilised (Schlensker and Distl, 2013). Developing and using reliable radiographic screening programs is however associated with several challenges. The prevalence of thoracic hemivertebra is extremely high in neurologically normal French and English Bulldogs, while the associated prevalence of neurological signs is low in these two breeds (Ryan et al., 2017). These factors complicate designing a meaningful screening program for these two specific breeds.

Although it seems logical to allocate a more severe grade of hemivertebra to dogs with a higher number of hemivertebra along the vertebral column (The Finnish Canine Expert, 2020), there is some evidence to suggest that fewer instead of more hemivertebrae are more likely to result in neurological signs (De Decker et al., 2019). Studies have suggested that a Cobb angle exceeding 35° is associated with an increased likelihood of clinical signs and measurement of Cobb angles on diagnostic imaging studies is relatively easy and reliable (Guevar et al., 2014; De Decker et al., 2019). It should however be considered that

1. Cobb angles exceeding 35° also occur in clinically normal dogs.
2. Some dogs with neurological signs associated with hemivertebra do not display kyphosis.
3. Spinal kyphosis should not be considered a static anatomical factor and can become progressively worse over time.

The latter is illustrated in a case report describing neurological signs and marked thoracic kyphosis in a 6-month-old Pug. Thoracic radiographs did however not disclose any kyphosis when the dog was only 2 months old (De Rycke et al., 2016). These findings illustrate that further research is needed to evaluate the reliability and clinical usefulness of radiologic screening programs for thoracic hemivertebrae in brachycephalic dogs. It is possible that existing and newly developed screening programs will need to be adapted when increasing evidence becomes available in this area.

TRANSITIONAL VERTEBRAE

Transitional vertebrae are anomalies found at the junction between two divisions of the vertebral column and occur commonly at the thoracolumbar and lumbosacral junction. This can result in variations in the number of vertebrae found within a vertebral segment and abnormal anatomy of the affected vertebra. Transitional vertebrae have symmetrical or asymmetrical morphological characteristics of two different vertebral types, such as the presence of one or two ribs on the first lumbar vertebra, absence of one or two ribs on the last thoracic vertebrae with rudimentary transverse processes, and a separate spinous process with rudimentary transverse processes on the first sacral vertebra (Westworth and Sturges, 2010).

Transitional vertebrae are common in brachycephalic dogs and especially in Pugs. Compared to French and English Bulldogs, Pugs have a higher prevalence of thoracolumbar and lumbosacral transitional vertebrae. Thoracolumbar and lumbosacral transitional vertebrae occur in, respectively, 30.9% and 54.2% of neurologically normal Pugs (Ryan et al., 2017; Bertram et al., 2019). Although transitional vertebrae are only rarely associated with clinical signs, they can complicate interpretation of diagnostic imaging studies and surgical planning and have the potential to alter the biomechanical characteristics of the vertebral column (Flückiger et al., 2006). This can result in acceleration of degenerative processes, such as intervertebral disc degeneration and herniation. A specific type of transitional vertebra has been described in Pugs, which is referred to as homeotic transformation of cervical into thoracic vertebrae. This is characterised by the presence of partial or complete cervical ribs on C7 (Figure 16.7). This anomaly occurs in 46% of neurologically normal Pugs, while it is rare in other dog breeds (Brocal et al., 2018b).

THORACIC VERTEBRAL CANAL STENOSIS

Thoracic vertebral canal stenosis secondary to hypertrophy of the dorsal lamina and articular processes has been reported in small brachycephalic dogs and English Bulldogs in particular. The cranial thoracic region is most commonly affected. Affected dogs are usually affected in the first year of life and English Bulldogs seem especially vulnerable. Caution should however be exercised as imaging findings compatible with cranial thoracic vertebral canal stenosis are common in French (31.6%) and English Bulldogs (39%) (Figure 16.8). Imaging findings suggestive for cranial thoracic stenosis are very rare in Pugs (Conte et al., 2020).

CAUDAL ARTICULAR PROCESS DYSPLASIA AND MENINGEAL FIBROSIS

The cranial and caudal vertebral articular processes of two adjacent vertebrae form together a facet joint, which provides stability and restricts motion of the vertebral column. Caudal articular process

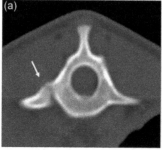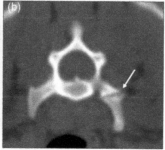

FIGURE 16.7 Transverse CT images at the level of (a) L1 and (b) C7 in two Pugs without neurological signs. (a) A thoracolumbar transitional vertebra with a rib at the right side of L1 (arrow). (b) Homeotic transformation of C7 with a cervical rib on the left side (arrow).

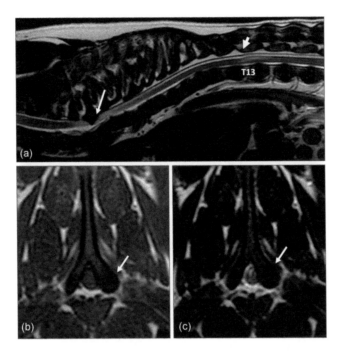

FIGURE 16.8   (a) sagittal T2-weighted, (b) transverse T1-weighted and (c) transverse T2-weighted MR images of a 7-year-old French Bulldog with progressive paresis and ataxia of the pelvic limbs. (a) A dorsal compression can be seen at the level of T2 (arrow), which is caused by (b and c) hypertrophy of the left articular process (arrow). This abnormality was however not considered the cause of the dog's clinical signs and highlights the difficulty of interpreting spinal imaging studies of small brachycephalic dogs. (a) The dog was also diagnosed with a spinal arachnoid diverticulum at the level of T12-T13 (arrowhead). His neurological signs resolved after surgical treatment of the spinal arachnoid diverticulum.

dysplasia is characterised by a complete (aplasia) or partial (hypoplasia) absence of one or more caudal vertebral articular processes. Although this malformation is strongly associated with the Pug breed (Fisher et al., 2013), it is also commonly observed on diagnostic imaging studies of other small brachycephalic breeds (Bertram et al., 2018). Similar to the malformations discussed above, caudal articular process dysplasia occurs commonly in neurologically normal brachycephalic dogs. Caudal articular process dysplasia has been observed in 70.4% of neurologically normal French Bulldogs, 81.4% of English Bulldogs and 97% of Pugs.

Caudal articular process dysplasia is more often seen in Pugs compared to other brachycephalic breeds (Bertram et al., 2018). In parallel to the discussion above regarding clinical relevance of hemivertebra, only a minority of dogs affected by caudal articular process dysplasia will develop neurological signs (Bertram et al., 2018; Rohdin et al., 2018b). The predisposition of Pugs for developing neurological signs is likely not only explained by the higher prevalence of caudal articular process dysplasia in this breed. Compared to French and English Bulldogs, Pugs demonstrate also a higher prevalence of caudal articular process aplasia (in contrast to hypoplasia), a higher prevalence of bilateral dysplasia, a higher number of affected vertebrae per dog and a different anatomical distribution along the vertebral column. While the cranial thoracic vertebral column is more often affected in French and English Bulldogs, the caudal thoracic region (T10 to T13) is more often affected in Pugs (Bertram et al., 2018). Compared to the cranial thoracic region, the articular facet joints of the caudal thoracic region have a more prominent role in providing vertebral stability. It has therefore been suggested that caudal articular process dysplasia can cause low-grade and repetitive vertebral instability (Fisher et al., 2013; Lourinho et al., 2020). This can lead to 'constrictive

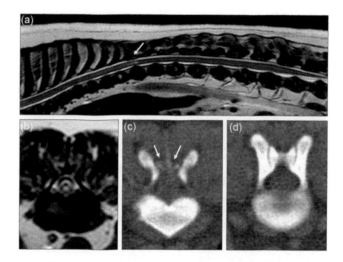

FIGURE 16.9　(a) Sagittal T2-weighted magnetic resonance image, (b) transverse T2-weighted magnetic resonance image and (c) CT image of an 8-year-old Pug with caudal articular process dysplasia and constrictive myelopathy. (a and b) The spinal cord is narrowed by a circumferential extradural spinal cord compression at the level of T11-T12. An intraparenchymal hyperintensity is also present at this site. (c) A CT image at this level reveals bilateral absence of the caudal articular processes (arrows). (d) Normal anatomy with intact cranial and caudal articular processes in the same dog at T12-T13.

myelopathy' or 'meningeal fibrosis', which is characterised by the formation of a dense, circumferential band of fibrotic tissue causing intradural adhesions between the arachnoid and pia mater and subsequent constriction of the spinal cord (Figure 16.9) (Fisher et al., 2013; Lourinho et al., 2020). Although meningeal fibrosis has traditionally been associated with caudal articular process dysplasia, this condition has also been reported in Pugs without concurrent vertebral anomalies (Rohdin et al., 2020). The causative relationship between meningeal fibrosis and caudal articular process dysplasia is therefore unclear. Meningeal fibrosis in Pugs could potentially represent a multifactorial neurological syndrome or a disease entity in itself (Rohdin et al., 2020).

If dogs develop clinical signs of meningeal fibrosis, this is usually later in life (median 7.7 years) with the most common clinical presentation being a slowly progressive paraparesis with ataxia of the pelvic limbs. Faecal and urinary incontinence can be present, and spinal hyperaesthesia is only rarely noted (Fisher et al., 2013; Tauro et al., 2019; Rohdin et al., 2020; Lourinho et al., 2020).

DIAGNOSIS AND TREATMENT OF CAUDAL ARTICULAR PROCESS DYSPLASIA

Because of the high prevalence of caudal articular process dysplasia in neurologically normal Pugs, the low prevalence of clinical signs and the uncertain causative relationship between this malformation and other spinal abnormalities, survey radiographs or CT cannot be used to obtain a clinical diagnosis of caudal articular process dysplasia. Obtaining a definitive diagnosis of caudal articular process dysplasia with concurrent meningeal fibrosis or other spinal pathology requires MRI (Driver et al., 2019; Lourinho et al., 2020). It is common for multiple spinal conditions to be present at the site of caudal articular process dysplasia (Figure 16.10). One study demonstrated compressive intervertebral disc protrusion in 56% and SAD in 22% of Pugs with caudal articular process dysplasia at the same site (Driver et al., 2019). Although a causal relationship has been hypothesised, the relationship between caudal articular process dysplasia and other spinal conditions is currently unclear (Driver et al., 2019; Nishida et al., 2019).

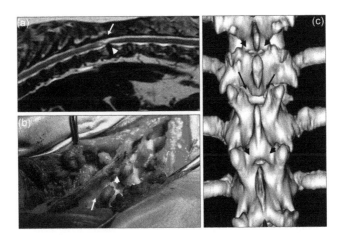

FIGURE 16.10   (a) Sagittal T2-weighted magnetic resonance image of an 8-year-old Pug demonstrating a spinal arachnoid diverticulum at the level of T11-T12 (arrow). An intervertebral disk protrusion causing moderate spinal cord compression is also present (arrowhead). (b) A 3D reconstructed CT image of the same dog demonstrates bilateral aplasia of the caudal articular process at T11-T12 (arrows). The caudal articular processes are present at T10-T11 and T12-T13 (arrowheads). (c) Intraoperative image of the same dog illustrating caudal articular process aplasia at T11-T12 (arrow). The caudal articular process is normally situated medial from the cranial articular process, which is normally formed. The caudal articular process is present at T12-T13 (arrowhead). Although vertebral malformations, such as caudal articular process dysplasia, are not often the primary cause of clinical signs, they have anecdotally been associated with development of other spinal disorders, such as intervertebral disk disease and spinal arachnoid diverticulum.

Little is known about the natural progression or results of medical management associated with caudal articular process dysplasia or meningeal fibrosis. Surgical treatment consisting of decompressive surgery in combination with dissection of the pia-arachnoid adhesions has been associated with a poor outcome (Fisher et al., 2013). Surgical treatment consisting of vertebral stabilisation with or without additional decompression has been associated with neurological improvement in the majority of cases (Tauro et al., 2019). Concurrent urinary or faecal incontinence is however expected to resolve in only less than half of affected cases (Tauro et al., 2019). Another study reported favourable outcomes after subarachnoid–subarachnoid shunt placement to bridge the site of cerebrospinal fluid flow obstruction in dogs with meningeal fibrosis (Meren et al., 2017). Objective comparison of outcome data is challenging however, because of the variety of imaging findings, pathological abnormalities and the substantial proportion of dogs with concurrent spinal disorders alongside caudal articular process dysplasia, such as intervertebral disc protrusion and SAD (Meren et al., 2017; Tauro et al., 2019).

SPINAL ARACHNOID DIVERTICULA

SAD, formerly called spinal arachnoid cysts, are cerebrospinal fluid filled focal dilations of the subarachnoid space resulting in progressive neurological signs. Pugs and French Bulldogs are predisposed to SAD (Mauler et al., 2014). This increasingly recognised disorder can be congenital or acquired. Underlying causes for acquired SAD include concurrent spinal conditions such as vertebral malformations, trauma, IVDD and inflammatory disease. Compared to other breeds, Pugs and French Bulldogs have more often concurrent spinal conditions with IVDD and vertebral malformation being the most common ones (Mauler et al., 2014, Mayousse et al., 2017). Although SAD can occur over the entire length of the vertebral column, in small breed dogs they occur most commonly in the thoracolumbar region (Mauler et al., 2014). The occurrence of cervical SAD, with a suspected hereditary aetiology, has however been reported in Pugs (Rohdin et al., 2014).

Affected dogs most commonly demonstrate a slowly progressive ataxia and paresis. Spinal hyperaesthesia and urinary or faecal incontinence are both present in around 20% of cases. There is a wide variation in the age of onset of clinical signs, which can be explained by the multifactorial aetiology of this condition. Compared to other breeds, Pugs develop clinical signs later in life (Mauler et al., 2014), which contributes to the assumption that Pugs suffer most often from acquired SAD (Flegel et al., 2013).

Diagnosis and Treatment of Spinal Arachnoid Diverticula

A diagnosis of SAD can be obtained by myelography, CT myelography or MRI (Figure 16.11). Medical treatment consists of administration of anti-inflammatory doses of prednisolone. The rationale for administering prednisolone is to decrease the rate of cerebrospinal fluid production. Several surgical techniques have been described, which include a durotomy, durectomy or marsupialisation (Skeen et al., 2003). There is some evidence that surgery is associated with better long-term outcomes compared to medical management. One study reported that 82% of dogs treated surgically experienced a long-term neurological improvement, while 3% remained stable and 16% deteriorated. After medical management, 30% experienced a long-term neurological improvement, while 30% remained stable and 40% deteriorated (Mauler et al., 2017). Although rare, late-onset recurrence of clinical signs is possible. The median time to recurrence of neurological signs is 20.5 months after surgery (Alcoverro et al., 2018).

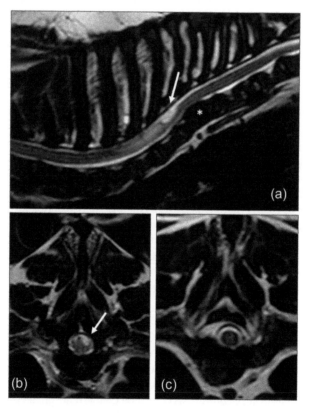

FIGURE 16.11   (a) sagittal T2-weighted, (b) transverse T2-weighted and (c) transverse T2-weighted MR images of a 6-year-old French Bulldog with a T4-T5 spinal arachnoid diverticulum. Although the dog has a hemivertebra just caudal (asterisk) from this lesion, the causative relationship between both abnormalities is unclear. (a) Tear-drop-shaped expansion of the dorsal subarachnoid space (arrow). (b) The focal dilatation of the dorsal subarachnoid space is lateralised to the left (arrow). (c) An unaffected side for comparison. Note how the normal subarachnoid space has an even width around the spinal cord.

Although surgery has been associated with good outcomes in most dogs, it has been suggested that Pugs have a poor long-term prognosis after surgery for thoracolumbar SAD. One study demonstrated that although 80% of Pugs experienced a good short-term outcome after surgery, 85% experienced late onset recurrence of neurological signs (Alisauskaite et al., 2019). Although it is currently unclear why Pugs have a more unfavourable prognosis, it possible that thoracolumbar SAD formation in this breed is associated with a multifactorial aetiology and variety of pathological mechanisms.

SPINA BIFIDA, MENINGOCELE AND MENINGOMYELOCELE

Spina bifida is characterised by embryologic failure of the vertebral arches to fuse over the spinal cord. This malformation is only rarely associated with clinical signs and is commonly seen in neurologically normal brachycephalic dogs (Ryan et al., 2017; Bertram et al., 2019). Diagnostic imaging studies can demonstrate a cleft in the lamina or spinous process without any other abnormalities. This is also referred to as spina bifida occulta. Spina bifida should therefore be considered an incidental finding in the majority of cases and is most often seen at the first thoracic vertebra or the sacrum (Figure 16.12) (Ryan et al., 2017; Bertram et al., 2019). Studies evaluating CT scans of neurologically normal French Bulldogs, English Bulldogs and Pugs have demonstrated that 38.2% of Pugs without neurological signs have spina bifida at the level of T1 and approximately 10% of French Bulldogs, English Bulldogs and Pugs demonstrate spina bifida of the sacrum (Ryan et al., 2017; Bertram et al., 2019).

Concurrent spinal abnormalities can occasionally be seen in animals with, especially sacral, spinal bifida. Clinical signs might develop when neural tissue protrudes through the bony defect (Song et al., 2016). Meningocele is characterised by protrusion of the meninges, while meningomyelocele is characterised by protrusion of both the meninges and nervous tissue through the vertebral arch defect (Westworth and Sturges, 2010; Song et al., 2016). The protruding meninges may extend dorsally to the skin. Skin and hair coat changes may include abnormal streaming of the hair, palpable indentation or 'dimple', and crusting cutaneous lesions associated with leakage of cerebrospinal fluid (Figure 16.13).

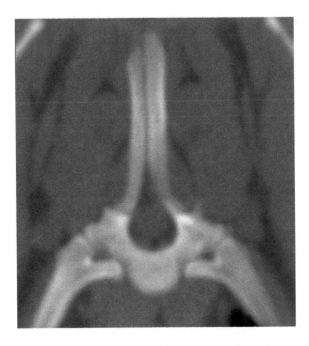

FIGURE 16.12   Transverse CT image at the level of T1 in a 4-year-old Pug without neurological signs. Spina bifida (occulta) is often seen at this level in neurologically normal Pugs and should probably be considered an incidental finding.

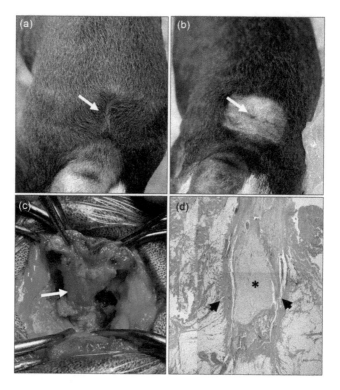

FIGURE 16.13 (a and b) Clinical, (c) surgical and (d) histopathological images of a 5-month-old French Bulldog with sacral spina bifida and meningomyelocele. (a) Abnormal streaming of the hair can be noticed (arrow). (b) Clipping of the area reveals an indentation or 'dimple' (arrow). (c) Surgical exploration revealed a fibrotic tract connecting the skin defect with the vertebral canal (arrow). (d) The surgically removed tissue contained spinal cord tissue (asterisk) and meninges (arrow heads), confirming the diagnosis of spina bifida with meningomyelocele. Image Courtesy of Dr Joe Fenn, Royal Veterinary College.

French and English Bulldogs are predisposed to clinically relevant lumbosacral spina bifida with varying degrees of meningocele and meningomyelocele (Martin Muñiz et al., 2020). Clinical signs occur in the first months after birth and include variable degrees of paraparesis, a bunny-hopping gait, lower motor neuron signs and faecal and urinary incontinence. Although the bony defect can be observed on radiographs or CT, MRI is necessary to characterise concurrent meningocele or meningomyelocele (Figure 16.14) (Song et al., 2016). Surgical resection of the protruding tract may halt progression or even improve the severity of neurological signs (Martín Muñiz et al., 2020).

INTERVERTEBRAL DISC DISEASE

Acute type I intervertebral disc extrusion (IVDE) is the most common canine spinal disorder and is also one of the most common spinal problems in small brachycephalic dogs, in particular the French Bulldog. Acute intervertebral disc extrusion is associated with specific characteristics in the French Bulldog, and there is evidence that an association exists between thoracic vertebral malformations and thoracolumbar IVDE. Cervical and thoracolumbar IVDE are especially prevalent in the French Bulldog (Aikawa et al., 2014; Mayousse et al., 2017). French Bulldogs are typically affected by IVDE at a young age, and several studies have demonstrated relevant associations between hemivertebra with kyphosis and thoracolumbar IVDE. Hemivertebra with kyphosis has been associated with early degeneration of adjacent intervertebral discs (Faller et al., 2014) and a higher risk of thoracolumbar IVDE. French Bulldogs with hemivertebra and kyphosis have twice

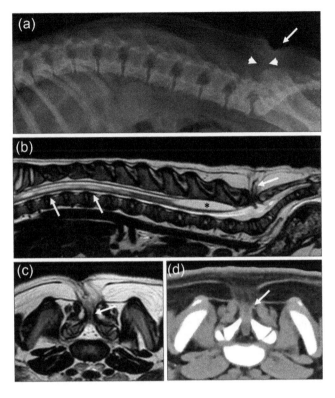

FIGURE 16.14 (a) Lateral survey radiograph, (b) T2-weighted sagittal MR image, (c) T2-weighted transverse MRI images and (d) CT image of the same dog as Figure 16.10. (a) A skin indentation or 'dimple' can be observed (arrow). An increased opacity can be seen ventral from this indentation, which likely represent the meningomyelocele (arrowheads). (b) There is dorsal displacement of the conus medullaris and a string with soft tissue intensity extending dorsally from the vertebral canal towards the skin. Just cranial from this lesion, a homogenous well-demarcated hyperintense lesion can be seen in the dorsal spinal cord parenchyma (asterisk), possibly representing syringomyelia or myelodysplasia. More cranially, there is also syringomyelia present (yellow arrows). (c) and (d) Dorsal displacement of neural tissue and meninges can be seen extending through the bony defect in the lamina (arrow).

the odds of being diagnosed with thoracolumbar IVDE compared to those without kyphosis (Inglez de Souza et al., 2018). French Bulldogs with kyphosis or scoliosis also have and altered distribution of IVDE with the caudal lumbar region being more commonly affected (Aikawa et al., 2014, Inglez de Souza et al., 2018). Compared to Dachshunds, another breed commonly affected by IVDE, French Bulldogs have an increased risk for a potentially fatal complication of severe thoracolumbar IVDE, ascending progressive myelomalacia (Aikawa et al., 2014). A recent report has suggested that a substantial proportion of French Bulldogs that successfully recovers from surgery for cervical or thoracolumbar intervertebral disc extrusion will suffer from a late onset recurrence of clinical signs. The severity of these recurrences is typically mild to moderate and occurs in up to 51% of French Bulldogs that initially recovered successfully after surgery (Kerr et al., 2021).

CONCLUSIONS

Spinal disorders, and congenital vertebral malformations in particular, occur commonly in small brachycephalic dogs. Such malformations occur however commonly in neurologically normal dogs, and differentiating between clinically relevant and irrelevant radiological findings can therefore be challenging. Pugs are more commonly clinically affected by complex vertebral malformations compared to other brachycephalic breeds, such as the French Bulldog. Although vertebral malformations are rarely the direct cause of clinical signs in French Bulldogs, they can contribute to the development of other spinal conditions, such as IVDD. Further research is necessary to evaluate the reliability and clinical usefulness of radiographic screening programs for thoracic vertebral malformations in brachycephalic dogs.

REFERENCES

Aikawa T., Kanazono S., Yoshigae Y., Sharp N.J., Muñana K.R. 2007. Vertebral stabilization using positively threaded profile pins and polymethylmethacrylate, with or without laminectomy, for spinal canal stenosis and vertebral instability caused by congenital thoracic vertebral anomalies. *Veterinary Surgery* 36:432–441.

Aikawa T., Shibata M., Asano M., et al. 2014. A comparison of thoracolumbar intervertebral disk extrusion in French Bulldogs and Dachshunds and association with vertebral anomalies. *Veterinary Surgery* 202:267–273.

Alcoverro E., McConnell J.F., Sanchez-Masian D., De Risio L., De Decker S., Gonçalves R. 2018. Late-onset recurrence of neurological deficits after surgery for spinal arachnoid diverticula. *Veterinary Record* 182:380.

Alisauskaite N., Cizinauskas S., Jeserevics J., et al. 2019. Short- and long-term outcome and magnetic resonance imaging findings after surgical treatment of thoracolumbar spinal arachnoid diverticula in 25 Pugs. *Journal of Veterinary Internal Medicine* 33:1376–1383.

Bertram S., Ter Haar G., De Decker S. 2018. Caudal articular process dysplasia of thoracic vertebrae in neurologically normal French bulldogs, English bulldogs, and Pugs: prevalence and characteristics. *Veterinary Radiology and Ultrasound* 59:396–404.

Bertram S., Ter Haar G., De Decker S. 2019. Congenital malformations of the lumbosacral vertebral column are common in neurologically normal French Bulldogs, English Bulldogs, and Pugs, with breed-specific differences. *Veterinary Radiology and Ultrasound* 60:400–408. doi:10.1111/vru.12753.

Brocal J., De Decker S., José-López R., et al. 2018a. Evaluation of radiography as a screening method for detection and characterisation of congenital vertebral malformations in dogs. *Vet Record* 182:573.

Brocal J., De Decker S., José-López R., et al. 2018b. C7 vertebra homeotic transformation in domestic dogs – are Pug dogs breaking mammalian evolutionary constraints? *Journal of Anatomy* 233:255–265.

Charalambous M., Jeffery N.D., Smith P.M., et al. 2014. Surgical treatment of dorsal hemivertebrae associated with kyphosis by spinal stabilisation with or without decompression. *The Veterinary Journal* 202:267–273.

Conte A., Bernardini M., De Decker S., et al. 2020. Thoracic vertebral canal stenosis associated with vertebral arch anomalies in small brachycephalic screw-tail dog breeds. *Vet Comp Orthop Traumatol.* doi: 10.1055/s-0040-1721375.

De Decker S, Packer RMA, Capello R, et al. 2019. Comparison of signalment and computed tomography findings in French bulldogs, Pugs, and English bulldogs with and without clinical signs associated with thoracic hemivertebra. *J Vet Intern Med* 33: 2151–2159.

De Rycke L.M., Crijns C., Chiers K., et al. 2016. Late-onset wedge-shaped thoracic vertebrae in a six-month-old pug. *Veterinary Record Case Reports* 4:e000317.

Driver C.J., Rose J., Tauro A., Fernandes R., Rusbridge C. 2019. Magnetic resonance imaging in pug dogs with thoracolumbar myelopathy and concurrent caudal articular process dysplasia. *BMC Veterinary Research* 15:182.

Dewey C.W., Davies E., Bouma J.L. 2016. Kyphosis and kyphoscoliosis associated with congenital malformations of the thoracic vertebral bodies in dogs. *Veterinary Clinics of North America: Small Animal Practice* 46:295–306.

Elford J.H., Oxley B., Behr S. 2020. Accuracy of placement of pedicle screws in the thoracolumbar spine of dogs with spinal deformities with three-dimensionally printed patient-specific drill guides. *Veterinary Surgery* 49: 347–353.

Faller K., Penderis J., Stalin C., Guevar J., Yeamans C., Gutierrez-Quintana R. 2014. The effect of kyphoscoliosis on intervertebral disc degeneration in dogs. *The Veterinary Journal* 200:449–451.

Fisher S.C., Shores A., Simpson S.T. 2013. Constrictive myelopathy secondary to hypoplasia or aplasia of the thoracolumbar caudal articular processes in Pugs: 11 cases (1993–2009). *Journal of the American Veterinary Medical Association* 242:223–229.

Flegel T., Müller M.K., Truar K., Löffler C., Oechtering G. 2013. Thoracolumbar spinal arachnoid diverticula in 5 pug dogs. *Canadian Veterinary Journal* 54:969–973.

Flückiger M.A., Damur-Djuric N., Hässig M., Morgan J.P., Steffen F. 2006. A lumbosacral transitional vertebra in the dog predisposes to cauda equina syndrome. *Veterinary Radiology and Ultrasound* 47:39–44.

Guevar J., Penderis J., Faller K., Yeamans C., Stalin C., Gutierrez-Quintana R. 2014. Computer- assisted radiographic calculation of spinal curvature in brachycephalic "screw-tailed" dog breeds with congenital thoracic vertebral malformations: Reliability and clinical evaluation. *PloS One.* 9:e106957.

Gutierrez-Quintana R. Guevar J., Stalin C., Faller K., Yeamans C., Penderis J. 2014. A proposed radiographic classification scheme for congenital thoracic vertebral malformations in brachycephalic "screw-tailed" dog breeds. *Veterinary Radiology and Ultrasound.* 55:585–591.

Gutierrez-Quintana R., De Decker S. 2021. The tail end of the brachycephalic – How to work up and treat spinal malformations. *In Practice.* 43:124–134.

Inglez de Souza M.C.C.M., Ryan R., Ter Haar G., Packer R.M.A., Volk H.A., De Decker S. 2018. Evaluation of the influence of kyphosis and scoliosis on intervertebral disc extrusion in French bulldogs. *BMC Veterinary Research* 14:5.

Kerr S., Crawford A.H., De Decker S. 2021. Late onset recurrence of clinical signs after surgery for intervertebral disc extrusion in French bulldogs. *J Small Anim Pract.* DOI: 10.1111/jsap.13331.

Lourinho F., Holdsworth A., McConnell J. F., et al. 2020. Clinical features and MRI characteristics of presumptive constrictive myelopathy in 27 pugs. *Vet Radiol Ultrasound* 61: 545–554.

Mansour T.A., Lucot K., Konopelski S.A., et al. 2018. Whole genome variant association across 100 dogs identifies a frame shift mutation in DISHEVELLED 2 which contributes to Robinow-like syndrome in Bulldogs and related screw tail dog breeds. *PLOS Genetics* 14:e1007850.

Martin Muñiz L., Del Magno S., Gandini G., et al. 2020. Surgical outcomes of six bulldogs with spinal lumbosacral meningomyelocele or meningocele. *Veterinary Surgery* 49: 200–206.

Mathiesen C.B., de la Puerta B., Groth A.M., Rutherford S., Capello R. 2018. Ventral stabilization of thoracic kyphosis through bilateral intercostal thoracotomies using SOP (String of Pearls) plates contoured after a 3-dimensional print of the spine. *Veterinary Surgery* 47:843–851.

Mauler D.A., De Decker S., De Risio L., et al. 2014. Signalment, clinical presentation, and diagnostic findings in 122 dogs with spinal arachnoid diverticula. *Journal of Veterinary Internal Medicine* 28:175–181.

Mauler D.A., De Decker S., De Risio L., et al. 2017. Spinal arachnoid diverticula: outcome in 96 medically or surgically treated dogs. *Journal of Veterinary Internal Medicine* 31:849–853.

Mayousse V., Desquilbet L., Jeandel A., Blot S. 2017. Prevalence of neurological disorders in French bulldog: a retrospective study of 343 cases (2002–2016). *BMC Veterinary Research* 13:212.

Meren I.L., Chavera J.A., Alcott C.J., Barker A.K., Jeffery N.D. 2017. Shunt tube placement for amelioration of cerebrospinal fluid flow obstruction caused by spinal cord subarachnoid fibrosis in dogs. *Veterinary Surgery* 46:289–296.

Moissonnier P., Gossot P., Scotti S. 2011. Thoracic kyphosis associated with hemivertebra. *Veterinary Surgery* 40:1029–1032.

Nishida H., Nakata K., Maeda S., Kamishina H. 2019. Prevalence and pattern of thoracolumbar caudal articular process anomalies and intervertebral disk herniations in pugs. The *Journal of Veterinary Medical Science* 81:906–910.

Parker H.G. 2012. Genomic analyses of modern dog breeds. *Mammalian Genome* 23:19–27.

Rohdin C., Nyman H.T., Wohlsein P., Hultin Jäderlund K. 2014. Cervical spinal intradural arachnoid cysts in related, young pugs. *Journal of Small Animal Practice* 55:229–234.

Rohdin C., Jäderlund K.H., Ljungvall I., Lindblad-Toh K., Häggström J. 2018a. High prevalence of gait abnormalities in pugs. *Veterinary Record* 182:167.

Rohdin C., Häggström J., Ljungvall I., et al. 2018b. Presence of thoracic and lumbar vertebral malformations in pugs with and without chronic neurological deficits. *The Veterinary Journal* 241:24–30.

Rohdin C., Ljungvall I., Häggström J., et al. 2020. Thoracolumbar meningeal fibrosis in pugs. *Journal of Veterinary Internal Medicine* 34: 797–807.

Ryan R., Gutierrez-Quintana R., Ter Haar G., De Decker S. 2017. Prevalence of thoracic vertebral malformations in French bulldogs, Pugs and English bulldogs with and without associated neurological deficits. *The Veterinary Journal.* 221:25–29.

Ryan R., Gutierrez-Quintana R., Ter Haar G., De Decker S. 2019. Relationship between breed, hemivertebra subtype, and kyphosis in apparently neurologically normal French Bulldogs, English Bulldogs, and Pugs. *American Journal of Veterinary Research* 80:189–194.

Schlensker E., Distl O. 2016. Heritability of hemivertebrae in the French bulldog using an animal threshold model. *The Veterinary Journal* 207:188–189.

Skeen T.M., Olby N.J., Muñana K.R., Sharp N.J. 2003. Spinal arachnoid cysts in 17 dogs. *Journal of the American Animal Hospital Association* 39:271–282.

Song R.B., Glass E.N., Kent M. 2016. Spina bifida, meningomyelocele, and meningocele. *Veterinary Clinics of North America: Small Animal Practice* 46:327–345.

Tauro A., Rose J., Rusbridge C., Driver C.J. 2019. Surgical management of thoracolumbar myelopathies in Pug dogs with concurrent articular facet dysplasia. *Veterinary and Comparative Orthopaedics and Traumatology Open* 2:e60–e72.

The Finnish Canine Expert - The Finnish Kennel Club [Internet]. 2020. Anu Lappalainen. Perinnölliset selkämuutokset. Nikamaepämuodostumat [accessed 24 April 2020]. Available from: https://www.kennelliitto.fi/kasvatus-ja-terveys/koiran-terveys/perinnolliset-sairaudet-ja-koiran-hyvinvointi/perinnolliset-selkamuutokset.

Westworth D.R., Sturges B.K. Congenital spinal malformations in small animals. 2010. *Veterinary Clinics of North America: Small Animal Practice* 40: 951–981.

Wyatt S., Gonçalves R., Gutierrez-Quintana R., De Decker S. 2018. Results of non-surgical treatment for congenital vertebral body malformations: 13 dogs (2009–2016). *Journal of the American Veterinary Medical Association* 253:768–773.

Wyatt S., Lafuente P., Ter Haar G., Packer R.M.A., Smith H., De Decker S. 2019. Gait analysis in French bulldogs with and without vertebral kyphosis. *Veterinary Journal* 244:45–50.

17 Obesity and Weight Management of Brachycephalic Animals

Eleanor Raffan
University of Cambridge

CONTENTS

Introduction...312
What Causes Obesity? 312
Management of Feeding and Exercise...312
Risk Factors Independent of Owner Management...312
Obesity as a Condition of Disordered Homeostasis 313
Genetic Variation Influences Obesity Susceptibility..314
Practical Implications of Genetic Variation in Obesity Susceptibility...314
Pathophysiology of Obesity 315
Insulin Resistance Affects Metabolism and Cell Growth 316
Obesity-Associated Inflammation Acts Systemically...316
Increased Fat Mass Exerts a Mechanical Effect 316
Obesity and the Respiratory System 317
Obesity Presents Particular Problems for Brachycephalic Animals 317
Management of Obesity 317
    The Key to Successful Weight Loss is Food Restriction...317
Defining Ideal and Target Weights..318
Selecting a Weight Loss Diet 318
Calculating Energy Ration for Weight Loss 319
Exercise as Part of the Weight Management Plan..320
Do Weight Management Efforts Succeed? 320
How Can We Provide Effective Support for Owners of Overweight Pets? 321
Improving Motivation 321
Improving Owner Effectiveness..322
Creating an Environment That Promotes Behaviour Change 322
Implementing Change – Goal Setting, Action Planning and Monitoring.....................................323
Preventing Obesity 324
Overview..324
References...324

INTRODUCTION

As many as 65% of pet dogs and cats are overweight or obese (German 2006, German et al. 2018, Larsen and Villaverde 2016). Obesity is also a common problem in rabbits (Courcier et al. 2012). Obesity is commonly defined at the excess accumulation of adipose tissue, although precise cut-offs for what constitutes normal, overweight and obese levels of 'excess' fat are debated given the lack of appropriate, accurate population-level data (Day 2017).

Animals gain weight where there is a chronic imbalance of energy intake in food and output for maintenance, thermogenesis and exercise. Being obese is associated with an increased incidence of many diseases and clinical problems that shorten lifespan and reduce quality of life in affected animals (Yam et al. 2016, Weeth 2016, Ruchti et al. 2018, Salt et al. 2019, Lawler et al. 2008). In brachycephalic animals, obesity-associated morbidities can exacerbate conformation-related health problems, and treatment of obesity may be complicated by pre-existing medical issues.

This chapter will discuss the causes and pathophysiological consequences of overweight and obesity (henceforth 'obesity') with particular reference to how this impacts on brachycephalic animals. The focus is on dogs and cats, since the evidence base for those species is greatest. Rabbits are mentioned briefly where relevant peer-reviewed information exists.

WHAT CAUSES OBESITY?

Simplistically, an individual gains weight because there is a mismatch between energy intake and expenditure. That balance is most commonly affected by alterations in food intake and energy expended during exercise but may also be influenced by clinical conditions that affect the efficiency of substrate utilisation (e.g. diabetes mellitus) or factors that alter metabolic rate (e.g. thyroid disease, ambient temperature). Obesity risk is therefore highly complex and is influenced by owner management and other environmental factors in combination with physiological variation due to factors including genetics, concurrent disease, sex hormones and age (Raffan 2013, Loftus and Wakshlag 2015).

MANAGEMENT OF FEEDING AND EXERCISE

The veterinary literature to date has mainly focussed on how owners manage their pets' weight and has identified many food- and exercise-related risk factors which are logical to those working with dogs. Overweight dogs and cats are more likely to receive human food, titbits and other treats outside of regular meals and to have owners who don't recognise obesity as a disease. Feeding *ad libitum* and feeding a dry or premium diet is associated with feline obesity. Dogs which are regarded as 'a member of the family' or otherwise humanised by their owners are more prone to weight gain (Munoz-Prieto et al. 2018, German 2006, Larsen and Villaverde 2016). Rabbits given access to concentrate food or inadequate access to grass/hay are predisposed to obesity (Prebble, Shaw, and Meredith 2015).

Obesity is consistently associated with less frequent, shorter and lower intensity exercise and, in cats, limited or no outdoor access. While it is compelling to blame lack of exercise for development of obesity, it is also possible that obesity and its comorbidities cause or exacerbate the inactivity (German et al. 2017, Robertson 2003, Courcier et al. 2010, Larsen and Villaverde 2016).

RISK FACTORS INDEPENDENT OF OWNER MANAGEMENT

Despite the importance of those 'human-controlled' risk factors, breed, age, sex and neuter status are also consistently identified as affecting obesity risk (Larsen and Villaverde 2016). Their effect is physiological and independent of owners. So, while owners should be able to tailor their dogs' food and exercise to keep them lean, it is disingenuous to dismiss the role of physiology in obesity.

OBESITY AS A CONDITION OF DISORDERED HOMEOSTASIS

Rather than blaming owners for obesity, it is better to view it as the failure of owners to curb a physiological drive towards weight gain in their pets. The drive to eat is governed by a homeostatic process, summarised in Figure 17.1.

Food intake and, to a lesser extent, energy expenditure are regulated by the hypothalamus, which coordinates peripheral signals about short-term energy intake in food and long-term energy storage in adipose tissue (Raffan 2013, Loftus and Wakshlag 2015). Gut-derived hormones released in response to changing nutrient concentrations in the gut and bloodstream signal to central receptors to trigger hunger after a period without food (e.g. ghrelin) and satiety after eating (e.g. cholecystokinin and peptide YY), acting in parallel with input from the visceral nerves.

Those short-term signals act against a background drive to promote hunger when fat stores are scant or reduce food intake when adipose tissue contains lots of energy stored as lipid. That background drive is mainly down to the action of leptin, a fat-derived hormone released in greater amounts after fat mass gain and that acts in the brain to reduce hunger. These neuro-hormonal

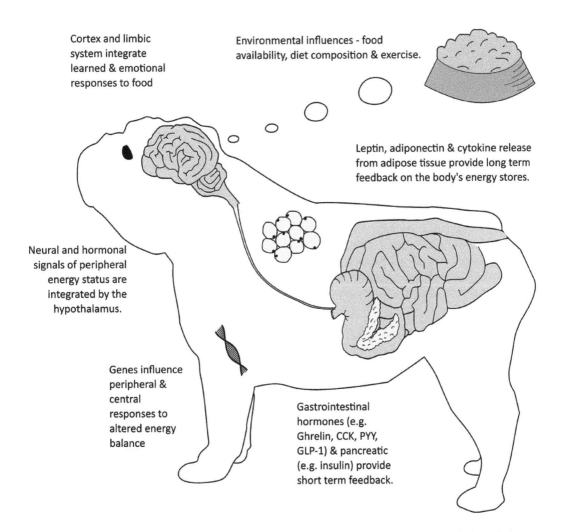

FIGURE 17.1 Peripheral signals of short- and long-term energy status are integrated by the hypothalamus. Neuro-hormonal effector responses affect appetite, satiety and energy expenditure to regulate bodyweight.

inputs converge upon the hypothalamus, specifically the leptin-melanocortin signalling pathway, to regulate hunger and satiety dynamically.

There is a wealth of evidence that mammals regulate fat mass so that increases or reductions in adipose tissue mass activate responses that favour return to their original weight. Thus, weight loss following caloric restriction is associated with reduced metabolic rate and increased appetite, and weight gain the opposite. Data from humans and laboratory animals suggest that mammals will defend a set point (or range) of bodyweight, determined for each individual by their genetic make-up (Speakman et al. 2011).

If body mass is homeostatically controlled, why should detrimental obesity develop? In reality, the homeostatic drives are a background on which other neurological inputs are superimposed. Cues related to food availability, taste and smell interact with hedonic pathways, integrating with inputs from memory and decision-making centres (van der Klaauw and Farooqi 2015). Obtaining the pleasurable effects of food is a powerful motivating force that can override homeostatic satiety signals. Therefore, exposure to, and availability of, highly palatable food, as is common for pets, is a powerful driver to increasing food intake beyond what is needed for maintenance of healthy bodyweight. Alongside the restricted exercise experienced by many pets, this likely drives the high prevalence of obesity seen.

GENETIC VARIATION INFLUENCES OBESITY SUSCEPTIBILITY

There is ample evidence from humans and rodents that susceptibility to obesogenic pressures is governed by genetics and that, in the main, genetic variation exerts its effect by altering eating behaviour (van der Klaauw and Farooqi 2015). Clear breed predispositions to obesity are evidence that genetics are important in dogs too. Among brachycephalic dog breeds, Pugs and English Bulldogs are particularly obesity-prone, but obesity can affect any dog of any breed (Larsen and Villaverde 2016, Corbee 2013, O'Neill et al. 2019, 2016, Edney and Smith 1986). In cats, obesity is similarly seen in all breeds and mixed breed cats; British shorthair and Persian cats are brachycephalic breeds reported to be predisposed to obesity (Larsen and Villaverde 2016, Corbee 2014).

The best canine example of how genes alter obesity susceptibility is a mutation in the gene POMC which has been identified in Labrador Retrievers. The mutation disrupts signalling through the leptin-melanocortin signalling pathway and predisposes affected dogs to obesity by increasing their food-seeking behaviour (Raffan et al. 2016). This is rare example of a single mutation having a large effect. Obesity is more commonly a complex genetic trait in which individual's susceptibility is governed by the 'dose' of risk alleles across many genetic loci (van der Klaauw and Farooqi 2015). Although each has a small effect, cumulatively they increase or decrease an individual's susceptibility to obesity.

PRACTICAL IMPLICATIONS OF GENETIC VARIATION IN OBESITY SUSCEPTIBILITY

Dogs and cats that are highly food-motivated are prone to obesity (Sallander et al. 2010, Ohlund, Palmgren, and Holst 2018, Raffan et al. 2015), and some of the most highly food-motivated dog breeds (e.g. Labrador Retrievers, Pugs) are also those which are particularly prone to obesity (Raffan et al. 2015). This supports the importance of genetically driven differences in food motivation to governing obesity susceptibility. In practical terms, it means that some breeds and individuals are particularly prone to overeating, manifested as increased food seeking, scavenging and begging in the home. They may also appear more 'grateful' for food when owners provide it. For owners of such animals, it is much harder to effectively restrict access to food and to keep them at a healthy weight than it would be for the owner of a 'picky' animal.

Owners of overweight dogs have been reported as actually making a greater effort to restrict their dogs' food intake than those of normal weight dogs – owners do recognise the need to slim their pets but are ineffective in doing so (Raffan et al. 2015). From a veterinary perspective, it means that owners of overweight pets need support on how to effectively restrict their dogs' food, rather than to be blamed for the problem.

PATHOPHYSIOLOGY OF OBESITY

Obesity is recognised as a disease in humans and animals (Day 2017). Increased fat mass leads to physiological disruptions that promote increased incidence of metabolic, endocrine, respiratory, orthopaedic, dermatological, neoplastic and other diseases in obese dogs, cats and rabbits (Day 2017, German 2006, Ruchti et al. 2018) (Table 17.1) At the most fundamental level, obesity shortens lives, with obese dogs dying on average 1–2 years earlier than dogs fed to maintain a lean body habitus (Salt et al. 2019, Kealy et al. 2002). There is evidence that the quality of life of overweight dogs is lower than for lean dogs and that quality of life improves after weight loss (Yam et al. 2016, Flanagan et al. 2017, German et al. 2012).

TABLE 17.1

Diseases and Clinical Signs That Occur with Greater Frequency in Overweight Pets For a Comprehensive Review of Obesity-Associated Conditions, Readers are Referred to Weeth (2016).

Shorter lifespan[a](dog)

Poor quality of life[a](dog)

Hyperthermia/heat stroke[a](dog)

Anaesthetic complications[a](dog)

Osteoarthritis and other joint diseases (dog), pododermatitis (rabbit)

Metabolic

Increased liver enzymes[a](dog)

Insulin resistance (dog), diabetes mellitus (cat, dog)

Dyslipidaemia (hypertriglyceridemia, hypercholesterolemia) (dog, cat)

Respiratory disorders

Tracheal Collapse[a](dog)

BOAS[a](dog, cat)

Laryngeal paralysis (dog)

Inflammatory airway disease (human)

Obstructive sleep apnoea[a] (human)

Hiatus hernia, gastroesophageal reflux[a] (human)

Increased blood pressure (dog, cat)

Neoplasia (mammary and transitional cell carcinoma) (dog)

Dystocia[a](dog)

Urinary disorders

Urethral sphincter mechanism incompetence (dog)

Urolithiasis (cat)

Other lower urinary tract disorders

[a]*Indicates conditions for which brachycephalic animals are already inherently predisposed, regardless of the additional risk following obesity. 'Human' in parentheses indicates associations that are well established in humans but not yet evidenced in dogs, cats or rabbits. 'Cat' or 'dog' in parentheses denotes species in which this has been reported in the veterinary literature, although readers should note that an absence of evidence does not mean this does not occur in reality.*

INSULIN RESISTANCE AFFECTS METABOLISM AND CELL GROWTH

Trying to identify a single inciting event for obesity at a molecular level is arguably unhelpful, but there are now multiple lines of evidence supporting what is known as the *adipose expandability theory*. In essence, it appears that individuals have a limit to their ability to store fat healthily. Once that limit is met, adipose tissue ceases to store energy efficiently and lipids accumulate in tissues where they don't belong leading to insulin resistance and inflammation (Virtue and Vidal-Puig 2010).

The homeostatic drive to maintain normoglycaemia means the initial response to insulin resistance is to increase circulating insulin concentration. Initially, this allows maintenance of normoglycaemia, but over time, pancreatic beta cells can be damaged leading to intermittent (post-meal) or persistent hyperglycaemia (diabetes mellitus). In dogs, disordered glucose regulation has been documented in obesity, but it is not recognised as causing frank diabetes although there is some epidemiological evidence that obesity may contribute to the development of canine diabetes mellitus (Wejdmark et al. 2011, Mattin et al. 2014, Raffan 2013). In cats, obesity-related insulin resistance is the most common inciting cause for diabetes mellitus (Osto et al. 2013).

A cluster of physiological changes reminiscent of human metabolic syndrome (reported in dogs as obesity-related metabolic dysfunction) are seen in a subset of overweight dogs; these include dyslipidaemia, increased postprandial blood glucose, increased blood pressure and cardiac remodelling. Blood pressure changes are usually mild (not clinically significant) but have been well documented in overweight dogs and are likely to related to the cardiac remodelling that has also been observed (Tvarijonaviciute et al. 2012, Clark and Hoenig 2016).

Dyslipidaemia is reported in overweight dogs, manifesting as hypertriglyceridemia and increased cholesterol, although atherosclerosis tends not to develop in dogs (probably due to differences in the ratio of high to low-density lipoprotein cholesterol between dogs and humans). Obesity-related hepatic lipid accumulation likely plays a role in the increased liver enzymes commonly observed in overweight dogs (Tvarijonaviciute et al. 2019). In dogs with respiratory compromise due to tracheal collapse, increased liver enzymes have been documented (Bauer et al. 2006). That is reminiscent of the hypoxic hepatitis recognised in humans and may mean that obese brachycephalic dogs need to deal with multiple hepatic challenges. In cats, dyslipidaemia and hepatic lipid accumulation are common sequelae to obesity, insulin resistance and diabetes mellitus (Biddinger and Kahn 2006).

Insulin is an anabolic hormone and promotion of cell growth and proliferation due to hyperinsulinaemia is implicated in the association between obesity and cancer, which is well established in humans and for which there is emerging evidence in dogs (Weeth 2016).

OBESITY-ASSOCIATED INFLAMMATION ACTS SYSTEMICALLY

Obesity is associated with chronic low-grade inflammation, primarily due to release of pro-inflammatory cytokines from adipose tissue macrophages. Obesity-associated inflammation exacerbates insulin resistance and is implicated in the obesity–cancer association and in exacerbating other obesity-associated diseases such as (human) asthma or osteoarthritis (Weeth 2016).

INCREASED FAT MASS EXERTS A MECHANICAL EFFECT

Subcutaneous fat reduces the surface area-to-volume ratio and has an insulating action, increasing the risk of heat stroke in overweight dogs, a condition to which brachycephalic breeds are predisposed. Higher body mass increases the load on joints and is implicated, along with inflammation, in causing or exacerbating cruciate disease, osteoarthritis and other orthopaedic disorders (Frye, Shmalberg, and Wakshlag 2016).

In humans, increased abdominal pressure due to obesity is a well-recognised risk factor for oesophageal reflux and hiatal hernia (both of which are commonly reported in some brachycephalic breeds), and it is likely that this can exacerbate the same problems in dogs (Tack and Pandolfino 2018).

Obesity raises the risk of dystocia. This is in part a mechanical effect; fat deposits can impede passage of the foetus through the birth canal. It is also likely that as in other species, enlarged foetal size and other consequences of metabolic dysfunction also contribute to this increased risk.

OBESITY AND THE RESPIRATORY SYSTEM

In otherwise healthy overweight dogs, there is a lower arterial partial pressure of oxygen, although not below the normal reference range; this reduction was shown to reverse after weight loss (Pereira-Neto et al. 2018). Overweight dogs have reduced tidal volume and functional residual capacity, increased resting respiratory rates and increased airway resistance. Tracheal collapse is exacerbated by obesity, as is sleep apnoea and airway inflammation (in humans). In brachycephalic breeds, obesity is a risk factor for developing clinical signs of BOAS (Packer et al. 2015, Manens et al. 2012, Chandler 2016, Pereira-Neto et al. 2018).

Obesity causes those changes by a combination of its mechanical, inflammatory and metabolic effects. Fat deposits in the thorax and abdomen mechanically reduce ventilatory capacity and increase the pressure within the thorax. Human obesity is well documented as reducing lung compliance, decreasing respiratory muscle strength, altering ventilation distribution and promoting airway inflammation (e.g. in asthma and chronic obstructive pulmonary disease), and it is likely the same is true in dogs (Chandler 2016).

Obstructive sleep apnoea has been well documented in the English Bulldog and is also recognised in other brachycephalic breeds (Packer et al. 2015). In humans, obesity and increased neck girth are well documented as risk factors for sleep apnoea, and the same is reported anecdotally in dogs. Finally, overweight dogs also have higher rates of anaesthetic complication, in part due to respiratory compromise (Clutton 1988).

OBESITY PRESENTS PARTICULAR PROBLEMS FOR BRACHYCEPHALIC ANIMALS

In summary, obesity is a particular problem in brachycephalic dogs for two main reasons. First, Pugs and English Bulldogs are particularly prone to obesity, with high food motivation in those breeds. Second, obesity can lead to a 'double whammy' of physiological challenges in brachycephalic dogs when obesity-related physiological or mechanical compromise worsen or increase the already elevated risk of morbidities related to conformational extremes such as BOAS, dystocia or hyperthermia.

Although less well documented, the negative effects on respiratory function are also likely to impact on BOAS in cats and rabbits. Finally, brachycephaly-associated dental disease in rabbits may lead to them avoiding hay/grass in favour of energy dense pelleted/muesli foods that promote obesity.

MANAGEMENT OF OBESITY

THE KEY TO SUCCESSFUL WEIGHT LOSS IS FOOD RESTRICTION

To lose weight, an animal needs to eat fewer calories than it expends over a sustained period. The most effective and sustainable way to alter energy 'ins and outs' is by food restriction. That is because the proportion of daily energy expended during exercise is usually no more than 25%, so making a large impact on energy balance by increasing exercise is unrealistic for most pet-owner dyads (Figure 17.2).

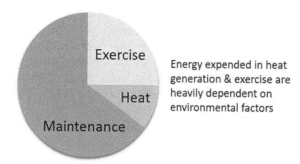

FIGURE 17.2   Maintenance of essential bodily functions accounts for most of daily energy expenditure. The remainder is used for heat production (approximately 10%) and exercise (approximately 25%). Heat loss is in part obligate – the thermic effect of food – and in part responsive to the environment – thermogenesis to maintain body temperature. Since energy expenditure during exercise only accounts for approximately 25% of daily energy expenditure, even big increases in exercise have only a modest effect on overall energy expenditure.

The basics of energy restriction are straightforward. After estimating an ideal weight and settling on a realistic target for weight loss, calculate the amount of energy required to produce weight loss and select a diet that will provide essential nutrients while still restricting energy. Monitor progress at least monthly, and alter the feeding ration as required to maintain steady weight loss.

Below I will expand on how best to implement the basics of weight loss management in dogs and cats but also discuss why dieting often fails and how acknowledging the challenges that owners of highly food-motivated dogs and cats face can lead to more effective management of obesity. Readers should consult species-specific text books for advice on managing obesity in rabbits (Blas and Wiseman 2020).

DEFINING IDEAL AND TARGET WEIGHTS

Ideally, clinical records of both weight and body condition score (BCS) should be kept throughout dog and cat lives. This allows owners and veterinarians to know the baseline for their pet's healthy, lean weight. In reality, clinical records are often incomplete, but a reasonable estimate can be made using the BCS. For the best validated 9-point BCS scale, each 1 point increase over the ideal of 4–5/9 equates to an approximately 10% increase in body mass, allowing estimation of a lean bodyweight (Mawby et al. 2004, German et al. 2006).

However, especially for profoundly overweight pets, weight loss to the ideal weight can be very difficult. Fortunately, more modest weight loss commonly still has a positive impact on health and welfare (German 2016b). In one study of dogs with osteoarthritis, weight loss of as little as 8% bodyweight was sufficient to produce a measurable and clinically important improvement in lameness (Marshall et al. 2010). Clinical experience suggests this is also the case in alleviating respiratory or other clinical signs.

SELECTING A WEIGHT LOSS DIET

It is vital to pay attention to the macro- and micronutrient composition of the diet offered during weight loss. If calorie restriction is achieved by simply reducing the volume of a standard maintenance diet, there is a risk the diet may fail to provide sufficient of both macronutrients (particularly protein) and micronutrients (e.g. selenium and choline). This risk is alleviated by feeding a diet specially formulated for weight loss as these are designed to have maintenance amounts of macro- and micronutrients while still restricting calorie density (Linder et al. 2013, 2012, German, Holden, et al. 2015).

The formulation of specialist weight loss diets is founded on our understanding of energy homeostasis. Satiety is promoted in response to nutrients within the gut lumen and circulation, physical stretch of the gut wall and neuronal stimuli. Moderating the diet can increase the intensity and duration of satiety after a meal. Broadly, this can be achieved by altering diet composition to slow gastrointestinal transit (high fibre); promote mechanical gut fill (high fibre); prolong the period over which food is absorbed (complex nutrients, high fibre); and increase nutrients which promote satiety (high protein) (Raffan 2013, Loftus and Wakshlag 2015).

There is plentiful evidence that diets designed for weight loss successfully improve satiety as well as the overall success of weight loss – on this basis, they are recommended by this author and others (Weber et al. 2007, Flanagan et al. 2017, German et al. 2010, German, Titcomb, et al. 2015). Offering a diet with higher protein content (than standard maintenance diets) is also important because they promote maintenance of lean muscle mass during weight loss (Linder and Parker 2016).

Concurrent morbidities may complicate the choice of food. For instance, brachycephalic breeds are at increased risk of hiatal hernia and gastrointestinal inflammation which could benefit from diets other than those for weight management. In such cases, it is possible to calculate whether nutrient density is adequate on a calorie-restricted diet – clinicians might prefer to contact a nutrition specialist for advice.

CALCULATING ENERGY RATION FOR WEIGHT LOSS

At the start of a weight loss programme, two options are available to decide on how much to feed. The first is to take an accurate current dietary history and reduce food intake by 10–20% regardless. Alternatively (and possibly better because dietary histories are notoriously inaccurate), feed the resting energy requirement for the estimated ideal bodyweight. After 2–4 weeks, assess progress and alter the ration to ensure a rate of weight loss of 0.5%–2% body mass per week.

Realistically, many owners will struggle to stop giving 'treats' to their pets. This is discussed in more detail below, but here it is important to remember that titbits are unlikely to be nutritionally balanced so they should not make up more than 10% of the daily ration. They should, however, be included in dietary calculations – finding the calories provided by different titbits can be difficult, so encourage owners to use standard foods or titbits to help with rationing.

BOX 17.1   A WORKED EXAMPLE OF HOW TO PLAN A WEIGHT LOSS INTERVENTION

A 4-year-old, female neutered, markedly overweight Pug is showing marked signs of respiratory distress due to BOAS. You believe that weight loss is indicated to reduce dyspnoea. Other than BOAS and obesity, the dog is clinically well. Current bodyweight is 11 kg and BCS 8/9.

First, calculate ideal (lean) weight:

$$Ideal\ bodyweight = Current\ weight \times (100/(100 + 10 \times (current\ BCS - 5)))$$
$$= 11*(100/(100 + (10*3)))$$
$$= 8.5\,kg$$

This can be simplified: Ideal bodyweight if BCS   *6 = Current weight/1.1*
7 = Current weight/1.2
8 = Current weight/1.3
9 = Current weight/1.4

However, you suspect 10% weight loss will improve clinical signs.
Target weight = 11*0.9 = 9.9 kg

Dietary history reveals the dog is already on a specialist weight management diet, and the owner weighs the food daily. However, the dog gets human food at mealtimes and treats during walks. The owner is reluctant to cut out treats. Increasing exercise is not feasible given the respiratory signs.

Next, calculating maintenance energy requirements for the estimated lean body mass (a good starting point for weight loss).

$$Energy\ requirement\ for\ lean\ body\ mass = 70*(lean\ body\ mass)^{0.75}$$
$$= 70 \times 8.5^{0.75}$$
$$= 347\ kcal/day$$

Calculate how this can be divided to meet the wishes of the owner:

- 40 g *kibble as meals twice a day* @ 310 kcal/100 g   = 248 kcal
- 20 g *kibble reserved for 'treats' during the day*   = 62 kcal
- 25 g *chicken breast at human mealtimes* @ 150 kcal/100 g   = 37 kcal

Total 347 kcal/day
Give clear written instructions:

- An 'if… then…' planning sheet
- Information leaflet about obesity to share with other family members
- Weight and feeding record sheet.

In first month, you hope to see weight loss of 0.5%–2% a week (e.g. 220–880g for this example). Tailor food ration depending on progress, and use the recheck as an opportunity to provide ongoing support.

EXERCISE AS PART OF THE WEIGHT MANAGEMENT PLAN

Meaningfully shifting to a negative energy balance using exercise alone is unrealistic both theoretically and in practice (Chapman et al. 2019). However, improving the quantity and intensity of exercise during weight loss has been shown to reduce the degree of muscle loss observed during weight loss, and one study showed the addition of exercise increased the rate of weight loss (Chauvet et al. 2011, Vitger et al. 2016).

Consequently, it is sensible to encourage increased exercise alongside food restriction for weight loss, but its importance should not be overemphasised. This is particularly important where increasing exercise is problematic. That might be because an owner finds management change difficult, or there is a physical reason not to increase exercise. This is of particular relevance to brachycephalic animals with airway compromise where introducing vigorous exercise might risk hypoxia or hyperthermia. In those cases, it is legitimate to focus primarily on dietary modification. In cats, encouraging play or increasing the time the cat spends outside can promote increased activity.

DO WEIGHT MANAGEMENT EFFORTS SUCCEED?

In experimental conditions, weight loss will proceed steadily at 1%–2% of bodyweight a week following that approach. In real life, however, weight loss programmes are often unsuccessful, even when owners profess enthusiasm for the project. From the literature, three clear and somewhat chastening conclusions can be made: (1) although some pets do effectively lose weight in the home setting, there is a high rate of drop-out, (2) owner compliance with feeding and exercise recommendations are commonly poor, and (3) rebound weight gain is common even after successful weight loss (Flanagan et al. 2017, German 2016a, German et al. 2010, Vitger et al. 2016, Yaissle, Holloway, and Buffington 2004, German, Titcomb, et al. 2015).

Rebound weight gain after loss is a particularly disappointing outcome, but the homeostatic drive to defend a 'set point' of bodyweight is a well-studied phenomenon. The same occurs in dogs and cats with maintenance requirements commonly being just 10% more than that required for weight loss. Therefore, continued energy restriction, maintenance of exercise and feeding for satiety are all sensible approaches to improving long-term weight management (German et al. 2011).

More complex are the reasons why owners who initially express enthusiasm for reducing their pet's weight later give up, although poor owner compliance almost always explains why diets fail. Reasons cited in the veterinary literature include dogs developing other disease that supersedes obesity as the focus of care, owners' personal reasons, and decreased satisfaction with progress as weight loss slows over time. Dogs that are initially more overweight (and arguably need to lose weight more) are less likely to reach target weight. Faster weight loss, a longer duration of weight loss, feeding a dried weight management diet and being female or entire are predictors of greater success (German, Titcomb, et al. 2015, Flanagan et al. 2017, Porsani et al. 2019). Those are limited, measurable factors, however, and in the next section I will take a broader view of why owners fail to make the changes required to slim their pets.

HOW CAN WE PROVIDE EFFECTIVE SUPPORT FOR OWNERS OF OVERWEIGHT PETS?

By definition, owners of obese animals have already failed to make the original small adjustments needed to keep their pet slim; it follows that they will find making the more meaningful changes required later to reduce their dog or cat's weight difficult. Although there is evidence for owner 'blindness' to obesity, there is also data that suggests owners do try to restrict food in overweight pets, they just aren't effective at doing so (described above). Therefore, professionals need to provide effective support that acknowledges the challenges faced by owners of obesity-prone pets.

The essence of 'feed less, do more exercise' is simple but for owners of overweight animals, that often means making many small changes to their daily routines. Changing habits is difficult. There is a large body of literature in human psychology on methods to implement human behavioural change (Cane, O'Connor, and Michie 2012). With few exceptions (Rohlf et al. 2010, Webb et al. 2018), these lessons have not been formally applied to promote canine weight loss.

One well-characterised framework describes the central pillars required to promote human behaviour change as 'motivation, ability and opportunity' (Michie, van Stralen, and West 2011). As animal care professionals, we should take time to communicate effectively with owners so as (a) to improve owners' **motivation** to change their behaviour, (b) to improve owner effectiveness by giving practical suggestions (**ability**) and (c) to create an environment that promotes behaviour change (**opportunity**). In Figure 17.3, I outline some of the questions that should be considered to better understand where an individual owner stands in relation to each of those pillars. Additionally, there is also evidence that some specific techniques promote behaviour change. Broadly, they fall into the categories of **goal setting, action planning and monitoring**.

Below and in Table 17.2, I comment on each element and how this might translate practically into supporting owners of overweight dogs and cats.

IMPROVING MOTIVATION

If weight loss is to happen, owners need to be motivated to change. Improving their motivation first requires an understanding of why they may not already be motivated. Positive messages are more successful – although there is some merit in warning owners of the negative implications of obesity,

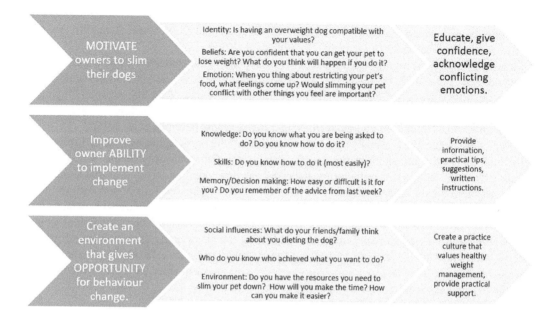

FIGURE 17.3 **Supporting human behaviour change requires motivation, ability and opportunity.** To promote the human behaviour changes required for pet weight loss, ask owners about the reasons why they struggle to change and consider how each aspect can be supported by the actions you take as a clinician.

owners are more likely to be motivated to change by an assurance that weight loss can rapidly have a beneficial effect on quality of life and clinical signs. Motivation can wane, particularly as weight loss slows, so maintaining a positive message and reassuring owners that a drop-off is to be expected can help maintain their enthusiasm.

IMPROVING OWNER EFFECTIVENESS

Many owners are already trying to restrict their dog's food intake, although ineffectively, when they embark on a weight loss programme. Commonly, owners do not understand that diet should be tailored to the dog (rather than feeding what it says on a packet), fail to realise the contribution small treats can make to their dog's energy intake, are sloppy in measuring the food or fail to notice the reduced intensity of their dog's activity. By helping them identify and rectify these barriers to effectively implementing a weight management plan, you will improve outcomes.

CREATING AN ENVIRONMENT THAT PROMOTES BEHAVIOUR CHANGE

People are more likely to make a change if others that they trust agree it is a good idea. Professionals can engender a veterinary practice culture that values dogs being kept at a healthy weight and highlight this by including weight monitoring in preventative healthcare visits.

One of the most potent forces that stops owners making the required behaviour changes is their pet itself – if we can reduce the 'pester power' of highly food motivated dogs or cats during calorie restriction, weight loss is more likely to succeed. The best evidence base for this comes from feeding a weight loss diet designed to promote satiety (Weber et al. 2007, German et al. 2010), but owners also report that complex puzzle feeders or scattering food for dogs to hunt through the day is helpful at reducing begging behaviour.

TABLE 17.2

Pet Owners Commonly Have Practical or Personal Concerns That Act as Barriers to Change. Identifying These Concerns Can Help Tailor Make a Support Plan Which Will Be More Effective for an Individual Pet and Owner. Some Examples are Given Here.

Barrier to Change	Potential Solutions	Practical Tips
Owners underestimate energy from 'treats'. Owners use approximate measures for kibble.	• Educate owners on the energy content of common treats. • Weigh food.	• Make a list of the energy content of common treats, and their human equivalents. • Weigh out meals for a week into tubs to save time day-to-day.
Owner is worried giving treats is essential for bonding.	• Reassure owner that making a fuss of their Pet will be an effective substitute. • Provide suggestions for low calorie treats. • Factor in measured and regular treats into the daily calorie allowance.	• Low-calorie treats include ice cubes and vegetable chunks. Alternatively, take treats of kibble from the daily ration. • Use the treat calorie counter to subtract regular treats from the ration.
Dog constantly seeking food when on restricted rations	• Feed a satiety-promoting diet • Provide toys and exercise as distraction. • Consider using a complex/time consuming puzzle toy to provide an outlet for food-seeking behaviour without many calories.	• Tough rubber toys can be stuffed with canned food and frozen before being given to the dog as part of their daily ration. • Feeding the entire ration scattered in a garden can provide a free roaming dog with hours of 'work'.
Owner doesn't feel they will succeed	• Provide positive examples.	• Waiting room displays or an album of successful slimmers may provide inspiration.
Owner doesn't think obesity is a problem	• Point to respected authorities who care. • Maintain a practice culture of regular weight checks, recording and early intervention.	• Many pet welfare charities have webpages on obesity. • Prepare waiting room displays and encourage owners to drop in and weigh between consultations.
Dog steals food	• Remove or avoid regular sources.	• e.g. move cats' bowls out of reach, get a secure bin.

IMPLEMENTING CHANGE – GOAL SETTING, ACTION PLANNING AND MONITORING

Having considered the motivation and barriers to implementing the management changes required to get their pets to become slimmer, owners are likely to be more successful if that is followed by goal setting, action planning and monitoring. Importantly, owners should be actively involved in generating goals and plans – a didactic approach from clinicians is unlikely to be successful.

Measurable and achievable goals are important, and this is the rationale behind defining both an ideal (lean) weight and target weight (e.g. 10% loss) at the start of a weight loss programme. Monitoring needs to be objective and likely to show a small difference – hence, weight is a more valuable metric than BCS. Simple record sheets help owners record not only weight loss progress but perhaps also success and failure at making feeding changes (although avoid over-facing owners with paperwork).

Action planning can be facilitated by 'if…then…' planning as applied to overweight dogs by Webb et al. (2018). That paper has a helpful supplementary table showing common problems and potential solutions. For instance, if owners might relate to the statement 'if my dog begs for food at table …,' they could choose one or more solutions from '…I will keep him out of the dining room at mealtimes', '… I will feed him before we eat so he is distracted' or '… I will tell him to lie in his bed'. Owners should record their plans and keep them accessible.

Despite all this, owners are likely to find it difficult to sustain the changes needed to slim their pets and to require ongoing support from veterinary professionals. That is best done face to face, but it may be more sustainable with telephone or email consultations. Expectation management is important – it is unlikely that even modest 10% weight loss will happen in less than 3 months, but it is worth the effort – seeing the health and welfare improvement in a slimming pet can be very rewarding.

PREVENTING OBESITY

Given that obesity treatment is so often ineffective, it is sensible to focus on preventing dogs gaining weight in the first place. I hypothesise that better owner education about the risks of gaining weight and how to recognise it at an early stage might reduce obesity rates in the target population. Training owners in how to use BCS charts might empower them to monitor their dogs' health at home. Regular monitoring and reporting of weight and condition scores from puppyhood can provide objective evidence of weight gain; if this becomes routine at every practice visit and all staff are trained to raise the issue if a dog has gained weight, it might lead to more frequent and effective interventions.

Brachycephalic dogs with airway compromise are at particular risk if they become overweight so targeting them as a 'high risk' group for weight monitoring seems sensible. Since neutering is commonly associated with weight gain, there is a strong logic to post-neutering counselling on the risks of obesity.

OVERVIEW

Obesity is a common disorder in dogs of all breeds and is associated with numerous comorbidities. In brachycephalic animals, obesity presents a particular problem because it can worsen respiratory compromise, dystocia risk and other conditions that occur with high frequencies in these breeds.

While risk factors for obesity are well recognised, we are not good at preventing or treating obesity. This is likely because an overly simplistic view of its pathogenesis has led to dogmatic and overly simplistic treatment efforts, reliant on changing the ingrained habits of owners. It is likely that more thoughtful treatment interventions that acknowledge the impact that genetically driven variability in animal behaviour has on obesity risk will continue to develop; testing which are beneficial experimentally will be important.

In combination with well-established evidence-based protocols that use dietary restriction and specialist diets to calorie restrict while providing essential nutrients and promoting satiety, this offers the hope of continued improvement in our clinical management of obesity.

REFERENCES

Bauer, N. B., M. A. Schneider, R. Neiger, and A. Moritz. 2006. "Liver disease in dogs with tracheal collapse." *J Vet Intern Med* 20 (4):845–9. doi:10.1892/0891-6640(2006)20[845:ldidwt]2.0.co;2.

Biddinger, S. B., and C. R. Kahn. 2006. "From mice to men: insights into the insulin resistance syndromes." *Annu Rev Physiol* 68:123–58. doi:10.1146/annurev.physiol.68.040104.124723.

de Blas, C., and J. Wiseman. 2020. *Nutrition of the Rabbit*. 3rd edition. Wallingford, Oxfordshire; Boston, MA: CABI Publishing.

Cane, J., D. O'Connor, and S. Michie. 2012. "Validation of the theoretical domains framework for use in behaviour change and implementation research." *Implement Sci* 7:37. doi:10.1186/1748-5908-7-37.

Chandler, M. L. 2016. "Impact of obesity on cardiopulmonary disease." *Vet Clin N Am Small Anim Pract* 46 (5):817–30. doi:10.1016/j.cvsm.2016.04.005.

Chapman, M., G. R. T. Woods, C. Ladha, C. Westgarth, and A. J. German. 2019. "An open-label randomised clinical trial to compare the efficacy of dietary caloric restriction and physical activity for weight loss in overweight pet dogs." *Vet J* 243:65–73.doi:10.1016/j.tvjl.2018.11.013.

Chauvet, A., J. Laclair, D. A. Elliott, and A. J. German. 2011. "Incorporation of exercise, using an underwater treadmill, and active client education into a weight management program for obese dogs." *Can Vet J* 52 (5):491–6.

Clark, M., and M. Hoenig. 2016. "Metabolic effects of obesity and its interaction with endocrine diseases." *Vet Clin N Am Small Anim Pract* 46 (5):797–815. doi:10.1016/j.cvsm.2016.04.004.

Clutton, R. E. 1988. "The medical implications of canine obesity and their relevance to anaesthesia." *Br Vet J* 144 (1):21–8. doi:10.1016/0007-1935(88)90149-2.

Corbee, R. J. 2013. "Obesity in show dogs." *J Anim Physiol Anim Nutr (Berl)* 97 (5):904–10. doi:10.1111/j.1439-0396.2012.01336.x.

Corbee, R. J. 2014. "Obesity in show cats." *J Anim Physiol Anim Nutr (Berl)* 98 (6):1075–80. doi:10.1111/jpn.12176.

Courcier, E. A., D. J. Mellor, E. Pendlebury, C. Evans, and P. S. Yam. 2012. "Preliminary investigation to establish prevalence and risk factors for being overweight in pet rabbits in Great Britain." *Vet Rec* 171 (8):197. doi:10.1136/vr.100792.

Courcier, E. A., R. M. Thomson, D. J. Mellor, and P. S. Yam. 2010. "An epidemiological study of environmental factors associated with canine obesity." *J Small Anim Pract* 51 (7):362–7. doi:10.1111/j.1748-5827.2010.00933.x.

Day, M. J. 2017. "One health approach to preventing obesity in people and their pets." *J Comp Pathol* 156 (4):293–5. doi:10.1016/j.jcpa.2017.03.009.

Edney, A. T., and P. M. Smith. 1986. "Study of obesity in dogs visiting veterinary practices in the United Kingdom." *Vet Rec* 118 (14):391–6.

Flanagan, J., T. Bissot, M. A. Hours, B. Moreno, A. Feugier, and A. J. German. 2017. "Success of a weight loss plan for overweight dogs: The results of an international weight loss study." *PLoS One* 12 (9):e0184199. doi:10.1371/journal.pone.0184199.

Frye, C. W., J. W. Shmalberg, and J. J. Wakshlag. 2016. "Obesity, exercise and orthopedic disease." *Vet Clin North Am Small Anim Pract* 46 (5):831–41. doi:10.1016/j.cvsm.2016.04.006.

German, A. J. 2006. "The growing problem of obesity in dogs and cats." *J Nutr* 136 (7 Suppl):1940S–6S.

German, A. J. 2016a. "Outcomes of weight management in obese pet dogs: What can we do better?" *Proc Nutr Soc* 75 (3):398–404. doi:10.1017/S0029665116000185.

German, A. J. 2016b. "Weight management in obese pets: The tailoring concept and how it can improve results." *Acta Vet Scand* 58 (Suppl 1):57. doi:10.1186/s13028-016-0238-z.

German, A. J., E. Blackwell, M. Evans, and C. Westgarth. 2017. "Overweight dogs exercise less frequently and for shorter periods: Results of a large online survey of dog owners from the UK." *J Nutr Sci* 6:e11. doi:10.1017/jns.2017.6.

German, A. J., S. L. Holden, T. Bissot, P. J. Morris, and V. Biourge. 2010. "A high protein high fibre diet improves weight loss in obese dogs." *Vet J* 183 (3):294–7. doi:10.1016/j.tvjl.2008.12.004.

German, A. J., S. L. Holden, N. J. Mather, P. J. Morris, and V. Biourge. 2011. "Low-maintenance energy requirements of obese dogs after weight loss." *Br J Nutr* 106 (Suppl 1):S93–6. doi:10.1017/s0007114511000584.

German, A. J., S. L. Holden, G. L. Moxham, K. L. Holmes, R. M. Hackett, and J. M. Rawlings. 2006. "A simple, reliable tool for owners to assess the body condition of their dog or cat." *J Nutr* 136 (7 Suppl):2031S–3S.

German, A. J., S. L. Holden, S. Serisier, Y. Queau, and V. Biourge. 2015. "Assessing the adequacy of essential nutrient intake in obese dogs undergoing energy restriction for weight loss: A cohort study." *BMC Vet Res* 11:253. doi:10.1186/s12917-015-0570-y.

German, A. J., S. L. Holden, M. L. Wiseman-Orr, J. Reid, A. M. Nolan, V. Biourge, P. J. Morris, and E. M. Scott. 2012. "Quality of life is reduced in obese dogs but improves after successful weight loss." *Vet J* 192 (3):428–34. doi:10.1016/j.tvjl.2011.09.015.

German, A. J., J. M. Titcomb, S. L. Holden, Y. Queau, P. J. Morris, and V. Biourge. 2015. "Cohort study of the success of controlled weight loss programs for obese dogs." *J Vet Intern Med* 29 (6):1547–55. doi:10.1111/jvim.13629.

German, A. J., G. R. T. Woods, S. L. Holden, L. Brennan, and C. Burke. 2018. "Dangerous trends in pet obesity." *Vet Rec* 182 (1):25. doi:10.1136/vr.k2.

Kealy, R. D., D. F. Lawler, J. M. Ballam, S. L. Mantz, D. N. Biery, E. H. Greeley, G. Lust, M. Segre, G. K. Smith, and H. D. Stowe. 2002. "Effects of diet restriction on life span and age-related changes in dogs." *J Am Vet Med Assoc* 220 (9):1315–20.

Larsen, J. A., and C. Villaverde. 2016. "Scope of the problem and perception by owners and veterinarians." *Vet Clin North Am Small Anim Pract* 46 (5):761–72. doi:10.1016/j.cvsm.2016.04.001.

Lawler, D. F., B. T. Larson, J. M. Ballam, G. K. Smith, D. N. Biery, R. H. Evans, E. H. Greeley, M. Segre, H. D. Stowe, and R. D. Kealy. 2008. "Diet restriction and ageing in the dog: major observations over two decades." *Br J Nutr* 99 (4):793–805. doi:10.1017/s0007114507871686.

Linder, D. E., L. M. Freeman, S. L. Holden, V. Biourge, and A. J. German. 2013. "Status of selected nutrients in obese dogs undergoing caloric restriction." *BMC Vet Res* 9:219. doi:10.1186/1746-6148-9-219.

Linder, D. E., L. M. Freeman, P. Morris, A. J. German, V. Biourge, C. Heinze, and L. Alexander. 2012. "Theoretical evaluation of risk for nutritional deficiency with caloric restriction in dogs." *Vet Q* 32 (3–4):123–9. doi:10.1080/01652176.2012.733079.

Linder, D. E., and V. J. Parker. 2016. "Dietary aspects of weight management in cats and dogs." *Vet Clin N Am Small Anim Pract* 46 (5):869–82. doi:10.1016/j.cvsm.2016.04.008.

Loftus, J. P., and J. J. Wakshlag. 2015. "Canine and feline obesity: a review of pathophysiology, epidemiology, and clinical management." *Vet Med (Auckl)* 6:49–60. doi:10.2147/VMRR.S40868.

Manens, J., M. Bolognin, F. Bernaerts, M. Diez, N. Kirschvink, and C. Clercx. 2012. "Effects of obesity on lung function and airway reactivity in healthy dogs." *Vet J* 193 (1):217–21. doi:10.1016/j.tvjl.2011.10.013.

Marshall, W. G., H. A. Hazewinkel, D. Mullen, G. De Meyer, K. Baert, and S. Carmichael. 2010. "The effect of weight loss on lameness in obese dogs with osteoarthritis." *Vet Res Commun* 34 (3):241–53. doi:10.1007/s11259-010-9348-7.

Mattin, M., D. O'Neill, D. Church, P. D. McGreevy, P. C. Thomson, and D. Brodbelt. 2014. "An epidemiological study of diabetes mellitus in dogs attending first opinion practice in the UK." *Vet Rec* 174 (14):349. doi:10.1136/vr.101950.

Mawby, D. I., J. W. Bartges, A. d'Avignon, D. P. Laflamme, T. D. Moyers, and T. Cottrell. 2004. "Comparison of various methods for estimating body fat in dogs." *J Am Anim Hosp Assoc* 40 (2):109–14. doi:10.5326/0400109.

Michie, S., M. M. van Stralen, and R. West. 2011. "The behaviour change wheel: A new method for characterising and designing behaviour change interventions." *Implement Sci* 6:42. doi:10.1186/1748-5908-6-42.

Munoz-Prieto, A., L. R. Nielsen, R. Dabrowski, C. R. Bjornvad, J. Soder, E. Lamy, I. Monkeviciene, B. B. Ljubic, I. Vasiu, S. Savic, F. Busato, Z. Yilmaz, A. F. Bravo-Cantero, M. Ohlund, S. Lucena, R. Zelvyte, J. Aladrovic, P. Lopez-Jornet, M. Caldin, C. Lavrador, B. Karveliene, V. Mrljak, J. Mazeikiene, and A. Tvarijonaviciute. 2018. "European dog owner perceptions of obesity and factors associated with human and canine obesity." *Sci Rep* 8 (1):13353. doi:10.1038/s41598-018-31532-0.

O'Neill, D. G., E. C. Darwent, D. B. Church, and D. C. Brodbelt. 2016. "Demography and health of Pugs under primary veterinary care in England." *Canine Genet Epidemiol* 3:5. doi:10.1186/s40575-016-0035-z.

O'Neill, D. G., A. M. Skipper, J. Kadhim, D. B. Church, D. C. Brodbelt, and R. M. A. Packer. 2019. "Disorders of dogs under primary veterinary care in the UK in 2013." *PLoS One* 14 (6):e0217928. doi:10.1371/journal.pone.0217928.

Ohlund, M., M. Palmgren, and B. S. Holst. 2018. "Overweight in adult cats: A cross-sectional study." *Acta Vet Scand* 60 (1):5. doi:10.1186/s13028-018-0359-7.

Osto, M., E. Zini, C. E. Reusch, and T. A. Lutz. 2013. "Diabetes from humans to cats." *Gen Comp Endocrinol* 182:48–53. doi:10.1016/j.ygcen.2012.11.019.

Packer, R. M., A. Hendricks, M. S. Tivers, and C. C. Burn. 2015. "Impact of facial conformation on canine health: Brachycephalic obstructive airway syndrome." *PLoS One* 10 (10):e0137496. doi:10.1371/journal.pone.0137496.

Pereira-Neto, G. B., M. A. Brunetto, P. M. Oba, T. Champion, C. Villaverde, T. H. A. Vendramini, J. C. C. Balieiro, A. C. Carciofi, and A. A. Camacho. 2018. "Weight loss improves arterial blood gases and respiratory parameters in obese dogs." *J Anim Physiol Anim Nutr (Berl)* 102 (6):1743–8. doi:10.1111/jpn.12963.

Porsani, M. Y. H., F. A. Teixeira, A. R. Amaral, V. Pedrinelli, V. Vasques, A. G. de Oliveira, T. H. A. Vendramini, and M. A. Brunetto. 2019. "Factors associated with failure of dog's weight loss programmes." *Vet Med Sci.* doi:10.1002/vms3.229.

Prebble, J. L., D. J. Shaw, and A. L. Meredith. 2015. "Bodyweight and body condition score in rabbits on four different feeding regimes." *J Small Anim Pract* 56 (3):207–12. doi:10.1111/jsap.12301.

Raffan, E. 2013. "The big problem: Battling companion animal obesity." *Vet Rec* 173 (12):287–91. doi:10.1136/vr.f5815.

Raffan, E., R. J. Dennis, C. J. O'Donovan, J. M. Becker, R. A. Scott, S. P. Smith, D. J. Withers, C. J. Wood, E. Conci, D. N. Clements, K. M. Summers, A. J. German, C. S. Mellersh, M. L. Arendt, V. P. Iyemere, E. Withers, J. Söder, S. Wernersson, G. Andersson, K. Lindblad-Toh, G. S. Yeo, and S. O'Rahilly. 2016. "A deletion in the canine POMC gene is associated with weight and appetite in obesity-prone labrador retriever dogs." *Cell Metab* 23 (5):893–900. doi:10.1016/j.cmet.2016.04.012.

Raffan, E., S. P. Smith, S. O'Rahilly, and J. Wardle. 2015. "Development, factor structure and application of the Dog Obesity Risk and Appetite (DORA) questionnaire." *PeerJ* 3:e1278. doi:10.7717/peerj.1278.

Robertson, I. D. 2003. "The association of exercise, diet and other factors with owner-perceived obesity in privately owned dogs from metropolitan Perth, WA. *Prev Vet Med* 58 (1–2):75–83.

Rohlf, V. I., S. Toukhsati, G. J. Coleman, and P. C. Bennett. 2010. "Dog obesity: Can dog caregivers' (owners') feeding and exercise intentions and behaviors be predicted from attitudes?" *J Appl Anim Welf Sci* 13 (3):213–36. doi:10.1080/10888705.2010.483871.

Ruchti, S., A. R. Meier, H. Wurbel, G. Kratzer, S. G. Gebhardt-Henrich, and S. Hartnack. 2018. "Pododermatitis in group housed rabbit does in Switzerland-Prevalence, severity and risk factors." *Prev Vet Med* 158:114–21. doi:10.1016/j.prevetmed.2018.06.011.

Sallander, M., M. Hagberg, A. Hedhammar, M. Rundgren, and J. E. Lindberg. 2010. "Energy-intake and activity risk factors for owner-perceived obesity in a defined population of Swedish dogs." *Prev Vet Med* 96 (1–2):132–41. doi:10.1016/j.prevetmed.2010.05.004.

Salt, C., P. J. Morris, D. Wilson, E. M. Lund, and A. J. German. 2019. "Association between life span and body condition in neutered client-owned dogs." *J Vet Intern Med* 33 (1):89–99. doi:10.1111/jvim.15367.

Speakman, J. R., D. A. Levitsky, D. B. Allison, M. S. Bray, J. M. de Castro, D. J. Clegg, J. C. Clapham, A. G. Dulloo, L. Gruer, S. Haw, J. Hebebrand, M. M. Hetherington, S. Higgs, S. A. Jebb, R. J. Loos, S. Luckman, A. Luke, V. Mohammed-Ali, S. O'Rahilly, M. Pereira, L. Perusse, T. N. Robinson, B. Rolls, M. E. Symonds, and M. S. Westerterp-Plantenga. 2011. "Set points, settling points and some alternative models: Theoretical options to understand how genes and environments combine to regulate body adiposity." *Dis Model Mech* 4 (6):733–45. doi:10.1242/dmm.008698.

Tack, J., and J. E. Pandolfino. 2018. "Pathophysiology of gastroesophageal reflux disease." *Gastroenterology* 154 (2):277–88. doi:10.1053/j.gastro.2017.09.047.

Tvarijonaviciute, A., R. Baric-Rafaj, A. Horvatic, A. Munoz-Prieto, N. Guillemin, E. Lamy, A. Tumpa, J. J. Ceron, S. Martinez-Subiela, and V. Mrljak. 2019. "Identification of changes in serum analytes and possible metabolic pathways associated with canine obesity-related metabolic dysfunction." *Vet J* 244:51–9. doi:10.1016/j.tvjl.2018.12.006.

Tvarijonaviciute, A., J. J. Ceron, S. L. Holden, D. J. Cuthbertson, V. Biourge, P. J. Morris, and A. J. German. 2012. "Obesity-related metabolic dysfunction in dogs: A comparison with human metabolic syndrome." *BMC Vet Res* 8:147. doi:10.1186/1746-6148-8-147.

van der Klaauw, A. A., and I. Sadaf Farooqi. 2015. "The hunger genes: Pathways to obesity." *Cell* 161 (1):119–32. doi:10.1016/j.cell.2015.03.008.

Virtue, S., and A. Vidal-Puig. 2010. "Adipose tissue expandability, lipotoxicity and the Metabolic Syndrome—an allostatic perspective." *Biochim Biophys Acta* 1801 (3):338–49. doi:10.1016/j.bbalip.2009.12.006.

Vitger, A. D., B. M. Stallknecht, D. H. Nielsen, and C. R. Bjornvad. 2016. "Integration of a physical training program in a weight loss plan for overweight pet dogs." *J Am Vet Med Assoc* 248 (2):174–82. doi:10.2460/javma.248.2.174.

Webb, T. L., M. Krasuska, Z. Toth, H. R. du Plessis, and L. Colliard. 2018. "Using research on self-regulation to understand and tackle the challenges that owners face helping their (overweight) dogs lose weight." *Prev Vet Med* 159:227–31. doi:10.1016/j.prevetmed.2018.08.017.

Weber, M., T. Bissot, E. Servet, R. Sergheraert, V. Biourge, and A. J. German. 2007. "A high-protein, high-fiber diet designed for weight loss improves satiety in dogs." *J Vet Intern Med* 21 (6):1203–8.

Weeth, L. P. 2016. "Other risks/possible benefits of obesity." *Vet Clin N Am Small Anim Pract* 46 (5):843–53. doi:10.1016/j.cvsm.2016.04.007.

Wejdmark, A. K., B. Bonnett, A. Hedhammar, and T. Fall. 2011. "Lifestyle risk factors for progesterone-related diabetes mellitus in elkhounds - a case-control study." *J Small Anim Pract* 52 (5):240–5. doi:10.1111/j.1748-5827.2011.01052.x.

Yaissle, J. E., C. Holloway, and C. A. Buffington. 2004. "Evaluation of owner education as a component of obesity treatment programs for dogs." *J Am Vet Med Assoc* 224 (12):1932–5. doi:10.2460/javma.2004.224.1932.

Yam, P. S., C. F. Butowski, J. L. Chitty, G. Naughton, M. L. Wiseman-Orr, T. Parkin, and J. Reid. 2016. "Impact of canine overweight and obesity on health-related quality of life." *Prev Vet Med* 127:64–9. doi:10.1016/j.prevetmed.2016.03.013.

18 Reproduction in Brachycephalic Companion Animal Species

Aoife Reid
Vets Now Ltd.

Laura Cuddy
Veterinary Specialists Ireland

Dan G. O'Neill
The Royal Veterinary College London

CONTENTS

Introduction...330
Fertility, Gestation and predicting parturition...331
Canine Artificial Insemination (AI) 331
Predicting Gestational Age and Planning Elective
Pre-parturient C-Section in Brachycephalic Bitches 331
Ovulation, LH Peak, Serum Progesterone Levels, Dioestrous and Body Temperature in Dogs.......332
Ultrasonography and Radiography for Gestational Aging in Dogs 333
Gestational Considerations for Pregnant Brachycephalic Dogs 334
Birth and C-section in the Brachycephalic Bitch...335
Assessment..335
C-Section: Indications...335
C-Section: Technique 336
    Anaesthesia 336
Standard Protocol..337
Surgical Preparation 338
Surgical Technique 339
Post-natal Care of the Brachycephalic Dam and Her Puppies...339
    The Brachycephalic Puppy 339
Neonatal Resuscitation..341
Common Canine Brachycephalic Congenital Malformations 342
Orofacial Clefts...342
Foetal Anasarca 343
The Brachycephalic Dam...344
The Brachycephalic Cat and Breeding 344
Fertility and Gestation...344
Birth 344
Postnatal Care of the Brachycephalic Kitten 345
Conclusions...346
References..346

INTRODUCTION

The popularity of some brachycephalic dog breeds has soared worldwide over the last decade (Maclennan & Smith 2019, Teng et al., 2016, Packer et al., 2019), despite many reported predispositions to disorders intrinsically related to their conformation. Linked to this popularity, canine fertility clinics offering assisted breeding techniques, such as artificial insemination (AI) and elective pre-parturient caesarean section (C-section), are becoming more common (Loeb, 2020). This chapter discusses some reproduction principles with special relevance for brachycephalic companion animals, including fertility, gestation, birth and postnatal care. Managing breeding in brachycephalic companion animals can be both clinically and professionally challenging for veterinary teams, and also high stakes for owners of much-loved pets or valuable genetic lines (Smith, 2007).

Several common brachycephalic dog breeds including French Bulldogs, Boston Terriers and Pugs are predisposed to dystocia (O'Neill et al., 2017). The typical shapes of brachycephalic breeds, including their distinctive flat-faced, large cranium, combined with a narrow maternal pelvis, create a cephalopelvic disproportion which promotes dystocia (Packer, Kennel Club). Mortality rates following dystocia of >20% for puppies and >1% for dams are reported in the general population of dogs (O'Neill et al., 2017). This suggests that there is significant scope for better management of dystocia to improve welfare and clinical outcomes, especially in brachycephalic breeds predisposed to dystocia.

Reliance on AI and C-section to allow some brachycephalic breeds to reproduce (Bannasch et al., 2010) raises many welfare and ethical issues (Quartuccio et al., 2020). A European owner-reported study focusing on the English Bulldog recorded an AI rate of 74.4%, with a corresponding C-section rate of 94.8% (Wydooghe et al., 2013). Ethical breeding practices are being debated more generally by breeders, veterinary governing bodies, animal welfare legislation, veterinary organisations and veterinary professionals (Ravetz, 2017, Fawcett et al., 2019), especially in relation to brachycephalic breeds that routinely rely on veterinary intervention to facilitate breeding (Fawcett et al., 2019). In the UK, animal welfare regulations provide guidance for breeding dogs (DEFRA, 2020), as does The Kennel Club in the UK (Kennel Club, 2021), in an effort to discourage perpetuating breeding which routinely relies on C-section. The UK RCVS Code of Professional Conduct for veterinary surgeons releases veterinary surgeons from the professional obligation to maintain client confidentiality in order to report C-sections (RCVS, 2019). But many challenges relating to reproduction remain, especially because many breeding dogs do not fall under the umbrella of formal kennel clubs or government registration.

Brachycephalic breeds also feature highly among commonly registered pedigree cat breeds (Plitman et al., 2019). The Persian, and related breeds such as the Exotic Shorthair, are characterised by their more extreme brachycephalic faces (Bertolini et al., 2016, Lipinski et al., 2008). A questionnaire survey of cat breeders in the UK reported that brachycephalic cat types have substantially higher dystocia levels than mesocephalic cats (Gunn-Moore & Thrusfield, 1995). Some brachycephalic breeds, including Persian and Exotic Shorthair cats, have been reported with high stillbirth rates; the Persian breed had the highest kitten mortality rate in a large UK cat breeder-based questionnaire study (Sparkes et al., 2006).

Although rabbits are popular companion animals in Europe, the USA, parts of Australia and some Asian countries (De Mello, 2016), little is published about companion rabbit breeding in the UK or the USA (Gosling et al., 2018, De Mello, 2014). This paucity of evidence pertaining to pet rabbit breeding seems at odds with their position as the third most popular companion animal in the UK, following dogs and cats (PDSA, 2019). Although there are a few isolated single case reports of dystocia in rabbits within the literature (Dickie, 2011, Sankar et al., 2017, Singh et al., 2000), obstetrical problems are reportedly rare in rabbits (Harcourt-Brown, 2002, Quesenberry & Carpenter, 2011). There is no reported association in the literature between dystocia and brachycephalism in companion rabbits at the time of writing.

FERTILITY, GESTATION AND PREDICTING PARTURITION

The number and range of fertility services offered for brachycephalic dogs has risen in recent years, and now includes pregnancy diagnosis, semen collection, semen analysis, progesterone testing and artificial insemination. This increase has been linked with the rising levels of ownership of some popular brachycephalic breeds, in particular the English Bulldog (Loeb & Evans, 2020). In the UK, some fertility facilities are staffed with non-veterinary personnel, which has raised regulatory issues including uncertainty about which procedures should be carried out by veterinary professionals versus competent laypeople, and also releases these facilities from a requirement to comply with Royal College of Veterinary Surgeons guidelines (Loeb, 2020). Rules and legislation can vary between countries, so owners should familiarise themselves with relevant regulations in their location to avoid contravening the law.

CANINE ARTIFICIAL INSEMINATION (AI)

AI is an assisted reproductive technique that may be used because of inability to attain a natural mating, on owner request, or excessive geographical distance between the dam and sire (Farstad, 2010, Mason, 2018, Smith-Carr, 2006). Data from the UK Kennel Club reports increasing numbers of puppies being born as a result of AI (Loeb, 2020). An increase in canine fertility services has been linked with the increased ownership of some brachycephalic breeds, in particular the English Bulldog, and may be heavily contingent on the frequent use of AI in that breed (Loeb & Evans, 2020). One owner-reported study on the English Bulldog reported an AI rate of 74.4% (Wydooghe et al., 2013). Reliance on AI has been linked to heritable anatomical traits of both dams and sires that preclude natural mating, such as an overloaded front end in combination with a smaller pelvis, along with respiratory compromise associated with a prolonged 'tie' phase during natural breeding (Gaytán et al., 2020, Heise, 2012, Wydooghe et al., 2013).

Poor timing of insemination is reportedly the most common cause of pregnancy failure following AI in the bitch (Mason, 2018). Appropriate use of AI requires good understanding of the physiology and anatomy of the bitch, coupled with careful semen collection and management. There are two general methods adopted for non-surgical insemination and two methods for surgical insemination. The first non-surgical insemination method is intravaginal, where semen is placed directly into the vaginal canal; this may be undertaken (and it commonly is) by competent laypersons, depending on local governing veterinary regulations. The second non-surgical method is transcervical, using a catheter and/or an endoscope; this generally may be undertaken only by a veterinary surgeon. Although transcervical insemination can be performed without sedation, light sedation may sometimes be required (Farstad, 2010). Surgical methods of insemination include via laparoscopy or surgical laparotomy. Surgical methods require general anaesthesia and are prohibited in some countries, such as the UK (RCVS, 2019).

The rules on registration of puppies produced by AI differ from country to country. Registration by the UK Kennel Club requires that the AI must be performed using a non-surgical method. (The Kennel Club, 2021) however requirements from other kennel clubs may vary.

PREDICTING GESTATIONAL AGE AND PLANNING ELECTIVE PRE-PARTURIENT C-SECTION IN BRACHYCEPHALIC BITCHES

Gestation describes the period from conception to birth (parturition). Accurate prediction of parturition dates is especially important in brachycephalic breeds because it facilitates better management of the birthing process, reduces neonatal mortality and is particularly important when an elective pre-parturient C-section is planned.

Bitches of many brachycephalic breeds are predisposed to dystocia: French Bulldogs had 16 times the odds of dystocia compared with crossbred bitches in the UK (O'Neill et al., 2017). Many dystocic bitches will require surgical intervention; 87% of dystocic English Bulldogs underwent C-section in the UK (O'Neill et al., 2019). Several brachycephalic breeds including the Boston Terrier, English Bulldog and French Bulldog have C-section rates reported in excess of 80% (Evans and Adams, 2010). To avoid the risk of an emergency C-section, the pet owner in collaboration with veterinary professionals may choose an elective pre-parturient C-section performed at a convenient time (Ryan & Wagner, 2006). An elective pre-parturient C-section may be requested or advised for reasons including previous dystocia, a breed or individual bitch at known high risk of dystocia, the value of the dam and litter, and differing ease of access to veterinary services to manage the elective versus emergency procedures (Ryan &Wagner, 2006). Elective pre-parturient C-section is considered a safe, effective and often justified intervention in the bitch (De Cramer & Nothling, 2019). Criteria for optimal timing of a pre-parturient C-section require being early enough to prevent associated risks of dystocia via spontaneous labour, but late enough to allow sufficient time for foetal maturation in order to promote long-term survival of the puppies. Optimal timing should minimise general anaesthetic and surgical risks by enabling appropriate fasting, pre-anaesthetic examination and diagnostics for the dam, and access to adequate resources for the veterinary team.

Methods used to predict optimal timing for elective C-section include direct and indirect determinations of ovulation at breeding, and ultrasonographic measurements of extra-foetal and foetal structures. Combining different methods can increase the prediction accuracy of any one method (Beccaglia et al., 2016). The most reliable methods available to determine foetal readiness for birth are detection of the luteinising hormone (LH) surge, ovulation timing and detection of day 1 of dioestrous (Lopate, 2008).

An intermediate approach is to allow an initial attempt at spontaneous unassisted parturition, although this can be problematic as it can result in foetal distress and/or losses depending on the circumstances. One North American study reported that the likelihood of having all puppies alive at birth following an emergency C-section is much lower (one-third as likely) compared with an elective pre-parturient C-section performed just after the drop of progesterone (Moon et al., 2000). See Table 18.3 for a summary of the tools that can be used to determine optimal timing for elective pre-parturient C-section. Below is a list of the parameters that can help determine optimal timing.

OVULATION, LH PEAK, SERUM PROGESTERONE LEVELS, DIOESTROUS AND BODY TEMPERATURE IN DOGS

The gestation period in dogs lasts on average 65 days but may range between 57 and 72 days from known mating dates. Parturition usually occurs 65 ± 1 day after the LH peak and 63 ± 1 day after ovulation (Concannon et al., 1983). However, detection of the LH peak and ultrasonographic identification of ovulation necessitates daily determinations that may prove impractical for many pet owners. The LH peak is the most accurate means of identifying ovulation timing; ovulation usually occurs 2 days afterwards. This gives flexibility to plan the best date for mating, as well as determine the expected date of whelping if the mating is successful. However, the LH peak may occur within a 24-hour period, so daily serum samples are drawn from the 4th or 5th day of proestrous, or when vaginal cytology approaches 50% cornification. The LH measurement is usually interpreted in conjunction with serum progesterone levels because only qualitative LH tests are currently readily available and LH levels can vary at the peak of the LH surge (Mason, 2018).

Serum progesterone ranges of 2–3 ng/mL occur during the LH peak and 4–10 ng/mL occur during ovulation (Kutzler et al., 2003). Additionally, serum progesterone levels can be used to predict parturition; a drop below 2 ng/mL has been reported to signify impending parturition within 24 hours (Concannon et al., 1978). However, a more recent study reports that a single progesterone determination has low diagnostic value because of low test sensitivity, although it can be useful as a first screening (Rota et al., 2015). The use of aglepristone, a competitive progesterone antagonist, as a primer prior to planned C-section, has been described in bitches in which progesterone has not yet dropped, but are 60–62 days post-ovulation, without increased neonatal mortality (Roos et al., 2018). Additionally, another recent study reported it was not a prerequisite for the level of serum progesterone to decrease below 2 ng/mL to obtain high survival ratios at birth (De Cramer & Nothling, 2019). Other notable exceptions to the decrease in serum progesterone levels to predict impending parturition include singleton pregnancies or pregnancies where all the foetuses have already died (Johnson, 2008). Absence of signs of parturition in conjunction with a low progesterone level should prompt a more comprehensive evaluation of the dam, including ultrasonography to assess for foetal distress.

Although it has been suggested that elective pre-parturient C-sections can be scheduled and performed from 57 days after onset of cytological dioestrus with documented puppy survival rates of 99%, more research is required to establish if this is routinely safe for all breeds (De Cramer & Nothling, 2020).

A drop in rectal temperature has also been associated with decreasing serum progesterone levels and can be a useful, indirect marker for breeders and veterinary professionals of impending parturition (Geiser et al., 2014, Johnston et al., 2001). Temperature generally drops by at least 1°C within 12–24 hours prior to parturition. However, because rectal temperatures in late pregnancy tend to be around 38°C, it is advisable to measure temperatures several times daily as the expected whelping date draws close, to detect a significant decrease beyond what is considered 'normal' for each individual bitch. Although a rectal temperature below 37.6°C has been cited as an indicator of impending parturition within 12–24 hours (Linde-Forsberg & Eneroth, 2000), there is some conflicting evidence regarding the association between a decline in rectal temperature and impending parturition (Veronesi et al., 2002). For this reason, it may be prudent to use temperature in combination with other parameters when assessing a bitch for impending parturition.

ULTRASONOGRAPHY AND RADIOGRAPHY FOR GESTATIONAL AGING IN DOGS

Ultrasonographic evaluation of organ development and foetal biometry can be used to confirm pregnancy, estimate gestational age and ascertain foetal viability. When the date of the LH surge and/or ovulation is unknown, ultrasonographical gestational aging may not be sensitive or specific enough to safely schedule an elective pre-parturient surgery. Because the accuracy of information provided by ultrasound examination may be limited by the sonographer's level of experience and the resolution level of the ultrasound equipment used, this needs to be factored into clinical decision-making. However, foetal organ evaluation can provide information to help support decision-making for optimal surgical intervention with elective pre-parturient C-section.

Two small Brazilian studies reported a strong correlation between gestational age and ultrasonographic evaluation of foetal kidney length (Gil et al., 2018) and foetal intestinal development (Gil et al., 2015). Although both studies were small, they did include several brachycephalic breeds. However, until more information is available, neither evaluations should be used as a sole indicator for foetal maturity and subsequent elective pre-parturient C-section planning. See Table 18.1 for further information on gestational aging based on ultrasonographic detection of foetal features.

Other foetal parameters that can be measured and inputted into a formula to predict gestational age include crown-to-rump length (CRL) measurement and gestational sac diameter (GSD); these can be useful up to day 40 after the LH surge to estimate gestational age. However, beyond day 40, use of the biparietal diameter (BPD) from multiple foetuses is reported as more accurate (Lopate, 2008). The BPD is the distance between the parietal bones. However, recently reported variation in BPD within and among litters has questioned its reliability to determine readiness for C-section (De Cramer & Nothling, 2018). Additionally, there is limited evidence on BPD reference values for specific brachycephalic breeds (Borella et al., 2019, Almeida et al., 2003, Camargo, 2012, Feliciano et al., 2015). It has been reported that care should be taken when BPD measurements are performed in the last week of pregnancy, or with small or large litters, because the variability among littermates and within litters is so large that it is unsuitable to predict foetal readiness for delivery. However, a Spanish study focusing exclusively on two brachycephalic breeds reported that elective

TABLE 18.1

Gestational Aging Based on Ultrasonographic Detection of Canine Foetal Features

Foetal Feature	First Detection (Days) After LH Surge
Heartbeat	23–25
Skeleton	33–39
Bladder	35–39
Stomach	36–39
Lung hyperechoic compared to liver	38–42
Liver hyperechoic compared to abdomen	39–47
Kidney	39–47
Eyes	39–47
Intestine	57–63
Intestinal peristalsis	62–64

Source: Data from Lopate, C. Estimation of gestational age and assessment of canine fetal maturation using radiology and ultrasonography: A review. *Theriogenology* 70, no. 3 (2008): 397–402.

TABLE 18.2

Approximate Radiographic Canine Gestational Aging Using Traditional Radiography

Feature	Detection (Days) After LH Surge	
	Mean	Range
Spherical uterine swellings	35	31–38
Ovoid uterine swellings	41	38–44
First evidence of skull mineralisation	45	43–46
Scapula, humerus and femur mineralised	48	46–51
Radius, ulna and tibia mineralised	52	50–53
Pelvis and all ribs mineralised	54	53–59
Coccygeal vertebrae, fibula, calcaneus and distal extremities mineralised	61	55–64
Teeth detectable	61	58–63

Source: Data from Lopate, C. Estimation of gestational age and assessment of canine fetal maturation using radiology and ultrasonography: A review. *Theriogenology* 70, no. 3 (2008): 397–402.

pre-parturient C-sections can be scheduled safely when the foetal BPD reaches a minimum value of 27.0 and 29.5 mm for French and English Bulldog breeds, respectively (Batista et al., 2014).

Other diagnostic imaging methods such as radiography are inadequate to determine foetal maturity but may provide an approximate estimate of gestational age (see Table 18.2).

Radiography offers little assistance to plan the timings of elective pre-parturient C-section. The foetal skeletal is not distinct until days 44–45 after the LH surge. Foetal mineralisation may be completed as early as day 58 after the LH surge, but survival of neonates delivered at this stage via C-section would be unlikely (Concannon, 2000, Rendano 1983, Rendano et al., 1984).

GESTATIONAL CONSIDERATIONS FOR PREGNANT BRACHYCEPHALIC DOGS

The main gestational disorders described across all pregnant bitches include metabolic diseases (e.g. diabetes mellitus, gestational toxaemia), infections (brucellosis and herpes virus infection) and non-infectious disturbances such as trauma, genetic and foetal congenital disorders, inadequate nutrition, hypothyroidism and hypoluteoidism (Kustritz, 2006). A fuller description of each of these disorders is beyond the scope of this chapter.

However, it is worth noting that brachycephalic bitches have certain pathological predispositions which may be exacerbated by physiological changes associated with pregnancy. Many brachycephalic breeds are affected by the progressive and lifelong respiratory effects of brachycephalic upper airway syndrome (BOAS) (Packer & Tivers, 2015). These effects may be compounded by physiological changes associated with late pregnancy, particularly an enlarged uterus causing cranial displacement of the diaphragm that further decreases functional residual capacity (FRC) and total lung volume (TLV), thus lowering oxygen reserve capacity (Pascoe & Moon, 2001).

Aspiration pneumonia, often associated with BOAS, is commonly reported in brachycephalic breeds (Meola, 2013). Pregnancy increases the risk of aspiration pneumonia because delayed gastric emptying, decreased oesophageal sphincter tone and increased gastrin levels all promote vomiting and regurgitation (Pascoe & Moon, 2001). The combination of these anatomical, physiological and pathological changes may potentially be associated with an increased risk of respiratory compromise in late pregnancy and during parturition of brachycephalic breeds; however, there is a paucity of evidence on the true extent of these (Harðardóttir et al., 2019).

BIRTH AND C-SECTION IN THE BRACHYCEPHALIC BITCH

Whelping (parturition) is a critical time for the brachycephalic dam and puppies. Dystocia leads to maternal and neonatal losses, and the majority of dystocia cases lead to C-section (Bergström et al., 2006, Münnich & Küchenmeister, 2009). French Bulldogs are 15.9 times more likely to have dystocia and are 2.4 times more likely to require C-section than mixed breeds (O'Neill et al., 2017). Medical management of dystocia may be considered, if appropriate. This would ideally require diagnostic imaging to exclude obstruction, and bloodwork to screen for metabolic factors that may require correction. This chapter will not discuss medical management of dystocia and will instead focus on surgical management, given the predilection of several brachycephalic breeds to surgical dystocia.

ASSESSMENT

Clinical examination of the dam should include a full physical examination with emphasis on the major body systems as well as the urogenital systems. This should include a sterile digital vaginal examination to assess the size of the pelvic canal and for the presence of a foetus in the birth canal. If dystocia is suspected, radiographs should be performed to assess for maternal pelvic obstruction, and number of foetuses. Foetal death may be identified by the presence of foetal intravascular gas or intrauterine free gas, depending on the duration of the post-mortem period (Barr, 1988); however, ultrasound is a more useful tool to assess foetal viability, and hypoxaemic distress which manifests as foetal heart rate deceleration (Gil et al., 2014). If dystocia is suspected, bloodwork to assess the dam's status is recommended, as many dystocia cases may present dehydrated, fatigued or in weak condition; therefore, a basic database including PCV/refractometric total solids, glucose, ionised calcium (and potentially other electrolytes) as well as venous blood gas analysis is advised. A full panel including biochemistry may be warranted for dams that present systemically unwell.

C-SECTION: INDICATIONS

Dystocia may result from maternal or foetal causes. Although primary uterine inertia is the most common cause in non-brachycephalic breeds, a small pelvic canal relative to the foetal head is not uncommon in brachycephalic breeds (Runcan and da Silva, 2018). Reliable estimates of gestational

TABLE 18.3

Tools for Timing of Elective Pre-parturient C-section

An elective pre-parturient C-section may be indicated if any of the following are present:

- High risk brachycephalic breed
- Nulliparous bitch ≥6 years of age
- Litter size ≤2 or ≥8
- Previous history of dystocia
- Owner lives in an area with limited access to emergency veterinary care

Use as many tools below as possible to increase the accuracy of prediction

Maternal Assessment Tools	Foetal Assessment Tools

Early term
Hormonal assays
• LH
• Progesterone
Vaginal cytology
Mating dates
	Early term (<day 40)
	Ultrasonography[a]
	• Approximate gestational aging through:
	  • Detection of foetal features
	  • CRL measurement
	  • GSD measurement
	Radiography
	• Approximate gestational aging through skeletal structures visible
	Late term (>day 40)
	Radiography
	• Approximate gestational aging through skeletal structures visible (up to day 58)
	Ultrasonography[a]
	• Approximate gestational aging through BPD

Impending parturition **Hormonal assays** • Progesterone **Bitch physical examination** • Rectal temperature 37.6 °C (99.7 °F) or lower • Signs of first-stage labour	**Impending parturition** • **Ultrasonography**[a] • Foetal heart rate deceleration (foetal distress) • Foetal intestinal peristalsis • Foetal renal cortex observation

[a] Machine and operator dependent.

timing are important in these high-risk pregnancies to establish a due date that helps to recognise dystocia or to plan for elective pre-parturient C-section (Table 18.3).

Emergency C-section should be considered in cases of prolonged gestation, failure to progress from stage 1 to stage 2 labour within 12–24 hours, >4 hours between puppies or signs of toxaemia. Puppy and maternal mortality have been reported to be higher in emergency C-section compared with elective pre-parturient C-section (Moon et al., 2000). Bitches undergoing an emergency C-section may require preoperative stabilisation prior to induction of anesthesia. See Table 18.4 for some common indications for emergency surgical intervention in a dystocia case.

C-SECTION: TECHNIQUE

ANAESTHESIA

Key goals of anaesthesia for C-section are to provide analgesia and minimise morbidity to the dam, while reducing the effect of anaesthesia on neonates. A pre-anaesthetic checklist should be completed for each patient to open lines of communication among the team regarding goals, specific

TABLE 18.4

Common Indications for Emergency Surgical Intervention in a Dystocia Case

Emergency C-section is typically indicated by any of the following:

Maternal indications:
- Failure of medical management for non-obstructive uterine inertia
- Intense non-productive contractions for >30 minutes with no expulsion of foetus
- Failure of onset of Stage 2 labour (primary uterine inertia)
- Prolonged Stage 2 labour with >4 foetuses in absence of obstruction
- Systemic maternal illness
- Suspicion of uterine torsion, rupture, prolapse or herniation
- Gross abnormalities of the maternal pelvis (e.g. fractures)
- Uteroverdin (greenish/black vulval discharge) >30 minutes **prior** to the first foetal delivery[a]

Foetal indications:
- Foeto-maternal disproportion
- Evidence of marked foetal distress (FHR < 150/minute or FHR <180 if more than one foetus remaining in-utero)
- Foetal death/putrefaction
- Foetal malposition that cannot be corrected

[a] If the goal is to have as many live puppies as possible.

---

patient risks and plans in advance. C-section carries increased anaesthetic risk due to pregnancy - associated physiologic changes (Hay Kraus, 2016). Overall mortality rates of the dams of up to 3% have been reported for C-section (Conze et al., 2020, Doebeli et al., 2013. Metcalfe et al., 2014).

There are various anaesthetic protocols described for C-section in brachycephalic bitches (Batista et al., 2014, Vilar et al., 2018). Most drugs used in traditional anaesthetic protocols readily cross the placenta and will affect the foetus (Pascoe and Moon, 2001). Puppy mortality rates are reported to be lower with protocols that utilise propofol and isoflurane (Funkquist et al., 1997, Moon et al., 2000). The use of regional anaesthesia via epidural administration of a local anaesthetic agent has been described (Luna et al., 2004). Although theoretically local anaesthesia offers advantages, when used alone in a brachycephalic dam, the airway would be unprotected in a patient at high risk for regurgitation and aspiration and is therefore contraindicated. Furthermore, the use of local anaesthesia has not demonstrated improved puppy survival or vigour (Cruz et al., 1997).

Label doses and market authorisation for specific drugs vary between countries, so veterinary practitioners are advised to be familiar with the prescribing cascade and recommendations in their location. Gaining explicit informed consent prior to off label administration from the pet owner may be necessary in some countries (RCVS, 2020).

No single anaesthetic protocol is suitable for every C-section. Protocols should be tailored to each patient, considering various factors (i.e., condition of the dam and foetuses, experience and familiarity with various anaesthetic agents and techniques, available resources and facilities, and emergency versus elective presentation) (Ryan & Wagner, 2006a).

STANDARD PROTOCOL

1. **Premedication:** The need for premedication depends on the status and temperament of the dam. The administration of opioids prior to, or at the time, of induction reduces the amount of induction and inhalant agent required. If administered prior to delivery, the administration of opioids to the dam will result in placental transfer to the foetuses, but the amount differs depending on the specific drug. Opioids may be withheld until after delivery of neonates but should be given prior to dam recovery. Maropitant is helpful to try to limit

regurgitation and aspiration in brachycephalic breeds. Maropitant (1 mg/kg SC) can be administered 30 minutes prior to opioid administration, or diluted and administered IV over 5 minutes if urgent (Robertson, 2016).

Other drugs routinely used in anaesthetic protocols should be avoided. Low dose acepromazine can be used if the dam is very anxious or stressed, but should not be used routinely as it cannot be reversed and has a long duration of action.

2. **Ocular protection**: Brachycephalic breeds are predisposed to corneal ulcerative disease (O'Neill et al., 2017), which may be further exacerbated during anaesthesia. Perioperative eye protection strategies and postoperative ophthalmic examination are advised to reduce the occurrence of corneal ulcers (Downing et al., 2018, Park et al., 2013).

3. **Antibiosis:** An uncomplicated caesarean section is classified as a 'clean-contaminated' procedure and perioperative antibiosis should be administered prior to induction (Goericke-Pesch et al., 2018).

4. **Fluid therapy:** Intravenous crystalloid fluid therapy should be initiated preoperatively and maintained to support maternal blood pressure throughout surgery.

5. **Preoxygenation:** The dam should be provided with at least 3–5 minutes of oxygen (100 mL/kg/min) by face mask prior to induction to optimise oxygenation during the induction process (McNally et al., 2009, Ryan & Wagner, 2006)). This will permit preservation of adequate blood oxygenation in the event of difficult induction and intubation, or post-induction apnoea.

6. **Induction:** Rapid induction with intravenous propofol or alfaxalone is used widely. Propofol is rapidly metabolised but can cause cardiopulmonary depression in puppies. Although there is no reported difference in mortality, alfaxalone may offer advantages as an induction agent, with higher Agpar (puppy vigour) scores at 5, 15 and 60 minutes (Doebeli et al., 2013). The dam's airway should be established and protected as rapidly as possible with an appropriately inflated endotracheal tube cuff. Oesophageal suction and lavage should be performed in cases where passive regurgitation occurs.

7. **Maintenance**: The dam should be maintained with isoflurane or sevoflurane titrated to the lowest effective level. Electrocardiogram, blood pressure, pulse oximetry and end-tidal CO_2 are monitored closely intraoperatively. Atropine can be administered if the foetal heart rates are low; glycopyrrolate should not be used to increase foetal heart rate as it does not cross the placental barrier. It is vital to maintain maternal systemic blood pressure, as there is a direct correlation between uterine blood flow and systemic blood pressure. Therefore, decreased systemic blood pressure will indirectly result in reduced foetal oxygen delivery. Intraoperative haemorrhage should initially be treated with crystalloids, or colloids where necessary (Kraus, 2016). Hypotension that is not responsive to fluid therapy can be treated with an ephedrine bolus (0.03–0.1 mg/kg IV) to increase blood pressure while simultaneously maintaining uterine blood flow (Kraus, 2016).

SURGICAL PREPARATION

Although there is no strong evidence that prolonged anaesthesia or delivery time affects outcomes, a swift induction to delivery time is preferred. Good preparation and a team effort is essential to achieve this goal. A full abdominal clip for exploratory laparotomy should be performed from xiphoid to pubis and lateral to the mammary glands, ideally with the patient awake and standing if tolerated without undue stress. Although supine hypotension has not been demonstrated in pregnant dogs and cats, it is prudent to keep the dam in lateral recumbency until surgery is ready to begin (Robertson, 2016). An initial sterile skin preparation can be performed prior to induction if the dam is amenable and not overly stressed. The dam should be moved to the operating room and induced when the team is ready, and a surgeon already scrubbed in. It can be helpful to tilt the operating table upwards 10°–15° at the cranial end to relieve pressure from the diaphragm and improve respiration.

SURGICAL TECHNIQUE

The entire abdomen should be draped from xiphoid to pubis, and the same distance laterally again from midline to the mammary glands. A ventral midline coeliotomy should be performed from halfway between xiphoid and umbilicus down to the pubis. Care should be taken when incising the stretched and thin linea alba to avoid unintentional uterine incision. Adhesions may be present if previous C-section has been performed. The uterus should be exteriorised and isolated with moistened laparotomy sponges. A ventral midline incision is made in the body of the uterus, and the most caudal foetus is removed first, prior to milking foetuses in each horn in alternating order caudally towards the incision. The placenta should release from the attachment zone and can be left to pass naturally, if the cervix is open at the time of surgery. However, if the cervix is not open, the membranes are removed gently and slowly with uterine massage to prevent haemorrhage (Traas, 2008b). The foetus should be handed off in a sterile towel to a non-sterile assistant for care. The umbilical cord is clamped 2–6 cm from the puppy's abdominal wall to allow access to the umbilical vein if needed for resuscitation. The whole uterus should be palpated for contents, and an assistant should examine the vaginal vault for the presence of a foetus. The hysterotomy should be closed with 3-0/4-0 monofilament absorbable suture on taper needle, in either a single layer continuous and or in a 2-layer closure. A brief lavage of the abdominal cavity can be performed prior to routine closure. Ideally, intradermal sutures are used to avoid puppy confusion with external sutures during nursing. Local anaesthetic can be infiltrated into the area surrounding the incision site, prior to incision (lidocaine) and at closure (bupivacaine 1.5–2 mg/kg). Oxytocin IM or IV (2 IU/kg, maximum 20 IU) may be administered to promote uterine involution if there has been excess bleeding or the placentae could not be removed (Ryan & Wagner, 2006b).

With client consent, ovariohysterectomy may be elected for in conjunction with C-section. This can be performed by either en bloc resection of the uterus with foetuses, or a standard ovariohysterectomy after the puppies have been removed. En bloc resection requires a larger support team as all puppies are simultaneously retrieved, rather than the team being handed one neonate at a time (Robertson, 2016). In a UK retrospective general population-based study, 31% of dams had ovariohysterectomy at the time of C-section (O'Neill et al., 2019).

As soon as the dam has recovered sufficiently from the procedure, the neonates should be placed with her to nurse and allow bonding, and for thermoregulation. Typically, if healthy, the dam and neonates are discharged to their owner's care as early as possible. Many commonly used analgesics, including opioids and non-steroidal anti-inflammatories, are poorly excreted in milk; however, it is prudent to be cognisant of medicine label contraindications for use in lactation, prior to prescription of any medicines.

C-section does not seem to have a high impact on fertility, with one study reporting a fertility rate of 100% after C-section (Conze et al., 2020).

POST-NATAL CARE OF THE BRACHYCEPHALIC DAM AND HER PUPPIES

THE BRACHYCEPHALIC PUPPY

Many brachycephalic puppies (>80% for English Bulldogs, Boston Terriers and French Bulldogs) are delivered surgically via elective or emergency C-section (Evans & Adams, 2010). Several studies have reported reduced survival and vitality of neonates delivered via C-section from maternal brachycephalic breeds (Moon et al., 2000, Wydooghe et al., 2013). Therefore, high standards of clinical evaluation and management of brachycephalic neonates in the immediate post-natal period are critical. The Apgar (Appearance, Pulse, Grimace, Activity and Respiration) scoring system is a rapid, reliab2le and easy method for evaluating both human and animal neonates (Veronesi et al., 2009, Fusi et al., 2020). The canine Apgar scoring system provides objective measures of viability from combined status scores (0–2) for five parameters: mucous membrane colour, heart rate, reflex irritability, motility and respiratory efforts (Veronesi, 2016). This facilitates the identification and development of effective strategies for resuscitation of neonates after either C-section or dystocia (Veronesi et al., 2009). Studies of canine Apgar systems in Spain (Batista et al., 2014) and Italy (Castagnetti et al., 2017) evaluating prognostic parameters collected at birth for brachycephalic neonates born via vaginal or C-section identified significantly higher Apgar scores in surviving neonates compared with non-surviving neonates. The mean initial Apgar score of surviving neonates was 5.0±2.0 compared to 2.5±1.0 for non-surviving neonates. Neonates scoring above four at birth had survival rates approaching 100%. Routine use of canine Apgar scores for brachycephalic neonates, regardless of route of delivery, offers a simple, rapid tool for veterinary professionals to detect neonatal distress and decreased viability sooner and more accurately (see Table 18.5 for details). Table 18.6 interprets puppy viability and offers some suggested actions based on Apgar total scores.

Several other markers may also help evaluate puppy prognosis in the first 24 hours of life via blood and amniotic fluid. Lactate, in association with the assessment of Apgar score at birth, can usefully identify the most severely distressed brachycephalic neonates, in order to facilitate receipt of appropriate support (Castagnetti et al., 2017). That study used handheld point-of-care monitors on blood obtained from puncture of the neonate's footpad with a 30-gauge needle. Neonatal lactate levels may typically be higher following vaginal births than C-section. However, failure of lactate levels to normalise in the first 24 hours, in combination with a low Apgar score, is a concerning prognostic sign.

TABLE 18.5
Apgar Score for Canine Newborn Viability Evaluation

| | Score | | |
Parameter	0	1	2
Mucous colour	Cyanotic	Pale	Pink
Heart rate (bpm)	<180	180–220	>220
Reflex irritability	Absent	Grimace	Vigorous
Motility	Flaccid	Some flexions	Active motion
Respiratory efforts[a]	No crying/<6bpm	Mild crying/6–15 bpm	Clear crying/>15 bpm

[a] Vocalisation and respiratory rate; bpm=breaths per minute.

Source: Data from Veronesi et al. An Apgar score system for routine assessment of newborn puppy viability and short-term survival prognosis. *Theriogenology*, 72 (2009): 401–407.

TABLE 18.6
Canine Apgar Total Scores for Evaluation of Action Needed for Canine Neonates

Canine Apgar rating total score	Viability	Action suggested
Ratings of 0–3	Severely distressed or critical neonates	Immediate resuscitation
Ratings of 4–6	Moderately distressed or moderate viability neonates	Immediate resuscitation
Ratings of 7–10	Normal viability neonates	Monitor closely to ensure • Continued stability • Normal temperature • Normal suckling to receive colostrum

NEONATAL RESUSCITATION

The vitality of puppies at delivery can range from alert (squeaking, wriggling, breathing) to limp, regardless of vaginal or caesarean delivery. For a basic neonatal resuscitation equipment list, see Table 18.7. Initial neonatal resuscitation should focus on the respiratory system, the cardiovascular system and thermoregulation, which are particularly critical for neonates born following dystocia or C-section, where higher neonatal mortality rates have been reported (Moon et al., 2000, Titkova et al., 2017). See Table 18.8 for a summary of the steps recommended for immediate resuscitation of neonates born following dystocia and/or C-section.

The neonatal chest is naturally compressed during vaginal delivery, expelling fluid from the respiratory tract and stimulating the first breath through recoil of the chest wall upon birth. This compression/recoil does not occur with caesarean delivery, and so surgically delivered neonates should be rubbed vigorously to encourage respiration once their foetal membranes are removed. Ideally, there should be one assistant available for each neonate. Neonates should be assessed immediately on hand-off to a non-sterile assistant, and resuscitation started rapidly where necessary. The oral cavity and nostrils should be cleared of fluid manually or with a neonatal bulb syringe. Flow-by oxygen should be provided where needed. Respiratory effort can be stimulated further by insertion of a narrow gauge needle (25 g or smaller) into the Jen Chung acupuncture point at the nasal philtrum (Traas, 2008a). Traditional 'swinging' of neonates to remove fluid/mucus and encourage respiration is no longer recommended due to the risk of inducing significant brain trauma (Grundy et al., 2009). Doxapram has a short duration of action, decreased effect on the hypoxic brain and a tendency to decrease cerebral blood flow, so is no longer recommended as a respiratory stimulant for apnoeic, hypoxic neonates (Moon et al., 2001). Intubation can be performed carefully if needed using

---

TABLE 18.7

Basic Neonatal Resuscitation Equipment List

Neonatal resuscitation equipment list includes
- Supplemental oxygen source
- Small tight-fitting masks
- Heat source
- Clean warmed towels
- Neonatal bulb syringes
- Monitoring equipment
- Resuscitation drugs
- 1 mL insulin syringes/1 mL syringe and 25G needles

TABLE 18.8

Immediate Resuscitation Protocol for Neonates Born Following Dystocia or C-Section

- Remove foetal membranes and suction fluid from upper airway if necessary
- Rub vigorously with a clean, warm towel
  - To stimulate respiration
  - To dry off helping to maintain body temperature
- Inspect umbilicus and ligate umbilical cord >1cm from body wall
- Perform and record Apgar score
- Place in a warm environment if stable

a 14–18G intravenous catheter or a 2–3 O.D. endotracheal tube. The umbilicus should be examined for haemorrhage and ligated at least 1 cm from the body wall. Finally, good thermoregulation is critical for neonates to conserve energy and minimise oxygen demands in the early neonatal period. Neonates cannot shiver and therefore cannot maintain body temperature. As well as drying with towels, neonates should be placed in a warmed environment (ideally 30°C–32°C with humidity of 55%–65%) once they are breathing, vocalising and moving (Cavanagh, 2017). This is of particular importance for neonates born via C-section because the dam will not be immediately available to mother the neonates, and neonates can cool rapidly via evaporation and heat radiation after birth.

If there is a poor response to initial resuscitation, then supplemental oxygen, warming and mechanical stimulation 'as described above' take priority once the airways are cleared. Resuscitative drugs (atropine, adrenaline, lidocaine and naloxone) can be administered via sublingual, intraosseous, endotracheal or intravenous (umbilical vein) routes, depending on availability of access. Naloxone (0.002–0.02 mg/kg IV or 1–2 drops sublingual) may need to be administered to reverse opioids given to the dam. If a puppy is in asystole, epinephrine can be administered (0.1 ug/kg via umbilical vein) in conjunction with lateral thoracic compressions.

Each neonate's mouth should be carefully inspected for cleft lip, or cleft hard or soft palate, as this will impede nursing ability and often requires surgical correction. When identified, affected neonates often need assisted feeding until they are mature enough for surgical correction. Neonates should also have a general health check for congenital sternal malformation or hernias.

Mortality of 8%–15% has been reported in brachycephalic puppies during the first 24 hours after birth (Castagnetti et al., 2017, Batista et al., 2014, Wydooghe et al., 2013). Risk factors for perinatal death include the presence of another puppy that was stillborn at birth, primiparous bitches over 6 years old, emergency C-section, larger litter sizes and congenital malformations (Batista et al., 2014, Moon et al., 2000, Tønnessen et al., 2012, Veronesi et al., 2009). Trauma is a significant cause of mortality in the first 24 hours due to direct injuries from the dam or from the environment (Moon et al., 2001). For this reason, dams should be closely supervised post-whelping with their puppies until good mothering behaviour emerges. It is particularly important that bitches post-C-section are not left unsupervised with neonates until they have fully recovered from anaesthesia. Neonatal hypothermia (<34.4°C), bradycardia, periods of apnoea or weight loss should all raise concerns and trigger prompt intensive care and monitoring. Correction of hypothermia, dehydration and hypoglycaemia should be a priority, whenever present, along with aggressive supportive care.

COMMON CANINE BRACHYCEPHALIC CONGENITAL MALFORMATIONS

Congenital malformations are reported in up to 7%–8% of some brachycephalic breeds (Batista et al., 2014, Wydooghe et al., 2013). These two European studies identified the most common congenital defects as palatochisis (orofacial cleft) and foetal anasarca, with congenital malformations being more likely in larger litters over four puppies (Batista et al., 2014). Other abnormalities reported more frequently in brachycephalic breeds include Swimmer Puppy Syndrome (SPS), pectus excavatum and pulmonic stenosis (Ontiveros et al., 2019). SPS results in puppies lying in sternal recumbency, unable to walk (Suter, 1984). The cause of SPS is as yet unclear. Treatment is centred around muscle strengthening and physiotherapy (Verhoeven et al., 2006). SPS should not be confused with another condition seen more often in brachycephalic breeds, pectus excavatum (Boudrieau et al., 1990). In one study, English Bulldogs and French Bulldogs were 13 and 7 times, respectively, more likely to have pulmonic stenosis compared to a general population presenting to a university. Pulmonic stenosis appears to be autosomal recessive in English Bulldogs (Ontiveros et al., 2019).

OROFACIAL CLEFTS

Orofacial clefts include a range of anatomical and functional disorders where a fissure connects the oral and nasal cavities; these include cleft lip, failure of fusion of the maxillary prominence and the medial nasal process, and palatochisis (Lobodzinska et al., 2014). Brachycephalic breeds were over-represented among cases of orofacial clefts in an owner-reported American study (Roman et al., 2019). Brachycephalic dog breeds including Boston Terrier, Pekingese, Boxer, English Bulldog, French Bulldog and Cavalier King Charles Spaniel are predisposed to palatochisis (Nelson, 2003, Grellet, 2008, Gough and Thomas, 2010, Moura et al., 2012). Their typical broad head, that requires a greater distance to be bridged by the two palatal shelves during development, is reported to explain the predisposition to palatochisis in brachycephalic breeds (Warzee et al., 2001).

Orofacial clefts in puppies are characterised by a typical history of milk draining from the nares during or after feeding, gagging, coughing or sneezing while eating, and the clefts are usually confirmed by visual examination (Lobodzinska et al., 2014). Although it is possible to diagnose palatochisis by clinical examination alone, additional tests such as computed tomography can assist to diagnose concomitant disorders and their associated severity which may affect prognosis (Pankowski et al., 2018). Orofacial clefts have a significant impact on morbidity and mortality, as well as financial loss for breeders. Affected puppies should be eliminated from the breeding pool, and many cases result in euthanasia in the early neonatal period (Roman et al., 2019). Surgical repair may be possible and is usually undertaken once the puppy is ideally over 16 weeks old or has stopped growing, meaning that affected puppies need constant observation and care in the intervening period (Pankowski et al., 2018).

FOETAL ANASARCA

Foetal anasarca (also known as Hydrops foetalis) is characterised by an excessive accumulation of subcutaneous fluids, and varying amounts of fluid in other body cavities (see Figure 18.1). The condition has been described in the English Bulldog, Pug and French Bulldog (Cunto et al., 2015, Sridevi et al., 2016, Maronezi et al., 2018) and is attributed to cardiac malformations as well as a variety of other causes. The condition is heritable in the English Bulldog (Hopper et al., 2004), likely as a recessive trait (Zoldag et al., 2001). Anasarca may affect all or part of the litter and is associated with foetal dystocia and increased neonatal mortality rate in affected foetuses, called 'walrus puppies' or 'bulldog water puppies'. It is thought to develop in mid-late gestation and may be identified by ultrasonography, which improves pet owner counselling and management of high-risk

FIGURE 18.1 Foetal Anasarca in an English Bulldog puppy at birth. The subcutaneous accumulation of fluid gives a walrus-like appearance, hence the common description of 'walrus puppy'.

litters in the neonatal period (Hopper et al., 2004, Sridevi et al., 2016). It has been suggested that English Bulldog bloodlines with anasarcous pups should be identified and banned from breeding in order to eradicate their persistence (Wydooghe, 2013). Puppies affected by anasarca often need to be delivered by C-section. Anasarca is easily identifiable at birth as the subcutaneous accumulation of fluid. These puppies may be treated by administration of furosemide and potassium chloride, but often are euthanised shortly after birth.

THE BRACHYCEPHALIC DAM

It is generally advisable that the dam and her puppies are examined by the veterinary team within 24 hours of whelping to identify and discuss any potential issues that may have arisen. The care required during the post-partum period of brachycephalic bitches is similar to other breeds, and bitches should be monitored for signs of reproductive complications including inappropriate maternal behaviour, agalactia, mastitis, eclampsia, metritis and uterine prolapse. Conditions occur specific to brachycephalic dams include postpartum haemorrhage and aspiration pneumonia (Moon et al., 1998, Robertson, 2016).

THE BRACHYCEPHALIC CAT AND BREEDING

There are far fewer published reports covering the breeding of brachycephalic cats compared to dogs. However, some pertinent comparative information is detailed below.

FERTILITY AND GESTATION

Cats are induced ovulators, so progesterone cannot be used to predict the timing of ovulation or parturition (Beccaglia et al., 2016). This also means the progesterone-associated temperature decrease seen in dogs prior to parturition does not occur in cats. Although relaxin, a pregnancy-specific peptide hormone, can be measured from day 20 to 35, this does not allow accurate gestational timing (Beccaglia et al., 2016). Gestation in cats have been reported to last around 65–67 days (Lein, 1989, Root et al., 1995, Sparkes et al., 2006, England & Heimendahl, 2010), although larger litter sizes may have shorter gestations (Sparkes et al., 2006) and a range of 52–74 days from mating is accepted (England & Heimendahl, 2010).

In cats, pregnancy detection can be carried out via ultrasonographic examination from day 10 after mating (Beccaglia et al., 2016) and via radiographic evidence of foetal mineralisation from day 40 after mating (England & Heimendahl, 2010). Similar to dogs, ultrasonographic imaging of the feline reproductive tract may provide a more accurate method for early pregnancy detection compared to abdominal palpation or radiography. Delaying ultrasonographic examination until day 16 or later after mating may increase accuracy of initial pregnancy detection; false positives may be recorded earlier because of the similarity between empty loops of small intestine and early pregnancy (England & Heimendahl, 2010). Table 18.9 details proposed gestational aging utilising ultrasonography, and Table 18.10 details proposed gestational aging using radiography.

BIRTH

Early identification of dystocia and subsequent intervention should improve outcomes; mortality due to dystocia in cats is reported at 2% of pregnancies overall (Holst et al., 2017). However, over 85% of parturitions in cats are reported to occur without any medical or surgical intervention (Sparkes et al., 2006, Strom Holst & Frossling, 2009). Providing a calm environment is essential as excessive human interference can disrupt the normal kittening process and promote cannibalism (England & Heimendahl, 2010). Cats have a much shorter birthing process than their canine counterparts. First-stage labour normally lasts from 6 to 12 hours; however, first-stage labour was reported to last

TABLE 18.9

Gestational Aging Based on Ultrasonographic Detection of Feline Foetal Features

Foetal Features	First Detection (Days) After Mating
Small circular anechoic structure	10–11
Embryo	14
Heart beat	16–18
Bladder	29–32
Stomach	29–30
Kidney	38–41
Bowel	38–42

Source: Data from Zambelli, D., and F. Prati. Ultrasonography for pregnancy diagnosis and evaluation in queens. *Theriogenology* 66, no. 1 (2006): 135–144.

TABLE 18.10

Approximate Radiographic Gestational Age for the Feline Foetus Using Traditional Radiography

Feature	Detection (Days) After Mating
First evidence of skull mineralisation (skull, scapula, humerus, femur, vertebrae and ribs)	38–40 onwards
Tibia, fibula, ilium and ischium mineralised	43 onwards
Metatarsals and metacarpals mineralised	49 onwards
Digits and sternum mineralised	52–53 onwards
Molar teeth detectable	56–63 onwards

under 2 hours in > 80% of purebred cats in one large study (Sparkes et al., 2006). In cats, vulval discharge in late first-stage labour is a reddish-brown colour compared to the green colour observed in dogs. Second-stage labour in cats usually lasts 4–16 hours, although the interval from birth of the first kitten to the last kitten in cats that do not require veterinary assistance can be quite short, with one report detailing <6 hours for 85% of purebred cats (Sparkes et al., 2006). It is worth noting, however, that a prolonged interval between delivery of each kitten does not necessarily contribute to poor viability in the kitten (Axelsson, 2019). Although the Persian breed has a reported higher incidence of dystocia (Ekstrand & Linde-Forsberg, 1994, Gunn-Moore & Thrusfield, 1995), a more recent dystocia study, based on Swedish insurance data, did not identify an increased incidence in Persian or Exotic Shorthair cats compared to other purebred cats, although the study was relatively underpowered and should be interpreted with caution (Holst et al., 2017). Persians have a higher incidence of dystocia because their disproportionately high braincase cranial conformation acts as a mechanical obstacle during the birth process (Kunzel et al., 2003). In the cat, further studies are needed to determine whether foetal heart rates can be used to determine foetal distress at the time of parturition (Zambelli & Prati, 2006).

POSTNATAL CARE OF THE BRACHYCEPHALIC KITTEN

Some brachycephalic breeds, such as Persian and Exotic Shorthair cats, are reported to have the highest mean stillbirth rate; Persian cats had the highest kitten mortality rate at 25% of all purebred cats in a large UK owner-based questionnaire study (Sparkes et al., 2006). British Shorthair had an increased rate of kitten mortality in Sweden (Strom Holst & Frossling, 2009). Apgar scoring for kittens has been suggested as a method to potentially decrease neonatal mortality (Axelsson, 2019); this may be especially helpful in brachycephalic breeds with higher kitten mortality rates.

CONCLUSIONS

- Several brachycephalic dog breeds, including the English Bulldog, have become heavily reliant on artificial insemination for successful reproduction.
- Some brachycephalic dog breeds, including the English Bulldog, French Bulldog and Boston Terrier, are predisposed to dystocia and have high caesarean section rates in excess of 80%.
- Elective pre-parturient caesarean section is now commonplace in several brachycephalic dog breeds. Optimal timing of this intervention is critical for good neonatal survival. It is recommended to use as many tools as possible in order to increase the prediction accuracy for timing these surgical procedures.
- Canine brachycephalic neonates need to be assessed rapidly after birth as they may have decreased vigour compared to other breeds. Canine Apgar scoring is recommended as a useful framework.
- Brachycephalic cat types have higher dystocia levels than mesocephalic cats. Some brachycephalic breeds, including Persian and Exotic Shorthair cats, have been reported with high stillbirth rates.
- Reliance on significant intervention for reproduction in many brachycephalic breeds raises serious welfare and ethical issues for veterinary professionals and pet owners.

REFERENCES

Almeida, A.H., Sterman, F.A. and Miglino, M.A., 2003. Ultrasonographic mensuration of head dimensions in boxers normal gestations and its correlation with the gestational age. *Brazilian Journal of Veterinary Research and Animal Science* 40:136–140

Axelsson, R., 2019. APGAR score as a method for prediction of survival prognosis in newborn puppies and kittens. https://stud.epsilon.slu.se/14800/ [accessed 8/8/20].

Bannasch, D., Young, A., Myers, J., Truvé, K., Dickinson, P., Gregg, J., Davis, R., Bongcam-Rudloff, E., Webster, M.T., Lindblad-Toh, K. and Pedersen, N., 2010. Localization of canine brachycephaly using an across breed mapping approach. *PloS One*, 5(3).

Barr, F.J., 1988. Pregnancy diagnosis and assessment of fetal viability in the dog: A review. *Journal of Small Animal Practice*, 29(10):647–656.

Batista, M., Moreno, C., Vilar, J., Golding, M., Brito, C., Santana, M. and Alamo, D., 2014. Neonatal viability evaluation by Apgar score in puppies delivered by cesarean section in two brachycephalic breeds (English and French bulldog). *Animal Reproduction Science*, 146(3–4):218–226.

Beccaglia, M., Alonge, S., Trovo', C. and Luvoni, G.C., 2016. Determination of gestational time and prediction of parturition in dogs and cats: An update. *Reproduction in Domestic Animals*, 51:12–17.

Bergström, A., Nødtvedt, A., Lagerstedt, A.S., et al., 2006. Incidence and breed predilection for dystocia and risk factors for cesarean section in a Swedish population of insured dogs. *Vet Surg* 35:786–91.

Bertolini, F., Gandolfi, B., Kim, E.S. et al., 2016. Evidence of selection signatures that shape the Persian cat breed. *Mammalian Genome* 27:144–155.

Borella, C.M., da Silva, T.F.P., Nascimento, Í., Coimbra, A.D.L., de Magalhães, F.F. and da Silva, L.D.M., 2019. Fetal biparietal diameter by ultrasonography as a tool for estimating gestational age in bitches of the French Bulldog race. *Ciência Animal*, 29(Suppl. 1):41–44.

Boudrieau, R.J., Fossum, T.W., Hartsfield, M.S., Hobson, H.P. and Rudy, L.R., 1990. Pectus excavatum in dogs and cats. *Compendium on Continuing Education for the Practicing Veterinarian* 12:341–354.

Camargo, N.I., 2012. Estimativa da idade gestacional por biometria fetal através de exames ultrassonográficos em cadelas (Canis familiaris, Linnaeus, 1758) das raças Bulldog Francês, Pug e Shih Tzu. 66f. Dissertação (Mestrado em Ciência Veterinária), Universidade Federal de Pernambuco, Recife.

Castagnetti, C., Cunto, M., Bini, C., Mariella, J., Capolongo, S. and Zambelli, D., 2017. Time-dependent changes and prognostic value of lactatemia during the first 24 h of life in brachycephalic newborn dogs. *Theriogenology*, 94:100–104.

Cavanagh, A., 2017. Neonatal Resuscitation, Clinician's Brief https://www.cliniciansbrief.com/article/neonatal-resuscitation [accessed 25/5/20].

Conze, T., Jurczak, A., Fux, V., Socha, P., Wehrend, A. and Janowski, T., 2020. Survival and fertility of bitches undergoing caesarean section. *Veterinary Record*, 186(13):416–416.

Concannon, P.W., 2000. Canine pregnancy: Predicting parturition and timing events of gestation. In P.W. Concannon, G. England, J. Verstegen III and C. Linde-Forsberg (eds.), *Recent Advances in Small Animal Reproduction*. Ithaca, NY: International Veterinary Information Service. [A1202.0500 (http://www.ivis.org)].

Concannon, P.W., Butler, W.R., Hansel, W., Knight, P.J. and Hamilton, J.M., 1978. Parturition and lactation in the bitch: Serum progesterone, cortisol and prolactin. *Biology of Reproduction*, 19:1113–1118.

Concannon, P., Whaley, S., Lein, D., et al., 1983. Canine gestation length: Variation related to time of mating and fertile life of sperm. *American Journal of Veterinary Research* 44(10):1819–1821.

Cruz, M.L., Luna, S.P.L., Clark, R.M.O., Massone, F. and Castro, G.B., 1997. Epidural anaesthesia using lignocaine, bupivacaine or a mixture of lignocaine and bupivacaine in dogs. *Journal of Veterinary Anaesthesia*, 24(1):30–32.

Cunto, M., Zambelli, D., Castagnetti, C., Linta, N. and Bini, C., 2015. Diagnosis and treatment of foetal anasarca in two English Bulldog puppies. *Pakistan Veterinary Journal*, 35(2): 251–253.

De Cramer, K.G.M. and Nöthling, J.O., 2018. Is the biparietal diameter of fetuses in late gestation too variable to predict readiness for cesarean section in dogs? *Theriogenology*, 113:50–55.

De Cramer, K.G.M. and Nöthling, J.O., 2019. Curtailing parturition observation and performing preparturient cesarean section in bitches. *Theriogenology*, 124:57–64.

De Cramer, K.G. and Nöthling, J.O., 2020. Towards scheduled pre-parturient caesarean sections in bitches. Reproduction in Domestic Animals.

DEFRA, 2020. The Animal Welfare (Licensing of Activities Involving Animals) (England) Regulations 2018 Guidance notes for conditions for breeding dogs https://assets.publishing.service.gov.uk/government/uploads/system/uploads/attachment_data/file/880217/dog-breeding-guidance.pdf. [accessed 7/5/20].

De Mello, M., 2014. Rabbits in captivity. The ethics of captivity, pp.77–89.

De Mello, M., 2016. Rabbits multiplying like rabbits: The rise in the worldwide popularity of rabbits as pets. In *Companion Animals in Everyday Life*. New York: Palgrave Macmillan, pp. 91–107.

Dickie, E., 2011. Dystocia in a rabbit (Oryctolagus cuniculus). *The Canadian Veterinary Journal*, 52(1):80.

Doebeli, A., Michel, E., Bettschart, R., Hartnack, S. and Reichler, I.M., 2013. Apgar score after induction of anesthesia for canine cesarean section with alfaxalone versus propofol. *Theriogenology*, 80(8):850–854.

Downing, F. and Gibson, S., 2018. Anaesthesia of brachycephalic dogs. *Journal of Small Animal Practice*, 59(12):725–733.

Ekstrand, C. and Linde-Forsberg, C., 1994. Dystocia in the cat: a retrospective study of 155 cases. *Journal of Small Animal Practice*, 35(9):459–464.

England, G.C. and Heimendahl, A.V., 2010. *BSAVA Manual of Canine and Feline Reproduction and Neonatology* (2nd edn.). Quedgeley: British Small Animal Veterinary Association, pp. 98–105.

Evans, K.M. and Adams, V.J., 2010. Proportion of litters of purebred dogs born by caesarean section. *Journal of Small Animal Practice* 51:113–8.

Farstad, W.K., 2010. Artificial insemination in dogs. In BSAVA manual of canine and feline reproduction and neonatology (pp. 80–88). BSAVA Library.

Fawcett, A., Barrs, V., Awad, M., Child, G., Brunel, L., Mooney, E., Martinez-Taboada, F., McDonald, B. and McGreevy, P., 2019. Consequences and management of canine brachycephaly in veterinary practice: Perspectives from Australian veterinarians and veterinary specialists. *Animals*, 9(1):3.

Feliciano, M.A.R., Maciel, G.S., Coutinho, L.N., Almeida, V.T., Uscategui, R.R. and Vicente, W.R.R., 2015. Gestational echo biometry in brachycephalic pregnant bitches. *Ciência Animal Brasileira*, 16(3):419–427.

Funkquist, P.M., Nyman, G.C., Löfgren, A.J. and Fahlbrink, E.M., 1997. Use of propofol-isoflurane as an anesthetic regimen for cesarean section in dogs. *Journal of the American Veterinary Medical Association*, 211(3):313.

Fusi, J., Faustini, M., Bolis, B. and Veronesi, M.C., 2020. Apgar score or birthweight in Chihuahua dogs born by elective Caesarean section: which is the best predictor of the survival at 24 h after birth?. *Acta Veterinaria Scandinavica*, 62(1):39. doi:10.1186/s13028-020-00538-y.

Gaytán, L., Rascón, C.R., Angel-García, O., Véliz, F.G., Contreras, V. and Mellado, M., 2020. Factors influencing English Bulldog bitch fertility after surgical uterine deposition of fresh semen. *Theriogenology*, 142:315–319.

Geiser, B., Burfeind, O., Heuwieser, W., and Arlt, S., 2014. Prediction of parturition in bitches utilizing continuous vaginal temperature measurement. *Reproduction in Domestic Animals*, 49:109–114

Gil, E.M., Garcia, D.A. and Froes, T.R., 2015. In utero development of the fetal intestine: Sonographic evaluation and correlation with gestational age and fetal maturity in dogs. *Theriogenology*, 84(5):681–686.

Gil, E.M.U., Garcia, D.A.A., Giannico, A.T. and Froes, T.R., 2014. Canine fetal heart rate: Do accelerations or decelerations predict the parturition day in bitches? *Theriogenology*, 82(7):933–941.

Gil, E.M.U., Garcia, D.A.A., Giannico, A.T. and Froes, T.R., 2018. Early results on canine fetal kidney development: Ultrasonographic evaluation and value in prediction of delivery time. Theriogenology, 107:180–187.

Goericke-Pesch, S., Fux, V., Prenger-Berninghoff, E. and Wehrend, A., 2018. Bacteriological findings in the canine uterus during Caesarean section performed due to dystocia and their correlation to puppy mortality at the time of parturition. *Reproduction in Domestic Animals = Zuchthygiene*, 53(4):889–894. doi:10.1111/rda.13181.

Gosling, E.M., Vázquez-Diosdado, J.A. and Harvey, N.D., 2018. The status of pet rabbit breeding and online sales in the UK: A glimpse into an otherwise elusive industry. *Animals*, 8:199.

Gough A. and Thomas A., 2010. *Breed Predispositions to Disease in Dogs and Cats* (2nd edn.). Ames, IA: Wiley-Blackwell, pp. 30, 36, 41–42, 254.

Grellet A., 2008. La supplémentation en acide folique pourrait prévenir les fentes palatines chez le chien. La Dép.eche vétérinaire, 986.

Grundy, S.A., Liu, S.M. and Davidson, A.P., 2009. Intracranial trauma in a dog due to being "swung" at birth. *Topics in Companion Animal Medicine*, 24(2):100–103.

Gunn-Moore, D.A. and Thrusfield, M.V., 1995. Feline dystocia: prevalence, and association with cranial conformation and breed. *The Veterinary Record*, 136(14):350–353.

Harcourt-Brown, F. 2002. *Textbook of Rabbit Medicine.* Oxford: Butterworth- Heinemann, p. 348.

Harðardóttir, H., Thierry, F. and Murison, P.J. 2019. Anaesthesia management of a pug (in late-stage pregnancy) with lung lobe torsion. *Veterinary Record Case Reports* 7:e000765. doi:10.1136/vetreccr-2018-000765.

Hay Kraus, B., 2016. Anesthesia for cesarean section in the dog. *Veterinary Focus*, 26(1):24.

Heise, A., 2012. Artificial insemination in veterinary science. *A Bird's-Eye View of Veterinary Medicine*, 17–33.

Holst, B.S., Axnér, E., Öhlund, M., Möller, L. and Egenvall, A., 2017. Dystocia in the cat evaluated using an insurance database. *Journal of Feline Medicine and Surgery*, 19(1):42–47.

Hopper B.J., Richardson J.L. and Lester N.V., 2004. Spontaneous antenatal resolution of canine hydrops fetalis diagnosed by ultrasound. *Journal of Small Animal Practice* 45:2–8.

Johnson, C.A., 2008. High-risk pregnancy and hypoluteoidism in the bitch. *Theriogenology*, 70(9):1424–1430.

Johnston, S.D., Root Kustritz, M.V. and Olson, P.N.S., 2001. Canine parturition – eutocia and dystocia. In S.D. Johnston, M.V. RootKustritz, and P.N.S. Olson (eds.). *Canine and Feline Theriogenology*. Philadelphia, PA: WB Saunders, pp. 105–128.

The Kennel Club, 2021. Reporting C-sections and surgeries, https://www.thekennelclub.org.uk/dog-breeding/first-time-breeders/whelping-your-first-litter/reporting-c-sections-and-surgeries/ [accessed 01/05/2021].

The Kennel Club, 2021. When you are ready for your first litter, https://www.thekennelclub.org.uk/dog-breeding/first-time-breeders/when-you-are-ready-for-your-first-litter/ [accessed 01/05/21].

Kunzel, W., Breit, S. and Opel, M., 2003. Morphometric investigations of breed-specific features in feline skulls and considerations on their functional implications. *Anatomy, Histology, Embryology*, 32:218e23.

Kustritz M.V.R., 2006. Pregnancy diagnosis and abnormalities of pregnancy in the dog. Theriogenology, 64:755–765.

Kutzler, M.A., Mohammed, H.O., Lamb, S.V. and Meyers-Wallen, V.N., 2003. Accuracy of canine parturition date prediction from the initial rise in preovulatory progesterone concentration. *Theriogenology*, 60(6):1187–1196.

Lein, D.H., 1989. Female reproduction. In R.G. Sherding (ed.), *The Cat - Diseases and Clinical Management* (1st edn.). New York: Churchill Livingstone, pp. 1479–1497.

Linde-Forsberg, C. and Eneroth, A., 2000. Abnormalities in pregnancy, parturition and the periparturient period. In S.J. Ettinger and E.C. Feldman (eds.), *Textbook of Veterinary Internal Medicine* (5th edn.). Philadelphia, PA: Saunders, pp. 1527–1539

Lipinski, M.J., Froenicke, L., Baysac, K.C., Billings, N.C., Leutenegger, C.M., Levy, A.M., Longeri, M., Niini, T., Ozpinar, H., Slater, M.R. and Pedersen, N.C., 2008. The ascent of cat breeds: genetic evaluations of breeds and worldwide random-bred populations. *Genomics*, 91(1):12–21.

Lobodzinska, A., Gruszczynska, J., Max, A., Bartyzel, B.J., Mikula, M., Mikula Jr., I. et al. 2014. Cleft palate in the domestic dog, Canis lupus familiaris–etiology, pathophysiology, diagnosis, prevention, and treatment. *Acta Scientiarum Polonorum Zootechnica*. 13(3):5–28.

Loeb, J. and Evans, E., 2020. Puppy power: Fertility clinics on the rise. *Veterinary Record*, 186(5):140. doi:10.1136/vr.m394.

Loeb, J., 2020. Who is regulating fertility clinics? *Veterinary Record* 186(5):137.

Lopate, C., 2008. Estimation of gestational age and assessment of canine fetal maturation using radiology and ultrasonography: A review. *Theriogenology*, 70(3):397–402.

Luna, S.P.L., Cassu, R.N., Castro, G.B., Neto, F.T., Silva, J.R. and Lopes, M.D., 2004. Effects of four anaesthetic protocols on the neurological and cardiorespiratory variables of puppies born by caesarean section. *Veterinary Record*, 154(13):387–389.

Maclennan, T. and Smith, D., 2019. The influence of media on the ownership of brachycephalic breed dogs. *Veterinary Nursing Journal*, 34(12):302–306.

Maronezi, M.C., Madruga, G.M., Uscategui, R.A.R., Simões, A.P.R., Silva, P., Rodrigues, M.G.K., Cintra, C.A., Assis, A.R., Vicente, W.R.R. and Feliciano, M.A.R., 2018. Pulmonar ARFI elastography and ultrasonography of canine fetal hydrops: Case report. *Arquivo Brasileiro de Medicina Veterinária e Zootecnia*, 70(5):1409–1413.

Mason, S.J., 2018. Current review of artificial insemination in Dogs. *Veterinary Clinics: Small Animal Practice*, 48(4):567–580.

McNally, E.M., Robertson, S.A. and Pablo, L.S., 2009. Comparison of time to desaturation between preoxygenated and nonpreoxygenated dogs following sedation with acepromazine maleate and morphine and induction of anesthesia with propofol. *American Journal of Veterinary Research*, 70(11):1333–1338.

Meola, S.D., 2013. Brachycephalic airway syndrome. *Topics in Companion Animal Medicine*, 28:91–6.

Monteiro, C.L.B., Campos, A.I.M., Madeira, V.L.H., Silva, H.V.R., Freire, L.M.P., Pinto, J.N., De Souza, L.P. and Da Silva, L.D.M., 2012. Pelvic differences between brachycephalic and mesaticephalic cats and indirect pelvimetry assessment. *Veterinary Record*, 172:16.

Metcalfe, S., Hulands-Nave, A., Bell, M., Kidd, C., Pasloske, K., O'Hagan, B., Perkins, N. and Whittem, T., 2014. Multicentre, randomised clinical trial evaluating the efficacy and safety of alfaxalone administered to bitches for induction of anaesthesia prior to caesarean section. *Australian Veterinary Journal*, 92(9):333–338.

Moon, P.F., Erb, H.N., Ludders, J.W., Gleed, R.D., and Pascoe, P.J., 1998, Perioperative management and mortality rates of dogs undergoing cesarean section in the United States and Canada. *Journal of the American Veterinary Medical Association*, 213 (3):365–369.

Moon, P.F., Erb, H.N., Ludders, J.W., Gleed, R.D. and Pascoe, P.J., 2000. Perioperative risk factors for puppies delivered by cesarean section in the United States and Canada. *Journal of the American Animal Hospital Association*, 36(4):359–368.

Moon, P.F., Massat, B.J. and Pascoe, P.J., 2001. Neonatal critical care. *Veterinary Clinics: Small Animal Practice*, 31(2):343–367.

Moura, E., Cirio, S.M. and Pimpão, C.T., 2012. Nonsyndromic cleft lip and palate in boxer dogs: Evidence of monogenic autosomal recessive inheritance. *The Cleft Palate-Craniofacial Journal*, 49 (6):759–760.

Münnich, A. and Küchenmeister, U., 2009. Dystocia in numbers - evidence-based parameters for intervention in the dog: Causes for dystocia and treatment recommendations. *Reproduction in Domestic Animals*, 44 (Suppl 2):141–7.

Nelson, A.W., 2003. Cleft palate. In D.H. Slatter (ed.), *Textbook of Small Animal Surgery* (3rd edn.). Philadelphia, PA: Saunders, pp. 814–823.

Ontiveros, E.S., Fousse, S.L., Crofton, A.E., Hodge, T.E., Gunther-Harrington, C.T., Visser, L.C. and Stern, J.A., 2019. Congenital cardiac outflow tract abnormalities in dogs: Prevalence and pattern of inheritance from 2008 to 2017. *Frontiers in Veterinary Science*, 6:52.

O'Neill, D.G., O'Sullivan, A.M., Manson, E.A., Church, D.B., Boag, A.K., McGreevy, P.D. and Brodbelt, D.C., 2017. Canine dystocia in 50 UK first-opinion emergency-care veterinary practices: prevalence and risk factors. *Veterinary Record*, 181(4):88.

O'Neill, D.G., O'Sullivan, A.M., Manson, E.A., Church, D.B., McGreevy, P.D., Boag, A.K. and Brodbelt, D.C., 2019. Canine dystocia in 50 UK first-opinion emergency care veterinary practices: clinical management and outcomes. *Veterinary Record*, 184:409. doi:10.1136/vr.104944.

Packer, R.M., O'Neill, D.G., Fletcher, F. and Farnworth, M.J., 2019. Great expectations, inconvenient truths, and the paradoxes of the dog-owner relationship for owners of brachycephalic dogs. *PloS One*, 14(7):e0219918.

Packer, R.M. and Tivers, M.S., 2015. Strategies for the management and prevention of conformation-related respiratory disorders in brachycephalic dogs. *Veterinary Medicine: Research and Reports*, 6:219.

Pankowski, F., Paśko, S., Max, A., Szal, B., Dzierzęcka, M., Gruszczyńska, J., Szaro, P., Gołębiowski, M. and Bartyzel, B.J., 2018. Computed tomographic evaluation of cleft palate in one-day-old puppies. *BMC Veterinary Research*, 14(1):316.

Pascoe, P.J. and Moon, P.F., 2001. Periparturient and neonatal anesthesia. *Veterinary Clinics of North America: Small Animal Practice* 31:315–41.

Plitman, L., Černá, P., Farnworth, M.J., Packer, R. and Gunn-Moore, D.A., 2019. Motivation of owners to purchase pedigree cats, with specific focus on the acquisition of brachycephalic cats. *Animals*, 9(7):394.

PDSA (Peoples Dispensary for Sick Animals) PDSA Animal Wellbeing (PAW) Report. 2019. https://www.pdsa.org.uk/media/7420/2019-paw-report_downloadable.pdf [accessed online 17/5/20].

Quartuccio, M., Biondi, V., Liotta, L. and Passantino, A., 2020. Legislative and ethical aspects on use of canine artificial insemination in the 21st century. *Italian Journal of Animal Science*, 19(1):630–643.

Quesenberry, K. and Carpenter, J.W., 2011. *Ferrets, Rabbits and Rodents-E-Book: Clinical Medicine and Surgery*. New York: Elsevier Health Sciences.

Ravetz, G., 2017. Conformation-altering surgeries, caesareans and data–the veterinary team's role? *Veterinary Nursing Journal*, 32(1):22–24.

RCVS, 2019. Client confidentiality 14.4. Code of Professional Conduct for Veterinary Surgeons. https://www.rcvs.org.uk/setting-standards/advice-and-guidance/code-of-professional-conduct-for-veterinary-surgeons/supporting-guidance/client-confidentiality/ [accessed 17/5/20].

RCVS, 2019. Canine surgical artificial insemination 27.30 Miscellaneous procedures: legal and ethical considerations. Code of Professional Conduct for Veterinary Surgeons. https://www.rcvs.org.uk/setting-standards/advice-and-guidance/code-of-professional-conduct-for-veterinary-surgeons/supporting-guidance/miscellaneous/ [accessed 17/5/20].

RCVS, 2020. Veterinary Medicines. https://www.rcvs.org.uk/setting-standards/advice-and-guidance/code-of-professional-conduct-for-veterinary-surgeons/supporting-guidance/veterinary-medicines/ [accessed 2/9/20].

Rendano, V.T., 1983. Radiographic evaluation of fetal development in the bitch and fetal death in the bitch and queen. In R.W. Kirk (ed.), *Current Veterinary Therapy*. Philadelphia, PA: WB Saunders, vol. VIII, pp. 947–952.

Rendano, V.T., Lein, D.H. and Concannon, P.W., 1984. Radiographic evaluation of prenatal development in the Beagle: Correlation with time of breeding, LH release, and parturition. *Veterinary Radiology*, 25:132–141.

Robertson, S., 2016. Anaesthetic management for caesarean sections in dogs and cats. *In Practice*, 38(7):327–339.

Roman, N., Carney, P.C., Fiani, N. and Peralta, S., 2019. Incidence patterns of orofacial clefts in purebred dogs. *PLoS One*, 14(11):e0224574.

Roos, J., Maenhoudt, C., Zilberstein, L., Mir, F., Borges, P., Furthner, E., Niewiadomska, Z., Nudelmann, N. and Fontbonne, A., 2018. Neonatal puppy survival after planned caesarean section in the bitch using aglepristone as a primer: A retrospective study on 74 cases. *Reproduction in Domestic Animals* 53:85–95.

Root, M.V., Johnston, S.D. and Olson, P.N., 1995. Estrous length, pregnancy rate, gestation and parturition lengths, litter size, and juvenile mortality in the domestic cat. *Journal of the American Animal Hospital Association*, 31(5):429–433.

Rota, A., Charles, C., Starvaggi Cucuzza, A. and Pregel, P., 2015. Diagnostic efficacy of a single progesterone determination to assess full-term pregnancy in the bitch. *Reproduction in Domestic Animals*, 50(6):1028–1031.

Runcan, E.E. and da Silva, M.A.C., 2018. Whelping and dystocia: maximizing success of medical management. Topics in companion animal medicine, 33(1): 12–16.

Ryan, S.D. and Wagner, A.E., 2006a. Cesarean section in dogs: Anesthetic management. *Compendium on Continuing Education for the Practicing Veterinarian*, 28(1):44–56.

Ryan, S.D. and Wagner, A.E., 2006b. Cesarean section in dogs: Physiology and perioperative considerations. *Compendium on Continuing Education for the Practicing Veterinarian*, 28(1):34–43.

Sankar, P., Mandal, D., Kumar, V. and Mondal, M., 2017. Dystocia in rabbits and its surgical management. *Exploratory Animal and Medical Research*, 7:216–217.

Singh, M., Kumar, A., Sood, P., Varshney, A.C. and Sharma, A., 2000. Dystocia due to foetal anasarca in an Angora rabbit: Anaesthetic and surgical management. *Indian Journal of Veterinary Surgery*, 21(2):109–110.

Smith, F.O., 2007. Challenges in small animal parturition—Timing elective and emergency cesarian sections. *Theriogenology*, 68(3):348–353.

Smith-Carr, S., 2006. Canine artificial insemination. *Veterinary Technician*, 27(8):474.

Sparkes, A.H., Rogers, K., Henley, W.E., Gunn-Moore, D.A., May, J.M., Gruffydd-Jones, T.J. and Bessant, C., 2006. A questionnaire-based study of gestation, parturition and neonatal mortality in pedigree breeding cats in the UK. *Journal of Feline Medicine and Surgery*, 8(3):145–157.

Sridevi, P., Reena, D. and Safiuzamma, M., 2016. Diagnosis of fetal anasarca by real time ultrasonograpy in a pug bitch and its surgical management. *The Indian Journal of Animal Reproduction*, 37(2):65–66.

Strom Holst, B and Frossling, J., 2009. The Swedish breeding cat: Population description, infectious diseases and reproductive performance evaluated by a questionnaire. *Journal of Feline Medicine and Surgery*, 11:793–802.

Suter, P.F., 1984. "Swimmers", flat pup syndrome. In P.F. Suter (ed.), *Thoracic Radiography: A Text Atlas of Thoracic Diseases of the Dog and Cat*. Wettswil, Switzerland, pp. 164–165.

Teng, K.T., McGreevy, P.D., Toribio, J.A.L. and Dhand, N.K., 2016. Trends in popularity of some morphological traits of purebred dogs in Australia. *Canine Genetics and Epidemiology*, 3(1):2.

The Cat Fanciers' Association, 2020. The Cat Fanciers' Association Announces Most Popular Breeds for 2019. https://cfa.org/cfa-news-releases/top-breeds-2019/ [accessed online 18/5/20].

Titkova, R., Fialkovicova, M., Karasova, M. and Hajurka, J., 2017. Puppy Apgar scores after vaginal delivery and caesarean section. *Veterinární medicína*, 62(9):488–492.

Tønnessen, R., Borge, K.S., Nødtvedt, A. and Indrebø, A., 2012. Canine perinatal mortality: a cohort study of 224 breeds. *Theriogenology*, 77(9):1788–1801.

Traas, A.M., 2008a. Resuscitation of canine and feline neonates. *Theriogenology*, 70(3):343–348.

Traas, A.M., 2008b. Surgical management of canine and feline dystocia. Theriogenology, 70(3):337–342.

Verhoeven, G., De Rooster, H., Risselada, M., Wiemer, P., Scheire, L. and Van Bree, H., 2006. Swimmer syndrome in a Devon rex kitten and an English bulldog puppy. *Journal of Small Animal Practice*, 47(10):615–619.

Veronesi, M.C., Battocchio, M. and Marinelli, L., 2002. Correlations among body temperature, plasma progesterone, cortisol and prostaglandin F2alpha of the periparturient bitch. *Journal of Veterinary Medicine. A, Physiology, Pathology, Clinical Medicine* 49(5):264–268.

Veronesi, M., 2016. Assessment of canine neonatal viability-the Apgar score. *Reproduction in Domestic Animals* 51(Suppl 1):46–50.

Veronesi, M.C., Panzani, S., Faustini, M. and Rota, A., 2009. An Apgar score system for routine assessment of newborn puppy viability and short- term survival prognosis. *Theriogenology*, 72:401–407.

Vilar, J.M., Batista, M., Pérez, R., Zagorskaia, A., Jouanisson, E., Díaz-Bertrana, L. and Rosales, S., 2018. Comparison of 3 anesthetic protocols for the elective cesarean-section in the dog: Effects on the bitch and the newborn puppies. *Animal Reproduction Science*, 190:53–62.

Warzee, C.C., Bellah, J.R. and Richards, D., 2001. Congenital unilateral cleft of the soft palate in six dogs. *Journal of Small Animal Practice*, 42:338–340.

Wydooghe, E., Berghmans, E., Rijsselaere, T. and Van Soom, A., 2013. International breeder inquiry into the reproduction of the English Bulldog. *Vlaams Diergeneeskundig Tijdschrift*, 82(1):38–43.

Zambelli, D. and Prati, F., 2006. Ultrasonography for pregnancy diagnosis and evaluation in queens. *Theriogenology*, 66(1):135–144.

Zoldag, L., Albert, M., Fodor, Z., Padar, Z., Kontadakis, K. and Eszes, F. 2001. Hereditary and pathohistological study of anasarca (congenital edema) in Hungarian English bulldog population. *Magyar Allatorvosok Lapja* 123:335–342.

19 Anaesthesia for the Brachycephalic Patient

Fran Downing and Rebecca Robinson
Davies Veterinary Specialists

CONTENTS

Conditions that Influence Anaesthetic Safety 354
    Brachycephalic Obstructive Airway Syndrome (BOAS) 356
    Gastrointestinal Disorders 357
    Ocular and Other Conditions 357
Anaesthetic Management 357
    Pre-anaesthetic Assessment and Stabilisation 358
        History 358
        Clinical Examination 358
        Functional Tests 360
        Fasting and Regurgitation 360
    Premedication 361
        Sedation 362
        Analgesia 363
    Temperature Management 363
    Preoxygenation 364
    Anaesthetic Induction 365
    Anaesthetic Maintenance 365
        Ventilation and Lung Function 365
        Ocular Care 366
Anaesthetic Recovery and Post-operative Management 367
        Preparation for Recovery 367
        Provision of Analgesia 367
        Administration of Anti-inflammatory and Vasoconstrictive Drugs 368
        Equipment for Emergency Re-intubation 368
        Patient Positioning 368
        Recovery Following Tracheal Extubation 368
        Sedation 369
        Nebulisation and Physiotherapy 369
        Tracheostomy 370
Conclusion 371
References 371

An increase in the numbers of brachycephalic companion animals owned has been recently seen, including certain breeds of dogs, cats and rabbits. Brachycephaly and its associated problems mean that such animals often require special consideration when presented to veterinary surgeons, especially with regards to anaesthesia. It is worth noting that, whilst the principal aims and

ASA Grade	Definition
1	A normal healthy patient
2	A patient with mild systemic disease
3	A patient with severe systemic disease
4	A patient with severe systemic disease that is a constant threat to life
5	A moribund patient who is not expected to survive without surgery
E	Emergency surgery, where a delay in treatment would lead to a significant increase in threat to life.

FIGURE 19.1 American Society of Anaesthesiologists (ASA) physical status classification system which is used to quantify a patient's anaesthetic risk (American Society of Anesthesiologists, 2019).

considerations for anaesthesia are the same in all species, there are important practical differences in the management techniques used in the aforementioned species that are beyond the scope of this chapter to thoroughly appraise. Therefore, the focus of this chapter will be on discussing anaesthetic management in brachycephalic dogs.

As ownership of certain brachycephalic dog breeds such as French Bulldogs, Pugs and Bulldogs has risen in recent years, brachycephalic dogs are increasingly presented to veterinary practices for sedation and general anaesthesia to facilitate both routine and emergency procedures. General anaesthesia carries an inherent but relatively low risk of death in all species; this risk can be influenced by several factors including the reason for anaesthesia (e.g. emergency vs. elective surgery), pre-existing disease and other patient-related factors. Anaesthetic risk can be quantified using the American Society of Anaesthesiologists (ASA) grading system (Figure 19.1). The Confidential Enquiry into Perioperative Small Animal Fatalities (CEPSAF) by Brodbelt et al. (2008) reported death rate in low risk dogs (ASA 1–2) at 0.05% compared with 1.33% in higher risk (ASA 3–5) patients. Brachycephalic dogs have been reported to be at greater risk of experiencing complications in the perioperative period compared to their non-brachycephalic counterparts (Gruenheid et al. 2018); the postoperative/recovery period carries particular risk of complications including regurgitation, vomiting, aspiration pneumonia and stertorous breathing (Gruenheid et al. 2018). In order to reduce risk in brachycephalic patients, it is essential that clinicians are familiar with the common pre-existing conditions of these types of dogs that may necessitate alterations in anaesthetic management and how best to approach them.

CONDITIONS THAT INFLUENCE ANAESTHETIC SAFETY

Brachycephalic breeds are predisposed to many conditions that can affect anaesthetic safety and thus require alternations in its management. These morbidities associated with brachycephaly include respiratory problems (primarily but not limited to brachycephalic obstructive airway syndrome (BOAS)), gastrointestinal abnormalities including gastro-oesophageal reflux (GOR) and corneal ulceration (Figure 19.2).

Conditions associated with brachycephaly in dogs	Problems seen during the perioperative period.	Perioperative interventions
Brachycephalic Obstructive Airway Syndrome (BOAS); • Primary problems – stenotic nares, hypoplastic trachea, aberrant turbinates and elongated soft palate. • Secondary problems – excessive oropharyngeal soft tissue, everted laryngeal saccules, laryngeal collapse and lower airway dysfunction.	• URT obstruction, hypoventilation and hypoxaemia. • Difficult tracheal intubation. • Hyperthermia (due to decreased capacity to dissipate heat by panting)	• Minimise patient stress (may include appropriate sedation). • Preoxygenate patients. • Secure and protect a patent airway e.g. endotracheal intubation. • Monitor oxygenation using pulse oximetry +/- blood gas analysis during anaesthesia. • Monitor ETCO$_2$ using capnography. • Positive pressure ventilation (PPV) if needed. • Have equipment available to manage a difficult tracheal intubation (see Figure 19.7) • Actively cool patients if needed.
Gastrointesinal disease: • Gastroesophageal reflux (GOR) • Hiatal hernia	• Regurgitation and vomiting. • Aspiration pneumonia.	• Preanaesthetic fasting. • Gastroprotectants, e.g. omeprazole. • Placement of a cuffed ET tube. • Patient positioning e.g. head elevated.
Ocular disease: • Reduced corneal sensitivity • Corneal dryness and ulceration.	• Corneal ulceration.	• Regular ocular lubrication - during anaesthesia and for a minimum of 24 hours after recovery. • Care when positioning – avoid direct trauma.
Being overweight and/or Obese	• Increased risk of upper respiratory tract obstruction • Hypoventilation • Tachycardia • Hypertension • Left ventricular hypertrophy and diastolic dysfunction	• Take appropriate considerations as discussed for peri-operative BOAS management. • Care with patient positioning to avoid exacerbation of hypoventilation or upper respiratory tract obstruction
Susceptible to overheating	• Exacerbation of existing respiratory disease and distress • Uncontrolled hyperthermia	• Close monitoring of body temperature • Active cooling +/- judicious sedation when required
Hypercoagulable state, including shortened clotting times and delayed fibrinolysis.	• Unlikely to be clinically significant.	• Unlikely to require any intervention.
Relative hypertension.	• Unlikely to be clinically significant.	• Unlikely to require any intervention.

FIGURE 19.2   Common problems associated with brachycephalic dog breeds, how they may affect upon anaesthetic management and actions that can be taken to minimise their impact.

Brachycephalic Obstructive Airway Syndrome (BOAS)

Whilst brachycephaly does not automatically infer that a patient will have substantial airway obstruction, and thus BOAS, it is often not possible to definitively determine whether BOAS is present until after anaesthesia is performed. Additionally, significant airway obstruction may develop at any point during the peri-anaesthetic period, even if clinical signs were not previously reported. Therefore, when performing anaesthesia in brachycephalic patients which the clinician believes to have a high risk of suffering from BOAS, it is safest to assume BOAS is present until proven otherwise. As a result, all potential respiratory problems are considered under this heading. BOAS results from varied combinations of primary anatomical abnormalities including stenotic nares, elongated soft palate, hypoplastic trachea and hypertrophic/malformed nasal turbinates (Dupré and Heidenreich 2016, Emmerson 2014), with secondary changes including mucosal oedema, everted laryngeal saccules, tonsil eversion and laryngeal collapse. These conformational abnormalities can lead to both dynamic and static upper respiratory tract (URT) obstruction, increased work required for breathing, hypoxaemia and hypoventilation in affected patients. Surgical interventions to correct or improve BOAS are discussed more fully in Chapter 11. Surgical management of BOAS includes palatoplasty, laryngeal sacculectomy and rhinoplasty, which may improve the flow of air through the upper airways in up to 90% of dogs with clinical signs of BOAS (Poncet et al. 2006, Torrez and Hunt 2006, Fasanella et al. 2010, Riecks, Birchard, and Stephens 2007, Haimel and Dupré 2015).

In dogs, the URT obstruction and subsequent hypoxaemia caused by BOAS may be exacerbated by sedation and/or anaesthesia itself along with the stress and agitation associated with hospitalisation. The anatomical and conformational abnormalities associated with brachycephaly also make it more difficult to visualise, secure and maintain a patent airway in these patients during the perioperative period. This is exacerbated in patients with clinical airway obstruction (BOAS). In humans, obstructive sleep apnoea (OSA) (which shares many similar features with BOAS) is also associated with an inherent increase in anaesthetic risk triggered by the presence of concurrent problems including hypoxia, hypertension, arrhythmia, heart disease and certain endocrinopathies (Martinez and Faber 2011, Corso et al. 2017). Preoperative management of OSA, including the use of continuous positive airway pressure (CPAP) therapy and weight loss, helps reduce anaesthetic risk (Martinez and Faber 2011). It is reasonable to assume that similar benefits may apply in brachycephalic dogs, especially those diagnosed with BOAS, although the routine use of preoperative CPAP therapy is unlikely to be feasible in conscious veterinary patients. Interestingly, it has been shown that previous surgical correction of BOAS reduces the likelihood of post-anaesthetic complications (including regurgitation, prolonged time to extubation and hypoxaemia) in brachycephalic dogs that underwent subsequent anaesthetic events (Doyle et al. 2020).

In cases of severe BOAS, airway obstruction at any point can result in the development of acute pulmonary oedema, similar to post-obstructive or negative pressure pulmonary oedema, which can be rapidly fatal. The pathophysiology is considered multifactorial, resulting in an increase in hydrostatic pressure favouring the movement of fluid out of capillaries into interstitial tissue and alveoli with resultant pulmonary oedema (Senior 2005, Bashir et al. 2017). Management of post-obstructive pulmonary oedema requires its rapid identification and supportive care, including maintaining airway patency, oxygen supplementation and positive pressure ventilation (PPV) (Udeshi, Cantie, and Pierre 2010). The administration of diuretics, including furosemide, has been recommended by some authors in the management of post-obstructive pulmonary oedema in humans, especially in those having received aggressive intraoperative fluid therapy (Udeshi, Cantie, and Pierre 2010). However, diuretics can exacerbate hypovolaemia and hypoperfusion in some patients and should only be used when deemed clinically appropriate.

BOAS has also been associated with an increased susceptibility to heat stress, due to a reduced capacity for dissipating heat through panting. The action of panting itself can exacerbate both dynamic and static URT obstruction, meaning that maintaining normothermia during the perioperative period is essential.

GASTROINTESTINAL DISORDERS

Brachycephalic dogs are reported to have high incidence of oesophageal and gastrointestinal abnormalities, including GOR, hiatal hernia, gastritis, distal oesophagitis and pyloric hyperplasia, (Poncet et al. 2006, Poncet et al. 2005). Increased incidence of GOR is believed to result, in part, from the chronically high negative intrathoracic pressures generated to overcome URT obstruction (Shaver et al. 2017, Boesch et al. 2005). Presence of GOR is itself associated with an increased risk of oesophagitis and stricture formation, rhinitis and aspiration pneumonia (Shaver et al. 2017, Wilson and Walshaw 2004), which can contribute to increased anaesthetic risk and therefore may require specific management. The severity of gastrointestinal signs is positively correlated with the severity of respiratory disease (and vice versa) in French Bulldogs, males and heavy dogs (Poncet et al. 2005). Proactive management of gastrointestinal disease using gastroprotectant, prokinetic and anti-emetic drugs can help to reduce vomiting, regurgitation and GOR in the perioperative period and may contribute to improved recoveries. Gastrointestinal disorders in brachycephalics are discussed comprehensively in Chapter 10.

OCULAR AND OTHER CONDITIONS

Brachycephalic breeds are predisposed to corneal ulcerative disease (O'Neill et al. 2017), which may be further exacerbated during sedation and general anaesthesia. Several factors, such as prominent eyes, corneal drying and reduced corneal sensitivity in brachycephalic breeds, may contribute to their heightened risk of corneal disease (Barrett et al. 1991); Ophthalmological conditions in brachycephalics are discussed in more detail in Chapter 12.

The presence of a hypercoagulable state (shortened clotting times and delayed fibrinolysis) measured by thromboelastography is reported in brachycephalic dogs (Crane et al. 2017, Hoareau and Mellema 2015), with the degree of hypercoagulability increasing with the severity of BOAS (Crane et al. 2017). However, there is currently no indication that the reported hypercoagulability is clinically significant and there is no evidence to suggest that the routine use of anticoagulant medications in necessary in brachycephalic dogs.

Brachycephalic dog breeds are reported to have lower arterial blood oxygen concentrations, higher blood CO_2 concentrations and hypertension compared with mesocephalic and dolichocephalic dog breeds (Hoareau et al. 2012). These findings are mirrored in human patients with OSA and contribute to their increased anaesthetic risk (Martinez and Faber 2011). The clinical significance of the observed hypertension in brachycephalic dogs is, however, unclear.

Obesity is now considered as a distinct disease entity in veterinary medicine (Ward, German, and Churchill 2018) with prevalence estimates ranging from 6% to 41% in the general canine population (O'Neill et al. 2016, Lund et al. 2006, McGreevy et al. 2005, O'Neill et al. 2014). Obesity is reportedly associated with increased risk for several diseases including endocrinopathies, metabolic abnormalities, joint disease and neoplasia (Lund et al. 2006, German 2006). Crucially, in the context of brachycephaly, obesity is associated with cardiorespiratory disease and functional alterations such as exercise and heat intolerance (German 2006). Increasing levels of excess body-weight are associated with worsening respiratory disease in brachycephalic dogs (Packer et al. 2015). Anaesthetic risk is also increased with obesity (Brodbelt 2009). Body condition scores of 7/9 or higher result in a pro-inflammatory state with links to numerous potential anaesthetic problems including upper respiratory tract obstruction and hypoventilation, tachycardia, hypertension and left ventricular hypertrophy with diastolic dysfunction (Love and Cline 2015).

ANAESTHETIC MANAGEMENT

Brachycephalic dogs require sedation and general anaesthesia for an array of elective and emergency diagnostic and surgical procedures. Brachycephalic patients have particular risk of complications including URT obstruction, GOR and aspiration following sedation/premedication, induction of anaesthesia, recovery and the immediate post-operative period (Gruenheid et al. 2018). This section discusses the general considerations for planning and managing anaesthesia, with particular emphasis on areas pertinent to brachycephalic patients (Figure 19.3). While the information in this chapter is primarily aimed at the canine brachycephalic patient, some of the basic principles can be applied to other species such as cats. However, clinicians need to be mindful of the additional species differences present which require specific management.

PRE-ANAESTHETIC ASSESSMENT AND STABILISATION

All patients should be fully assessed prior to anaesthesia to identify and address potential problems proactively. Following this assessment, it may be beneficial to postpone the procedure for animals where the anaesthetic risk is deemed too high for an elective procedure until any reversible risk factors are reduced. Risk reduction could include weight loss, improving fitness and providing pre-operative gastro-protective agents.

History

Assessment begins with acquiring a full history, including both details of the presenting complaint and pre-existing disease. For brachycephalic patients, even when presenting for reasons other than airway surgery, it is important to specifically question owners on respiratory and gastrointestinal signs (Figure 19.4). Literature suggests that owners report a high prevalence of exercise intolerance (up to 88%), heat intolerance (50%) and sleep disturbances (56%) in brachycephalic dogs when specifically questioned (Roedler, Pohl, and Oechtering 2013) but equally that many owners may consider these as 'normal' for these breeds and hence not offer information unless specifically asked (Packer et al. 2015). Prior warning of respiratory or gastrointestinal issues can highlight potential difficulties during anaesthesia for individual patients, allowing formulation of mitigation plans in advance for prevention or effective management. Pertinent considerations for all brachycephalic breeds include difficult tracheal intubation, URT obstruction and desaturation of the patient, regurgitation and aspiration, and corneal ulceration.

Clinical Examination

In addition to patient history, clinical examination is an essential component of pre-anaesthetic assessment. The aim should be to perform this without causing undue stress on the patient; beginning with an initial 'hands-off' assessment is often useful. Particular attention should be paid to the respiratory system, evaluating for the presence of abnormalities including dyspnoea, tachypnoea, hyperpnoea and decreased nasal airflow. Many brachycephalic dogs will be affected with some degree of URT noise at rest, the degree of which can give an indication as to the underlying severity of disease although such URT noise will tend to worsen with patient stress during transport and admission to hospital. Laryngeal auscultation may provide further information pre-anaesthesia on severity of respiratory compromise. Auscultation of a laryngeal stridor to predict laryngeal collapse is reported with a specificity of 100% and sensitivity of 60% prior to exercise and 70% following exercise (Riggs et al. 2019). Should laryngeal stridor be auscultated, the clinician can be alerted to the presence of more advanced airway disease; however, the absence of laryngeal stridor does not reliably exclude laryngeal collapse.

Thoracic auscultation can be challenging in brachycephalic patients because referred URT sounds are commonly detected and can obstruct cardiac and more subtle lower respiratory sounds. Although very crude, assessment of mucous membrane colour is an easy way to assess some features of cardiovascular and, notably for brachycephalic patients, respiratory function. Colour assessment can give a quick indication on whether urgent intervention is required; cyanosis indicates severe hypoxaemia, with the concentration of blood deoxyhaemoglobin exceeding 15 g/L (Goss, Hayes, and Burdon 1988). This should prompt immediate oxygen supplementation and addressing the

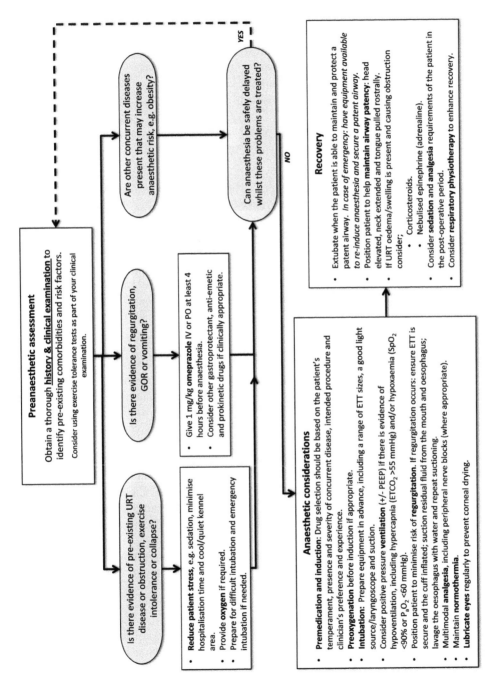

FIGURE 19.3   The decision-making process and common considerations when planning and managing anaesthesia in brachycephalic dogs. (Taken from Downing & Gibson, 2018.)

- Is there any breathing noise such as stertor/stridor during rest or exercise?
- Does the patient snore when sleeping?
- What is the length of time spent sleeping during the day?
- What is the patient's exercise tolerance?
- What is the patient's heat tolerance?
- Has there ever been a history of cyanosis and syncopy?
- Does the patient ever suffer from gagging and retching?
- Does the patient ever regurgitate, choke or vomit?
- Has the patient ever suffered with ocular problems?

FIGURE 19.4   Questions to ask the owner of a brachycephalic dog prior to anaesthesia.

underlying cause rapidly where possible. However, it is important to note that detection of cyanosis is a poorly sensitive method for determining hypoxaemia – crucially the absence of cyanosis does not guarantee normal blood oxygenation. Pulse oximetry, while recognised as challenging to gain accurate readings in the conscious patient, offers superior oxygenation monitoring. In humans, cyanosis may not always be detectable visually despite hypoxaemia ($PaO_2 < 80$ mmHg [10.7 kPa] and/or $SpO_2 < 94\%$ when breathing room air). Ambient light additionally influences apparent mucous membrane colour; for example, use of heat lamps renders colour interpretation almost impossible. Pigmentation of membranes and concurrent disease states resulting in icterus will also affect interpretation.

Functional Tests

Basic exercise tests can provide useful information. Assessment of factors such as respiratory noise, inspiratory and expiratory effort, and presence of dyspnoea, cyanosis and syncope pre-, intra and post-exercise test allows estimation of respiratory compromise (Liu et al. 2016, Riggs et al. 2019). A 3-minute trot test is easy and inexpensive and can improve the sensitivity of the clinical examination for BOAS diagnosis from 56.7% to 93% (Riggs et al. 2019). This can greatly assist the clinician to accurately identify higher risk patients. Given the risk for acute decompensation during such a test, it is recommended that exercise tests are performed within close proximity to additional support should it be required, including staff members, supplementary oxygen, sedative drugs and intubation kits. The UK Kennel Club now has a respiratory function assessment scheme available for owners of Pugs, French Bulldogs and Bulldogs which involves such exercise tolerance tests. While such assessments can provide valuable information, one must be mindful of the timeframe as it is recommended these assessments are performed at least every 2 years (The Kennel Club 2020). Whole body plethysmography can also aid identification of patients with clinically significant BOAS (Liu et al. 2016), although this is not readily available or applicable in general clinical practice.

Fasting and Regurgitation

Withholding food and water prior to anaesthesia is associated with a reduced risk of mortality in dogs (Brodbelt 2006). Therefore, it is generally recommended that canine patients are fasted prior to general anaesthesia. The rationale is that decreased volumes of food and liquid in the stomach may reduce the risk of vomiting or regurgitating during the peri-operative period, which should therefore reduce the aspiration risk. Current guidelines in human medicine for fasting are for withdrawal of solid food 6 hours prior, and clear fluids 2 hours prior, to anaesthesia (Smith et al. 2011). In veterinary medicine, research has shown that prolonged fasting (>12 hours) is associated with decreased gastric pH and increased incidence of reflux (Galatos and Raptopoulos 1995). Feeding a small meal of canned food 3 hours prior to canine anaesthesia increased gastric pH with minimal effect on gastric volume (Savvas, Rallis, and Raptopoulos 2009) and significantly decreased the incidence of reflux compared to feeding 10 hours prior to anaesthesia (Savvas, Raptopoulos, and Rallis 2016). Therefore, prior to elective anaesthesia, whilst it is common to advise withdrawal of food after midnight (this may be the only practical approach), it may be more appropriate to reduce the fasting period in healthy patients to allow a small meal 3–6 hours before anaesthesia. Although some suggest water withdrawal 2–4 hours prior to anaesthesia, the authors commonly allow water to be available until the time of premedication.

Brachycephalic patients commonly tend to show gastrointestinal abnormalities, with increasing severity of airway compromise associated with increased gastrointestinal disease (Poncet et al. 2005). Up to 97% of brachycephalic cases presented for airway disease have oesophageal, gastric or duodenal changes (although often subclinical) with regurgitation likely, (Poncet et al. 2005). If identified, medical management of the gastrointestinal disease (see Chapter 10) prior to anaesthesia can decrease the complication rate associated with URT surgery and improve prognosis (Poncet et al. 2006). Although we lack definitive evidence, it is reasonable to assume similar benefits for cases anaesthetised for other reasons. While there are a number of gastro-protectant and prokinetic drugs available, oral omeprazole has been shown to be superior to famotidine (Tolbert et al. 2011). Administration of 1 mg/kg omeprazole at least 4 hours prior to the induction of anaesthesia reduces the incidence of GOR (Panti et al. 2009) and has been recommended for routine use before anaesthetic induction in brachycephalic dogs (Downing and Gibson 2018).

Gastro-oesophageal reflux during anaesthesia carries a risk of aspiration of gastric contents and subsequent pneumonia, oesophagitis, irritation of pharyngeal tissues and oesophageal stricture formation. If regurgitation is observed, it is useful to,

1. Check the ETT cuff is inflated and secure - this protects the airway and reduce the risk of aspiration.
2. Position the patient with the head elevated slightly and nose angled downwards to facilitate drainage of fluid away from the larynx while discouraging further regurgitation. *This may not be feasible if regurgitation occurs and the patient cannot be moved, e.g. during surgery*
3. Remove any fluid from the oesophagus and pharyngeal area by gentle suction as soon as practically possible.
4. Lavage the oesophagus with water and repeat suctioning until clear fluid is retrieved. This increases oesophageal pH after GOR and may reduce the risk of the associated complications (Wilson and Evans 2007). *N.B. some sources also advise instilling dilute sodium bicarbonate solution into the oesophagus but this is still controversial.*

PREMEDICATION

While not mandatory, premedication of any patient prior to general anaesthesia is highly recommended. The terms 'sedation' and 'premedication' are often used interchangeably and there is often a high degree of crossover regarding the overall aims of each (Figure 19.5). However, it is important to appreciate that premedication agents include not only drugs used for pre-anaesthetic sedation, but also include any other drug therapy given in the lead up to anaesthesia. For example, anti-emetic drugs are considered a premedication agent in human anaesthesia (Steeds and Orme 2006) and the use of pre-anaesthetic omeprazole in brachycephalic breeds as previously discussed can be considered part of the premedication plan. There is no 'perfect' premedication (and induction) drug protocol for brachycephalic dogs, with no robust peer-reviewed clinical evidence to support the use of one technique over another. Drug selection depends upon numerous factors including:

- Reduce anxiety
- Improve patient handling
- Provide analgesia
- Provide muscle relaxation
- Reduce doses of other drugs or agents
- Reduce unwanted side effects due to excessive drug doses
- Improve anaesthetic induction and recovery quality
- Pre-emptively manage likely potential complications such as regurgitation and vomiting

FIGURE 19.5   Broad aims and requirements of an anaesthetic premedication.

- Patient temperament
- The presence and severity of concurrent disease
- The diagnostic or surgical procedure to be carried out under anaesthesia
- Availability of different drugs
- The clinician's familiarity with a particular protocol

In general, drug related effects are less important to anaesthetic risk than procedure, patient and management factors (Brodbelt et al. 2007). This highlights the value of protocol familiarity and it may be better using routine drugs regardless of availability of an alternative that may be touted as 'safer'.

Sedation

Despite the above caveats, premedication often include sedative agents because reducing patient anxiety can vastly improve the overall patient experience and our ability to handle the patient. This can facilitate necessary procedures such as intravenous cannula placement and reduce the required dose of anaesthetic induction agents, thus the risk of post-induction apnoea. The considered use of sedative agents in brachycephalic patients can be a vital part of management during all stages of hospitalisation; vocalisation, stress and panting are all reduced which can lead to improved airflow and reduced risk of hyperthermia and airway oedema. However, excessive sedation in brachycephalic breeds can be problematic, resulting in hypoventilation and exacerbation of airway obstruction due to relaxation of pharyngeal musculature and recumbency (Mosing 2016, Murrell 2016, Ellis and Leece 2017). Therefore sedated brachycephalic patients must be closely monitored at all times for signs of URT obstruction, including increased stertor/stridor, increased inspiratory effort and exaggerated movement of the chest wall. It is recommended that patients clinically affected with respiratory compromise receive low doses of sedative agents where required and appropriate. Commonly used sedative drugs and their dosages are highlighted in Figure 19.6 with acepromazine (ACP) and alpha-2 agonists such as medetomidine or dexmedetomidine being the most common sedative agents incorporated into premedication or sedative plans. Other drugs, notably trazodone and gabapentin, have recently become accepted in veterinary medicine for their sedative actions, particularly for hospitalised patients showing signs of stress or anxiety. Trazodone has been shown to reduce stress related signs and behaviours in hospitalised dogs (Gilbert-Gregory et al. 2016), and aid calmness during necessary confinement in post-surgical patients (Gruen et al. 2014). It is advisable to avoid using trazodone in patients who are prescribed tramadol, as concurrent administration may result in serotonin syndrome (Borland, Clutton, and Bradbrook 2015). Anecdotally, the authors have used trazodone with success in brachycephalic patients during the hospitalisation period. Despite anecdotal reports of use of gabapentin, there is currently no evidence to support its routine use as a sedative in dogs.

Analgesia

For patients with pre-existing pain or those scheduled for a painful procedure, the provision of analgesia forms a large component of the premedication. Several analgesic options are available for management of acute surgical pain in dogs (Bradbrook and Clark 2018a, b). Opioids are often a reasonable analgesic choice, particularly for premedication of surgical patients, with methadone or buprenorphine licensed in dogs in the UK and commonly used. While providing excellent analgesia, opioids also carry potential adverse effects. The notable respiratory depressant effect of opioids seen in humans is less commonly a problem in dogs but sedation, vomiting, regurgitation and panting are post-opioid effects of particular concern for brachycephalic breeds. The sedative effects of certain opioids can be advantageous, especially when used as part of a premedication, although care needs to be taken to avoid excessive levels as discussed previously. Panting is most commonly associated with the administration of methadone (Monteiro et al. 2008), which might be deleterious in brachycephalic breeds, resulting in altered airflow and increased risk of airway oedema. Vomiting and regurgitation is more commonly associated with morphine (Lefebvre, Willems, and Bogaert 1981); however it has been shown that the indiscriminate use of methadone postoperatively in dogs following orthopaedic surgery, regardless of clinical need or pain score, dramatically increased the risk of vomiting and vocalisation compared with those receiving the drug based on individual pain scoring (Bini et al. 2018). Therefore, while analgesia should never be denied, additional analgesia following premedication should always be based upon individual needs.

While not strictly considered part of the patient's premedication, it is wise to consider other forms of analgesia, taking a multi-modal approach (Bradbrook and Clark 2018a, b, Gurney 2012). It is reasonable to assume that this could reduce the required dose of opioids and therefore reduce associated adverse effects. This may include use of additional systemic drugs such as paracetamol, ketamine and anti-inflammatory drugs when deemed appropriate. Locoregional anaesthetic techniques can provide excellent perioperative analgesia in a variety of surgical patients. However it is worth noting that the techniques are not without risk, with potential complications including nerve damage, prolonged motor blockade, vascular puncture and toxicity associated with local anaesthetic drugs (Gurney and Leece 2014).

TEMPERATURE MANAGEMENT

Efforts should be made to keep the patient's temperature within the normal range throughout the whole hospitalisation period. Stress and agitation, for example as a result of transportation, admittance and confinement within the hospital, can contribute to hyperthermia and exacerbate existing respiratory disease and distress. If patients are hyperthermic (but not pyrexic), active cooling may be required including the use of fans, cold intravenous (IV) fluid therapy, cold water baths and in extreme cases, cold water enemas. Anxious and stressed patients may benefit from sedation as previously discussed (Figure 19.6) but observation and management for URT obstruction in sedated patients is essential. Sedated patients also require regular temperature monitoring because they may subsequently develop hypothermia. This can be exacerbated by general anaesthesia, which impairs normal homeostatic thermoregulatory mechanisms. Hypothermia can result in an increased incidence of post-operative infections, impaired coagulation and increased intraoperative blood loss, altered drug metabolism (most commonly an increased duration of action) and potentially a prolonged recovery (Frank 2001). Shivering in the post-operative period can also increase oxygen demand by up to 400% (Mosing 2016), which is worthy of particular note in brachycephalic dogs who already have greater inherent risk of hypoxaemia. Methods for conserving heat in patients include passive warming techniques such as use of blankets, airway humidification through use of heat-moisture exchangers (HME) and active warming techniques including heat pads and forced warm air blankets.

Drug	Dose	Route of administration
Acepromazine	**Premedication:** 5-20 µg/kg*	IM or IV
	Post-operative sedation: 5-10 µg/kg q. 3-6 hours	IV
Medetomidine	**Premedication:** 1-10 µg/kg	IM or IV
	Post-operative sedation: 1-3 µg/kg IV up to q. 1 hour if needed +/- CRI 1-2 µg/kg/hr	IV
Dexmedetomidine	**Premedication:** 1-5 µg/kg mcg/kg	IV or IM
	Post-operative sedation: 0.5-1 µg /kg IV up to q. 1 hour +/- CRI 0.5-1 µg/kg/hr	IV
Methadone	0.1-0.3 mg/kg	IM or IV
Buprenorphine	0.01-0.02 mg/kg	IM or IV
Butorphanol	0.1-0.3 mg/kg	IM or IV
Propofol	1-4 mg/kg (given slowly to effect)	IV
Alfaxalone	2 mg/kg (given slowly to effect)	IV
Trazadone	2-10 mg/kg ^	PO

FIGURE 19.6   Common drugs available for use for premedication, induction of anaesthesia and post-operative sedation in **dogs.** (Adapted from Downing & Gibson, 2018.) The doses specified are those commonly used by the authors and lie within licensed ranges in the UK. In dogs with clinical signs of BOAS the authors recommend using low doses of sedative drugs where required and appropriate. (* Lower doses of acepromazine (10 µg/kg or less) are generally recommended in Boxer breeds (Murrel, 2016). ^ To avoid excessive sedation, the authors recommend starting with the lower end of the dose range and increasing the dose and frequency of administration as required in accordance with the needs of the individual patient. Trazodone can be given up to twice daily but is <u>not</u> a licensed veterinary medicine in the UK.)

PREOXYGENATION

Preoxygenation using a tight-fitting facemask is routinely performed in human medicine prior to anaesthetic induction. The associated increase in arterial oxygen saturation reduces the rate of desaturation should post-induction apnoea occur (Nimmagadda, Salem, and Crystal 2017). Preoxygenation with 100% oxygen via a facemask for 3 minutes prior to anaesthetic induction also increased the time to desaturation in dogs (McNally, Robertson, and Pablo 2009). This can be particularly useful in brachycephalic breeds with their propensity for URT obstruction and increased risk of hypoxaemia during anaesthetic induction, combined with the challenges often encountered when attempting tracheal intubation (Downing and Gibson 2018). Brachycephalic dogs often have lower arterial oxygen concentrations than non-brachycephalic dogs when conscious (Hoareau et al. 2012) which means that desaturation may occur more rapidly in brachycephalic breeds during periods of apnoea.

Despite its benefits, preoxygenation is not a benign procedure. While potential problems reported in humans such as absorption atelectasis, production of reactive oxygen species and undesirable haemodynamic effects (Nimmagadda, Salem, and Crystal 2017) are unlikely to be of clinical consequence in dogs due to the brevity of preoxygenation, the act of preoxygenation can cause significant patient stress. This may arise due to physical restraint, application of a facemask (which may also result in corneal damage) and the noise or sensation of oxygen flow (Downing and Gibson 2018).

Therefore it is important to consider on a dog-by-dog basis whether the benefits of preoxygenation outweigh the risks of exacerbating patient respiratory difficulties and increased oxygen requirement as a result of increased stress levels.

Anaesthetic Induction

Anaesthetic induction should ideally be performed using injectable agents, which allow rapid transition to unconsciousness and can be titrated to effect. Propofol and alfaxalone are both licensed and commonly used for this purpose. Prior to tracheal intubation, if possible, it is advisable to assess the presence and severity of BOAS and laryngeal collapse in all brachycephalic dogs. Along with providing a reference point to determine progression of pre-existing URT disease, airway examination can help to identify higher risk patients, which may develop problems during anaesthetic recovery or during subsequent anaesthetics. This will allow appropriate preparations to be made to prevent or manage potential difficulties more successfully. While airway assessments may also be of benefit in brachycephalic cats, clinicians need to be aware that the clinical picture of feline BOAS is different, with stenotic nares being the prominent feature; airway obstruction rarely occurs due to other aspects that are common to dogs, including an elongated soft palate and laryngeal collapse.

Tracheal intubation can be challenging in brachycephalic dogs and prior consideration should be given to additional equipment that may be required (Figure 19.7). Access to, and use of, a laryngoscope is highly recommended, with the possibility of an additional light source being useful. A wide range of sizes of cuffed endotracheal tubes (which may need to be longer with a narrow diameter) should be readily available. A rigid stylet is sometimes useful to aid visualisation of the larynx. In severe cases of tracheal hypoplasia, a urinary catheter can act either as a stylet to 'rail-road' an endotracheal tube over the top, or it can be modified to connect to a breathing system to allow direct oxygen delivery. Availability of a suction catheter and machine to allow removal of regurgitated material or saliva can be useful.

Anaesthetic Maintenance

Unless contra-indicated, maintenance of anaesthesia is best achieved using inhalational agents. Both isoflurane and sevoflurane are licensed for dogs and either is appropriate for brachycephalic patients. They are rapidly cleared from the body via the lungs upon recovery, thus expediting the recovery process. Depending upon the surgical or diagnostic procedure, additional injectable agents may be needed to provide balanced anaesthesia. If required, use of short-acting or reversible agents such as fentanyl and medetomidine, or lower doses of drugs including ketamine (5–10 mcg/kg/min IV) and lidocaine (30–50 mcg/kg/min IV), used as intravenous infusions may reduce unwanted effects during the anaesthetic recovery period.

Ventilation and Lung Function

Hypoventilation and hypercapnia are common in the conscious brachycephalic dog (Hoareau et al. 2012). The increased airway resistance and work of breathing typical of brachycephalic dogs and which contributes to hypoventilation may worsen during general anaesthesia because of tracheal hypoplasia and the resultant narrow diameter endotracheal tubes required. Sedation and general anaesthesia are likely to exacerbate hypoventilation and hypercapnia due to a combination of drug effects, patient positioning and excessive body condition score. Capnography to monitor end tidal carbon dioxide ($ETCO_2$) levels will allow detection of significant hypercapnia following hypoventilation. This can be managed using positive pressure ventilation (PPV). While $ETCO_2$ values of 35–45 mmHg (4.7–6.0 kPa) are reported as typical of conscious dogs, $ETCO_2$ up to 55 mmHg can be acceptable in healthy dogs of any breed that are stable under anaesthesia before interventions including ventilatory support are required.

- Calm environment
- Appropriate sedation and analgesia if necessary.
- Equipment for emergency tracheal intubation:
  - IV induction agent.
  - A good light source +/- a rigid stylet to aid visualisation of the larynx, e.g. a laryngoscope with appropriate sized blades.
  - A selection of clean, appropriately sized ET tubes.
  - Tie to secure the ET tube.
- Suction machine (or similar) and catheter for managing regurgitation.
- Ready access to an oxygen supply.
- +/- Nebuliser machine.
- An experienced assistant

FIGURE 19.7   Equipment and requirements for anaesthetic induction/recovery.

As brachycephalic breeds have lower resting arterial oxygen levels compared to other dogs (Hoareau et al. 2012) and often suffer from pre-existing respiratory disease (De Lorenzi, Bertoncello, and Drigo 2009, Roedler, Pohl, and Oechtering 2013, Packer et al. 2015), all options to preserve or improve lung function during anaesthesia should be considered. During anaesthesia, atelectasis is common in part resulting from reduced ventilation, recumbency and absorption atelectasis due to the high concentrations of inspired oxygen. Atelectasis leads to decreased lung compliance, increased pulmonary vascular resistance and in severe cases may trigger development of acute lung injury (ALI) (Ray, Bodenham, and Paramasivam 2014). Absorption atelectasis can be slowed, thus maintaining better lung aeration and gas exchange, by routinely using lower inspired concentrations of oxygen; this is achieved by using a medical air/oxygen mix during anaesthesia. Dogs receiving an inspired oxygen concentration (FiO_2) of 40% have been shown to have better lung aeration and gas exchanged then those receiving 100% oxygen (Staffieri et al. 2007). However, a reduced FiO_2 should only be used if there is no evidence of hypoxaemia during anaesthesia. This can be determined through use of pulse oximetry to measure the saturation of haemoglobin with oxygen (SpO_2), ensuring that levels >90% are achieved and maintained.

When available, the use of advanced ventilation strategies including PPV, continuous positive airway pressure (CPAP) and positive-end expiratory pressure (PEEP) can reduce the development of atelectasis and help preserve lung function during anaesthesia (Ray, Bodenham, and Paramasivam 2014). However, use of such strategies should be balanced with the potential for the negative effects they may have on haemodynamic variables such as blood pressure (Downing and Gibson 2018).

Ocular Care

Protective ocular reflexes are depressed or absent under anaesthesia, increasing the risk of corneal drying and ulceration. Corneal abrasions can occur from direct trauma, chemical injuries and excessive drying as a result of warming devices. Steps should be taken to avoid injury including considered patient positioning and regular eye lubrication with either carbomer eye gels or sodium hyaluronate to minimise corneal drying. Commonly used anaesthetic agents including some opioids, inhalation agents and medetomidine have been shown to reduce tear production. General anaesthesia reduces tear production for at least 24 hours (Herring et al. 2000) and therefore it is commonly advised to use corneal lubricants for at least 48 hours following both sedation and anaesthesia. Brachycephalic breeds often have protruding eyes, which combined with a reduced corneal sensitivity result in a delayed return of palpebral reflex during anaesthetic recovery (Park et al. 2013, Barrett et al. 1991), putting them at even greater risk of ocular damage.

ANAESTHETIC RECOVERY AND POST-OPERATIVE MANAGEMENT

General anaesthesia results in many physiological alterations that the patient must normalise during recovery. The recovery period is reported as the time when most anaesthetic mortality occurs in dogs (Brodbelt et al. 2008). The anatomical and physiological anomalies in brachycephalic breeds can further increase mortality risks at this time when ventilatory and respiratory effects from residual anaesthetic drugs, hypothermia and increased work of breathing make hypoventilation and hypoxia more likely. Additionally, a major concern with any brachycephalic dog during anaesthetic recovery is URT obstruction, particularly in patients that have undergone airway surgery. With the relative excess of URT tissue in brachycephalics, it is imperative that patients are sufficiently alert and have enough muscle tone to maintain airway patency themselves **before** attempting tracheal extubation. The high prevalence of gastrointestinal disorders and GOR (Poncet et al. 2006, 2005) means aspiration of regurgitant material during the recovery period is an increased concern. Risks are amplified following airway surgery, as post-surgical mucosal oedema, bleeding or blood clots may be present exacerbating URT obstruction and risk of aspiration.

It is crucial that the veterinary team are well prepared for the recovery period. Appropriate planning and a proactive approach can reduce the chance that a worsening cycle whereby URT obstruction and oedema, altered airflow and dynamic airway collapse may potentially culminate in pulmonary oedema (Downing and Gibson 2018).

Preparation for Recovery

Some degree of forethought and preparation for anaesthetic recovery is essential to ensure the smoothest possible progress. As tracheal re-intubation is more likely in brachycephalic dogs, it is advisable that two suitably qualified people (e.g. veterinary surgeon and RVN) are present throughout the recovery period. General considerations prior to commencing anaesthetic recovery should include:

- Provision of adequate analgesia
- Provision of anti-inflammatory drugs
- Access to necessary equipment for emergency re-intubation
- Patient positioning
- Provision of additional sedation

Provision of Analgesia

For any surgical procedure, aside from the moral obligation to provide analgesia, it is prudent to ensure that brachycephalic patients have appropriate analgesia prior to, and throughout the post-operative period. Painful patients may show numerous behavioural changes that may further the already compromised respiratory function in brachycephalic dogs. Associated agitation and vocalisation can exacerbate URT oedema and collapse. Reduced movement and recumbency can lead to reduced lower airway and lung function due to poor lung expansion, atelectasis and reduced mobilisation or clearance of secretions (Pathmanathan, Beaumont, and Gratrix 2014). Pain can impact on respiratory function, tissue oxygen supply and normal gastrointestinal function, causing further compromise to the patient. Effects can include an altered ventilation (either tachypnoea or hypoventilation) and increased sympathetic response resulting in increased blood pressure, heart rate and myocardial oxygen demand. Pain can result in gastrointestinal signs ranging from reduced motility and delayed gastric emptying to nausea and vomiting, compounding any pre-existing clinical signs. As a result, painful patients may have prolonged hospitalisation periods that may ultimately impact upon the speed of recovery and incidence of adverse effects. This can also carry financial implications for the pet owner. However, as discussed previously, all analgesic drugs have the potential for unwanted side effects including vomiting, vocalisation and panting; therefore, analgesia must always be used appropriately based on the patient's need.

Administration of Anti-inflammatory and Vasoconstrictive Drugs

Specifically for airway surgery, anti-inflammatory drugs are recommended prior to commencing recovery. Administration of intravenous glucocorticoids (usually dexamethasone) is most usual and may need to be repeated at intervals throughout the recovery period depending upon the clinical picture. Some clinicians may prefer to use non-steroidal anti-inflammatory drugs (NSAIDs), although these are usually reserved for less severely affected patients. Use of a topical vasoconstrictor such as xylometazoline (Otrivine; GSK, UK) prior to anaesthetic recovery may help manage nasal oedema and improve airflow following instillation into the nares.

Equipment for Emergency Re-intubation

Equipment and additional personnel that may be required (Figure 19.7) should be readily accessible prior to commencing the recovery process, with the clinician ready to perform an emergency re-intubation as required. While the prophylactic administration of gastroprotectant drugs such as omeprazole should reduce the incidence of GOR during anaesthesia (Panti et al. 2009), the recovery team should still be prepared to clear the mouth and pharynx and provide gentle suction of the oesophagus should GOR occur. However, care should be taken as depending upon the stage of recovery and the patient's level of consciousness - attempts at suctioning may initiate the gag reflex and further regurgitation in the later stages of recovery.

Patient Positioning

Wherever possible, brachycephalic patients should be placed in sternal recumbency for recovery. Elevating the head slightly (e.g. using a folded towel) and ensuring the patient's neck is extended will minimise the potential for pharyngeal soft tissue to cause airway obstruction. Pulling the patient's tongue rostrally out of the mouth will also help ensure optimal airflow through the narrowed pharynx. Some patients will benefit from having their mouth held open; if tolerated, a roll of cohesive bandage material makes for a good temporary mouth gag, although the patient must never be left unattended with this in place.

Care should be taken regarding the timing of tracheal extubation. While extubation of non-brachycephalic patients is usually recommended at the point when the swallowing reflex returns (Bednarski et al. 2011), it is widely accepted that brachycephalic breeds will tolerate endotracheal tubes for much longer during the recovery period. Whilst we need to be mindful that delayed tracheal extubation could precipitate gagging and regurgitation, it is possible for brachycephalic patients to tolerate the endotracheal tube up to the point of full consciousness. This is advantageous by allowing respiratory and ventilatory function more time to recover from the anaesthetic effects and ensuring a patent airway which is protected against potential aspiration of regurgitate material, blood or fluid. However, even if the endotracheal tube is removed at this late stage, laryngeal collapse or airway compromise due to post-surgical oedema can still occur, resulting in URT obstruction and necessitating emergency re-intubation.

Where possible, manual expression of the patient's urinary bladder prior to recovery can aid with patient comfort levels and contribute to a smoother recovery.

Recovery Following Tracheal Extubation

Patients must be monitored closely and regularly following tracheal extubation, with intravenous access maintained until the point of hospital discharge where possible. This section of the perioperative period has one of the highest risks for upper respiratory tract obstruction. It is sensible to ensure the person nominated for patient monitoring is prepared and competent for emergency re-intubation if required (Figure 19.7). The aim for anaesthetic recovery should be to keep the patient **cool and calm.** While hypothermia should ideally be prevented and treated if present, brachycephalic patients are particularly prone to heat stress. Close observation of body temperature is required to avoid over-heating, which can contribute to worsening of upper respiratory tract signs. Efforts should also be made to minimise stress and anxiety levels in recovering patients as excessive vocalisation and panting can cause deterioration of airway swelling.

Sedation

Sedation (Figure 19.6) may be required in the post-operative period to aid with patient stress management. The aim is to have a calm patient without causing excessive sedation that may result in pharyngeal tissue relaxation, thus contributing to airway obstruction. Knowledge of the patient's response to a prior premedication can help guide both what drugs and dose are most likely to benefit the patient.

Nebulisation and Physiotherapy

In severe URT obstruction in both dogs and humans, nebulisation with epinephrine (adrenaline) can effectively reduce laryngeal oedema, likely due to localised vasoconstriction. This can alleviate obstruction and may avoid the need for tracheostomy tube placement (Ellis and Leece 2017, MacDonnell, Timmins, and Watson 1995; Franklin, Liu, and Ladlow 2021). However, nebuliser use may be stressful for some patients due to the associated noise and physical restraint required, causing exacerbation of URT obstruction; in such cases nebulisation may have to be avoided. Quiet and compact nebuliser units such as the Flexineb C1 nebuliser (Nortev, Ireland) can negate some of these problems and are generally very well tolerated (Figure 19.8).

Literature suggests that a total dose of 0.05 mg/kg epinephrine diluted into 5 mL of 0.9% saline and nebulised for 10 minutes can form part of a successful management plan for pre- and post-surgical URT obstruction (Ellis and Leece 2017; Franklin, Liu, and Ladlow 2021). It is important that epinephrine dilution is achieved using an isotonic solution because water can cause bronchoconstriction (Beasley, Rafferty, and Holgate 1988). Anecdotally, higher doses of epinephrine (Figure 19.9) have also been effective without noted untoward effects, although it may be advisable to use an ECG to monitor the cardiac rhythm during treatment with these higher doses.

Use of physiotherapy, specifically respiratory physiotherapy, is commonplace in human critical care medicine (Pathmanathan, Beaumont, and Gratrix 2014) where recumbency, tracheal intubation and drug therapies can impair clearing mechanisms and lung function. While many techniques reported for use in humans are not directly applicable for veterinary species, early mobilisation of patients can assist with clearance of secretions and pulmonary re-expansion; this reduces atelectasis and improves respiratory function by decreasing the work of breathing and improving oxygenation. This may simply involve encouraging the patient to stand or gently ambulate, with or without assistance. Whilst the potential benefits of incorporating respiratory physiotherapy in the postoperative period are not exclusive to brachycephalic breeds, the higher frequency of underlying respiratory disease in these

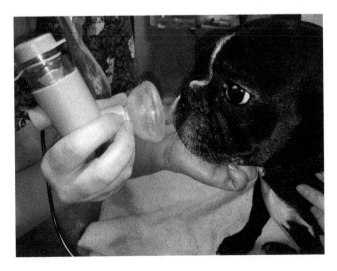

FIGURE 19.8 The Flexineb C1 Nebuliser (Nortev, Ireland), which can be used to administer nebulised medication, such as epinephrine (adrenaline).

Patient weight (kg)	Dose of 1:1000 (1 mg/mL) epinephrine	Volume of 0.9% saline (mL)
2.5 – 10	1 mg (1 mL)	4
10 – 20	2 mg (2 mL)	3
> 20	3 mg (3 mL)	2

FIGURE 19.9   Anecdotal epinephrine (adrenaline) doses and 0.9% saline volumes used for nebulisation in dogs. Nebulisation is carried out for 5–10 minutes and can be repeated after 30 minutes if no improvement is seen.

dogs mean that they may be more susceptible to anaesthetic-related decreases in lung function such as absorption atelectasis. Physiotherapy should always be conducted in conjunction with adequate analgesia; good pain control can help minimise complications by promoting earlier ambulation and enabling patients to take deeper breaths (Ray, Bodenham, and Paramasivam 2014).

Tracheostomy

In patients where sustained tracheal extubation is not successful, for example due to severe BOAS, laryngeal collapse or post-surgical oedema, a temporary or permanent tracheostomy may be required. Tracheostomy placement and management is described in more detail in Chapter 11.

It has been reported that up to 58% of BOAS patients suffering concurrent laryngeal collapse required temporary tracheostomy placement following surgical intervention (White 2012). Increasing patient age has been shown to be a risk factor for placement; the odds of dogs requiring a temporary tracheostomy tube increases by 30% for each 1 year increase in age (Worth et al. 2018). Additional risk factors for tracheostomy placement identified in the same study included the treatment of pneumonia and corticosteroid administration post-operatively, although it is considered that these were not causal associations, merely a reflection of disease severity.

The placement of a tracheostomy tube can be life-saving in some patients. However, once placed, the patient will require intensive nursing, which can have both practical and financial implications (see Chapter 3). There are significant risks associated with temporary tracheostomy placement, with reported complication rates of 86% and mortality rates of 19%, (Nicholson and Baines 2012). Brachycephalic breeds were over-represented in those patients with complications, the most frequent of which were airway or tube obstruction, dislodgement of the tube, swelling and discharge from the stoma and aspiration pneumonia. It is also noted that patients requiring temporary tracheostomy following BOAS surgery had increased hospitalisation periods (Worth et al. 2018), although this retrospective study showed the mortality did not differ significantly between patients requiring temporary tracheostomy and those that did not.

A similarly high complication rate has been noted from permanent tracheostomy due to severe laryngeal collapse in association with BOAS. A median survival time of 100 days was reported with major complications occurring in 80% of dogs, resulting in death or euthanasia in a little over half of the total study population. However, it was also reported that one-third of this study population had long term survival (> 5 years), deemed to be with a good quality of life (Gobbetti et al. 2018).

CONCLUSION

Ultimately, there is not a single, perfect anaesthetic protocol for brachycephalic patients. When anaesthetising these patients, it is important to consider their management during the entire perioperative period, including:

- Pre-anaesthetic assessment and stabilisation.
- Induction and maintenance of general anaesthesia.
- Recovery from anaesthesia and postoperative management.

Management during these periods will be influenced by the patient's temperament, the presence and severity of concurrent disease, the procedure to be carried out and the individual veterinary surgeon involved. Where concurrent conditions that increase in anaesthetic risk are present in a patient, e.g. gastro-oesophageal reflux or obesity, consider addressing these prior to anaesthesia where practical and possible.

REFERENCES

American Society of Anesthesiologists 2019. ASA Physical Status Classification System [online] Available at https://www.asahq.org/standards-and-guidelines/asa-physical-status-classification-system Accessed on 05/2/2020.

Barrett, P. M., R. H. Scagliotti, R. E. Merideth, P. A. Jackson, and F. L. Alarcon. 1991. "Absolute corneal sensitivity and corneal trigeminal nerve anatomy in normal dogs." *Prog. Vet Comp Ophthalmol.* 1 (4):245–254. https://eurekamag.com/research/002/017/002017472.php.

Bashir, A., S. Q. Ahmad, J. Silverman, E. Concepcion, and H. Lee. 2017. "Post-obstructive pulmonary edema from aspirated nuts." *SAGE Open Med Case Rep* 5. https://doi.org/10.1177/2050313X17717391. https://www.ncbi.nlm.nih.gov/pubmed/28717510.

Beasley, R., P. Rafferty, and S. T. Holgate. 1988. "Adverse reactions to the non-drug constituents of nebuliser solutions." *Br J Clin Pharmacol* 25 (3):283–287. https://doi.org/10.1111/j.1365-2125.1988.tb03305.x. http://www.ncbi.nlm.nih.gov/pubmed/3358893.

Bednarski, R., K. Grimm, R. Harvey, V. M. Lukasik, W. S. Penn, B. Sargent, and K. Spelts. 2011. "AAHA anesthesia guidelines for dogs and cats." *J Am Anim Hosp. Assoc.* 47 (6):377–385. https://doi.org/10.5326/jaaha-ms-5846. https://www.jaaha.org/doi/abs/10.5326/JAAHA-MS-5846.

Bini, G., E. Vettorato, C. De Gennaro, and F. Corletto. 2018. "A retrospective comparison of two analgesic strategies after uncomplicated tibial plateau levelling osteotomy in dogs." *Vet Anaesth Analg* 45 (4):557–565. https://doi.org/10.1016/j.vaa.2017.11.005.

Boesch, R. P., P. Shah, M. Vaynblat, M. Marcus, M. Pagala, S. Narwal, and M. Kazachkov. 2005. "Relationship between upper airway obstruction and gastroesophageal reflux in a dog model." *J Invest Surg* 18 (5):241–245. https://doi.org/10.1080/08941930500248656. https://www.ncbi.nlm.nih.gov/pubmed/16249167.

Borland, K., R.E. Clutton, and C. Bradbrook. 2015. "Possible association of tachycardia and hypertension with perioperative trazodone administration in a dog." *Vet Rec Case Rep* 3 (1):e000231. https://doi.org/10.1136/vetreccr-2015-000231. https://vetrecordcasereports.bmj.com/content/vetreccr/3/1/e000231.full.pdf.

Bradbrook, C. A., and L. Clark. 2018a. "State of the art analgesia- recent developments in pharmacological approaches to acute pain management in dogs and cats. Part 1." *Vet J* 238:76–82. https://doi.org/10.1016/j.tvjl.2018.06.003. http://www.ncbi.nlm.nih.gov/pubmed/29907456.

Bradbrook, C., and L. Clark. 2018b. "State of the art analgesia-Recent developments pharmacological approaches to acute pain management in dogs and cats: Part 2." *Vet J* 236:62–67. https://doi.org/10.1016/j.tvjl.2018.04.009. http://www.ncbi.nlm.nih.gov/pubmed/29871752.

Brodbelt, D. 2009. "Perioperative mortality in small animal anaesthesia." *Vet J* 182 (2):152–161. https://doi.org/10.1016/j.tvjl.2008.06.011. http://www.ncbi.nlm.nih.gov/pubmed/18658000.

Brodbelt, D. C., K. J. Blissitt, R. A. Hammond, P. J. Neath, L. E. Young, D. U. Pfeiffer, and J. L. Wood. 2008. "The risk of death: The confidential enquiry into perioperative small animal fatalities." *Vet Anaesth Analg* 35 (5):365–373. https://doi.org/10.1111/j.1467-2995.2008.00397.x. http://www.ncbi.nlm.nih.gov/pubmed/18466167.

Brodbelt, D. C., D. U. Pfeiffer, L. E. Young, and J. L. Wood. 2007. "Risk factors for anaesthetic-related death in cats: Results from the confidential enquiry into perioperative small animal fatalities (CEPSAF)." *Br J Anaesth* 99 (5):617–623. https://doi.org/10.1093/bja/aem229. http://www.ncbi.nlm.nih.gov/pubmed/17881744.

Brodbelt, D. C. 2006. "The Confidential Enquiry into Perioperative Small Animal Fatalities." Doctor of Philosophy, University of London. http://www.rvc.ac.uk/Staff/Documents/dbrodbelt_thesis.pdf

Corso, R., V. Russotto, C. Gregoretti, and D. Cattano. 2017. "Perioperative management of obstructive sleep apnea: A systematic review." *Minerva Anestesiol.* https://doi.org/10.23736/S0375-9393.17.11688-3. https://www.ncbi.nlm.nih.gov/pubmed/28402089.

Crane, C., E. A. Rozanski, A. L. Abelson, and A. deLaforcade. 2017. "Severe brachycephalic obstructive airway syndrome is associated with hypercoagulability in dogs." *J Vet Diagn Invest* 29 (4):570–573. https://doi.org/10.1177/1040638717703434. https://www.ncbi.nlm.nih.gov/pubmed/28381131.

De Lorenzi, D., D. Bertoncello, and M. Drigo. 2009. "Bronchial abnormalities found in a consecutive series of 40 brachycephalic dogs." *J Am Vet Med Assoc* 235 (7):835–840. https://doi.org/10.2460/javma.235.7.835. http://www.ncbi.nlm.nih.gov/pubmed/19793013.

Downing, F., and S. Gibson. 2018. "Anaesthesia of brachycephalic dogs." *J Small Anim Pract* 59 (12):725–733. https://doi.org/10.1111/jsap.12948. http://www.ncbi.nlm.nih.gov/pubmed/30374971.

Doyle, C. R., T. K. Aarnes, G. A. Ballash, E. L. Wendt-Hornickle, C. F. Baldo, R. A. Johnson, T. E. Wittum, and M. A. McLoughlin. 2020. "Anesthetic risk during subsequent anesthetic events in brachycephalic dogs that have undergone corrective airway surgery: 45 cases (2007-2019)." *J Am Vet Med Assoc* 257 (7): 744–749. https://doi.org/10.2460/javma.257.7.744. https://www.ncbi.nlm.nih.gov/pubmed/32955391.

Dupré, G., and D. Heidenreich. 2016. "Brachycephalic Syndrome." *Vet Clin North Am Small Anim Pract* 46 (4):691–707. https://doi.org/10.1016/j.cvsm.2016.02.002. https://www.ncbi.nlm.nih.gov/pubmed/27012936.

Ellis, J., and E. A. Leece. 2017. "Nebulized adrenaline in the postoperative management of brachycephalic obstructive airway syndrome in a pug." *J Am Anim Hosp Assoc* 53 (2):107–110. https://doi.org/10.5326/JAAHA-MS-6466. http://www.ncbi.nlm.nih.gov/pubmed/28282230.

Emmerson, T. 2014. "Brachycephalic obstructive airway syndrome: A growing problem." *J Small Anim Pract* 55 (11):543–544. https://doi.org/10.1111/jsap.12286. https://www.ncbi.nlm.nih.gov/pubmed/25352108.

Fasanella, F. J., J. M. Shivley, J. L. Wardlaw, and S. Givaruangsawat. 2010. "Brachycephalic airway obstructive syndrome in dogs: 90 cases (1991–2008)." *J Am Vet Med Assoc* 237 (9):1048–1051. https://doi.org/10.2460/javma.237.9.1048. https://www.ncbi.nlm.nih.gov/pubmed/21034343.

Frank, S. M. 2001. "Consequences of hypothermia." *Curr Anaesth Crit Care* 12 (2):79–86. https://doi.org/http://dx.doi.org/10.1054/cacc.2001.0330. http://www.sciencedirect.com/science/article/pii/S0953711201903308.

Franklin, P. H., N. C. Liu, and J. F. Ladlow. 2021. "Nebulization of epinephrine to reduce the severity of brachycephalic obstructive airway syndrome in dogs." *Vet Surg* 50 (1): 62–70. https://doi.org/10.1111/vsu.13523. https://www.ncbi.nlm.nih.gov/pubmed/33044024.

Galatos, A. D., and D. Raptopoulos. 1995. "Gastro-oesophageal reflux during anaesthesia in the dog: the effect of preoperative fasting and premedication." *Vet Rec* 137 (19):479–483. http://www.ncbi.nlm.nih.gov/pubmed/8578660.

German, A. J. 2006. "The growing problem of obesity in dogs and cats." *J Nutr* 136 (7 Suppl):1940S–1946S. https://doi.org/10.1093/jn/136.7.1940S.

Gilbert-Gregory, S. E., J. W. Stull, M. R. Rice, and M. E. Herron. 2016. "Effects of trazodone on behavioral signs of stress in hospitalized dogs." *J Am Vet Med Assoc* 249 (11):1281–1291. https://doi.org/10.2460/javma.249.11.1281.

Gobbetti, M., S. Romussi, P. Buracco, V. Bronzo, S. Gatti, and M. Cantatore. 2018. "Long-term outcome of permanent tracheostomy in 15 dogs with severe laryngeal collapse secondary to brachycephalic airway obstructive syndrome." *Vet Surg* 47 (5):648–653. https://doi.org/10.1111/vsu.12903. https://onlinelibrary.wiley.com/doi/abs/10.1111/vsu.12903.

Goss, G. A., J. A. Hayes, and J. G. Burdon. 1988. "Deoxyhaemoglobin concentrations in the detection of central cyanosis." *Thorax* 43 (3):212–213. https://doi.org/10.1136/thx.43.3.212. http://www.ncbi.nlm.nih.gov/pubmed/3406907.

Gruen, M. E., S. C. Roe, E. Griffith, A. Hamilton, and B. L. Sherman. 2014. "Use of trazodone to facilitate postsurgical confinement in dogs." *J Am Vet Med Assoc* 245 (3):296–301. https://doi.org/10.2460/javma.245.3.296. https://www.ncbi.nlm.nih.gov/pubmed/25029308. https://www.ncbi.nlm.nih.gov/pmc/articles/PMC4414248/.

Gruenheid, M., T. K. Aarnes, M. A. McLoughlin, E. M. Simpson, D. A. Mathys, D. F. Mollenkopf, and T. E. Wittum. 2018. "Risk of anesthesia-related complications in brachycephalic dogs." *J Am Vet Med Assoc* 253 (3):301–306. https://doi.org/10.2460/javma.253.3.301. https://www.ncbi.nlm.nih.gov/pubmed/30020004.

Gurney, M. A. 2012. "Pharmacological options for intra-operative and early postoperative analgesia: An update." *J Small Anim Pract* 53 (7):377–386. https://doi.org/10.1111/j.1748-5827.2012.01243.x. http://www.ncbi.nlm.nih.gov/pubmed/22747730.

Gurney, M. A., and E. A. Leece. 2014. "Analgesia for pelvic limb surgery. A review of peripheral nerve blocks and the extradural technique." *Vet Anaesth Analg* 41 (5):445–458. https://doi.org/10.1111/vaa.12184.

Haimel, G., and G. Dupré. 2015. "Brachycephalic airway syndrome: A comparative study between pugs and French bulldogs." *J Small Anim Pract* 56 (12):714–719. https://doi.org/10.1111/jsap.12408. https://www.ncbi.nlm.nih.gov/pubmed/26563910.

Herring, I. P., J. P. Pickett, E. S. Champagne, and M. Marini. 2000. "Evaluation of aqueous tear production in dogs following general anesthesia." *J Am Anim Hosp Assoc* 36 (5):427–430. https://doi.org/10.5326/15473317-36-5-427. http://www.ncbi.nlm.nih.gov/pubmed/10997519.

Hoareau, G. L., G. Jourdan, M. Mellema, and P. Verwaerde. 2012. "Evaluation of arterial blood gases and arterial blood pressures in brachycephalic dogs." *J Vet Intern Med* 26 (4):897–904. https://doi.org/10.1111/j.1939-1676.2012.00941.x.

Hoareau, G., and M. Mellema. 2015. "Pro-coagulant thromboelastographic features in the bulldog." *J Small Anim Pract* 56 (2):103–117. https://doi.org/10.1111/jsap.12299. https://www.ncbi.nlm.nih.gov/pubmed/25482453.

Lefebvre, R. A., J. L. Willems, and M. G. Bogaert. 1981. "Gastric relaxation and vomiting by apomorphine, morphine and fentanyl in the conscious dog." *Eur J Pharmacol* 69 (2):139–145. https://doi.org/10.1016/0014-2999(81)90408-8.

Liu, N. C., V. J. Adams, L. Kalmar, J. F. Ladlow, and D. R. Sargan. 2016. "Whole-body barometric plethysmography characterizes upper airway obstruction in 3 brachycephalic breeds of dogs." *J Vet Intern Med* 30 (3):853–865. https://doi.org/10.1111/jvim.13933.

Love, L., and M. G. Cline. 2015. "Perioperative physiology and pharmacology in the obese small animal patient." *Vet Anaesth Analg* 42 (2):119–132. https://doi.org/10.1111/vaa.12219. http://www.ncbi.nlm.nih.gov/pubmed/25131820.

Lund, E., J. Armstrong, C. Kirk, and J. S. Klausner. 2006. "Prevalence and risk factors for obesity in adult dogs from private US veterinary practices." *Intern J Appl Res Vet Med* 4:177–186.

MacDonnell, S. P., A. C. Timmins, and J. D. Watson. 1995. "Adrenaline administered via a nebulizer in adult patients with upper airway obstruction." *Anaesthesia* 50 (1):35–36. https://doi.org/10.1111/j.1365-2044.1995.tb04510.x. http://www.ncbi.nlm.nih.gov/pubmed/7702142.

Martinez, G., and P. Faber. 2011. "Obstructive sleep apnoea." *Continuing Educ Anaesth Crit Care Pain* 11 (1):5–8.

McGreevy, P. D., P. C. Thomson, C. Pride, A. Fawcett, T. Grassi, and B. Jones. 2005. "Prevalence of obesity in dogs examined by Australian veterinary practices and the risk factors involved." *Vet Rec* 156 (22):695–702. https://doi.org/10.1136/vr.156.22.695.

McNally, E. M., S. A. Robertson, and L. S. Pablo. 2009. "Comparison of time to desaturation between preoxygenated and nonpreoxygenated dogs following sedation with acepromazine maleate and morphine and induction of anesthesia with propofol." *Am J Vet Res* 70 (11):1333–1338. https://doi.org/10.2460/ajvr.70.11.1333. http://www.ncbi.nlm.nih.gov/pubmed/19878015.

Monteiro, E. R., C. D. Figueroa, J. C. Choma, D. Campagnol, and C. M. Bettini. 2008. "Effects of methadone, alone or in combination with acepromazine or xylazine, on sedation and physiologic values in dogs." *Vet Anaesth Analg* 35 (6):519–527. https://doi.org/10.1111/j.1467-2995.2008.00412.x.

Mosing, M. 2016. "General principles of perioperative care." In *BSAVA Manual of Canine and Feline Anaesthesia and Analgesia*, edited by T. Duke-Novakovski, M. de Vries and C. Seymour, 13–23. British Small Animal Veterinary Associationm, Quedgeley.

Murrell, J. C. 2016. "Pre-anaesthetic medication and sedation." In *BSAVA Manual of Canine and Feline Anaesthesia and Analgesia*, edited by T. Duke-Novakovski, C. de Vries and C. Seymour, 170–189. British Small Animal Veterinary Association, Quedgeley.

Nicholson, I., and S. Baines. 2012. "Complications associated with temporary tracheostomy tubes in 42 dogs (1998 to 2007)." *J Small Anim Pract* 53 (2):108–114. https://doi.org/10.1111/j.1748-5827.2011.01167.x. https://onlinelibrary.wiley.com/doi/abs/10.1111/j.1748-5827.2011.01167.x.

Nimmagadda, U., M. R. Salem, and G. J. Crystal. 2017. "Preoxygenation: Physiologic basis, benefits, and potential risks." *Anesth Analg* 124 (2):507–517. https://doi.org/10.1213/ane.0000000000001589.

O'Neill, D. G., D. B. Church, P. D. McGreevy, P. C. Thomson, and D. C. Brodbelt. 2014. "Prevalence of disorders recorded in dogs attending primary-care veterinary practices in England." *PLoS One* 9 (3):e90501. https://doi.org/10.1371/journal.pone.0090501.

O'Neill, D. G., E. C. Darwent, D. B. Church, and D. C. Brodbelt. 2016. "Demography and health of Pugs under primary veterinary care in England." *Canine Genet Epidemiol* 3:5. https://doi.org/10.1186/s40575-016-0035-z.

O'Neill, D. G., M. M. Lee, D. C. Brodbelt, D. B. Church, and R. F. Sanchez. 2017. "Corneal ulcerative disease in dogs under primary veterinary care in England: Epidemiology and clinical management." *Canine Genet Epidemiol* 4:5. *Epidemiol* 4: 5. https://doi.org/10.1186/s40575-017-0045-5. https://www.ncbi.nlm.nih.gov/pubmed/28630713.

Packer, R. M., A. Hendricks, M. S. Tivers, and C. C. Burn. 2015. "Impact of facial conformation on canine health: brachycephalic obstructive airway syndrome." *PLoS One* 10 (10):e0137496. https://doi.org/10.1371/journal.pone.0137496. http://www.ncbi.nlm.nih.gov/pubmed/26509577.

Panti, A., R. C. Bennett, F. Corletto, J. Brearley, N. Jeffery, and R. J. Mellanby. 2009. "The effect of omeprazole on oesophageal pH in dogs during anaesthesia." *J Small Anim Pract* 50 (10):540–544. https://doi.org/10.1111/j.1748-5827.2009.00818.x. http://www.ncbi.nlm.nih.gov/pubmed/19796313.

Park, Y. W., W. G. Son, M. B. Jeong, K. Seo, L. Y. Lee, and I. Lee. 2013. "Evaluation of risk factors for development of corneal ulcer after nonocular surgery in dogs: 14 cases (2009–2011)." *J Am Vet Med Assoc* 242 (11):1544–1548. https://doi.org/10.2460/javma.242.11.1544. http://www.ncbi.nlm.nih.gov/pubmed/23683020.

Pathmanathan, N., N. Beaumont, and A. Gratrix. 2014. "Respiratory physiotherapy in the critical care unit." *BJA Education* 15 (1):20–25. https://doi.org/10.1093/bjaceaccp/mku005. https://doi.org/10.1093/bjaceaccp/mku005.

Poncet, C. M., G. P. Dupre, V. G. Freiche, and B. M. Bouvy. 2006. "Long-term results of upper respiratory syndrome surgery and gastrointestinal tract medical treatment in 51 brachycephalic dogs." *J Small Anim Pract* 47 (3):137–142. https://doi.org/10.1111/j.1748-5827.2006.00057.x.

Poncet, C. M., G. P. Dupre, V. G. Freiche, M. M. Estrada, Y. A. Poubanne, and B. M. Bouvy. 2005. "Prevalence of gastrointestinal tract lesions in 73 brachycephalic dogs with upper respiratory syndrome." *J Small Anim Pract* 46 (6):273–239. http://www.ncbi.nlm.nih.gov/pubmed/15971897.

Ray, K, A. Bodenham, and E. Paramasivam. 2014. "Pulmonary atelectasis in anaesthesia and critical care." *Continuing Educ Anaesth Crit Care Pain* 14 (5):236–245. https://doi.org/10.1093/bjaceaccp/mkt064. https://doi.org/10.1093/bjaceaccp/mkt064.

Riecks, T. W., S. J. Birchard, and J. A. Stephens. 2007. "Surgical correction of brachycephalic syndrome in dogs: 62 cases (1991–2004)." *J Am Vet Med Assoc* 230 (9):1324–1328. https://doi.org/10.2460/javma.230.9.1324. https://www.ncbi.nlm.nih.gov/pubmed/17472557.

Riggs, J., N. C. Liu, D. R. Sutton, D. Sargan, and J. F. Ladlow. 2019. "Validation of exercise testing and laryngeal auscultation for grading brachycephalic obstructive airway syndrome in pugs, French bulldogs, and English bulldogs by using whole-body barometric plethysmography." *Vet Surg* 48 (4):488–496. https://doi.org/10.1111/vsu.13159.

Roedler, F. S., S. Pohl, and G. U. Oechtering. 2013. "How does severe brachycephaly affect dog's lives? Results of a structured preoperative owner questionnaire." *Vet J* 198 (3):606–610. https://doi.org/10.1016/j.tvjl.2013.09.009.

Savvas, I., T. Rallis, and D. Raptopoulos. 2009. "The effect of pre-anaesthetic fasting time and type of food on gastric content volume and acidity in dogs." *Vet Anaesth Analg* 36 (6):539–540. https://doi.org/10.1111/j.1467-2995.2009.00495.x. http://www.ncbi.nlm.nih.gov/pubmed/19845925.

Savvas, I., D. Raptopoulos, and T. Rallis. 2016. "A "Light Meal" three hours preoperatively decreases the incidence of gastro-esophageal reflux in dogs." *J Am Anim Hosp Assoc* 52 (6):357–363. https://doi.org/10.5326/JAAHA-MS-6399. http://www.ncbi.nlm.nih.gov/pubmed/27685364.

Senior, M. 2005. "Post-anaesthetic pulmonary oedema in horses: A review." *Vet Anaesth Analg* 32 (4):193–200. https://doi.org/10.1111/j.1467-2995.2005.00186.x. https://www.ncbi.nlm.nih.gov/pubmed/16008716.

Shaver, S. L., L. A. Barbur, D. A. Jimenez, B. M. Brainard, K. K. Cornell, M. G. Radlinsky, and C. W. Schmiedt. 2017. "Evaluation of gastroesophageal reflux in anesthetized dogs with brachycephalic syndrome." *J Am Anim Hosp Assoc* 53 (1):24–31. https://doi.org/10.5326/JAAHA-MS-6281. https://www.ncbi.nlm.nih.gov/pubmed/27841676.

Smith, I., P. Kranke, I. Murant, A. Smith, G. O'Sullivan, E. Søreide, C. Spies, and B. in't Veld. 2011. "Perioperative fasting in adults and children: guidelines from the European Society of Anaesthesiology." https://www.aagbi.org/publications/publications-guidelines/M/R, Accessed 12/12/16. https://www.aagbi.org/publications/publications-guidelines/M/R.

Staffieri, F., D. Franchini, G. L. Carella, M. G. Montanaro, V. Valentini, B. Driessen, S. Grasso, and A. Crovace. 2007. "Computed tomographic analysis of the effects of two inspired oxygen concentrations on pulmonary aeration in anesthetized and mechanically ventilated dogs." *Am J Vet Res* 68 (9):925–931. https://doi.org/10.2460/ajvr.68.9.925.

Steeds, C., and R. Orme. 2006. "Premedication." *Anaesth Intensive Care Med* 7 (11):393–396. https://doi.org/10.1053/j.mpaic.2006.08.002. https://doi.org/10.1053/j.mpaic.2006.08.002.

The Kennel Club. 2020. "Respiratory Function Grading Scheme." Accessed 03/02/2020. https://www.thekennelclub.org.uk/rfgs.

Tolbert, K., S. Bissett, A. King, G. Davidson, M. Papich, E. Peters, and L. Degernes. 2011. "Efficacy of oral famotidine and 2 omeprazole formulations for the control of intragastric pH in dogs." *J Vet Intern Med* 25 (1):47–54. https://doi.org/10.1111/j.1939-1676.2010.0651.x.

Torrez, C. V., and G. B. Hunt. 2006. "Results of surgical correction of abnormalities associated with brachycephalic airway obstruction syndrome in dogs in Australia." *J Small Anim Pract* 47 (3):150–154. https://doi.org/10.1111/j.1748-5827.2006.00059.x. https://www.ncbi.nlm.nih.gov/pubmed/16512847.

Udeshi, A., S. M. Cantie, and E. Pierre. 2010. "Postobstructive pulmonary edema." *J Crit Care* 25 (3):508.e1–508.e5. https://doi.org/10.1016/j.jcrc.2009.12.014. https://www.ncbi.nlm.nih.gov/pubmed/20413250.

Ward, E., A. J. German, and J. A. Churchill. 2018. "The Global Pet Obesity Initiative Position Statement." Accessed 03/02/2020. https://static1.squarespace.com/static/597c71d3e58c621d06830e3f/t/5da311c5519bf62664dac512/1570968005938/Global+pet+obesity+initiative+position+statement.pdf.

White, R. N. 2012. "Surgical management of laryngeal collapse associated with brachycephalic airway obstruction syndrome in dogs." *J Small Anim Pract* 53 (1):44–50. https://doi.org/10.1111/j.1748-5827.2011.01156.x. https://onlinelibrary.wiley.com/doi/abs/10.1111/j.1748-5827.2011.01156.x.

Wilson, D. V., and A. T. Evans. 2007. "The effect of topical treatment on esophageal pH during acid reflux in dogs." *Vet Anaesth Analg* 34 (5):339–343. https://doi.org/10.1111/j.1467-2995.2006.00340.x. http://www.ncbi.nlm.nih.gov/pubmed/17565573.

Wilson, D. V., and R. Walshaw. 2004. "Postanesthetic esophageal dysfunction in 13 dogs." *J Am Anim Hosp Assoc* 40 (6):455–460. https://doi.org/10.5326/0400455. https://www.ncbi.nlm.nih.gov/pubmed/15533965.

Worth, D. B., J. A. Grimes, D. A. Jiménez, A. Koenig, and C. W. Schmiedt. 2018. "Risk factors for temporary tracheostomy tube placement following surgery to alleviate signs of brachycephalic obstructive airway syndrome in dogs." *J Am Vet Med Assoc* 253 (9):1158–1163. https://doi.org/10.2460/javma.253.9.1158. https://doi.org/10.2460/javma.253.9.1158.

20 Conclusions
Where Are We Now? And What Should or Could We Do Next?

Dan G. O'Neill and Rowena M.A. Packer
Royal Veterinary College

CONTENTS

A Beginning to the End… or an End to the Beginning 377
Chapter-Specific Conclusions 378
The Brachycephalic Crisis: What Have We Learnt and Where Can We Go from Here? 382
References 385

A BEGINNING TO THE END… OR AN END TO THE BEGINNING

We hope that you have enjoyed your journey through the complex, concerning and fascinating world of brachycephalism in companion animals that we have all shared in this book and that this is just the beginning of your journey towards a deeper understanding of brachycephaly. You may have opted to read each chapter in the order presented in the book and therefore followed the story from a broader overview to the specific medical topics, or you may have preferred to pick'n'mix chapters to create your own unique journey. Either way, we hope that this book has encouraged you to reflect deeply on the many faces of brachycephalism that our expert authors have introduced and discussed. A critical requirement for learning and personal growth is to be open to new ideas and to explore problems from several viewpoints in order to gain a more holistic perspective. It is fine to disagree with some of the evidence and opinions presented in the book so long as you have listened and reflected on these first; it is the nature of complex problems that we are each wedded most strongly to our own unique perspectives based on our own life experiences and beliefs. This book is aimed at a diverse readership, and we anticipate that consensus on all views expressed is unlikely. In fact, these topics on brachycephaly with conflicting views between and within stakeholders are hugely important to identify and explore because understanding and resolving these issues helps us all to move forward towards new and more inclusive worldviews. With this personal growth in mind, we hope that you have challenged many of your prior beliefs and have 'tried on' some new viewpoints, like a new suit of clothes, to see how comfortable they feel for future use. If we have helped you to update even one of your prior beliefs about brachycephalism, we will feel that this book has been a success.

We began the book with a wide-ranging introduction that aimed to set the scene in relation to brachycephalism in companion animals. The complexity of the brachycephalic problem was described and then further expanded throughout the book. However, by definition, complex issues cannot be resolved with single simple solutions, and therefore, many contentious questions on brachycephaly still remain. For example, should we try to 'save' all brachycephalic breeds given the inherently severe health problems that many face? Many brachycephalic dog types, have existed in some form for centuries and have become part of our national and personal histories, but is 'tradition' a sufficient reason to justify their problematic futures? Would it not be better overall for humans to experience these breeds as historical entities in books and films without requiring living sentient animals to endure welfare-challenged lives just to reinforce this historical role? Given what we now know about the health issues associated with brachycephalism, can humanity justify future deliberate development of new breeds and types of animals that are brachycephalic, not just in dogs, but also in cats and rabbits? So many problematic questions still exist.

Complexities are limited not just to the 'solutions' side of the brachycephaly problem; even defining brachycephalism itself is challenging. Brachycephalism is more than just a binary feature (yes/no) but exists along a skull-length continuum with progressive shortening to a point where humankind has removed any nasal protuberance at all in these animals. At what point along this scale should we consider an animal as 'brachycephalic', and at what point should we consider this extreme brachycephalic conformation as unacceptable? As yet, we have no definitive answers to these questions and debate rages on; ultimately, it may require the courts and legislation to deliver final rulings if the key stakeholders in dog health and welfare abdicate their responsibilities and refuse to reach mutually acceptable agreement.

Our understanding of human attraction to, and interactions with, brachycephalic dogs is still in its relative infancy and also is underscored by complexity and confusion. How can so many owners love these dogs so deeply, but seemingly yet be so blind to the suffering brought on by their brachycephalism? Is it an innate human flaw that we prioritise our own pleasure from loving brachycephalic dogs above the lifetimes of suffering that brachycephalic animals may endure by receiving this 'love'? Regardless of the reasons for this paradoxical love, there is no doubting the strength of the dog–owner bonds between brachycephalic dogs and their owners, and the associated cognitive dissonance on the welfare impacts of their dogs much loved appearance. This then begs the question: can hearts and minds about owning brachycephalic animals be changed at all? The complexity of these and other questions relating to brachycephaly demands substantial ongoing research applying both social and natural sciences approaches to further unravel the breadth of issues here and to develop resilient mitigation strategies that can protect animal welfare.

CHAPTER-SPECIFIC CONCLUSIONS

So, what have we learned on our travels through this book? The first section of the book offers a range of chapters applying both the social and natural sciences. Work based on social sciences focusses on human elements contributing to the brachycephalic crisis. A fascinating trip through the history of brachycephalism in dog breeds since the Victorian times in Chapter 2 highlights that the problems facing brachycephalic breeds today were also present and recognised over a hundred years ago. Many efforts to redress these issues are explored, but human factors including disharmony between breeders and veterinarians have limited our progress. It is clear that the brachycephalic crisis continues to be, and always has been, a predominantly human issue. Consequently, welfare progress requires an alliance of the willing among key stakeholders to reach a human resolution; alternatively, a century of failure to resolve these issues may force more dramatic measures, such as legislative reforms, that circumvent the desires of some stakeholders. Whichever outcome happens is ultimately our choice, as stakeholders, based on our actions (or inactions). With great power comes great responsibility.

The brachycephalic situation we face right now is hugely paradoxical. More is known than ever before about the health issues of brachycephalic dogs, but yet their popularity is at an all-time high. This raises the critically important question of why owners are drawn to brachycephalic breeds and, even in the face of chronic or severe health problems, continue to show loyalty towards these breeds. Chapter 3 examines the growing literature on human attraction to brachycephalic animals, considering both biological explanations (e.g. *kindchenschema* and 'the cute phenomenon') and recent cultural explanations (e.g. fashion and lifestyle motivations). This literature portrays a complex relationship between humans and brachycephalic animals, rife with biased perceptions of dog health, but underscored by a deep dog–owner bond. This complexity may explain why 'simple' educational messaging aimed at informing owners about their inherent health problems has not led to reductions in their popularity. With millions of brachycephalic dogs currently being kept as companions, research aimed at improving their welfare (e.g. developing better diagnostic methods and treatments) remains extremely important. However, to move beyond 'firefighting' the brachycephalic issue with veterinary interventions and instead move towards reducing the number of brachycephalic dogs bred and kept as companions internationally, we need a much better understanding of the human side of this issue to ensure that canine welfare is improved in a lasting and meaningful way.

Ethical issues for veterinary professionals caring for brachycephalic animals are explored in Chapter 4. The complexity of these issues is highlighted by competing demands between the perspectives of caring for individual animals compared with ethical demands to improve the welfare at breed or species levels. Dealing with this complexity can lead to severe moral distress among veterinary team members who are motivated intrinsically by a desire to improve animal welfare at all levels of abstraction. However, awareness and explicit acknowledgement of competing ethical demands on veterinary teams at the levels of the patient and its breed/species, the clients and their communities, their practice and professional bodies as well as their core personal ethics as human beings can help to produce the healthiest decision-making for the animals and for the veterinary professionals.

Most brachycephalic animals have at least one human owner at some point in their life. Consequently, consideration of the motivations, emotions and beliefs of these owners is critical to improving communication between veterinary professionals and the owners of brachycephalic animals. Chapter 5 considers the importance of veterinary discussions with owners across a range of key aspects relating to current or future ownership of brachycephalic animals. The diversity of both acknowledged and subconscious preferences and beliefs across owners is highlighted as a key challenge for effective veterinary communications. Veterinary teams have an important role to play in crafting the ownership choices of the wider animal-owning public and improving the welfare of brachycephalic patients presented to them, but, for greatest effect, veterinary professionals need to move away from 'talking at' and towards 'discussing with' their clients.

Veterinary nurses play vital roles in caring for companion animals as well as their owners, and these roles continue to expand. With this background, Chapter 6 undertook the challenging task of defining the breadth of veterinary nurse activities in relation to brachycephalism in companion animals. Their advanced nursing training means that veterinary nurses can effectively support veterinary care of brachycephalic animals both generally, e.g. weight control and pre-purchase consultations, and specifically, e.g. running brachycephalic-specific clinics to promote skin fold and ocular care, and breathing assessment. From this chapter, it is clear that veterinary nurses have a unique opportunity to advise and guide owners using their communication skills, compassion, empathy and education, and to play a key role bridging between clients and veterinary surgeons. However, it may be that the full contribution of veterinary nurses as an advocate for brachycephalic animal welfare has remained a relatively untapped resource to date.

The first section of the book then moves towards an exploration of the quantitative evidence base underpinning our understanding and concern for brachycephalic health. Looking at brachycephalic animals from a population perspective, Chapter 7 offers the reader a condensed guide to epidemiology as a science. Epidemiological evidence is presented to show that almost one-fifth of all dogs in the UK are now brachycephalic. Escalating popularity of some key dog breeds, such as the French Bulldog, Pug and English Bulldog, means that common disorders of these breeds substantially skew the health estimates for brachycephalic dogs overall. This chapter highlights the importance of generating and applying good epidemiological evidence to promote meaningful and planned welfare improvements. We need to move beyond 'kicking the welfare can down the road' with statements such as 'more research is needed' that just promote further decades of deferred decision-making while animals continue to suffer. Instead, this chapter advocates that we twin the application of reliable evidence with concerted efforts on human behaviour change in order to reshape our human mindsets on brachycephalism.

Genetics underpins our human power to select for, and fix, phenotypic characteristics such as brachycephaly that humans may deem as desirable at some point in time but that nature may consider as an evolutionary disadvantage across all time. Chapter 8 offers a detailed exposition of the current state of knowledge on the genetics of brachycephaly at the levels of both the individual dog and the population. The chapter explores some genetics approaches that offer potential options to breed away from the worst of these welfare harms and concludes that genetic approaches may need to be developed uniquely for each breed rather than applied to all brachycephalic dogs. Genetic methods may offer opportunities for broader brachycephalic welfare impacts in the future but are currently dominated by efforts to improve selection in respiratory health. Progress here uses scoring systems that are reliant on phenotype testing such as exercise-based testing or respiratory grading assessment.

Brachycephalic health issues have been recognised for over a century, with a long history of efforts at mitigation that have been poorly successful. Chapter 9 undertakes a comprehensive review of the national and international activities that have been, and continue to be, tried to resolve the brachycephalic crisis. A lengthy list of such activities, especially in Europe and North America, are presented along with information on the key stakeholders involved and a summary of the outcomes. A sobering conclusion from this review is that, despite undoubtedly high activity in this area, these efforts have often been disjointed, poorly strategised and seldom multilateral, and rarely generated measurable outcomes that showed meaningful welfare improvements. The value of national and international activities that engage with the spectrum of stakeholders is acknowledged as an important route for democratic resolution of many brachycephalic issues.

Building on these broader topics, the second part of the book focuses on specific clinical topics. Given the wide interest in breathing issues as a contributor to reduced welfare in brachycephalic animals, Chapters 10 and 11 are dedicated to exploring clinical and surgical aspects of brachycephalic obstructive airway syndrome (BOAS), respectively. Rather than being a single disease, BOAS is a syndromic presentation generated by cumulative effects from several diseases affecting several respiratory tract sites. The relative contributions from these underlying diseases varies between dogs and breeds. Consequently, each BOAS case is unique and requires its own clinical workup to define a bespoke medical and surgical plan. BOAS is a progressive condition that generally worsens with aging. This aging effect carries serious welfare implications as our currently youthful population of (especially) French Bulldogs and Pugs becomes aged and the prevalence of respiratory problems will inevitably rise. Although success rates following medical and surgical therapy for BOAS have improved over recent years, it is clear that breeding away from phenotypes associated with breathing problems remains the wisest long-term approach. BOAS treatment should not be considered as curative, and it is important to remember that treated animals retain the potential to perpetuate this disorder to future generations if used for breeding.

For many humans, the eyes of brachycephalic animals are a hugely appealing feature. Paradoxically, these large and exposed eyes lead to some of the worst welfare outcomes for animals with brachycephaly. Chapter 12 takes a deep and thoughtful look at ophthalmology of brachycephalic animals. Many deviations from healthy ocular conformation and its surrounding adnexa are highlighted that have been either deliberately or inadvertently selected in these animals. Painful ocular conditions such as corneal ulcers are both common and highly predisposed in brachycephalic animals. Veterinary ophthalmological expertise can now offer medical and surgical management for many of the eye disorders that result from the brachycephalic conformation. Nonetheless, it is hard to avoid the conclusion that what we, as humans, view externally as 'cute puppy dog eyes' instead represent potential lifetime sources of pain for many brachycephalic animals from the inside.

Brachycephalic dogs are predisposed to a long list of dermatological conditions. Chapter 13 delves under the skin of these issues to dissect some common dermatological disorders of dogs. These include general disorders such as otitis externa and interdigital dermatitis, disorders with an inherited aetiology such as demodicosis and also disorders that are firmly linked to the brachycephalic phenotype such as intertriginous dermatitis (skin fold dermatitis). Given that the pruritus and pain associated with skin disease is often long term and/or recurrent, this chapter makes it clear that owners and veterinary teams need to recognise and manage these conditions as early as possible.

Decades of selection for deviation in the lengths of upper and lower jaws in brachycephalic animals have ensured that dental problems are commonplace in these animals. Chapter 14 expands on a range of dental problems associated with the brachycephalic conformation. Despite a high prevalence of dental disorders in companion animals overall, and especially in brachycephalic animals, it is proposed that teaching on dentistry is inadequate within most undergraduate veterinary curricula and that this deficit promotes further welfare issues whereby dental suffering often goes unrecognised. A thorough dental examination is recommended during all veterinary visits, along with additional value from whole-mouth intra-oral dental radiography. As similarly acknowledged throughout this book, this chapter highlights the welfare risks from the 'normalisation phenomenon' whereby serious health problems are considered 'normal' just because they are almost ubiquitous in a breed. Dental examples of normalisation include overshot and undershot jaws and dental crowding that may define a breed but are *de facto* diseases in their own right. The message from this chapter is that 'normal' should be measured against typical and desirable health characteristics of dogs overall rather than what is typical for individual breeds.

Structural changes to the brain cavity due to brachycephaly results in conformational changes to the brains of brachycephalic animals that can lead to neurological disorders. Chapter 15 identifies that much of the neurological impact from brachycephaly follows disruption of cerebrospinal fluid (CSF) movement and absorption. Brachycephalic breeds can also show brain disorders with suspected inherited tendency. Many of these brain disorders are associated with pain that can often be severe. Given the limitations to current medical and surgical therapies to address the disorders discussed in this chapter, prevention by breeding away from extreme phenotypes rather than striving for cure of clinical cases is advocated as the wiser approach for brain disorders associated with brachycephaly.

Typical body shapes for many popular brachycephalic breeds such as the French Bulldog and the Pug are short and 'cobby', often with short, absent or screw tails. Emerging evidence suggests that these conformations are associated with certain spinal and vertebral disorders in these breeds that can be progressive as these dogs age. Chapter 16 reinforces the importance of breed-specific approaches to health reforms because of inherent differences between brachycephalic breeds in their predisposition to spinal malformations and subsequent clinical status. This chapter concludes that there is still much to learn about the true clinical and welfare importance of the vertebral malformations commonly recorded in these breeds.

Obesity has become a silent epidemic in our modern companion animals. Chapter 17 highlights that a high prevalence of obesity, combined with implications such as cumulative respiratory compromise, means that obesity holds particularly severe welfare issues in certain brachycephalic breeds. Although the genetics of obesity are beginning to be elucidated, a core imbalance between energy intake and energy expenditure remains at the heart of the obesity issue. A key conclusion is that successful weight control requires collaboration between the veterinary team with a committed owner who understands the importance of preventing or redressing unnecessary weight gain and who is motivated to maintain these changes once achieved.

Dramatically increasing popularity of certain breeds of brachycephalic dogs means that demand for optimal reproduction has never been greater. However, Chapter 18 identifies that the typically large head and stubby body conformation of brachycephalic animals is associated with predisposition services to reduced fertility and many reproductive disorders. In consequence, market forces related to consumer demand for brachycephalic puppies have promoted expansion of veterinary reproduction for companion animals as a growing industry. Some brachycephalic dog breeds, such as the English Bulldog, have become heavily reliant on artificial insemination for successful reproduction, while elective caesarean surgical procedures are now commonplace in many brachycephalic breeds. Reproductive limitations in many brachycephalic breeds raise serious welfare and ethical issues for veterinary professionals and pet owners that have yet to be fully addressed. For example, should humans be promoting breeds of animals that are intrinsically incapable of breeding without human intervention?

Many of the earlier clinical chapters described diagnostic and therapeutic procedures that require general anaesthesia. Anaesthesia carries quantifiable risks for any companion animal. However, Chapter 19 explains that these risks escalate in brachycephalic animals due to direct challenges of reaching, maintaining and exiting a safe plane of anaesthesia in conjunction with underlying predispositions to many conditions that intrinsically increase anaesthetic risk, e.g. pre-existing respiratory compromise. It is clear that there is value to veterinary teams from regular experience of anaesthetising brachycephalic animals, but that these procedures will always remain challenging and risky in these animals.

THE BRACHYCEPHALIC CRISIS: WHAT HAVE WE LEARNT AND WHERE CAN WE GO FROM HERE?

To finish, we present some broader thoughts here that are synthesised from the book overall and that we hope will give the interested reader food for further reflection. The fact that this book required 20 chapters across a diversity of social and natural sciences to cover the brachycephalic issue is evidence of the complexity underlying these welfare issues. Consequently, given this complexity, efforts at some form of successful resolution of the brachycephalic issue are likely to require a systems thinking approach (Arnold and Wade 2015). This means that despite being well-meaning and appearing logical, many individual recommendations from the chapters in this book could lead to serious unintended consequences for other aspects of the brachycephalic issue if fully enacted as standalone solutions. For example, many chapters advocated phenotypic improvements within current breeds using breeding programmes with high selection intensity to push towards higher average craniofacial ratios (i.e. longer muzzles), better respiratory function grading or less folded skin. While these actions may appear sensible to reduce the specific targeted problem, the genetics chapter in this book highlights that limited genetic diversity within many brachycephalic breeds means that strong selection pressures are likely to reduce genetic diversity even further, with consequent genetic harms to wider health in future generations. Consequently, it is critical that actions aimed at improving the health and welfare of brachycephalic animals are viewed in a very wide context and should always consider potential unintended consequences before implementation.

A corollary from the previous paragraph, and one that was emphasised in Chapter 9 on national and international approaches, is that the most effective actions to improve brachycephalic welfare overall will require collaboration between multiple stakeholders. The complexity of the issues surrounding brachycephaly in companion animals is such that unilateral actions from individual stakeholders are likely to offer limited overall gains and could even result in a contrarian effect whereby overall welfare may be reduced. For example, prioritisation of one clinical domain, such as ophthalmology or respiratory function as a panacea for the overall brachycephalic syndrome, may take the focus off other equally or even more important clinical domains, with long-term detrimental effects on overall welfare. For the best effects, decisions and actions on brachycephaly should be multilateral and strategic, and need to engage the diversity of stakeholders who come together as an 'alliance of the willing'.

From the information presented right across this book, and especially Chapter 3, it is hard to avoid the realisation that the brachycephalic issue for companion animals is not truly an animal issue at all, but is instead a human issue that affects animals. It is humans who have deliberately selected towards the brachycephalic phenotype in species such as dogs, cats and rabbits despite prior evolutionary pressures over many millennia having promoted mesocephalic survival advantages (Onar et al. 2012). It is humans who invented the breed concept and have created a structure of closed populations based on breed standards that have changed little over a century with the aim of tightly fixing certain genotypes and phenotypes within groups of dogs while preventing recourse to remedial actions such as outcrossing or welfare-responsive changes (Farrell et al. 2015, Worboys et al. 2018). It is humans who have suddenly popularised current fad dog breeds such as the French Bulldog, Pug and English Bulldog. However, despite all this human activity, the animals are the ones that bear the bulk of the consequent welfare issues highlighted throughout this book. A logical conclusion, therefore, is that solutions to the brachycephalic crisis will necessarily require substantial human behaviour change; for example, to prioritise animal welfare over our current human preconceptions of what is cute or desirable.

A follow-on thought experiment from the previous point is to ask: 'Who owns the animal welfare issues relating to brachycephaly?' Throughout this book, an array of stakeholders have accepted some responsibility for resolving various aspects of the brachycephalic problem. These stakeholders include veterinarians, welfare scientists, charities, academics, breeders, kennel clubs and many more. Each intersection of the complex brachycephalic problem seems to have its own logical 'welfare controller' who can direct the changes in relation to that point. For example, veterinary professionals can control aspects related to diagnosis and clinical management of brachycephalic disorders, while breeders can control aspects related to selection away from extremes of conformation. However, there appears to be insufficient recognition of the responsibility of owners themselves who ultimately 'own' many decisions, such as choosing which breed to own, followed by subsequent decisions on feeding, exercise, presentation for veterinary care, compliance with veterinary advice and even euthanasia. Numerically, owners are by far the largest human stakeholder group in animal welfare. A natural conclusion therefore is that owners of animals are the primary owners of the welfare of brachycephalic animals. With this viewpoint in mind, one approach for the future would be to support increasing self-efficacy in owners by raising public awareness of their consumer power as puppy buyers to end the brachycephalic crisis by, for example, refusing to purchase brachycephalic animals or by demanding dramatically reduced conformational exaggeration in these breeds. Owners could opt to 'Stop and think before buying a (flat-faced) brachycephalic dog'. A very human problem needs a very human solution.

Continuing with this human theme, it is plain especially from Chapter 3 that development of effective strategies to resolve the cognitive and emotional dissonance on brachycephaly in the minds of many of the general public will be critical to reduce the brachycephalic crisis. Many humans experience deep appeal for the brachycephalic features of companion animals at an *emotional* level while at the same time being tugged rationally by their awareness that brachycephaly carries significant health risks for these animals at an *objective* level. Mental conflict between the 'act of sentiment' in loving the look of brachycephalic animals and the 'act of reason' in accepting that this conformation harms the welfare of affected animals can be lead to deep emotional turmoil in many owners who fail to accept one or other perspective as dominant. To reduce this conflict, we may need to accept the existence of our innate tendency to 'love' the brachycephalic look but simultaneously accept that holding this 'love' does not give us an automatic right to 'own' a brachycephalic animal.

From reading this book, it is clear that precision of language and choice of words are critical for effective communication, especially for such an emotive topic as brachycephaly. For example, the word 'normal' is used regularly in relation to breed health, e.g. 'it is normal for English Bulldogs to have deep skin folds on their faces' or 'it is normal for Pugs to snore, even when they're awake'. In these cases, 'normal' is used a synonym for 'typical,' but the cognitive effect is an inference suggesting a level of acceptability, desirability, naturalness, wholesomeness or even healthiness. Some chapters in this book have highlighted the hazards from the normalisation phenomenon whereby harmful physical features such as skin folds that are typical for some brachycephalic breeds are reported as 'normal' and therefore have become accepted as falling within the domain of 'good health' for these breeds, whereas these same features would represent poor health for dogs in general. Words describe meaning but also create meaning; when sharing our meanings, we need to think carefully and use the most appropriate words to ensure that what we mean to say is what we mean to be heard.

Complications relating to the hierarchies (levels) of abstraction used in discussions about brachycephalic health represent another recurring theme throughout this book. For example, we can discuss brachycephaly at various levels such as species, breed, type within a breed or even in relation to physical features within a type. While there are obvious commonalities between inference at each of these levels, there are also clear differences. At a species level, we are likely to be subconsciously comparing between brachycephalic and non-brachycephalic animals whereas within a brachycephalic breed, we are likely to be comparing between varying degrees of brachycephaly. However, discussions in relation to broader brachycephalic topics frequently end in stalemate and frustration, even between well-motivated and informed advocates, because levels of abstraction are often not stated explicitly in advance. Consequently, the advocates are unknowingly discussing concepts at differing levels. For example, while we might confidently state that brachycephaly in dogs is associated with high risk of severe breathing problems (species level of abstraction) (Chapters 10 and 11), we commit the cardinal epidemiological sin of ecological fallacy if we extend this knowledge to claim that all brachycephalic dog breeds (breed level of abstraction) have a similarly high risk. Pursuant to the previous paragraph, as well as using precise language and words, we must be precise in our inference also. To paraphrase Alice in Wonderland, it is important to both say what we mean as well as to also mean what we say.

This book has highlighted repeatedly that a conflict exists between a duality of apparently logical pathways that could end the current brachycephalic crisis. On the one hand, the clinical section of the book highlights several severe health issues associated with the brachycephalic conformation or brachycephalic breeds. Many of these chapters commend selectively breeding away from extreme brachycephaly as a logically ameliorative approach to improve the health of these breeds to a point where they become acceptable for promotion as companion animals by most stakeholders. This is the approach taken by The Kennel Club's 'Breed Health and Conservation Plans' project (The Kennel Club 2020). On the other hand, many chapters also explain that the brachycephalic crisis is predicated on escalating popularity for some brachycephalic breeds such as the French Bulldog, the Pug and the English Bulldog. Although these breeds may have had equally poor disorder profiles a decade ago, a brachycephalic crisis did not exist then because these breeds were less fashionable at that time so the cumulative welfare impact from brachycephaly on dogs overall was much lower at that time. With this latter perspective, the most sensible welfare approach would be to discourage prospective owners from purchasing brachycephalic animals of any type. This latter approach has been applied by the UK Brachycephalic Working Group with the widespread use of their strapline 'Stop and think before buying a flat-faced dog' that is aimed at the wider public (BWG 2020). In reality, we need to make progress on both of these fronts contemporaneously. In the current absence of effective legislation to stem the breeding of brachycephalic dogs in the UK, and given our inherent human attraction to the brachycephalic conformation, it is unlikely that brachycephaly will disappear as a conformational feature in companion animals in the near future. Even if brachycephalic populations are successfully reduced to smaller numbers over time, the potential for suffering in the remaining animals bred for extreme brachycephalic conformation cannot be ignored. Consequently, the logical conclusion is that all stakeholders need to continue working towards solutions for each of these two core issues: dramatically reducing numbers and popularity while also promoting a healthier phenotype of the remaining animals.

Despite the poor welfare pictures painted across many of the chapters in this book, it is also clear from this book that there is tremendous enthusiasm from many stakeholders for actions aimed at reducing the welfare issues associated with brachycephaly. A deluge of research over the past decade has provided overwhelming evidence on the harms from brachycephaly and has galvanised individuals and organisations both locally, nationally and internationally into action. There is increasing acceptance that extremes of exaggerated conformation in our companion animal species are unacceptable morally, ethically and from a welfare perspective. Many recent strategies aimed at resolving the brachycephalic crisis are moving towards the maxim laid down by Socrates over two thousand years ago; "the secret of change is to focus all of your energy not on fighting the old, but on building the new". Alliances of the willing are moving forwards from conflict between stakeholders and instead focusing on fighting poor welfare. There are many reasons for optimism that progress can be made to limit the negative welfare impacts from brachycephalism.

And now we end our journey in this book into the world of brachycephaly in companion animals. It has been our pleasure to share this time with you. We hope this is just the beginning of your personal journey towards a deeper understanding of brachycephaly. It is a complex story. It has elements of extreme love, extreme suffering, extreme despair but also extreme optimism for a better future for all companion animals with brachycephaly. We hope that we have stimulated you to challenge your prior beliefs and to reflect deeply on the issues surrounding brachycephaly from many perspectives. We sincerely hope that you are now equipped with sufficient factual and reflective knowledge to make informed decisions about what you can do, individually and within human collectives, to help resolve the brachycephalic crisis. This is a human-created crisis that needs solutions aimed at humans; brachycephalic animals need activity from veterinary and animal professionals along with owners and the wider public now more than ever for this crisis to be resolved.

REFERENCES

Arnold, R.D., and J.P. Wade. 2015. "A definition of systems thinking: A systems approach." *Procedia Computer Science* 44:669–678.

BWG. 2020. "The Brachycephalic Working Group." The Brachycephalic Working Group, accessed July 4. http://www.ukbwg.org.uk/.

Farrell, L., J. Schoenebeck, P. Wiener, D. Clements, and K. Summers. 2015. "The challenges of pedigree dog health: approaches to combating inherited disease." *Canine Genetics and Epidemiology* 2 (1):3.

Onar, V., C. Çakırlar, M. Janeczek, and Z. Kızıltan. 2012. "Skull typology of byzantine dogs from the theodosius harbour at yenikapı, Istanbul." *Anatomia, Histologia, Embryologia* 41 (5):341–352. doi:10.1111/j.1439-0264.2012.01143.x.

The Kennel Club. 2020. "Breed health and conservation plans (BHCPs)." The Kennel Club Limited, accessed November 2. https://www.thekennelclub.org.uk/health/breed-health-and-conservation-plans/.

Worboys, M., Strange, J.-M. and Pemberton, N. 2018. *The invention of the modern dog: breed and blood in Victorian Britain*. 1st ed. Baltimore, Maryland: Johns Hopkins University Press.

Index

Note: **Bold** page numbers refer to tables; *italic* page numbers refer to figures.

AAHA *see* American Animal Hospital Association
    (AAHA)
acepromazine (ACP) 190, 338, 362
active warming techniques 363
acute lung injury (ALI) 366
adipose expandability theory 316
advertisements 10
aggressive intraoperative fluid therapy 356
AI *see* artificial insemination (AI)
alapexy 183
alarplasty/rhinoplasty 182, *183*
ALI *see* acute lung injury (ALI)
Allgoewer's medial canthoplasty technique 205
alpha-2 agonists 362
ameliorative approach 384
American Animal Hospital Association (AAHA) 236
American bullies 18
American Kennel Club 140
American Society of Anaesthesiologists (ASA) 354, *354*
American Veterinary Medical Association (AVMA) 140
anaesthesia 336–337
    BOAS 356
    factors 354
    gastrointestinal disorders 357
    management (*see* anaesthetic management)
    ocular conditions 357
    problems 354, *355*
    recovery and post-operative management
        anti-inflammatory and vasoconstrictive drugs 368
        equipment, emergency re-intubation 368
        nebulisation and physiotherapy 369–370
        patient positioning 368
        preparation 367
        provision 367
        sedation 369
        tracheal extubation 368
        tracheostomy 370
    veterinary surgeons 353
anaesthetic considerations 179
anaesthetic management
    complications 357–358
    induction 365, *366*
    maintenance
        ocular care 366
        ventilation and lung function 365–366
    pre-anaesthetic assessment and stabilisation
        clinical examination 358, 360
        decision-making process *359*
        fasting and regurgitation 360–361
        functional tests 360
        history 358
    premedication
        aims and requirements *362*
        analgesia 363
        factors, drug selection 361
        sedation 361, 362, *364*

preoxygenation 364–365
    temperature 363
analgesia 363
anatomical monstrosities 15
anecdotal epinephrine (adrenaline) *370*
Animal Health Law 142
Animal Health Trust (AHT) 14
animal welfare 41
animal welfare change 16, 17
Animal Welfare Law 141
anti-inflammatory 368
aquaporins 256–257
artificial insemination (AI) 330, 331
arytenoid lateralisation 186
ASA *see* American Society of Anaesthesiologists (ASA)
aspiration pneumonia 190–191
assisted breeding techniques 4
atopic dermatitis/adverse food reactions 223–224
atrophic ventriculomegaly *262*
AVMA *see* American Veterinary Medical Association
    (AVMA)
awareness campaigns 138

baby schema effect 26–28
BCS *see* body condition score (BCS)
Beauchamp and Childress' Focal Virtues for Medical
    Professionals 46, **47**
bilateral subdural haematoma *268*
biparietal diameter (BPD) 333
Bmp3 antisense morpholinos 112
BMP3 F425L mutation 114
BMP3 (bone morphogenetic protein 3) signalling
    111, 112
BOAS *see* brachycephalic obstructive airway syndrome
    (BOAS)
BOAS functional grading 173
BOCCS *see* brachycephalic obstructive cerebrospinal fluid
    channel syndrome (BOCCS)
body condition score (BCS) 318
BOS *see* brachycephalic ocular syndrome (BOS)
BPD *see* biparietal diameter (BPD)
brachycephalic activism in twenty-first century 16–17
brachycephalic behaviour 28–29
brachycephalic bitch with dystocia, fetopelvic
    disproportion 49
brachycephalic breed health reforms
    actions and efforts
        breed-specific regulatory approaches 142–143
        commissions, working groups, statements...
            138–139
        decision-makers 137
        education and raising awareness 138
        epidemiological research programmes 137
        holistic scientific approach 137
        international multi-stakeholder 144
        qualitative research 137

brachycephalic breed health reforms (*cont.*)
  quality, dogs supplied 140
  supply, demand, trade and transport dogs 140–143
  historical perspective and stakeholders
    action types, benefits and challenges 134, **135**
    challenges 128–129
    showing and breeding dogs 134–136
    timeline, international focus 129, **130–133,** 133
    veterinarians and the veterinary profession
      136–137
  issues 127–128
  societal level 128
  veterinary and cynological organisations 128
brachycephalic breeds 9
brachycephalic conformation 43
brachycephalic conformation and reformers 17
brachycephalic demography and popularity 94–98, *96, 97*
brachycephalic disease
  and welfare activism 7
brachycephalic exaggeration 17
brachycephalic health
  epidemiology of 85–101
brachycephalic nurse clinics 70
brachycephalic obstructive airway syndrome (BOAS) 57,
      74, 90, 128, 136, 139, 143, 222, 335
  anaesthesia 356
  anatomical abnormalities, nasal cavity *170*
  clinical and surgical aspects 380
  clinical conditions 155
  clinical examination 166
  clinical signs 165–167, 172
  complications 187–188
  conformational risk factors **159,** *159*
  diagnosis 166
  disease characteristics 158–159
  disease recognition 165
  everted laryngeal saccules 164–165
  gastrointestinal signs 165
  genetic estimated breeding values 118
  genetics of 113–114
  genetic tests for 118–120
  history 166
  hyperplastic palate 163
  imaging 168–169
  laryngeal collapse 113, 164, *164*
  lesion sites 160, **160**
  macroglossia 163
  medical management 178–179
  nasal turbinate hypertrophy 161–162, *162*
  nasopharyngeal restriction and collapse 163
  postoperative care 188–191
  pre-surgical considerations 179
  prognosis/long-term outcome 193–196, **194**
  respiratory function testing 168
  skull and muzzle 156
  stenotic nostrils 160–161, *161*
  surgery (*see* surgery, BOAS)
  temporary tracheostomy *191,* 191–192
  tonsillar hypertrophy 165
  tracheal hypoplasia 163–164
  treatment decision-making 171–172
  upper respiratory disorders 156
brachycephalic obstructive cerebrospinal fluid channel
      syndrome (BOCCS) 270

brachycephalic ocular syndrome (BOS) 143
brachycephalic syndrome 108, 109
Brachycephalic Working Group (BWG) 17, 137, 139
brachycephalism 16
  binary feature 378
  biological explanations 378
  companion animals 379
  domestic dog breeds (skull and head shortening)
      109–113, **111**
  health issues 378
  history 378
  human mindsets 379
  learning and personal growth 377
  optimism 384
  resilient mitigation strategies 378
brain disorders
  brachycephalic skull 251
  cerebrospinal fluid physiology 256–257
  clinical syndromes 257–282
  differential diagnosis **282–284**
  morphogenesis 253–254
  nervous system 252–253
  neuroparenchymal and skull changes *252*
  skull base suture craniosynostosis **255**
  skull bones **253,** *254*
breathing assessment 74–76
breed health 14, 26
breeding programmes 8, 382
breed loyalty, development of 35, 56
breed-related disease 14
breeds of brachycephalic dogs 7
breed-specific instructions (BSI) 136
British dog breeders' specialist newspapers and veterinary
      journals 7–8
British Small Animal Veterinary Association
      (BSAVA) 14
British Veterinary Association (BVA) 14, 57, **58**
British Veterinary Association's 2016 Animal Welfare
      Strategy 45
BSAVA survey of veterinarians 15
BSI *see* breed-specific instructions (BSI)
bull-baiting 12
bulldogs 8, 9
  for elective caesarean section 49, **49**
  population 115
  to racial degeneration 12
  water puppies 343
BVA/KC screening schemes and gene tests 18
BWG *see* Brachycephalic Working Group (BWG)

CAD *see* canine atopic dermatitis (CAD)
Campaign for the Responsible Use of Flat-Faced Animals
      (CRUFFA) 137
Canadian Kennel Club 136
Canadian Veterinary Medical Association (CVMA) 140
canine Apgar systems 340, **340**
canine atopic dermatitis (CAD) 223–224
canine characteristics 26
canine faddists 13
cataracts 214–215
caudal articular process dysplasia 300–303, *302*
caudal buccal traumatic granulomas 240, *240*
Cavalier King Charles Spaniel (CKCS) 230
celebrity endorsement 31

CEPSAF *see* Confidential Enquiry into Perioperative
    Small Animal Fatalities (CEPSAF)
cerebrospinal fluid (CSF) 4, 381
    absorption and dynamics **258**
    glymphatic system and aquaporins 256–257
    hydrocephalus 256
    physiology **257**
    production and reabsorption 256
CFR *see* craniofacial ratio (CFR)
cherry eyes 63, 206
chiari malformation (CM)
    coronal suture craniosynostosis 272
    diagnosis 273, 275, 278
    genetic factors and breeding advice 280–282
    minimum MRI diagnostic protocol **277**
    MRI features **278**
    pathogenesis 271–272
    spectrum **270**, *271*
    surgical management **281**
CKCS *see* Cavalier King Charles Spaniel (CKCS)
CM *see* chiari malformation (CM)
CM-P *see* pain associated with CM (CM-P)
cognitive biases 85
cognitive dissonance 378
cognitive dissonance fuelling under recognition or
    underestimation 34
community-level advocacy 57, 58
companion animal health issues 85
    predispositions and protections to disorders 88–92,
        89, 91
companion animals
    assessment 335
    birth 344–345
    body temperature 332–333
    brachycephalic cat and breeding 344
    brachycephalic dam 344
    brachycephalic dental/oral issues 237
    brachycephalic puppies 339–340
    canine AI 331
    caudal buccal traumatic granulomas 240, *240*
    congenital malformations 342
    craniofacial exaggerations and deformities 236
    C-section (*see* elective pre-parturient caesarean section
        (C-section))
    dental crowding and rotation 241, *241*
    dental examination 237
    dioestrous 332–333
    emotional and objective level 383
    evolutionary survival mechanisms 236
    fertility and gestation 331, 344
    foetal anasarca *343,* 343–344
    furry palate 245
    gestational aging 333–334, **334**
    gingival hyperplasia 247
    LH peak 332–333
    malocclusion with abnormal contacts 238–240
    mandibular fracture 246–247
    mesaticephalic dog's mouth *238*
    neonatal resuscitation **341,** 341–342
    oral health care 235
    orofacial clefts 342–343
    ovulation 332–333
    postnatal care, brachycephalic kitten 345–346
    predicting parturition 331

pregnant, gestational considerations 334–335
rabbits 330
rostral mandible, Boxer dog *239*
serum progesterone levels 332–333
shapes 330
standard protocol 337–338
surgical preparation 338
surgical technique 339
tooth and periodontal tissues *242*
under-eruption and pericoronitis 241–243, *243*
unerupted teeth and dentigerous cysts 243–244
veterinarian's responsibility 236
veterinary dentistry 236
wide, loose symphysis 245–246
Confidential Enquiry into Perioperative Small Animal
    Fatalities (CEPSAF) 354
conformation-related disease 14, 25
confounding 88
congenital malformations 342
congenital vertebral malformations 291, 292, 295
consequentialism 46–47
constrictive myelopathy 301–302
continuous positive airway pressure (CPAP) 190, 356, 366
contractarianism 46
corneal conditions
    abscess 213
    corneal ulceration 211–212
    dermoid 211
    melting ulcer (keratomalacia) 212, *212*
    neoplasia 213–214
    pigmentation 213
corneal ulceration 11, 211–212, 357
corticosteroids 189
CPAP *see* continuous positive airway pressure (CPAP)
craniofacial ratio (CFR) 143, 157
critics of exaggerated conformation 12
CRL *see* crown-to-rump length (CRL)
crown-to-rump length (CRL) 333
CRUFFA *see* Campaign for the Responsible Use of Flat-
    Faced Animals (CRUFFA)
cuneiformectomy 186
CVMA *see* Canadian Veterinary Medical Association
    (CVMA)

Danish Dog Register 140
Danish study of breed motivations 79
decision-making of prospective dog owners 26
default orthodoxy in dog fancy 14
Demodex canis 221, *221*
Demodex injai 221, 221–223
demodicosis 221–222, *222*
dental crowding/rotation 241, *241*
dentigerous cysts 243–244
deontology 47, 49–51
dermatological problems
    atopic dermatitis 220
    atopic dermatitis/adverse food reactions 223–224
    clinical signs 222–223
    cutaneous abnormalities 220
    demodicosis 221–222, *222*
    diagnosis 223
    false pad formation 220
    hereditary skin disease 220–221
    interdigital draining tracts 224–226, *226, 227*

dermatological problems (*cont.*)
  intertriginous dermatitis 224, *225*
  management 223
  otitis externa 226–229
  pruritic behaviours 220
  PSOM 230–232
  VetCompass reports 219
dermoids 211
distichiasis 207, 209
DNA gene testing 16
Dog Attachment Questionnaire (DAQ) 28
dog fancy 8
dog legislation 142
dog–owner interactions 29
Doug the Pug 31
dry eye 210, *214*
Dutch Kennel Club 143
DVL2 (dishevelled 2) 111, 112
dyslipidaemia 316
dyspnoea 19, 71, 167, 186, 188, *189*, 191, 360
dystocia 10, 335, **337**, 341, 344

early brachycephalic reshaping and activism 8–13
early post-war veterinary efforts, pedigree dog
          health 14
ectopic cilia 209
educational efforts 19
elective pre-parturient caesarean section (C-section)
  birth 335
  breeding techniques 330
  client confidentiality 330
  indications 335–336
  planning elective pre-parturient 331–332
  predicting gestational age 331–332
  technique 336–337
  tools **336**
emphysema 205
endocrine disorders 231
energy expenditure *318*
enophthalmia 202
ethical and practical challenge, community-level
          advocacy 58
ethically challenging situations, veterinary practice
  brachycephalic breeds 42
  conformation-related conditions 42
  consequentialism 46–47
  contractarianism 46
  deontology 47, 49–51
  ethical frameworks 46
  mobilised veterinary individuals and organisations 42
  of treating brachycephalic dogs 43–46
  utilitarianism 46–47
  virtue ethics 46
ethics 46, 50
euryblepharon *204*, 204–205
evidence-based veterinary medicine 85
exaggerated brachycephalic conformation 18
excess adipose tissue 77
exercise and thermoregulation 63
exercise-based and exercise, respiratory assessment-based
          tests 119
exercise-based testing 380
existing health problems 64–65
extreme brachycephalic phenotypes

costs and benefits 47, **48**
veterinarians breeding, selling and surgery on dogs 50
eye examination 71–73
eyelid abnormalities
  cilia abnormalities 207–209, *208*
  diamond eye/pagoda eye 207, *208*

farm and laboratory animal welfare 14
FECAVA *see* Federation of Companion Animal
          Veterinary Associations (FECAVA)
Federation Cynologic Internationale 136
Federation of Companion Animal Veterinary Associations
          (FECAVA) 136, 137, 141
Federation of Veterinarians of Europe (FVE) 137, 141
feeding process 1
feline foetus **345**
five freedoms of animal welfare 49
  from discomfort 49
  to express normal behaviour 49
  from fear and distress 49
  from hunger and thirst 49
  from pain, injury and disease 49
Five Welfare Needs 59, **59, 65**
flatter-faced conformation 13
Flexineb C1 nebuliser *369*
foetal anasarca *343*, 343–344
food restriction 317–318
foreign-body-induced ulcerative palatitis 245
formalised screening schemes
  for HD 16
  for PRA 16
FRC *see* functional residual capacity (FRC)
French Bulldog 29
  heat stress, signs of 42, *42*
French Bulldogs and Pugs 30
functional residual capacity (FRC) 335
furry palate 245
FVE *see* Federation of Veterinarians of Europe (FVE)

gait abnormality 291
gastrointestinal abnormalities 178–179
gastrointestinal disorders 357
gastro-oesophageal reflux 361
gene tests 20
genetic diversity
  in brachycephalic breeds 114–116, *117*
  maintenance and restoration 117–118
genetic health 14
genetics of brachycephaly 109–113
genome-wide association study (GWAS) techniques
          116, *117*
Genome-wide Complex Trait Analysis (GCTA)
          software 113
genomic STR loci in bulldog 116
gestational aging 333–334, **334**
gestational sac diameter (GSD) 333
gingival hyperplasia 247
glaucoma 215
glymphatic system 256–257
good health for brachycephalic dog 33–35
GSD *see* gestational sac diameter (GSD)

haemorrhage 182, 189
health issues, refashioned animals 10

health issues of brachycephalic animals 88
health of brachycephalic dogs 56
health testing for brachycephalic breeds 107–121
healthy commercialism 12
heat apoplexy 10
heat-moisture exchangers (HME) 363
hemangiosarcoma 214
hemivertebra
    clinical signs 295–297
    Cobb angle 293
    degenerative spinal processes 298
    diagnosis and treatment 297, 297–298
    embryologic development 292
    prevalence 294–295
    screening programs 299
    screw-tail morphology 292, 293
    spinal hyperaesthesia and paraparesis 295
    subtypes 292
    survey radiographs 296
    thoracolumbar intervertebral disc extrusion 298
    3D-printed models 299
    3-year-old neurologically French Bulldog 293
hereditary skin disease 220–221
HME see heat-moisture exchangers (HME)
homeotic transformation 300, 300
human behaviour 4
human power 380
100 m walk test 119
husbandry and breeding practices 14
hydops foetalis 343
hydrocephalus
    clinical signs 258–259
    defined 257
    diagnosis
        magnetic resonance imaging 261, **262, 263,** 264, 266
        ultrasonography 264
    differentials 261, 262
    dysmorphia 259
    pathophysiology 259
    SM 260
    treatment and prognosis
        intraventricular pressure and valve selection 267
        medical management 264, **265**
        surgical management 264, 266–267
        VP shunt 267
hypercapnia 365
hyperplastic palate 163
hypoventilation 365
hypoxaemia 358, 360

ichthyosis 220–221
IDDSTs see interdigital draining sinus tracts (IDDSTs)
ideal companion dog 29
IDHWs see International Dog Health Workshops (IDHWs)
immoral ring of dog fanciers 11
individual and organisational welfare activism 17
insulin 316
insurance 63
interdigital dermatitis 380
interdigital draining sinus tracts (IDDSTs) 220
interdigital draining tracts 224–226, 226, 227
International Dog Health Workshops (IDHWs) 133, 137, 144

international multi-stakeholder 141, 144
International Partnership for Dogs (IPFD) 133
International Working Group on Extremes of Conformation in Dogs (IWGECD) 137
interstitial-fluid-lymphatic-vessel-centric model 256
intertriginous dermatitis 224, 225
intervertebral disc disease (IVDD) 292
intervertebral disc extrusion (IVDE) 306–307
intranasal obstruction 178
intraocular pressure (IOP) 215
intra-oral dental radiography 244, 244, 246, 381
intra-oral surgery 179, 180
intraventricular pressure (IVP) 267
IOP see intraocular pressure (IOP)
IPFD see International Partnership for Dogs (IPFD)
IPFD International Collaborative on Extreme Conformation in Dogs 136
IVDD see intervertebral disc disease (IVDD)
IVDE see intervertebral disc extrusion (IVDE)
IVP see intraventricular pressure (IVP)
IWGECD see International Working Group on Extremes of Conformation in Dogs (IWGECD)

KC/BVA health schemes and joint committee 16
KCS see keratoconjunctivitis sicca (KCS)
Kennel Club (KC) 8
Kennel Club/University of Cambridge Functional Grading Scheme, The 167, 172
keratoconjunctivitis sicca (KCS) 210
keratomalacia 212
kyphoscoliosis 293
kyphosis 293, 295, 296

Labrador Retriever vs. BOAS-affected French Bulldog 157
laryngeal auscultation 358
    stethoscope placement during pre- and post-exercise 75, 75
laryngeal collapse 164, 164
laryngeal saccules 164–165
larynx/laryngeal collapse
    advanced (grades II and III) 185–186
    everted laryngeal saccules/grade I laryngeal collapse 184–185
laser-assisted turbinectomy (LATE) 186–187, 187
LATE see laser-assisted turbinectomy (LATE)
legislative approaches 36
lens abnormalities 214–215
LH see luteinising hormone (LH)
lifestyle breeds 30
longevity study 98–99, **100**
low-energy dogs 30
luteinising hormone (LH) 332
machine learning algorithm 275
macroglossia 163
macropalpebral fissure 204
major histocompatibility complex (MHC) diversity 107
mandatory desexing of dogs 50
Manny the Frenchie 31
melting ulcer 212, 212
Members of European Parliament (MEPs) 142
meningeal fibrosis 300–303, 302
meningocele 305–306
meningoencephalitis of unknown origin (MUO) 261
meningomyelocele 305–306, 306, 307

MEPs *see* Members of European Parliament (MEPs)
mesocephalic survival advantages 382
metabolic diseases 334
Microsporum canis 232
mid-century veterinary intervention 14
minimum minor allele frequency (MAF) 116
mobilised veterinary individuals and organisations 42
modern change theory 18
modern pedigree dog breeding 14
molecular methods 115
Monash Dog Owner Relationship Scale (MDORS) 28
moral dilemmas, veterinary practice 41–51
moral distress 41
moral stress/distress 45
Morgan and Moore technique 207
mucinosis 221
MUO *see* meningoencephalitis of unknown origin (MUO)

nasal turbinate hypertrophy 161–162, *162*
nasolacrimal system abnormalities
    description 209
    dry eye 210
    imperforated punctum 209
    micropunctum 209, *210*
national pedigree organisations 135
nebulization 369–370
neonatal resuscitation **341,** 341–342
neutering 63, 324
NKU *see* Nordic Kennel Union (NKU)
non-pruritic alopecia 231–232
nonsteroidal anti-inflammatory drugs (NSAIDs) 179, 368
Nordic Kennel Union (NKU) 139
normalisation of obesity 77
NSAIDs *see* nonsteroidal anti-inflammatory drugs
    (NSAIDs)

obesity 4
    barriers to change **323**
    defined 312
    diseases 357
    diseases and clinical signs **315**
    disordered homeostasis 313–314
    environment, behaviour change 322
    factors 312
    fat mass, mechanical effect 316–317
    feeding and exercise management 312
    genetic variation
        influences 314
        practical implications 314–315
    goal setting, action planning and monitoring 323–324
    hypothalamus *313*
    ideal and target weights 318
    improving motivation 321–322, *322*
    improving owner effectiveness 322
    inflammation 316
    insulin resistance affects metabolism and cell
        growth 316
    management 317–318
    overweight pets 321
    pathophysiology 315
    prevention 324
    problems, brachycephalic animals 317
    respiratory system 317
    risk factors independent, owner management 312

    weight loss (*see* weight loss)
    weight management plan 320–321
    welfare issues 381
obstructive sleep apnoea (OSA) 317, 356
occipital dysplasia 272
occupational stressor 41
OM *see* otitis media (OM)
online marketing 60
ophthalmology, brachycephalic breeds
    glaucoma 215
    lens abnormalities 214–215
    orbital, periocular and adnexal conditions 201–210
    retinal diseases 215
    scleral show *202*
orbital conditions
    euryblepharon *204,* 204–205
    lack of enophthalmia 202
    nasal folds 205–206, *206*
    pneumatosis 205, *205*
    PNMG 206–207
    proptosis 202–204, *203*
orbital pneumatosis 205, *205*
'Oriental' fashionable aesthetics 10
'Oriental' lapdogs 9
orofacial clefts 342–343
OSA *see* obstructive sleep apnoea (OSA)
otitis externa 380
    atopic dermatitis 227
    clinical signs 228–229
    factors 226–227
    inflammation and swelling, pinna and auditory
        orifice *229*
    lop-eared rabbits 227
    normal Shar Pei *228*
    topical therapy 228
    treatment 229
otitis media (OM) 228
overweight 4
overweight pets 321
owner psychology 33
oxygen supplementation 190

pain associated with CM (CM-P)
    clinical signs 272, **274**
    diagnosis 276
    management 279–280
    MRI features **278**
    pain face, 9-year-old male neutered CKCS *274*
    pathogenesis 272
    prognosis 279
    treatment algorithm *281*
    uncharacteristic aggression, 4-year-old female
        neutered CKCS *275*
palatal mucosa 245, *245*
palatoplasty *181*
Pannus/Uberreiter syndrome 213
parvovirus 44
PDSA PAW report (Peoples Dispensary for Sick Animals
    2019) 77, 78
pectus excavatum 342
Pedigree Dogs Exposed (PDE) 16
PEEP *see* positive-end expiratory pressure (PEEP)
People's Dispensary for Sick Animals (PDSA)
    59, 60

pericoronitis 241–243
pet attachment questionnaire (PAQ) 27–28
phacoemulsification of lens 214
phenotype scores 20
physical weakness or deformity 12
physiotherapy 369–370
PLR *see* pupillary light reflexes (PLR)
PNMG *see* prolapse of the nictitans membrane gland (PNMG)
Pocket technique 207
popularity crisis in brachycephalic breeds 32
population-based view of brachycephalic health 85
population genetics 107–121
positive behavioural traits 29
positive-end expiratory pressure (PEEP) 366
positive pressure ventilation (PPV) 356, 365
postoperative analgesia
    airway swelling 188–189, *189*
    aspiration pneumonia 190–191
    avoiding stress 190
    haemorrhage 189
    oxygen supplementation 190
    recovery from anaesthetic 188
    regurgitation and vomiting 190
PPV *see* positive pressure ventilation (PPV)
preconceptions of breed 'laziness' 30
preoxygenation 364–365
pre-purchase consultations 59–60, 77–79, *80*
primary secretory otitis media (PSOM)
    bulging pars flaccida *231*
    clinical signs 230
    dermatophytosis, persian cats 232
    mucous evacuation *231*
    non-pruritic alopecia 231–232
    treatment 230–231
principles of communicating with existing owners about brachycephaly 60–62
progressive exaggeration 10
progressive retinal atrophy (PRA) 14, 15, 16
prolapse of the nictitans membrane gland (PNMG) 206–207
proptosis 202–204, *203*
prostituted Bulldog breeding 13
PSOM *see* primary secretory otitis media (PSOM)
public engagement on national television and radio 15
pug on weighing scales at brachycephalic weight clinic 77, *78*
pug popularity 30
pulmonic stenosis 342
pupillary light reflexes (PLR) 203
Puppy Contract and Puppy Information Pack 59
puppy vaccine consult
    cherry eyes 63
    exercise and thermoregulation 63
    insurance 63
    neutering 63
    respiratory 62
    skin 63
    veterinary checks 63
    weight 62

QDA *see* quadratic component analysis (QDA)
quadratic component analysis (QDA) 168
quadrigeminal cistern expansion 267–270, *269*

racial degeneration 12
reacquisition desire 35
regurgitation 178, 190
reliable and representative longevity (lifespan) and mortality (causes of death) statistics 98–99, **100**
reproductive limitations 381
Respiratory Function Grading Scheme 143
respiratory grading assessment 380
respiratory system 62, 317
retinal diseases 215
RFG Scheme 167, 171
rhinomanometry 162
Rodney Stone 10, *11*
Royal Society for the Prevention of Cruelty to Animals (RSPCA) 65
Royal Veterinary College (RVC) 157
RSPCA *see* Royal Society for the Prevention of Cruelty to Animals (RSPCA)
RUNX gene 112, 113
RVC *see* Royal Veterinary College (RVC)

SAD *see* spinal arachnoid diverticula (SAD)
SCC *see* squamous cell carcinoma (SCC)
SCCEDs *see* spontaneous chronic corneal epithelial defects (SCCEDs)
Schirmer tear test (STT) 210
Schirmer tear test with gentle handling 71, *72*
scoliosis 293
Scottish Society for the Prevention of Cruelty to Animals (SSPCA) 65
screw-tailed brachycephalic dogs 292
'screw' tails 10
sedation 361, 362, *364,* 369
sedative drugs 179
Singleton's campaign, brachycephalic disease 16
Singleton's confrontational approach 15
skin 63
skin folds 73, *74*
skull dimensions, mesaticephalic *vs.* brachycephalic dog *156*
skull index ratio 157
sleep disorders 76
SM *see* syringomyelia (SM)
SMOC2 (SPARC-related modular calcium binding 2) 111
SMOC1 gene 111
social influences on breed choice 30–32
spina bifida *305,* 305–306, *306*
spinal arachnoid cysts 303
spinal arachnoid diverticula (SAD) 291, *303,* 303–305, *304*
spontaneous chronic corneal epithelial defects (SCCEDs) *211*
SPS *see* Swimmer Puppy Syndrome (SPS)
squamous cell carcinoma (SCC) 213
staphylectomy 180, *181*
starvation pre-anaesthetic 76
stenotic nostrils 158, 160–161, *161*
stethoscope placement during pre- and post-exercise for laryngeal auscultation 75, *75*
struggling to breathe 33
STT *see* Schirmer tear test (STT)
supply of dogs
    authorities, multi-national/European efforts 141–142
    European Union (EU) – European Parliament (EP) 142
    international multi-stakeholder approach 141
    regulation of dog sales 141

supracollicular fluid collections 267–270
surgery, BOAS
    aberrant nasal turbinates 186–187
    larynx/laryngeal collapse 184–186
    nostrils/nares 182–183
    positioning 179–180
    soft palate 180–182
    strategies 192–193
    tonsils 184
Swedish Kennel Club 135, 291
Swedish Veterinary Association 136
sweeping generalisations 15
Swimmer Puppy Syndrome (SPS) 342
syringomyelia (SM)
    clinical signs 273
    diagnosis 273, 275, 278
    genetic factors and breeding advice 280–282
    minimum MRI diagnostic protocol 277
    MRI features 280
    pathogenesis 273
    spectrum 270, 271
    surgical management 281
syringomyelia-specific signs (SM-S)
    behavioural and clinical signs 273, 274
    diagnosis 277–278
    management 279–280
    mid-cervical syringes and superficial dorsal horn
        involvement 276
    prognosis 279
    treatment algorithm 281
    weakness and proprioceptive deficits 277
System That Builds Brachycephalic Dogs, The 129

targeted DNA test 18
TFEU see Treaty on the Functioning of the European
        Union (TFEU)
The Kennel Club Assured Breeders Scheme 60
thermoregulation 76–77
thoracic auscultation 358
thoracic vertebral canal stenosis 300
thoracic vertebral malformations 292
TLV see total lung volume (TLV)
tonsillar hypertrophy 165
tonsillotomy 184
total lung volume (TLV) 335
Toy Spaniel puppy 11, 12
tracheal hypoplasia 163–164
tracheostomy 186, 191, 191–192, 370
trader technique 182, 184
transitional vertebrae 292, 300
trazodone 362
Treaty on the Functioning of the European Union
        (TFEU) 141
trichiasis 209

UK Animal Welfare Acts 65
UK Brachycephalic Working Group 57

under-eruption 241–243, 243
unerupted teeth 243–244
Union of European Veterinary Practitioners (UEVP) 141
upper airway auscultation 166
upper respiratory tract (URT) 356
urogenital systems 335
URT see upper respiratory tract (URT)
utilitarianism 46–47

vaccine denialism 60
vasoconstrictive drugs 368
vertebral body formation defects 292
VetCompass 90, 92, 156, 219
veterinary checks 63
veterinary communications 379
veterinary costs associated with vaccination 44
veterinary epidemiology 86–88
veterinary nurses
    in brachycephalic health, role of 69–80
    and brachycephalic patient 69–80
    breathing assessment 74–76
    eye examination 71–73
    history 71
    infographic of 69, 70
    pre-purchase consultations 77–79, 80
    Schirmer tear test with gentle handling 71, 72
    skin folds 73, 74
    in small animal practice 69
    weight and thermoregulation 76–77
veterinary oaths and codes of conduct 44
veterinary profession 136–137
veterinary responsibility, brachycephalic health 56–57
virtue ethics approach 46, 50
vomiting 190

walrus puppies 343
WBBP see whole-body barometric plethysmography
        (WBBP)
weight 62
weight and thermoregulation 76–77
weight loss
    calculating energy ration 319
    diet 318–319
    food restriction 317–318
    intervention 319–320
weight management plan 320–321
welfare activism 7
welfare controller 383
welfare relevance of disorders to breed health 92–94,
        93–95
whelping (parturition) 335
whole-body barometric plethysmography (WBBP) 113,
        120, 168, 168, 169, 187, 193, 195, 196
World Small Animal Veterinary Association (WSAVA)
        137, 236
WSAVA see World Small Animal Veterinary Association
        (WSAVA)